DATE DUE

	MAR 2 1 2007	

NONVIRAL VECTORS
for
GENE THERAPY

NONVIRAL VECTORS —— *for* —— GENE THERAPY

Edited by

Leaf Huang
University of Pittsburgh School of Medicine
Pittsburgh, Pennsylvania

Mien-Chie Hung
University of Texas M. D. Anderson Cancer Center
Houston, Texas

Ernst Wagner
Vienna University Biocenter
Vienna, Austria

ACADEMIC PRESS

San Diego   London   Boston   New York   Sydney   Tokyo   Toronto

This book is printed on acid-free paper. ∞

Academic Press
A Division of Harcourt, Inc.
525 B Street, Suite 1900, San Diego, California 92101-4495, USA
http://www.apnet.com

Academic Press
24-28 Oval Road, London NW1 7DX, UK
http://www.hbuk.co.uk/ap/

Library of Congress Catalog Card Number: 99-63092

International Standard Book Number: 0-12-358465-5

PRINTED IN THE UNITED STATES OF AMERICA
99 00 01 02 03 04 EB 9 8 7 6 5 4 3 2 1

To our families

CONTENTS

PART I
INTRODUCTION

CHAPTER 1
Introduction

Leaf Huang and Ekapop Viroonchatapan

PART II
CATIONIC LIPOSOMES

CHAPTER 2
Progress in Gene Delivery Research and Development
Philip L. Felgner

CHAPTER 3
Cationic Lipid–Mediated Gene Delivery to the Airways
*John Marshall, Nelson S. Yew, Simon J. Eastman, Canwen Jiang,
Ronald K. Scheule, and Seng H. Cheng*

CHAPTER 4

Structure and Structure–Activity Relationships of Lipid-Based Gene Delivery Systems

Dan D. Lasic

PART III
OTHER VECTORS

CHAPTER 7
Nuclear Transport of Exogenous DNA

Magdolna G. Sebestyén and Jon A. Wolff

CHAPTER 8
Particle-Mediated Gene Delivery: Applications to Canine and Other Larger Animal Systems

Ning-Sun Yang, Gary S. Hogge, and E. Gregory MacEwen

CHAPTER 9
Polyethylenimines: A Family of Potent Polymers for Nucleic Acid Delivery

Antoine Kichler, Jean-Paul Behr, and Patrick Erbacher

CHAPTER 10
Ligand–Polycation Conjugates for Receptor-Targeted Gene Transfer

Ernst Wagner

CHAPTER 11
The Perplexing Delivery Mechanism of Lipoplexes

Lee G. Barron and Francis C. Szoka, Jr.

CHAPTER 12
Biopolymer–DNA Nanospheres
Kam W. Leong

CHAPTER 13
Novel Lipidic Vectors for Gene Transfer
Song Li and Leaf Huang

PART IV
ANIMAL MODELS AND CLINICAL TRIALS

CHAPTER 14
Mechanisms of Cationic Liposome–Mediated
Transfection of the Lung Endothelium

Dexi Liu, Joseph E. Knapp, and Young K. Song

CHAPTER 15
Cystic Fibrosis Gene Therapy

Uta Griesenbach, Duncan M. Geddes, and Eric W. F. W. Alton

CHAPTER 18
A Novel Gene Regulatory System

Steven S. Chua, Mark M. Burcin, Yaolin Wang, and Sophia Y. Tsai

FOREWORD

Viruses may be considered very efficient formulations, developed over many years of biological evolution, for the transfer and expression of genes. All steps involved are of a chemical, molecular, or supramolecular nature. Therefore, it is quite clear that there is an important role to be played by chemistry, in particular in the development of synthetic vectors that would be able to perform similar gene transfer functions while not being subject to constraints resulting from the biological nature of viruses. In view of the potential medical applications in gene therapy, it is not a surprise that much attention has been given to the problem and that intense activity has been directed at the design, synthesis, and evaluation of a number of purely chemical gene transfer agents. The complexity of the processes involved justifies the multiplicity of the approaches and the variety of artificial vectors investigated. Thus, for instance, a novel cationic lipid may not be just a "me too" addition to an already extended list, but rather may open new possibilities for differently influencing the highly complex series of steps along the path that leads from the free plasmid outside the cell to its expression in the nucleus in a living organism. Very significant progress—from passage through the cell membrane *in vitro* to expression *in vivo*—has been made over a rather short period of time thanks to the efforts of numerous laboratories. However, the gene transfer efficiency still falls short of what viruses are able to achieve. For instance, passage through nuclear pores and targeting of the nucleus represent steps beyond penetration into the cytoplasm. Thus, continued investigation of the factors that influence the overall process coupled with a better understanding of the biological events involved will lead to further progress toward the development of viable protocols for use in gene therapy. At the same time it may also open new vistas on the ability to act on cell biology processes by means of suitably designed synthetic molecules.

From another point of view, it is also quite rewarding to see that, at least in some steps, one of the basic functions of supramolecular chemistry, namely transport processes, finds here an extension into the area of molecular biology and even therapy.

This book presents a panorama of the achievements and perspectives for further exploration as seen by recognized experts, major actors, and leaders in the field of nonviral vectors for gene therapy. The editors must be congratulated for bringing the book into existence. It is very timely and will serve as a basis for future developments in an area of great value for both basic science and its application to human therapy.

Jean-Marie Lehn

FOREWORD

Gene therapy has rightly captured the attention of the public and the biomedical research community. As an oncologist, I am especially attracted to this area of investigation because we now know the cause of cancer: Cancer results from a series of inherited or acquired mutations in a few of the groups of genes that regulate cell proliferation and protect the integrity of the genetic material.

This profound statement, built on the discoveries of thousands of scientists, creates focused targets for new anticancer therapies. We can attempt to fix (replace) the defective genes or take advantage of their presence to selectively damage the cancer cells carrying these genes.

When therapy with genes becomes a practical reality, the applications to inherited and chronic degenerative diseases hold promise for great improvements in how we manage many of the illnesses that plague us today. Research during the past few years has provided *proof of concept* that gene transfer can be achieved in human beings, with transient expression of the gene product resulting in changes in cellular and biochemical activities.

The challenges now are multifold. Perhaps the most important question facing us is the need for better gene transfer technology. To date, viral vectors have provided the most effective carriers of genes into patients' cells. A year ago the National Institutes of Health conducted a review of gene therapy and concluded that the major need was for additional laboratory research to investigate improved ways of carrying out transfer and controlled expression of genes in cells.

In this volume, a variety of innovative ways for accomplishing gene transfers are presented in up-to-date chapters by experts who are carrying the field forward. *Nonviral Vectors for Gene Therapy* is, therefore, a timely and much-needed resource that will help investigators achieve the goal of improving health by introducing new genes into cells that change molecular behavior.

John Mendelsohn

CONTRIBUTORS

Numbers in parentheses indicate the pages on which the authors' contributions begin.

Eric W. F. W. Alton (337), Imperial College School of Medicine, at the National Heart & Lung Institute, London SW3 6LR, England

Lee G. Barron (229), School of Pharmacy, University of California, San Francisco, California 94143

Jean-Paul Behr (191), Laboratoire de Chimie Génétique, UMR-7514 CNRS/Université Louis Pasteur, Faculté de Pharmacié, F-67401 Strasbourg-Illkirch, France

Mark M. Burcin (409), Department of Cell Biology, Baylor College of Medicine, Houston, Texas 77030

Seng H. Cheng (39), Genzyme Corporation, Framingham, Massachusetts 01701

Steven S. Chua (409), Department of Cell Biology, Baylor College of Medicine, Houston, Texas 77030

Simon J. Eastman (39), Genzyme Corporation, Framingham, Massachusetts 01701

Patrick Erbacher (191), Laboratoire de Chimie Génétique, UMR-7514 CNRS/Université Louis Pasteur, Faculté de Pharmacié, F-67401 Strasbourg-Illkirch, France

Philip L. Felgner (25), Gene Therapy Systems, Inc., San Diego, California 92121

Duncan M. Geddes (337), Imperial College School of Medicine, at the National Heart & Lung Institute, London SW3 6LR, England

Uta Griesenbach (337), Imperial College School of Medicine, at the National Heart & Lung Institute, London SW3 6LR, England

Gary S. Hogge (171), Department of Medical Sciences, School of Veterinary Sciences, University of Wisconsin, Madison, Wisconsin 53706

Gabriel Hortobagyi (357), Department of Breast Medical Oncology, University of Texas M. D. Anderson Cancer Center, Houston, Texas 77030

Leaf Huang (3, 289), Laboratory of Drug Targeting, Department of Pharmacology, University of Pittsburgh School of Medicine, Pittsburgh, Pennsylvania 15261

Mien-Chie Hung (357), Department of Cancer Biology, Section of Molecular Cell Biology, University of Texas M. D. Anderson Cancer Center, Houston, Texas 77030

Canwen Jiang (39), Genzyme Corporation, Framingham, Massachusetts 01701

Antoine Kichler (191), URA-CNRS 1923 — Genethon III, Groupe de Vectorologie, F-91002, Evry Cedex 2, France

Joseph E. Knapp (313), Department of Pharmaceutical Sciences, School of Pharmacy, University of Pittsburgh, Pittsburgh, Pennsylvania 15261

Ilya Koltover (91), Materials Department, Physics Department, and Biochemistry and Molecular Biology Program, University of California, Santa Barbara, California 93106

Dan D. Lasic (69), Liposome Consultations, Newark, California 94560

Kam W. Leong (267), Department of Biomedical Engineering, School of Medicine, The Johns Hopkins University, Baltimore, Maryland 21205

Song Li (289), Laboratory of Drug Targeting, Department of Pharmacology, University of Pittsburgh School of Medicine, Pittsburgh, Pennsylvania 15261

Dexi Liu (313), Department of Pharmaceutical Sciences, School of Pharmacy, University of Pittsburgh, Pittsburgh, Pennsylvania 15261

E. Gregory MacEwen (171), Department of Medical Sciences, School of Veterinary Sciences, University of Wisconsin, Madison, Wisconsin 53706

John Marshall (39), Genzyme Corporation, Framingham, Massachusetts 01701

Donald M. McDonald (119), Cardiovascular Research Institute and Department of Anatomy, University of California, San Francisco, California 94143

John W. McLean (119), Cardiovascular Research Institute and Department of Anatomy, University of California, San Francisco, California 94143

Gary H. Rhodes (377), Department of Medical Pathology, University of California, Davis, California 95616

Cyrus R. Safinya (91), Materials Department, Physics Department, and Biochemistry and Molecular Biology Program, University of California, Santa Barbara, California 93106

Ronald K. Scheule (39), Genzyme Corporation, Framingham, Massachusetts 01701

Magdolna G. Sebestyén (139), Department of Pediatrics, University of Wisconsin–Madison, Madison, Wisconsin 53705-2280

Young K. Song (313), Department of Pharmaceutical Sciences, School of Pharmacy, University of Pittsburgh, Pittsburgh, Pennsylvania 15261

Francis C. Szoka, Jr. (229), School of Pharmacy, University of California, San Francisco, California 94143

Gavin Thurston (119), Cardiovascular Research Institute and Department of Anatomy, University of California, San Francisco, California 94143

Sophia Y. Tsai (409), Department of Cell Biology, Baylor College of Medicine, Houston, Texas 77030

Ekapop Viroonchatapan (3), Laboratory of Drug Targeting, Department of Pharmacology, University of Pittsburgh School of Medicine, Pittsburgh, Pennsylvania 15261

Ernst Wagner (207), Institute of Biochemistry, Vienna University Biocenter, Vienna A-1030, Austria

Shao-Chun Wang (357), Department of Cancer Biology, Section of Molecular Cell Biology, University of Texas M. D. Anderson Cancer Center, Houston, Texas 77030

Yaolin Wang (409), Schering-Plough Research Institute, Kenilworth, New Jersey 07033

Jon A. Wolff (139), Departments of Pediatrics and Medical Genetics, University of Wisconsin–Madison, Madison, Wisconsin 53705-2280

Ning-Sun Yang (171), Comprehensive Cancer Center, University of Wisconsin Medical School, Madison, Wisconsin 53792; and Institute of Bioagricultural Sciences, Academia Sinica 11529, Taipei, Taiwan, Republic of China

Nelson S. Yew (39), Genzyme Corporation, Framingham, Massachusetts 01701

PART I

Introduction

CHAPTER 1

Introduction

Leaf Huang and Ekapop Viroonchatapan
Laboratory of Drug Targeting, Department of Pharmacology,
University of Pittsburgh School of Medicine, Pittsburgh, Pennsylvania

I. Gene Therapy
II. Virus vs Nonvirus as a Vector
   A. Viral Vectors
   B. Nonviral Vectors
III. Characteristics of a Nonviral Vector
   A. Self-Assembly Processes
   B. Interactions with DNA
   C. Target Specificity
   D. Stability
IV. Delivery Barriers That a Vector
   Must Overcome
   A. In the Test Tube
   B. Route of Administration
   C. During Transport to the Target Cell
   D. Uptake by the Target Cell
   E. Escaping the Endosome (Endosomal Release)
   F. Stability in Cytoplasm
   G. Importing into the Nucleus
V. Cytoplasmic Expression
VI. Hybrid Viral/Nonviral Vectors
VII. Conclusion
   References

The concerted efforts in gene therapy to date have provided fruitful achievements toward a new era of treating human diseases. A number of obstacles, however, still must be surmounted for successful clinical applications. In this chapter, the milestones of the development of gene therapy are summarized. The characteristics of nonviral vectors are described and the delivery barriers to efficient gene transfer are also discussed.

I. GENE THERAPY

Gene therapy has been progressively developed with the hope that it will be an integral part of medical modalities in the future. The ultimate goal of gene therapy is to cure both inherited and acquired disorders in a straightforward manner by removing their causes, that is, by adding, correcting, or replacing genes. This notion gave birth to a wide variety of possibilities for true therapeutic approaches for treating human pathological conditions.

The wealth of knowledge on the genetic basis of human diseases laid a solid foundation for gene therapy. Gene therapy protocols were originally designed to correct heritable disorders in humans, such as adenosine deaminase deficiency, cystic fibrosis, and Gaucher's disease. Some of the key events in the development of gene therapy are summarized in Table 1. Although in the past two decades gene therapy trials have been initiated worldwide, little has been achieved in terms of curing disease. A number of hurdles must be overcome for successful clinical applications. One of the major obstacles is inefficiency in gene delivery. Ideal gene delivery systems should be biodegradable, nontoxic, nonimmunogenic, stable during storage and after administration, able to access to target cells, and suitable for efficient gene expressions.

II. VIRUS VS NONVIRUS AS A VECTOR

Vectors proposed for gene delivery generally fall into two categories: viral and nonviral. They differ primarily in their assembling process. A viral vector is assembled in a cell, whereas a nonviral vector is constructed in a test tube. The in-depth understanding of the biological self-assembly process may provide an idea for constructing synthetic self-assembling systems.

A. VIRAL VECTORS

Viral vectors are replication-defective viruses with part or all of the viral coding sequence replaced by that of therapeutic genes. There is considerable interest in using viruses for gene therapy because they are highly efficient in gene delivery. However, several drawbacks associated with their practical use are evident. Depending on the type of vector, the problems include insufficient pharmaceutical quantities (viral titers), toxicity, and the potential replication of competent viruses. Major viral vectors are as follows:

Table I

Key Events in Gene Therapy Development

Year	Event	Reference
1932	At the Sixth International Congress of Genetics held in Ithaca, New York, the term *genetic engineering* was first used with reference to the application of genetic principles to animal and plant breeding.	Crow (1992)
1944	A gene can be transferred within nucleic acids.	Avery *et al.* (1944)
1952	Viruses can transmit genes to Salmonella.	Zinder and Lederberg (1952)
1954	Discovery of DNA structure and its genetic implications.	Watson and Crick (1954)
1956	Viral genomes can be permanently incorporated into cell genomes.	Lederberg (1956)
1961–2	Foreign DNA can integrate in a stable fashion into the mammalian cellular genome.	Kay (1961), Bradley *et al.* (1962)
1968	Papovaviruses SV40 and polyoma transform normal cells into tumor cells by integrating their DNA into the genome of an infected cell.	Sambrook *et al.* (1968), Westphal and Dulbecco (1968)
1970	Human genetic engineering was discussed and the feasibility and ethics of several procedures were explored.	Davis (1970)
1972	A discussion of gene therapy was offered.	Friedmann and Roblin (1972)
1979	Introduction of the human β-globin gene into murine bone marrow cells with chemical transfection techniques.	Cline *et al.* (1980), Mercola *et al.* (1980)
1981–2	Retroviral vectors were developed to transfer foreign genes to essentially 100% of exposed mammalian cells.	Shimotohno and Temin (1981), Wei *et al.* (1981), Tabin *et al.* (1982)
1983	A restoration of functional hypoxanthine guanine phophoribosyl transferase activity was successfully executed by using a retrovirus as a vector.	Miller *et al.* (1983)
1987	Synthesis of a cationic lipid, N-[1-(2,3-dioleyloxy)propyl]-N,N,N-trimethylammonium chloride (DOTMA).	Felgner *et al.* (1987)
1990	The pioneer gene transfer trial in patients with advanced melanoma using tumor-infiltrating lymphocytes modified by inserted gene-encoding neomycin phosphotransferase.	Rosenberg *et al.* (1990)
1990	The adenosine deaminase gene therapy trial was initiated.	Culver *et al.* (1992)
1992	First gene therapy trial using DC-chol/DOPE cationic liposome.	Nabel *et al.* (1993)
1990s	Other gene delivery systems were developed including adenovirus, herpes virus, adeno-associated virus, and synthetic vectors.	Brenner (1995), Gao and Huang (1995), Felgner (1997), Friedmann (1997), Tseng and Huang (1998)

1. Retrovirus

Retrovirus is an eukaryotic RNA virus in the family Retroviridae that has RNA as its genome. It uses viral enzymes to copy its genome into DNA and integrate into the host chromosome (Varmus, 1988). Long-term gene expression can be achieved with this vector, but only in the cycling cells (Anderson, 1992). This vector also suffers from difficulties in preparing lots with high titer. Recent advances in lentivirus vectors have enabled transduction of nondividing cells (Bukrinsky et al., 1993; Naldini et al., 1996). Since integration of the viral genome into the host genome is a random process, there is a risk of insertional mutagenesis.

2. Adenovirus

Adenovirus is an icosahedral virus that contains a large DNA genome. Different from the retrovirus, adenovirus can infect both dividing and nondividing cells (Grunhaus and Horwitz, 1992). High viral titers, greater than 10^{10} CFU/mL, can be obtained (Berkner, 1992; Stratford-Perricaudet, 1994). Gene expression using adenovirus lasts for only a short time because adenoviral genome is not integrated into the host DNA. An additional shortcoming of adenoviral vectors is that they induce inflammatory responses in the host tissues, which also limits the duration of gene expression and the frequency of repeat administration (Schneider and French, 1993).

3. Adeno-Associated Virus

Adeno-associated virus (AAV) is a small nonpathogenic DNA virus. It can only produce progenies in the presence of a helper virus, such as adenovirus or herpes virus. In the absence of a helper virus, AAV instead incorporates its genome into the chromosomes of the host cells (Muzyczka, 1992). Both dividing and nondividing cells can be infected by AAV. The main disadvantage of AAV is that sufficiently high viral titer is hard to obtain. The virus also has a limited packaging size of about 5 kb.

4. Other Viral Vectors

Herpes simplex virus (HSV) has recently received much attention. HSV is a double-stranded DNA virus with a genome measuring more than 150 kb (Roizman and Sears, 1990). The infection of peripheral nerve terminals and subsequent retrograde axonal transport provide the way for HSV to access the nervous system (Glorioso et al., 1994). Similar to the advantages of adenovirus, HSV can infect a variety of dividing and nondividing cells. A very high viral titer can also be produced. However, HSV causes some cytotoxicity (Glorioso et al., 1994).

Vaccinia virus is a poxvirus with double-stranded DNA genome (Lyerly and DiMaio, 1993). Its advantages in gene delivery are similar to those of other DNA viruses. An attenuated vaccinia virus was commonly used 20 years ago in smallpox immunization. The immunologic elimination of vaccinia vectors may be effected by previous exposure to vaccinia virus.

B. Nonviral Vectors

The limitations of viral vectors make synthetic vectors an attractive alternative. Advantages of nonviral vectors include their nonimmunogenicity, low acute toxicity, simplicity, and feasibility to be produced on a large scale. There are, however, some drawbacks with these nonviral vectors, including their lower efficiency than viral vectors in gene transfer and their transient gene expressions. Many types of nonviral vectors have been proposed. The following is a summary of these vectors and delivery methods, which are discussed in greater detail in various chapters of this book.

1. Naked DNA

Naked DNA, or free DNA, is useful in gene transfer to skeletal muscle (Jiao et al., 1992; Wolff et al., 1990), liver (Hickman et al., 1994), and heart muscle (Ardehali et al., 1995). Intratumoral injection of naked DNA can induce transgene expression at a level that is high enough to elicit therapeutic consequences (Vile and Hart, 1993; Yang and Huang, 1996; Budker et al., 1998). For systemic administration, however, the plasmid DNA needs to be protected from degradation by endonucleases during delivery from the site of administration to the site of gene expression.

2. Gene Gun

Shooting DNA into cells can be carried out by using a physical method such as bioballistic bombardment or gene gun. The gene gun uses gold particles coated with DNA, which are transferred to a mylar carrier sheet. An electric arc generated by a high-voltage discharge accelerates the DNA-coated gold particles to high velocity, enabling efficient penetration of target organs in vivo or single cell layers in vitro (Yang et al., 1990; Andree et al., 1994; Sun et al., 1995). The major drawback of this system is that the target tissues must be surgically exposed, although the DNA transfer is relatively efficient.

3. Liposome / DNA Complex (Lipoplex)

Because of charge interaction, a complex can be easily formed between cationic liposomes (positive) and DNA (negative) (Gershon *et al.*, 1993; Sternberg *et al.*, 1994; Gustafsson *et al.*, 1995; Liu *et al.*, 1996). Liposome/DNA complex, or lipoplex, must be sufficiently small to enter the cells. Liposomes mainly consist of a positively charged lipid (cationic lipid) and a helper lipid (colipid), for example, dioleoyl phosphatidylethanolamine (DOPE) or dioleoyl phosphatidylcholine (DOPC). Several different kinds of cationic lipid, including quaternary ammonium detergents, cationic derivatives of cholesterol and diacyl glycerol, and lipid derivative of polyamines, have been developed.

4. Polymer / DNA Complex (Polyplex)

Several cationic polymers have been shown to form complexes with DNA and thus facilitate gene transfer (Zhou *et al.*, 1991; Gao *et al.*, 1993; Kupfer *et al.*, 1994). The effect of polymer structure (e.g., linear, branched, and spheroidal) on transfection efficiency does not appear to be a major factor. Poly(L-lysine), polyethylenimine, and starburst dendrimer have linear, branched, and spheroidal structures, respectively.

5. Liposome / Polymer / DNA Complex (Lipopolyplex)

Incorporation of polylysine into lipoplex to form ternary complexes leads to a tight condensation of DNA. This prevents the complex from aggregation and nuclease degradation. Cationic and anionic lipopolyplexes called LPDI and LPDII, respectively, have been formulated and shown to be effective in gene transfer (Gao and Huang, 1996; Lee and Huang, 1996; Li and Huang, 1997; Li *et al.*, 1997, 1998).

III. CHARACTERISTICS OF A NONVIRAL VECTOR

A. SELF-ASSEMBLY PROCESSES

Phospholipids are amphiles that contain both hydrophilic and hydrophobic moieties. In aqueous media, these molecules spontaneously associate themselves to form ordered structures to minimize the unfavorable interaction between the bulk aqueous phase and the long hydrocarbon fatty acyl chains. This self-assembly process of lipid molecules results in different microscopic and macroscopic structures, depending on their concentrations. The structures include micelles, planar bilayer sheets, and lamellar vesicles.

Through a process of condensation, DNA molecules organize into highly ordered structures at higher concentrations or upon the addition of chemical agents, such as multivalent cations, alcohols, basic proteins, cationic polymers, or cationic liposomes, to DNA (Bloomfield, 1996; Lasic, 1996). The complex of cationic liposomes and/or cationic polymers with DNA can be employed to deliver and transfer genes into target cells. The process of complex formation is governed by multiple types of weak molecular forces including ionic, H-bonding, hydrophobic, and solvent exclusion. The process can be kinetically slow and does not reach equilibrium.

B. Interactions with DNA

Both cationic polymers and liposomes will form a complex with DNA through charge interaction since DNA is negatively charged. Structures and properties of polyamine/DNA complex have been extensively studied (Wilson and Bloomfield, 1979; Smirnov et al., 1988). The complex is largely toroidal in shape. DNA stoichiometrically binds to polyamine and the condensation process is highly cooperative. The important factor in condensing DNA is the structure of polyamine. The results of DNA compaction induced by diaminoalkanes have been reported (Yoshikawa and Yoshikawa, 1995). It appeared that all of the positive charges must interact simultaneously to achieve binding and that the position of these charges plays an important role in optimal interaction. It was also shown that about 90% of the phosphates must be neutralized for a complete compaction (Smirnov et al., 1988; Sen and Crothers, 1986).

The binding of cationic liposomes to DNA, similar to polyamine–DNA binding, is a highly cooperative process. The interaction between DNA and cationic liposomes is inadequately understood and it is therefore difficult to control the size distribution of the complex produced. Electron microscopy of liposome/DNA complex has revealed a shortening and thickening of DNA in the presence of cationic liposomes (Gershon et al., 1993). Globular structures can be occasionally seen lining the DNA strands in an aggregated fashion, similar to beads on a string. This was the first evidence of the complex formation, suggesting that a liposome coating on the DNA strands causes DNA to be condensed. Electron microscopy further illustrates that at a 1/1 lipid/DNA ratio, approximately half of the DNA molecules are bound to liposomes. When the liposome concentration is increased, all the DNA is covered by the lipids. At low liposome/DNA ratios, the liposome/DNA complex is nearly spherical; but at high ratios, the liposome/DNA complex turns into smooth rod-like structures. These structures contradict the popular belief that the liposomes bound DNA on their surface while maintaining their original size and shape. Observations by freeze–fracture microscopy (Sternberg et al., 1994) have revealed two types of structures: liposome/DNA aggregates (meatballs) and

DNA covered with a single lipid bilayer (spaghetti). The size of the complex increases with the elevated liposome concentration. At high concentrations, lipid membrane becomes discontinuous and allows complete encapsulation of DNA. Recently, Rädler *et al.* (1997) discovered that a liposome/DNA complex consists of a higher ordered multilamellar structure with DNA sandwiched between cationic bilayers.

C. Target Specificity

Mammalian cell surface is coated with negatively charged sialic acid (Vigneron *et al.*, 1996). These anionic biological surfaces can be targeted through charge interaction by a cationic delivery system in an efficient but nonspecific manner (Filgner and Ringold, 1989). Among the different methods of targeting of liposomes, attachment of a ligand on the surface of a vector is the most efficient method of delivery to a selective site. Both antibodies and low-molecular-weight ligands have been used to target vectors to cell surface receptors.

Antibodies and other site-specific proteins can be attached to the liposomes on their surface (Heath, 1987) or be incorporated in their membrane as the liposomes are formed (Huang *et al.*, 1984; Holmberg *et al.*, 1989). When these liposomes are injected, they bind specifically and efficiently to the target site, if the liposomes are prepared with a lipid formulation that confers low reticuloendothelial system (RES) uptake of liposomes (Maruyama *et al.*, 1990). A novel approach using a monoclonal antibody against an epitope expressed on the lung endothelial cells has been developed (Trubetskoy *et al.*, 1992 a,b). The antibody was covalently linked to the positively charged polylysine chains, which are a part of the ternary complex containing cationic liposomes and DNA. In another approach to targeted delivery of DNA, ligands to different cell surface receptors can be incorporated into the surface of the liposome by covalently modifying the ligand with a lipid group and adding it during the formation of liposomes (Stavridis *et al.*, 1986; Wang and Huang, 1987; Stripp *et al.*, 1990).

D. Stability

Li *et al.* (1998) reported that, after exposure to mouse serum, cationic lipidic vectors become negatively charged, significantly increase in size, and aggregate to each other. Disintegration, DNA release, and degradation of the lipidic vectors occur if further incubation persists (Li *et al.*, in press). The rate of vector disintegration depends on the lipid composition of the vector. Incorporating dioleoylphosphatidylethanolamine (DOPE) into the composition accelerates the rate of vector disintegration (Li *et al.*, in press). In order to prevent aggregation, the surface of the complex can be modified with polyethylene glycol (Plank *et al.,* 1996). This also decreases the interaction of the complex with blood components.

IV. DELIVERY BARRIERS THAT A VECTOR MUST OVERCOME

There are several steps required for accomplishing an effective gene expression. A vector must be properly prepared, delivered from the administered site to the target site, and then taken up by the target cell. Moreover, a vector must be released from the endosome and finally imported into the nucleus. These steps are schematically summarized in Fig. 1 and they are also discussed next.

A. IN THE TEST TUBE

The initial difficulty in gene delivery is the preparation of the vector complex, especially a lipoplex. A heterogeneous population of the complex is produced because the population is not prepared in a controlled manner. Binding DNA to cationic liposome is a multivalent charge interaction, which usually leads to further aggregation or flocculation of the complex. At low concentrations, a small complex

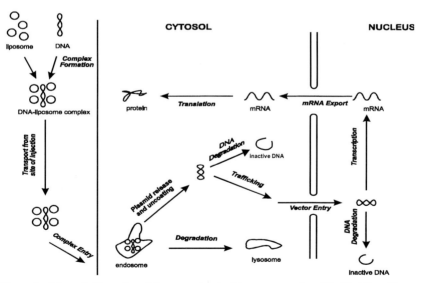

Figure 1 Schematic illustration of the process involved in gene expression. [Modified from *Pharmaceutical Science and Technology Today,* Volume 1, Tseng and Huang "Liposome-Based Gene Therapy" pages 206–213, Copyright 1998, with permission from Elsevier Science.]

(200–400 nm) is formed. In contrast, a large complex (>1 μm) is obtained at high concentrations. A two-vial formulation tentatively solves this problem. DNA and liposome are packaged in separate vials and then mixed together immediately before use. The entrance of a particulate matter into cells via endocytosis is a size-limiting process (Machy and Leserman, 1983; Zhou and Huang, 1994). Therefore, some of the complexes in the heterogeneous population are too large to be endocytosed, and thus are rapidly removed from blood circulation. This reduces the bioavailability of the transferred gene to be expressed in the cells. In general, a lipopolyplex, such as LPD, is more homogeneous in size and more stable upon storage than the lipoplex. The LPD can be stored at 4°C for at least 4 weeks (Li and Huang, 1997).

B. ROUTE OF ADMINISTRATION

Intravenous administration is the most effective way of systemically introducing liposome into a human body. The injected liposomes are mostly trapped in the liver and spleen (Hwang, 1987). To reduce or retard the uptake by RES, liposomes can be administered by other routes. For example, after injection of liposomes into the peritoneal cavity of mice or rats, the liposomes enter the lymphatic system and then the circulatory system, causing a delay in the RES uptake (Ellens et al., 1981; Parker et al., 1981). Similar effects were shown by intramuscular and subcutaneous injections (Tuner et al., 1983; Poste et al., 1984). Consequently, if the targeted cells reside in, or can be accessed from, the lymphatic system, liposomes may be able to reach them through the intramuscular or subcutaneous route.

Intratracheal instillation has been used to administer both liposomally encapsulated drugs (Fielding and Abra, 1992) and liposome/DNA complex (Brigham et al., 1989) to the lung. Yoshimura et al. (1992) showed the presence of mRNA of human cystic fibrosis transmembrane conductance (CFTR) regulator gene in mouse lungs after intratracheal administration of the corresponding plasmid with lipofectin, a cationic liposome formulation. Reporter gene activity for luciferase and β-galactosidase was also shown in the lung, suggesting that some of the airway epithelial cells were transfected. Brigham et al. (1989) also used lipofectin/DNA complex to transfect mouse lungs by intratracheal instillation as well as intravenous and intraperitoneal injections. The expression of the chloramphenicol acetyltransferase (CAT) gene has also been reported in mouse lung after administration of liposome/DNA complexes in aerosol form (Stribling et al., 1992). DC-chol liposome/DNA complexes have been directly injected into the tumor (Stewart et al., 1992) and also administered into arterial endothelial cells by catheter (Nabel et al., 1992).

Liposome/DNA complex and free antisense oligonucleotide have been used in clinical studies. Nabel et al. (1993) introduced a human HLA-B7 gene into subcutaneous melanoma tumors of HLA-B7-negative patients by direct gene transfer with liposome/DNA complex. The transferred gene was expressed and localized to the site of injection, and no apparent toxicity or anti-DNA antibodies were associated with this treatment. Hui *et al.* (1997) performed clinical trials in patients with different kinds of cancer. The patients were injected with a liposome/DNA complex directly into cutaneous nodules. Strong local responses were achieved in the treated nodules. Single or multiple intravenous infusion (0.06–2.00 mg/kg, infused over 2 h) of an intracellular adhesion molecule-1 antisense oligodeoxy-nucleotide (ISIS 2302) was given for the first time in healthy male volunteers in order to investigate its safety and pharmacokinetics (Glover *et al.*, 1997). ISIS 2302 was well tolerated and there were no clinical signs, symptoms, or changes in routine laboratory safety parameters.

C. DURING TRANSPORT TO THE TARGET CELL

Nonviral gene delivery systems rely on normal cellular uptake mechanisms. Before these systems can reach the cell surface, several hurdles must be overcome. For example, after intravenous administration, serum inactivation and DNA deg-radation must be avoided, or in the case of cystic fibrosis gene therapy (using airway administration) the delivery systems must penetrate the thick layer of mucus cov-ering the target cells. The simplest way of helping the delivery system reach the target cells is by local administration (Son and Huang, 1994; Caplen *et al.*, 1995; Aoki *et al.*, 1995; Takehara *et al.*, 1995; Schwartz *et al.*, 1995).

Gaining insight into the mechanisms involved in biodistribution and cellular uptake is the first step in developing targeted gene delivery vehicles for intravenous administration. Liu *et al.* (1997) and Song *et al.* (1997) demonstrated that a high cationic lipid/DNA ratio is essential for achieving better gene delivery efficiency *in vivo*. The increase of the cationic lipid/DNA ratio then results in a higher level of gene expression. One possible explanation is that an excess amount of positive charges may inhibit the neutralization effect of blood components and protect a transgene from degradation.

D. UPTAKE BY THE TARGET CELL

Uptake of a cationic, nonviral DNA delivery system is a nonspecific process. Due to its excess positive charge, the complex binds to the negatively charged cell membrane and is subsequently taken up by endocytosis. The rate of entry into cells varies with cell type and occurs relatively slowly. In treated COS and HeLa cells *in vitro*, less than 5% of the cells had taken up the complex after 30 min; maxi-mal uptake was reached after 6 h (Zabner *et al.*, 1995). After this 6-h exposure, approximately 50% of the cells had taken up the complex, accounting for as much as 80% of the delivered dose found within the cells. The specificity of gene expres-sion could be increased and toxicity might be reduced by adding a targeting ligand to the surface of the DNA delivery vehicle. After endocytosis, the DNA-containing particles are largely retained in perinuclear endosomes/lysosomes. The entrapment of the particles within these vesicles is one of the major barriers for transfection.

E. Escaping the Endosome (Endosomal Release)

Endocytosis is inefficient for gene transfer since most of the DNA is retained in the endosomes/lysosomes and is eventually degraded or inactivated (Zhou and Huang, 1994; Zabner et al., 1995). A number of strategies have been explored to enhance endosomal release. One involves using fusogenic lipids or peptides to disrupt the endosome membrane. These compounds form inverted hexagonal structures and/or pores in the membrane. DOPE is generally employed as a fusogenic helper lipid in a cationic lipid/DNA complex (Litzinger and Huang, 1992; Farhood et al., 1995). After binding of cationic lipids to anionic lipids in the membrane, phase separation may occur. This would initiate the inverted hexagonal phase formation and membrane destabilization in the DOPE-rich regions. The second approach involves using a DNA delivery system with a high buffering capacity and the flexibility to swell when protonated. Such a system reduces the acidification of the endosome and subsequently leads to rupture of the endosomal membrane (Remy et al., 1994). The rationale that liposome/DNA or polymer/DNA complexes are taken up by cells via endocytosis and that endosomes can be ruptured if the pH drop in the late endosome is inhibited by the buffering capacity of the formulation led to the use of polyethylenimine as a DNA delivery polymer (Boussif et al., 1995). In a similar manner, the pH-sensitive liposomes destabilize the endosome, thus allowing the DNA to escape into the cytoplasm before lysosomal degradation can take place (Wang and Huang, 1987).

Anionic lipids can displace DNA in a cationic liposome/DNA complex (Xu and Szoka, 1996). This mechanism of DNA release may be due to the multivalent nature of the anionic lipid surface and the collaborative effects of electrostatic interactions and hydrophilic–hydrophobic interactions of the lipid. Lipid mixing results in charge neutralization, which allows for the diffusion of the cationic lipids away from the DNA. Adding equal moles of anionic lipid to the cationic liposome/DNA complex leads to 80% release of the DNA. It has been hypothesized that anionic lipids, which are normally found in endosomal membranes, can efficiently dissociate the cationic lipids from the complex and release DNA into the cytoplasm.

F. Stability in Cytoplasm

Plasmid DNA must be released from the endosome to the cytoplasm before it can be imported into the nucleus. At this point, the stability of plasmid DNA in cytoplasm is a major concern of an efficient gene expression. Lechardeur et al. (in press) recently reported for the first time that naked plasmid DNA is rapidly degraded in the cytoplasm. Plasmid DNA was injected into the cytosol of COS and HeLa cells and it disappeared with an apparent half-life of 50–90 min. A cytosolic nuclease may have been responsible for the degradation.

G. Importing into the Nucleus

It was shown that the microinjection of liposome/DNA complex into the nucleus of oocytes resulted in no detectable level of expression when compared to free DNA that was also microinjected into the nucleus (Zabner *et al.*, 1995). This indicates that lipid coating of the DNA inhibits transcription. Furthermore, microinjection of free DNA directly into the nucleus has resulted in a much higher level of gene expression than injection of the DNA into the cytoplasm (Capecchi, 1980; Zabner *et al.*, 1995). These studies strongly indicated that only free DNA could be imported into the nucleus, although the process was very inefficient. However, a recent study by Coonrod *et al.* (1997), using BrdU-labeled DNA fragment of 1 kb in length, indicated that DNA can be readily imported into the nucleus. However, it is not clear whether the imported DNA is already degraded into small pieces and whether high-molecular-weight plasmid DNA is similarly imported. Dean (1997) showed that the import of plasmid DNA into the nucleus is a sequence-specific process. Cytoplasmically injected SV40 DNA was imported into the nucleus through the nuclear pore complex. The nuclear transport signal appears to include the region of SV40 DNA that contains the origin of replication and the early and late promoters. This is a promising approach; however, the relatively short half-life of DNA in the cytoplasm may be a hurdle.

V. CYTOPLASMIC EXPRESSION

DNA delivery into the cytoplasm of the cells is efficiently accomplished by various types of gene-transferring vectors. However, the ultimate expression of the DNA is still hindered by inefficient transport of DNA into the nucleus, where the transcription machinery resides. One of the strategies to circumvent the requirement for nuclear transport is given by a powerful cytoplasmic expression of a reporter gene (pT7-CAT) driven by the T7 promoter. T7 RNA polymerase is essential for the expression of this gene. It was shown that the purified T7 RNA polymerase could be codelivered with DNA using DC-chol liposomes (Gao and Huang, 1993). CAT reporter gene expression is transient. This may be due to rapid degradation or inactivation of the T7 RNA polymerase. To overcome the problem, several research groups (Dubendorff *et al.*, 1991; Deng *et al.*, 1994; Gao *et al.*, 1994; Chen *et al.*, 1995, 1998) developed T7 RNA polymerase autogenes (e.g., pT7

AUTO 2C⁻) to continuously produce high levels of T7 RNA polymerase in the cytoplasm. However, there are difficulties in using pT7 AUTO 2C⁻ and other autogenes. It is difficult and time consuming to amplify and purify these plasmids from bacteria. The initiation of expression of these autogenes also requires co-delivery of purified T7 RNA polymerase, which is costly and potentially immunogenic. A novel T7 RNA polymerase autogene, pCMV/ T7-T7pol, was recently reported (Brisson *et al.*, 1999). This autogene contains both CMV and T7 promoters. The CMV promoter allows the production of some T7 RNA polymerase, which then initiates the autocatalytic cycle of the autogene by using the T7 promoter. Although this dual-autogene approach is promising, the cytoplasmic nuclease degradation of DNA may be a hurdle to overcome.

VI. HYBRID VIRAL/NONVIRAL VECTORS

Some new formulations have combined the merits of both viral and nonviral vectors, such as virus–cationic liposome–DNA complex (Curiel *et al.*, 1991; Wu *et al.*, 1994; Kreuzer *et al.*, 1996; Saldeen *et al.*, 1996; Singh and Rigby, 1996; Noguiez-Hellin *et al.*, 1996). Plasmid DNA was encapsulated into liposomes containing the nuclear protein high mobility group and fused with inactivated hemagglutinating virus of Japan (HVJ). Expression of various genes could be obtained with this fusion hybrid of liposome and virus (Kaneda *et al.*, 1989; Tomita *et al.*, 1992, 1993; Kato *et al.*, 1993; Sawa *et al.*, 1995). Aoki *et al.* (1997) recently developed an efficient HJV/liposome complex for transferring the β-galactosidase gene to rat heart. Direct infusion of the complex into coronary artery resulted in the widespread presence of β-galactosidase in cardiac myocytes around the microvasculature. Curiel *et al.* (1991) successfully employed an adenovirus to promote the gene transfer of transferrin–polylysine conjugates. Adenovirus apparently disrupts the endosomal membrane to release the DNA into the cytoplasm and avoids lysosomal degradation of DNA.

VII. CONCLUSION

The concerted efforts in gene therapy to date have provided fruitful achievements toward a new era of curing human diseases. A number of obstacles, however, still must be surmounted for successful clinical applications. For that purpose, investigators from various areas of expertise are dealing with the problems, and the dream of exploiting gene therapy for medical modalities should become reality in the near future.

REFERENCES

Anderson, W. F. (1992). Human gene therapy. *Science* **256**, 808–813.

Andree, C., Swain, W. F., Page, C. P., Macklin, M. D., Slama, J., Hatzis, D., and Eriksson, E. (1994). *In vivo* transfer and expression of a human epidermal growth factor gene accelerates wound repair. *Proc. Natl. Acad. Sci. USA* **91**, 12188–12192.

Aoki, K., Yoshida, T., Sugimura, T., and Terada, M. (1995). Liposome-mediated *in vivo* gene transfer of antisense K-ras construct inhibits pancreatic tumor dissemination in the murine peritoneal cavity. *Cancer Res.* **55**, 3810–3816.

Aoki, M., Morishita, R., Muraishi, A., Moriguchi, A., Sugimoto, T., Maeda, K., Dzau, V. J., Kaneda, Y., Higaki, J., and Ogihara, T. (1997). Efficient *in vivo* gene transfer into the heart in the rat myocardial infarction model using the HVJ (Hemagglutinating Virus of Japan)–liposome method. *J. Mol. Cell Cardiol.* **29**, 949–959.

Ardehali, A., Fyfe, A., Laks, H., Drinkwater, D. C., Qiao, J. H., and Lusis, A. J. (1995). Direct gene transfer into donor hearts at the time of harvest. *J. Thorac. Cardiovasc. Surg.* **109**, 716–720.

Avery, O. T., MacLeod, C. M., and McCarty, M. (1944). Studies on the chemical nature of the substance inducing transformation of pneumococcal types. *J. Exp. Med.* **79**, 137–158.

Berkner, K. L. (1992). Expression of heterologous sequences in adenoviral vectors. *Curr. Top. Microbiol. Immunol.* **158**, 38–66.

Bloomfield, V. A. (1996). DNA condensation. *Curr. Opin. Struct. Biol.* **6**, 334–341.

Boussif, O., Lezoualc'h, F., Zanta, M. A., Mergny, M. D., Scherman, D., Demeneix, B., and Behr, J. P. (1995). A versatile vector for gene and oligonucleotide transfer into cells in culture and *in vivo:* Polyethylenimine. *Proc. Natl. Acad. Sci. USA* **92**, 7297–7301.

Bradley, T. R., Roosa, R. A., and Law, L. W. (1962). DNA transformation studies with mammalian cells in culture. *J. Cell. Comp. Physiol.* **60**, 127–138.

Brenner, M. K. (1995). Human somatic gene therapy: progress and problems. *J. Intern. Med.* **237**, 229–239.

Brigham, K. L., Meyrick, B., Christian, B., Magnuson, M., King, G., and Berry L. (1989). *In vivo* transfection of murine lungs with a functioning prokaryotic gene using a liposomes vehicle. *Am. J. Med. Sci.* **298**, 278–281.

Brisson, M., He, Y., Li, S., Yang, J-P., and Huang, L. (1999). A novel T7 RNA polymerase autogene for efficient cytoplasmic expression of target genes. *Gene Ther.* **6**, 263–270.

Budker, V., Zhang, G., Danko, I., Williams, P., and Wolff, J. (1998). The efficient expression of intravascularly delivered DNA in rat muscle. *Gene Ther.* **5**, 272–276.

Bukrinsky, M. I., Haggerty, S., Dempsey, M. P., Sharova, N., Adzhubel, A., Spitz, L., Lewis, P., Goldfarb, D., Emerman, M., and Stevenson, M. (1993). A nuclear localization signal within HIV-1 matrix protein that governs infection of non-dividing cells. *Nature* **365**, 666–669.

Capecchi, M. R. (1980) High efficiency transformation by direct microinjection of DNA into cultured mammalian cells. *Cell* **22**, 479–488.

Caplen, N. J., Alton, E. W., Middleton, P. G., Dorin, J. R., Stevenson, B. J., Gao, X., Durham, S. R., Jeffery, P. K., Hodson, M. E., Coutelle, C., Huang, L., Porteous, D. J., Williamson, R., and Geddes, D. M. (1995). Liposome-mediated CFTR gene transfer to the nasal epithelium of patients with cystic fibrosis. *Nature Med.* **1**, 39–46.

Chen, X., Li, Y., Xiong, K., Aizicovici, S., Xie, Y., Zhu, Q., Sturtz, F., Shulok, J., Snodgrass, R., Wagner, T. E., and Platika, D. (1998). Cancer gene therapy by direct tumor injections of a nonviral T7 vector encoding a thymidine kinase gene. *Human Gene Ther.* **9**, 729–736.

Chen, X., Li, Y., Xiong, K., Xie, Y., Aizicovici, S., Snodgrass, R., Wagner, T. E., and Platika, D. (1995). A novel nonviral cytoplasmic gene expression system and its implications in cancer gene therapy. *Cancer Gene Ther.* **2**, 281–289.

Cline, M. J., Stang, H., Mercola, K., Morse, L., Ruprecht, R., Brown, J., and Salser, W. (1980). Gene transfer in intact animals. *Nature* **284,** 422–425.

Coonrod, A., Li, F. Q., and Horwitz, M. (1997). On the mechanism of DNA transfection: Efficient gene transfer without viruses. *Gene Ther.* **4,** 1313–1321.

Crow, J. F. (1992). Anecdotal, historical and critical commentaries on genetics. *Genetics* **131,** 761–768.

Culver, K. W., Berger, M., Miller, A. D., Anderson, W. R., and Blaese, R. M. (1992). Lymphocyte gene therapy for adenosine deaminase deficiency. *Pediatr. Res.* **31,** 149A.

Curiel, D. T., Agarwal, S., Wagner, E., and Cotten, M. (1991). Adenovirus enhancement of transferrin-polylysine-mediated gene delivery. *Proc. Natl. Acad. Sci. USA* **88,** 8850–8854.

Davis, B. D. (1970). Prospect for genetic intervention in man. *Science* **170,** 1279–1283.

Dean, A. D. (1997). Import of plasmid DNA into the nucleus is sequence specific. *Exp. Cell Res.* **230,** 293–302.

Deng, H., and Wolff, J. A. (1994). Self-amplifying expression from the T7 promoter in 3T3 mouse fibroblasts. *Gene* **143,** 245–249.

Dubendorff, J. W., and Studier, F. W. (1991). Creation of a T7 autogene. Cloning and expression of the gene for bacteriophage T7 RNA polymerase under control of its cognate promoter. *J. Mol. Biol.* **219,** 61–68.

Ellens, H., Morselt, H., and Scherphof, G. (1981). *In vivo* fate of large unilamellar sphingomyelin-cholesterol liposomes after intraperitoneal and intravenous injection into rats. *Biochim. Biophys. Acta* **674,** 10–18.

Farhood, H., Serbina, N., and Huang, L. (1995). The role of dioleoyl phosphatidylethanolamine in cationic liposome mediated gene transfer. *Biochim. Biophys. Acta* **1235,** 289–295.

Felgner, P. L. (1997). Nonviral strategies for gene therapy. *Sci. Am.* **276,** 102–106.

Felgner, P. L., Barenholz, Y., Behr, J. P., Cheng, S. H., Cullis, P., Huang, L., Jessee, J. A., Seymour, L., Szoka, F., Thierry, A. R., Wagner, E., and Wu, G. (1997). Nomenclature for synthetic gene delivery systems. *Human Gene Ther.* **8,** 511–512.

Felgner, P. L., Gadek, T. R., Holm, M., Roman, R., Chan, H. W., Wenz, M., Northrop, J. P., Ringold, G. M., and Danielsen, M. (1987). Lipofection: A highly efficient, lipid-mediated DNA-transfection procedure. *Proc. Natl. Acad. Sci. USA* **84,** 7413–7417.

Felgner, P. L., and Ringold, G. M. (1989). Cationic-liposome mediated transfection. *Nature* **337,** 387–388.

Fielding, R. M., and Abra, R. M. (1992). Factors affecting the release rate of terbuline from liposome formulations after intratracheal instillation in the guinea pig. *Pharm. Res.* **9,** 220–223.

Friedmann, T. (1997) The road toward human gene therapy—a 25-year perspective. *Ann. Med.* **29,** 575–577.

Friedmann, T., and Roblin, R. (1972). Gene therapy for human genetic disease? *Science* **175,** 949–955.

Gao, X., and Huang, L. (1993). Cytoplasmic expression of a reporter gene by co-delivery of T7 RNA polymerase and T7 promoter sequence with cationic liposomes. *Nucleic Acids Res.* **21,** 2867–2872.

Gao, X., and Huang, L. (1995). Cationic liposome-mediated gene transfer. *Gene Ther.* **2,** 710–722.

Gao, X., and Huang, L. (1996). Potentiation of cationic liposome-mediated gene delivery by polycations. *Biochemistry* **35,** 1027–1036.

Gao, X., Jaffurs, D., Robbins, P. D., and Huang, L. (1994). A sustained, cytoplasmic transgene expression system delivered by cationic liposomes. *Biochem. Biophys. Res. Commun.* **200,** 1201–1206.

Gao, X., Wagner, E., Cotten, M., Agarwal, S., Harris, C., Romer, M., Miller, L., Hu, P., and Curiel, D. (1993). Direct *in vivo* gene transfer to airway epithelium employing adenovirus-polylysine-DNA complexes. *Human Gene Ther.* **4,** 17–24.

Gershon, H., Ghirlando, R., Guttman, S. B., and Minsky, A. (1993). Mode of formation and structural features of DNA-cationic liposome complexes used for transfection. *Biochemistry* **32,** 7143–7151.

Glorioso, J. C., Goins, W. F., Meaney, C. A., Fink, D. J., and DeLuca, N. A. (1994). Gene transfer to brain using herpes simplex virus vectors. *Ann. Neurol.* **35,** S28–34.

Glover, J. M., Leeds, J. M., Mant, T. G., Amin, D., Kisner, D. L., Zuckerman, J. E., Geary, R. S., Levin,

A. A., and Shanahan, W. R., Jr. (1997). Phase I safety and pharmacokinetic profile of an intercellular adhesion molecule-1 antisense oligodeoxynucleotide (ISIS 2302). *J. Pharmacol. Exp. Ther.* **282,** 1173–1180.

Grunhaus, A., and Horwitz, M. S. (1992). Adenoviruses as cloning vectors. *Semin. Virol.* **3,** 237–252.

Heath, T. D. (1987). Covalent attachment of proteins to liposomes. *Methods Enzymol.* **149,** 111–119.

Hickman, M. A., Malone, R. W., Lehmann-Buinsma, K., Sih, T. R., Knoell, D., Szoka, F. C., Walzem, R., Carlson, D. M., and Powell J. S. (1994). Gene expression following direct injection of DNA into liver. *Human Gene Ther.* **5,** 1477–1483.

Holmberg, E., Maruyama, K., Litzinger, D. C., Wright, S., Davis, M., Kabalka, G. W., Kennel, S. J., and Huang, L. (1989). Highly efficient immunoliposomes prepared with a method which is compatible with various lipid compositions. *Biochem. Biophys. Res. Commun.* **165,** 1272–1278.

Huang, L., Huang, A., and Kennel, S. (1984) Coupling of antibodies with liposomes. In G. Gregoriadis (Ed.), *Liposome technology* (Vol. III, pp. 51–62). Boca Raton, FL: CRC Press.

Hui, K. M., Ang, P. T., Huang, L., and Tay, S. K. (1997). Phase I study of immunotherapy of cutaneous metastases of human carcinoma using allogeneic and xenogeneic MHC DNA-liposome complexes. *Gene Ther.* **4,** 783–790.

Hwang, K. J. (1987). Liposome pharmacokinetics. In M. J. Ostro (Ed.), *Liposomes: From biophysics to therapeutics* (pp. 109–156). New York: Decker.

Jiao, S., Williams, P., Berg, R. K., Hodgeman, B. A., Liu, L., Repetto, G., and Wolff, J. A. (1992). Direct gene transfer into nonhuman primate myofibers *in vivo. Human Gene Ther.* **3,** 21–33.

Kaneda, Y., Iwai, K., and Uchida, T. (1989). Introduction and expression of the human insulin gene in adult rat liver. *J. Biol. Chem.* **264,** 12126–12129.

Kato, K., Dohi, Y., Yoneda, Y., Yamamura, K., Okada, Y., and Nakanishi, M. (1993). Use of the hemagglutinating virus of Japan (HVJ)-liposome method for analysis of infiltrating lymphocytes induced by hepatitis B virus gene expression in liver tissue. *Biochim. Biophys. Acta* **1182,** 283–290.

Kay, E. R. (1961). Incorporation of deoxyribonucleic acid by mammalian cells *in vitro. Nature* **191,** 387–388.

Kreuzer, J., Denger, S., Reifers, F., Beisel, C., Haack, K., Gebert, J., and Kubler, W. (1996). Adenovirus-assisted lipofection: Efficient *in vitro* gene transfer of luciferase and cytosine deaminase to human smooth muscle cells. *Atherosclerosis* **124,** 49–60.

Kupfer, J. M., Ruan, X. M., Liu, G., Matloff, J., Forrester, J., and Chaux, A. (1994). High-efficiency gene transfer to autologous rabbit jugular vein grafts using adenovirus-transferrin/polylysine-DNA complexes. *Human Gene Ther.* **5,** 1437–1443.

Lasic, D. D. (1996). Liposomes in gene therapy. *Adv. Drug. Del.* **20,** 221–266.

Lechardeur, D., Sohn, K-J., Haardt, M., Joshi, P. B., Monck, M., Graham, R. W., Beatty, B., Squire, J., O'Brodovich, H., and Lukacs, G. L. (in press). Metabolic instability of plasmid DNA in the cytosol: A potential barrier to gene transfer. *Gene Ther.*

Lederberg, J. (1956). Genetic transduction. *Am. Sci.* **44,** 264–280.

Lee, R. J., and Huang, L. (1996). Folate-targeted, anionic liposome-entrapped polylysine-condensed DNA for tumor cell-specific gene transfer. *J. Biol. Chem.* **271,** 8481–8487.

Li, S., Brisson, M., He, Y., and Haung, L. (1997). Delivery of a PCR amplified DNA fragment into cells: A model for using synthetic genes for gene therapy. *Gene Ther.* **4,** 449–454.

Li, S., and Huang, L. (1997). *In vivo* gene transfer via intravenous administration of cationic lipid-protamine-DNA (LPD) complexes. *Gene Ther.* **4,** 891–900.

Li, S., Rizzo, M. A., Bhattacharya, S., and Huang, L. (1998). Characterization of cationic lipid-protamine-DNA (LPD) complexes for intravenous gene delivery. *Gene Ther.* **5,** 930–937.

Li, S., Tseng, W-C., Stolz, D. B., Wu, S-P., Watkins, S. C., and Haung, L. (in press). Dynamic changes in the characteristics of cationic lipidic vectors after exposure to mouse serum: Implications for intravenous lipofection. *Gene Ther.*

Litzinger, D. C., and Huang, L. (1992). Phosphatidylethanolamine liposomes: Drug delivery, gene transfer and immunodiagnostic applications. *Biochim. Biophys. Acta* **1113,** 201–227.

Liu, F., Qi, H., Huang, L., and Liu, D. (1997). Factors controlling the efficiency of cationic lipid-mediated transfection *in vivo* via intraveneous administration. *Gene Ther.* **4,** 517–523.

Liu, F., Yang, J., Huang, L., and Liu, D. (1996). New cationic lipid formulations for gene transfer. *Pharm. Res.* **13,** 1856–1860.

Lyerly, H. K., and DiMaio, J. M. (1993). Gene delivery systems in surgery. *Arch. Surg.* **128,** 1197–1206.

Machy, P., and Leserman, L. D. (1983). Small liposomes are better than large liposomes for specific drug delivery *in vitro*. *Biochim. Biophys. Acta* **730,** 313–320.

Maruyama, K., Kennel, S. J., and Huang, L. (1990). Lipid composition is important for highly efficient target binding and retention of immunoliposomes. *Proc. Natl. Acad. Sci. USA* **87,** 5744–5748.

Mercola, K. E., Stang, H. D., Browne, J., Salser, W., and Cline, M. J. (1980). Insertion of a new gene of viral origin into bone marrow cells of mice. *Science* **208,** 1033–1035.

Miller, A. D., Jolly, D. J., Friedmann, T., and Verma, I. M. (1983). A transmissible retrovirus expressing human hypoxanthine phosphoribosyltransferase (HPRT): gene transfer into cells obtained from humans deficient in HPRT. *Proc. Natl. Acad. Sci. USA* **80,** 4709–4713.

Muzyczka, N. (1992). Use of adeno-associated virus as a general transduction vector for mammalian cells. *Curr. Top. Microbiol. Immunol.* **158,** 97–129.

Nabel, E. G., Gordon, D., Yang, Z-Y., Xu, L., San, H., Plautz, G. E., Wu, B-Y., Gao, X., Huang, L., and Nabel, G. J. (1992). Gene transfer *in vivo* with DNA-liposome complexes: Lack of autoimmunity and gonadal localization. *Human Gene Ther.* **3,** 649–656.

Nabel, G. J., Nabel, E. G., Yang, Z-Y., Fox, B. A., Plautz, G. E., Gao, X., Huang, L., Shu, S., Gordon, D., and Chang, A. E. (1993). Direct gene transfer with DNA-liposome complexes in melanoma: Expression, biologic activity, and lack of toxicity in humans. *Proc. Natl. Acad. Sci. USA* **90,** 11307–11311.

Naldini, L., Blomer, U., Gage, F. H., Trono, D., and Verma, I. M. (1996). Efficient transfer, integration, and sustained long-term expression of the transgene in adult rat brains injected with a lentiviral vector. *Proc. Natl. Acad. Sci. USA* **93,** 11382–11388.

Noguiez-Hellin, P., Meur, M. R., Salzmann, J. L., and Klatzmann, D. (1996). Plasmoviruses: Nonviral/viral vectors for gene therapy. *Proc. Natl. Acad. Sci. USA* **93,** 4175–4180.

Parker, R. J., Sieber, S. M., and Weinstein, J. N. (1981). Effect of liposome encapsulation of a fluorescent dye on its uptake by the lymphatics of the rat. *Pharmacology* **23,** 128–136.

Plank, C., Mechtler, K., Szoka, F. C., Jr., and Wagner, E. (1996). Activation of the complement system by synthetic DNA complexes: A potential barrier for intravenous gene delivery. *Human Gene Ther.* **7,** 1437–1446.

Poste, G., Kirsh, R., and Koestler, T. (1984) The challenge of liposome targeting *in vivo*. In G. Gregoriadis, (Ed.), *Liposome technology* (Vol. III, pp. 1–28). Boca Raton, FL: CRC Press.

Rädler, J. O., Koltover, I., Salditt, T., and Safinya, C. (1997). Structure of DNA-cationic liposome complexes: DNA intercalation in multilamellar membranes in distinct interhelical packing regimes. *Science* **275,** 810–814.

Remy, J-S., Sirlin, C., Vierling, P., and Behr, J. P. (1994). Gene transfer with a series of lipophilic DNA-binding molecules. *Bioconjugate Chem.* **5,** 647–654.

Roizman, B., and Sears, A. E. (1990). Herpes simplex viruses and their replication. In B. N. Fields and D. M. Knipe (Eds.), *Fields Virology* (2nd ed., pp. 1795–1842). New York: Raven Press.

Rosenberg, S. A., Aebersold, P., Cornetta, K., Kasid, A., Morgan, R. A., Moen, R., Karson, E. M., Lotze, M. T., Yang, J. C., Topalian, S. L., *et al.* (1990). Gene transfer into humans: Immunotherapy of patients with advanced melanoma using tumor infiltrating lymphocytes modified by retroviral gene transduction. *New Engl. J. Med.* **323,** 570–578.

Saldeen, J., Curiel, D. T., Eizirik, D. L., Andersson, A., Strandell, E., Buschard, K., and Welsh, N. (1996). Efficient gene transfer to dispersed human pancreatic islet cells *in vitro* using adenovirus-polylysine/DNA complexes or polycationic liposomes. *Diabetes* **45,** 1197–1203.

Sambrook, J., Westphal, H., Srivansan, P. R., and Dulbecco, R. (1968). The integrated state of viral DNA in SV40-transformed cells. *Proc. Natl. Acad. Sci. USA* **59,** 1288–1293.

Sawa, Y., Suzuki, K., Bai, H. Z., Shirakura, R., Morishita, R., Kaneda, Y., and Matsuda, H. (1995). Efficiency of *in vivo* gene transfection into transplanted rat heart by coronary infusion of HVJ liposome. *Circulation* **92**, 479–482.

Schneider, M. D., and French, B. A. (1993). The advent of adenovirus: Gene therapy for cardiovascular disease. *Circulation* **88**, 1937–1942.

Schwartz, B., Benoist, C., Abdallah, B., Scherman, D., Behr, J. P., and Demeneix, B. (1995). Lipospermine-based gene transfer into the newborn mouse brain is optimized by a low lipospermine/DNA charge ratio. *Human Gene Ther.* **6**,1515–1524.

Sen, D., and Crothers, D. M. (1986). Condensation of chromatin: Role of multivalent cations. *Biochemistry* **25**,1495–1503.

Shimotohno, K., and Temin, H. M. (1981). Formation of infectious progeny virus after insertion of herpes simplex tymidine gene into DNA of an avian retrovirus. *Cell* **26**, 67–77.

Singh, D., and Rigby, P. (1996). The use of histone as a facilitator to improve the efficiency of retroviral gene transfer. *Nucleic Acids Res.* **24**, 3113–3114.

Smirnov, I. V., Dimitrov, S. I., and Makarov, V. L. (1988). Polyamine-DNA interactions. Condensation of chromatin and naked DNA. *J. Biomol. Struct. Dyn.* **5**, 1149–1161.

Son, K., and Huang, L. (1994). Exposure of human ovarian carcinoma to cisplatin transiently sensitizes the tumor cells for liposome-mediated gene transfer. *Proc. Natl. Acad. Sci. USA* **91**, 12669–12672.

Song, Y. K., Liu, F., Chu, S., and Liu, D. (1997). Characterization of cationic liposome-mediated gene transfer *in vivo* by intravenous administration. *Human Gene Ther.* **8**, 1585–1594.

Stavridis, J. C., Deliconstantinos, G., Psallidopoulos, M. C., Armenakas, N. A., Hadjiminas, D. J., and Hadjiminas, J. (1986). Construction of transferrin-coated liposomes for *in vivo* transport of exogenous DNA to bone marrow erythroblasts in rabbits. *Exp. Cell. Res.* **164**, 568–572.

Sternberg, B., Sorgi, F. L., and Huang, L. (1994). New structures in complex formation between DNA and cationic liposomes visualized by freeze–fracture electron microscopy. *FEBS Lett.* **356**, 361–366.

Stewart, M. J., Plautz, G. E., Buono, L. D., Yang, Z. Y., Xu, L., Gao, X., Huang, L., Nabel, E. G., and Nabel, G. J. (1992). Gene transfer *in vivo* with DNA-liposome complexes: Safety and acute toxicity in mice. *Human Gene Ther.* **3**, 267–275.

Stratford-Perricaudet, L. D., and Perricaudet, M. (1994). Gene therapy: The advent of adenovirus. In J. A. Wolff, (Ed.), *Gene therapeutics: Methods and applications of direct gene transfer* Boston: Birkhäuser.

Stribling, R., Brunette, E., Liggitt, D., Gaensler, K., and Debs, R. (1992). Aerosol gene delivery *in vivo*. *Proc. Natl. Acad. Sci. USA* **89**, 11277–11281.

Stripp, B. R., Whitsett, J. A., and Lattier, D. L. (1990). Strategies for analysis of gene expression: Pulmonary surfactant proteins. *Am. J. Physiol.* **259**, 185–197.

Sun, W. H., Burkholder, J. K., Sun, J., Culp, J., Lu, X. G., Pugh, T. D., Ershler, W. B., and Yang, N. S. (1995). *In vivo* cytokine gene transfer by gene gun reduces tumor growth in mice. *Proc. Natl. Acad. Sci. USA* **92**, 2889–2893.

Tabin, C. J., Hoffman, J. W., Goff, S. P., and Weinberg, R. A. (1982). Adaptation of a retrovirus as a eucaryotic vector transmitting the herpes simplex thymidine kinase gene. *Mol. Cell. Biol.* **2**, 426–436.

Takehara, T., Hayashi, N., Miyamoto, Y., Yamamoto, M., Mita, E., Fusamoto, H., and Kamada, T. (1995). Expression of the hepatitis C virus genome in rat liver after cationic liposome-mediated *in vivo* gene transfer. *Hepatology* **21**, 746–751.

Tomita, N., Higaki, J., Kaneda, Y., Yu, H., Morishita, R., Mikami, H., and Ogihara, T. (1993). Hypertensive rats produced by *in vivo* introduction of the human renin gene. *Circ. Res.* **73**, 898–905.

Tomita, N., Higaki, J., Morishita, R., Kato, K., Mikami, H., Kaneda, Y., and Ogihara, T. (1992). Direct *in vivo* gene introduction into rat kidney. *Biochem. Biophys. Res. Commun.* **186**, 129–134.

Trubetskoy, V. S., Tochillin, V. P., Kennel, S., and Huang, L. (1992 a). Use of N-terminal modified poly-L-lysine-antibody conjugate as a carrier for targeted gene delivery in mouse lung endothelial cells. *Bioconjugate Chem.* **3**, 323–327.

Trubetskoy, V. S., Tochillin, V. P., Kennel, S., and Huang, L. (1992 b). Cationic liposomes enhance targeted delivery and expression of exogenous DNA mediated by N-terminal modified poly-L-lysine-antibody conjugate in mouse lung endothelial cells. *Biochim. Biophys. Acta* **1131**, 311–313.

Tseng, W.-C., and Huang, L. (1998). Liposome-based gene therapy. *Pharm. Sci. Tech. Today* **1**, 206–213.

Tuner, A., Kirby, C., Senior, J., and Gregoriadis, G. (1983). Fate of cholesterol-rich liposomes after subcutaneous injection into rats. *Biochim. Biophys. Acta* **760**, 119–125.

Varmus, H. (1988). Retroviruses. *Science* **240**, 1427–1435.

Vigneron, J-P., Oudrhiri, N., Fauquet, M., Vergely, L., Bradley, J-C., Basseville, M., Lehn, P., and Lehn, J-M. (1996). Guanidium-cholesterol cationic lipids: efficient vectors for the transfection of eukaryotic cells. *Proc. Natl. Acad. Sci. USA* **93**, 9682–9686.

Vile, R. G., and Hart, I. R. (1993). *In vitro* and *in vivo* targeting of gene expression to melanoma cells. *Cancer Res.* **53**, 962–967.

Wang, C-Y., and Huang, L. (1987). pH-sensitive immunoliposomes mediate target-cell-specific delivery and controlled expression of a foreign gene in mouse. *Proc. Natl. Acad. Sci. USA* **84**, 7851–7855.

Watson, J. D., and Crick, F. H. C. (1954). Genetical implications of the structure of deoxyribonucleic acid. *Nature* **171**, 737–739.

Wei, C., Gibson, M., Spear, P. G., and Scolnick, E. M. (1981). Construction and isolation of a transmissible retrovirus containing the src gene from Harvey Murine Sarcoma Virus and the thymidine kinase gene from herpes simplex virus type I. *J. Virol.* **39**, 935–944.

Westphal, H., and Dulbecco, R. (1968). Viral DNA in SV40 and polyoma-transformed cell lines. *Proc. Natl. Acad. Sci. USA* **59**, 1156–1162.

Wilson, R. W., and Bloomfield, V. A. (1979). Counterion-induced condensation of deoxyribonucleic acid: A light-scattering study. *Biochemistry* **18**, 2192–2196.

Wolff, J. A., Malone, R. W., Williams, P., Chong, W., Acsadi, G., Jani, A., and Felgner, P. L. (1990). Direct gene transfer into mouse muscle *in vivo*. *Science* **247**, 1465–1468.

Wu, G. Y., Zhan, P., Sze, L. L., Rosenberg, A. R., and Wu, C. H. (1994). Incorporation of adenovirus into a ligand-based DNA carrier system results in retention of original receptor specificity and enhances targeted gene expression. *J. Biol. Chem.* **269**, 11542–11546.

Xu, Y., and Szoka, F. C., Jr. (1996). Mechanism of DNA release from cationic liposome/DNA complexes used in cell transfection. *Biochemistry* **35**, 5616–5623.

Yang, J. P. and Huang, L. (1996). Direct gene transfer to mouse melanoma by intratumor injection of free DNA. *Gene Ther.* **3**, 542–548.

Yang, N-S., Burkholder, J., Roberts, B., Martinell, B., and McCabe, D. (1990). *In vivo* and *in vitro* gene transfer to mammalian somatic cells by particle bombardment. *Proc. Natl. Acad. Sci. USA* **87**, 9568–9572.

Yoshikawa, Y., and Yoshikawa, K. (1995). Diaminoalkanes with an odd number of carbon atoms induce compaction of a single double-stranded DNA chain. *FEBS Lett.* **361**, 277–281.

Yoshimura, K., Rosenfeld, M., Nakamura, H., Scherer, M. M., Pavirani, A., Lecocq, J-P., Crystal, R. G. (1992). Expression of the human cystic fibrosis transmembrane conductance regulator gene in the mouse lung after *in vivo* intracheal plasmid-mediated gene transfer. *Nucleic Acid Res.* **20**, 3233–3240.

Zabner, J., Fasbender, A. J., Moninger, T., Poellinger, K. A., and Welsh, M. J. (1995). Cellular and molecular barriers to gene transfer by a cationic lipid. *J. Biol. Chem.* **270**, 18997–19007.

Zhou, X., and Huang, L. (1994). DNA transfection mediated by cationic liposomes containing lipopolylysine: Characterization and mechanism of action. *Biochim. Biophys. Acta* **1189**, 195–203.

Zhou, X., Klibanov, A., and Huang, L. (1991). Lipophilic polylysines mediate efficient DNA transfection in mammalian cells. *Biochim. Biophys. Acta* **1065**, 8–14.

Zhou, X., Klibanov, A., and Huang, L. (1992). Improved encapsulation of DNA in pH-sensitive liposomes for transfection. *J. Liposome Res.* **2**, 125–139.

Zinder, N. D., and Lederberg, J. (1952). Genetic exchange in Salmonella. *J. Bacteriol.* **64**, 679–699.

PART II

Cationic Liposomes

CHAPTER 2

Progress in Gene Delivery Research and Development

Philip L. Felgner

Gene Therapy Systems, Inc., San Diego, California

I. Early *in Vitro* Transfection Studies
II. Molecular Biology and Biotechnology Applications for DNA Transfection
III. Application of Synthetic Vectors for Gene Therapy
IV. Developing More Efficient Synthetic Gene Delivery Systems
References

Although synthetic delivery systems entered the gene therapy repertoire somewhat later than viral vectors, there is a long history of nonviral gene delivery research. Some of the earliest scientifically controlled experiments go back more than 70 years, before the significance of DNA was even understood. Systematic *in vitro* transfection studies began in the late 1950s with results from several laboratories showing that purified viral nucleic acid could be transfected into cultured cells, leading to the proliferation of infectious virus particles. With the advent of recombinant DNA technology, transfection was used routinely and extensively in academic laboratories and industry to express recombinant proteins in cells and even to generate helper virus free recombinant virus particles for gene therapy applications. The discoveries that cationic lipids could greatly increase transfection in cultured cells and that naked DNA could be expressed in muscle following direct im injection helped to stimulate interest in the development of nonviral delivery systems for gene therapy applications. Although synthetic vectors have been applied slowly to gene therapy applications, today they are among the most widely tested vectors in human clinical trials.

Nonviral Vectors for Gene Therapy

I. EARLY *IN VITRO* TRANSFECTION STUDIES

In 1928, Fred Griffith was working with two different strains of the pneumococcus bacteria that exhibited different pathogenicities and colony morphologies. He showed that heat-killed pathogenic cells, when mixed with the live nonpathogenic cells, could transform a small percentage of the nonpathogenic cells into the pathogenic phenotype. Oswald Avery took up the task of defining the active principle responsible for this transformation. He originally believed that Griffith's "transforming factor" was a complex polysaccharide because the wild type and transformed colonies had different morphologies, and he reasoned that this characteristic difference might have been due to differences in the bacterial cell surface carbohydrates. In 1944, he showed that the transforming factor could be extracted and purified with the nucleic acid fraction from the heat-killed bacteria. It was stable to protease and RNase digestion but sensitive to DNase digestion. These results convinced most scientists that the active principle in the transforming factor was DNA. Although the significance of these experiments was not fully appreciated in this context, they probably represent the first scientific demonstration of DNA transfection. Today, bacterial cell transfection is a routine molecular biology technique that is essential to the application of recombinant DNA technology.

The earliest efforts to identify methods for enhancing delivery of functional purified polynucleotides into living mammalian cells (Table 1) were stimulated in the mid 1950s by the results of Alexander and Holland (Alexander *et al.*, 1958; Holland *et al.*, 1959) showing that purified poliovirus RNA was infectious in HeLa cells. Since the titer of this purified genomic RNA stock was extremely low (up to 1 million times less active than the original intact virus), it was questioned whether a vanishingly small quantity of intact virus had survived the purification procedure. Alexander and Holland showed that a hypertonic saline solution (1 M NaCl) could enhance infectivity of the purified RNA by about 100-fold. This result, which demonstrated that living virus particles were not required to produce living (i.e., infectious and replicating) virus, was subsequently confirmed and extended using

Table 1

**Methods Developed to Deliver
Infectious Viral Nucleic Acid *in Vitro***

Year	Author	Milestone
1958	Alexander *et al.*	Purified infectious poliovirus RNA
1959	Holland *et al.*	Purified infectious poliovirus RNA
1962	Smull *et al.*	Basic protein mediated
1965	Vaheri *et al.*	DEAE dextran mediated
1973	Graham *et al.*	Calcium phosphate method

improved methods for polynucleotide delivery, including calcium phosphate copre-cipitation (Graham and van der Eb, 1973) and DEAE dextran (Vaheri *et al.*, 1965).

II. MOLECULAR BIOLOGY AND BIOTECHNOLOGY APPLICATIONS FOR DNA TRANSFECTION

Broader utility for the calcium phosphate procedure was later demonstrated by experiments showing that transfection of a noninfectious fragment of the herpes simplex virus genome containing the thymidine kinase gene could transform thy-midine kinase-negative cells (Minson *et al.*, 1978). This result demonstrated that an infectious and proliferating virus was not necessary to induce cellular transforma-tion leading to a new cellular phenotype. The discovery that cells could be trans-fected with antibiotic resistence genes provided a convenient way to select for cells that were stably transduced with the new plasmid.

Transfection technology and the emerging recombinant DNA field con-verged in the late 1970s when Berg and colleagues applied calcium phosphate, DEAE dextran, and liposome-mediated transfection methodologies to the delivery and expression of recombinant plasmids in cultured mammalian cells (Mulligan *et al.*, 1979; Fraley *et al.*, 1980; Southern *et al.*, 1982). Mulligan showed that cells could be transfected with a plasmid encoding rabbit β-globin, leading to the ex-pression of authentic β-globin protein in the cells. Southern showed that cells could be transfected with the neomycin resistence gene, enabling the stably transduced cells to survive a neomycin antibiotic challenge. This finding provided a conve-nient way to select and expand stably transduced cells. A gene of interest, such as β-globin, could be cloned into a plasmid that also expressed the neomycin resis-tance gene; then all of the cells that survived after the neomycin challenge would also express β-globin. These findings led to widespread applications for transfection methodology in molecular and cellular biology and in the pharmaceutical industry. Today, plasmid transfection is usually the first step in producing stably transduced

Table 2

Synthetic Gene Delivery System Applications

Year	Author	Milestone
1978	Minson *et al.*	Thymidine Kinase Transformation
1979	Mulligen, Berg	Rabbit β-globin−Plasmid
1982	Southern, Berg	Neomycin (G418) Selection
1983	Mann *et al.*	Helper−Free Retrovirus Vector
1984	Hwang, Gilboa	Title: " . . . retroviral infection is more efficient than . . . DNA transfection."

cell lines that are used to manufacture recombinant proteins for the biotechnology industry.

Interestingly, plasmid transfection is also used in the production of recombinant viral particles for gene therapy applications. Theoretical support for the use of viruses for gene therapy initially arose from the knowledge that DNA and RNA tumor viruses were capable of introducing heritable changes into the genome of mammalian cells. Friedmann proposed that if the deleterious features of these viruses could be eliminated, one might safely deliver a desirable gene to correct cellular defects or genetic disorders. Mulligan and his co-workers (Mann *et al.*, 1983) first demonstrated the technical feasibility of this hypothesis by describing a method of producing recombinant infectious but nonreplicating retrovirus particles. This feat was accomplished by using a cell line that expressed all of the virus's structural genes but lacked a functional viral genome. When a plasmid encoding any gene of interest and a unique nucleotide sequence called the *packaging signal* are transfected into these cells, infectious virus particles containing the encapsulated transgene of interest are produced. These particles have a complete virus coat so that they can efficiently infect cells, but they lack a functional virus genome so they cannot replicate. The recognition and application of this elegant methodology greatly stimulated research in the gene therapy field; plasmid transfection is utilized to produce the packaging cell lines and also to introduce the recombinant gene of interest into the packaging cell.

III. APPLICATION OF SYNTHETIC VECTORS FOR GENE THERAPY

Enthusiasm for the use of nonviral gene delivery systems for gene therapy developed more slowly. As previously discussed, the earliest successful *in vitro* transfection experiments were done with purified infectious viral genomes and the end point was the demonstration of infectious virus particles or plaques (Alexander *et al.*, 1958; Holland *et al.*, 1959). Similarly, among the earliest reported *in vivo* transfections were experiments (Table 3) showing that virus replication could occur following injection of purified or cloned viral genomes *in vivo*. The highest levels of expression were observed with particulate formulations of calcium phosphate-precipitated DNA (reviewed in Felgner and Rhodes, 1991). Israel (Israel *et al.*, 1979) showed that polyoma virus DNA was infectious following intraperitoneal injection into mice and hamsters, but at a level 4 to 5 logs below that of intact polyoma virions. Dubensky and Bouchard showed that *in vivo* polyoma virus DNA replication efficiency could be improved by using calcium phosphate-precipitated DNA and by treating the tissue with hyaluronidase and collagenase (Dubensky *et al.*, 1984; Bouchard *et al.*, 1984). Both groups also showed that the infectious polyoma

Table 3

***In Vivo* Gene Transfer with Synthetic Systems
Using Infectious and Oncogenic Plasmids**[a]

Year	Author	Milestone
1979	Israel, *et al.*	Infectious polyoma virus plasmid
1983	Fung, *et al.*	Oncogenic SRC plasmid
1984	Bouchard *et al.*	Oncogenic polyoma plasmid
1984	Dubensky *et al.*	Infectious polyoma virus plasmid
1984	Seeger Varmus	Infectious hepatitis Virus plasmid

[a]Studies showing that infectious plasmids replicate in vivo and oncogenic plasmids form tumors in vivo.

virus plasmid produced tumors in a significant percentage of the treated animals. Fung showed that a noninfectious plasmid encoding the *src* oncogene could produce tumors in susceptible chickens (Fung *et al.,* 1983). Gould-Fogerite used liposomes to deliver a noninfectious, oncogenic papyloma virus plasmid DNA fragment to induce mouse tumors following direct subcutanteous injection (Gould-Fogerite *et al.,* 1989). Finally, Varmus and co-workers gave a single 20-μg intrahepatic injection of a cloned infectious hepatitis viral DNA (ground-squirrel-specific virus) into ground squirrels and produced seropositive animals at 11–18 weeks postinjection (Seeger, 1984). In all of these cases the observed effects were primarily attributed to the ability of the systems under study to be amplified by replication of an extremely rare, low-frequency infectious or oncogenic event. Discussion of how the apparent low level of expression obtained could be used to therapeutic benefit was limited. However, the results did demonstrate that infectious and oncogenic DNA sequences could be taken up and expressed, albeit at a low level, by cells following direct *in vivo* injection.

During the 1980s several reports showing *in vivo* expression from directly injected noninfectious, nononcogenic plasmid DNA sequences were published. (Table 4 and 5). *In vivo* gene expression using liposomes, calcium phosphate, a polylysine/protein conjugate, cationic lipids, and naked DNA was reported. Benvenisty detected chloramphenocol acetyltransferase (CAT) activity in newborn rat tissues, following intraperitoneal injection of calcium phosphate-precipitated plasmid DNA (Benvenisty and Reshef, 1996). CAT plasmid constructs containing different eukaryotic viral promoters, a proinsulin gene, and a human growth hormone gene were all reported to express messenger RNA and/or gene product.

Wu prepared a asialo-orosomucoid (AsOR)/polylysine conjugate that interacted spontaneously with pSV2 CAT plasmid DNA (Wu *et al.,* 1989). Twenty-four hours following intraveneous injection, CAT activity was detected in rat liver. Activity was detected only when the pSV2 CAT DNA was complexed with the

Table 4

Second-Generation Synthetic Gene Delivery Systems[a]

Year	Author	Milestone
1980	Fraley *et al.*	Liposomes
1987	Felgner *et al.*	Cationic lipids
1987	Wu *et al.*	Polylysine/receptor mediated
1988	Johnston *et al.*	Gene gun

[a] Systems developed to deliver plasmid *in vitro* and *in vivo*.

AsOR/PL conjugate; free plasmid DNA was not active *in vivo*. The activity declined to baseline levels by 96 h postdosing. Animals given a partial hepatectomy prior to dosing maintained high levels of CAT activity for up to 11 weeks postdosing. Analysis of DNA extracts from the tissues by Southern blot suggested that the CAT plasmid DNA was integrated into liver cell genomic DNA. These results led to the hypothesis that cells undergoing division in regenerating liver permit stable integration of episomal DNA into the host cell genome and that this newly integrated DNA is active and stable.

Table 5

Early Gene Transfer Studies with Noninfectious Plasmids

Year	Author	Milestone
1983	Nicolau *et al.*	Insulin gene lowers glucose levels
		Liposomes
1986	Benvenisty and Reshef	CAT plasmids in peritoneum
		Calcium phosphate
1987	Wang, Huang	CAT plasmid in peritoneal tumor
		Liposomes
1989	Wu, Wilson	CAT plasmids in liver
		Polylysine/orosomucoid
1989	Kaneda *et al.*	Reporter plasmids in liver
		Liposomes
1989	Brigham *et al.*	Lung plasmids in lung
		Cationic lipids
1990	Nabel *et al.*	β-Gal in blood vessel
		Cationic lipids
1990	Holt *et al.*	Luciferase in brain
		Cationic lipids
1990	Yang *et al.*	Reporter genes—liver, skin, muscle
		Gene gun
1990	Wolff *et al.*	Luciferase and CAT in muscle
		Naked DNA

The ability of liposomes to deliver genes *in vivo* was described in several reports. Nicholau reported a transient hypoglycemic effect and elevated insulin levels in the blood, spleen, and liver of rats, following intravenous injection of about 2 μg of a plasmid containing the rat preproinsulin gene (Nicholau *et al.,* 1983). Wang demonstrated CAT activity following intraperitoneal injection of pH-sensitive immunoliposomes containing a CAT plasmid into nude mice bearing an ascites tumor (Wang and Huang, 1987). Kaneda described experiments showing expression of SV40 large T antigen in the liver of rats following injection of a pBR SV40 plasmid into the portal vein (Kaneda *et al.,* 1989). The delivery vehicle was a complex mixture of components mixed in a particular order so as to produce complexes of phospholipid vesicles containing encapsulated DNA, ganglioside, Sendai virus fusion proteins, and red blood cell membrane proteins.

Cationic lipids were also studied for their ability to deliver genes *in vivo*. CAT gene expression was detected in the blood and lungs of mice following injection of 15 or 30 μg pSV2 CAT plasmid complexed with cationic lipid vesicles (Lipofectin reagent, BRL) (Brigham *et al.,* 1989). Injections were intraveneous, intraperitoneal, and intratracheal. Expression was observed for 6 days postinjection. Holt showed that direct injection of CAT and luciferase DNA and RNA into *Xenopus laevis* embryos resulted in expression of luciferase gene product (Holt *et al.,* 1990). Functional expression was shown to be dependent on the presence of the cationic lipid DOTMA, and expression could be enhanced by the coadministration of proteolytic enzymes. Luciferase activity in transfected embryos peaked during the first 48 h posttransfection and was still detectable 28 days later. Finally, Nabel showed that cationic lipids could enhance expression from plasmids administered into catheterized blood vessels (Nabel *et al.,* 1989).

Interest in the nonviral gene delivery approach was boosted by our results showing *in vivo* reporter gene expression from plasmids encoding luciferase, CAT, and β-galactosidase genes following direct injection into mouse muscle (Wolff *et al.,* 1990). Up to 1 ng of gene product could be isolated from tissues injected with $10-100$ μg of plasmid DNA, and expression was shown to persist for more than 6 months. The DNA did not integrate into the host genome, and plasmid essentially identical to the starting material could be recovered from the muscle months after injection. *In vitro* transcribed messenger RNA was also taken up and expressed in mouse muscle, but the duration of expression was much shorter due to enzymatic breakdown of the message. Interestingly, no special delivery system was required for these effects.

The practical potential of this nonviral approach to gene delivery was further substantiated by the demonstration that animals could be immunized following intramuscular injection of plasmids encoding heterologous antigens (Ulmer *et al.,* 1993). Gary Rhodes at Vical injected a plasmid encoding secreted HIV gp120, obtained from Nancy Haigwood at Chiron, into the skeletal muscle of mice and showed that antigen-specific antibodies and cytotoxic T-cells were generated.

Investigators at Vical and Merck subsequently showed that antibodies and CTL responses against influenza antigens could also be generated in this way, and the immune responses were sufficient to protect the plasmid-injected animals from a live virus challenge. Stephan Johnston showed that naked DNA plasmids encoding various different antigens could also be administered transdermally using the gene gun, leading to the generation of antibodies against the encoded antigens. These results led to a rapidly expanding interest in the new field of DNA vaccines, which is being actively investigated by immunologists, clinicians, and pharmaceutical companies today (Table 6). Human clinical trials evaluating DNA vaccines for HIV, influenza, malaria, and hepatitis are currently under way, and at least 10 other vaccine candidates are in advanced preclinical development.

Although naked DNA in the absense of any cationic lipids, is expressed reasonably well at the injection site following im administration, when the plasmid is administered by other routes, cationic lipids can frequently augment expression.

Table 6

Species in which DNA vaccine efficacy has been demonstrated		
Chicken	Rabbit	Cow
Trout	Ferret	Horse
Mouse	Cat	Monkey
Rat	Goat	Pig
Guinea Pig	Sheep	Chimpanzee

Effective DNA vaccine immunization against	
Viruses	
Avian and human influenza	LCMV
Bovine herpes	Measles
BVDV	Papilloma
Dengue fever	Rabies
Encephalitis	RSV
FELV	SHIV
Hepatitis B and C	SIV
Herpes simplex	SV-40
HIV-1	
Bacteria	
Lyme	*Rickettsia*
Moraxella bovis	*Salmonella*
Mycobacterium TB	Tetanus
Mycoplasma	
Parasites	
Cryptosporidium parvum	Malaria
Leishmania	*Schistosoma*

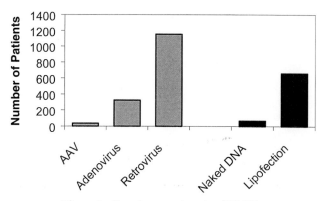

Figure 1   Gene therapy patients (as of 3/1/98).

Cationic lipid-mediated expression has been reported following direct intratracheal, intrahepatic, intracranial, and intratumor injection. Particularly high levels of cationic lipid-mediated expression can be obtained in tumor tissue that is growing within an accessible space such as the paritoneum or plural cavities. The ability of cationic lipids to enhance expression in tumor tissue *in vivo* has encouraged investigators to launch cancer gene therapy clinical trials. Figure 1 shows the number of patients administered synthetic gene delivery systems for cancer gene therapy applications compared to the numbers administered viral vectors.

IV. DEVELOPING MORE EFFICIENT SYNTHETIC GENE DELIVERY SYSTEMS

During the last 8 years we have witnessed substantial progress in the application of synthetic gene delivery systems. However, the clinical applications remain limited today because the *in vivo* expression levels obtainable remain too low. For this reason, investigators are exploring ways to improve the *in vivo* efficiency of these systems. The obstacles to more efficient *in vivo* gene delivery and expression can be described in terms of "barriers" (Table 7). There are extracellular barriers and intracellular barriers. The extracellular barriers refer to those obstacles that the injected gene encounters before it reaches its target cell. The four main extracellular barriers are opsinins, phagocytes, extracellular matrix, and degradative enzymes. Opsinins are proteins that attach themselves to the gene or the delivery system, inactivating the gene and its carrier. Phagocytes are cells that can seek out, engulf, and actively digest the delivery system. The extracellular matrix is a zone of polymerized protein and carbohydrate that is present between cells protecting the plasma membrane of

Table 7

Overcoming Barriers to Efficient Gene Delivery

Extracellular *in vivo* barriers
Opsonins
Phagocytic cells
Extracellular matrix
Degradative enzymes

Intracellular barriers
Plasma membrane
Endocome
Nuclear membrane

the target cell, and it can be difficult for a relatively large DNA carrier system to pass through this barrier. Finally, DNases present in the serum and extracellular fluid can rapidly digest unprotected DNA.

The principal intracellular barriers are the plasma membrane, the endosome, and the nuclear membrane. Once the gene delivery system reaches its target cell, it encounters the plasma membrane, which must be traversed before the gene can be expressed. Under the right circumstances, after the delivery system attaches to the plasma membrane it can be taken into the cell by endocytosis, but then the delivery system must have a mechanism for escaping from the endosome to avoid being degraded in the lysosomal compartment. Finally, if the gene is able to cross these barriers and enter the cell cytoplasm, it must still have a means of getting across the nuclear membrane.

Delivery system technology development directed at overcoming these barriers involves aspects of molecular biology, DNA condensation technology, and ligand conjugation chemistries (Table 8). With respect to molecular biology, inves-

Table 8

Improving Gene Delivery and Expression

Physical and structural aspects
Characterization and control

Identify the important biological barriers
In vitro and *invivo* mechanism studies

Modify the carrier
New condensing agents and ligand conjugates

Modify the DNA
Gene chemistry

Molecular biology aspects
Episomal vectors and Minichromosomes

tigators are looking for ways to increase the level of gene product expressed from plasmids by optimizing promoters, enhancers, introns, terminator sequences, and codon usage. Some investigators are also exploring ways to allow replication of the plasmid inside the nucleus of transfected cells. This approach would theoretically increase the number of plasmid copies per cell and also allow expression to persist at high levels within populations of rapidly dividing cells.

DNA condensation is another active area of scientific research. Today the principal ways of accomplishing condensation and packaging of plasmid DNA are with either hydrophobic cations or hydrophilic polycations. Hydrophobic cations form liposomes or micelles that can interact with DNA and reorganize into a cationic lipid/DNA complex called a "lipoplex." The hydrophilic polycations also form complexes when mixed with DNA and these complexes are called a "polyplex" (Table 9). The hydrophilic polymers are of two general types: the linear polymers such as polylysine and spermine; and the branched chain, spherical, or globular polycations such as polyethyleneimine and dendrimers. An active area of scientific research involves understanding and controlling the DNA condensation and packaging processes with these agents and determining the structure of the complexes. Following packaging of the DNA by these methods, investigators are interested in improving gene delivery efficiency by incorporating ligands into the complexes. These ligands are intended to introduce biological functions into the complexes in order to make them more effective at delivering genes into the target cells. Such ligands may include:

- *Peptides,* which have a specific cell surface receptor so that the complexes will be targeted to specific cells bearing this receptor.
- *Nuclear localization signals,* so that DNA can enter the nucleus more efficiently.
- *pH-sensitive ligands,* to encourage more efficient endosomal escape.
- *Steric stabilizing agents,* to avoid interaction with biological factors that would destabilize the complexes after introduction into the biological milieu.

Better methods for studying the biodistribution of plasmid *in vitro* and *in vivo* will facilitate synthetic gene delivery system research. A probe that would allow the

Table 9

Third–Generation Synthetic Gene Delivery Systems[a]

Year	Author	Milestone
1990	Wagner *et al.*	Polylysine/receptor mediated
1995	Boussif, Behr	Polyethylenimine
1996	Tang, Szoka	Dendrimer

[a] Improved systems for in vitro and in vivo gene delivery.

simultaneous localization of plasmid DNA and gene product expression at the cellular and subcellular level would offer a means for better understanding the cellular and molecular barriers to DNA delivery and should provide new insights leading to more effective plasmid delivery systems. To date, no technology has been reported that can enable the biodistribution of functionally and conformationally intact plasmid to be followed in real time in living cells, while simultaneously monitoring gene expression from the same plasmid. For this purpose, we recently developed a nonperturbing plasmid labeling procedure to generate a highly fluorescent plasmid DNA preparation.

In order to create this system, we used the property of peptide nucleic acids (PNA) to hybridize to nucleic acids in a high-affinity and sequence-specific manner. A fluorescent PNA conjugate was hybridized to its complementary sequence on a plasmid. The PNA binding site was cloned into a region of the plasmid that was not essential for transcription so that PNA binding would not interfere with expression. Fluorescent plasmid prepared in this way was neither functionally nor conformationally altered. PNA binding was sequence-specific, saturable, and extremely stable and did not influence the nucleic acid intracellular distribution. This method was utilized to study conformationally intact plasmid DNA biodistribution in living cells after cationic lipid-mediated transfection. A fluorescent plasmid expressing green fluorescent protein (GFP) enabled simultaneous localization of both plasmid and expressed protein in living cells and in real time. GFP was expressed only in cells containing detectable nuclear fluorescent plasmid. This detection method offers a way to simultaneously monitor the intracellular localization and expression of plasmid DNA in living cells. This tool should aid in the elucidation of the mechanism of plasmid delivery and its nuclear import with synthetic gene delivery systems, and in the development of more efficient synthetic gene delivery systems. In addition to providing a means for obtaining biologically active fluorescent DNA, this approach provides a way of introducing new physical and biological elements onto DNA without perturbing its transcriptional activity. This "gene chemistry" approach will be used in the future to couple ligands (e.g., nuclear localization signals, or other peptides) onto the DNA to improve its *in vivo* bioavailability and expression.

The science of synthetic gene delivery system discovery is still at an early stage and there are many different ways in which the current generation vectors can be improved. As these different approaches are explored and new developments are introduced, we can anticipate that improved delivery systems will emerge. Delivery systems that confer higher levels of *in vivo* gene product expression should yield increasingly successful clinical outcomes.

REFERENCES

Alexander, H. E., Koch, G., Moran-Mountain, I., Sprunt, K., and Van Damme, O. (1958). Infectivity of ribonucleic acid of poliovirus on HeLa cell monolayers. *J. Exp. Med.* **108,** 493–506.

Benvenisty, N., and Reshef, L. (1986). Direct introduction of genes into rats and expression of the genes. *Proc. Natl. Acad. Sci. USA* **83,** 9551–9555.

Bouchard, L., Gelinas, C., Asselin, C., and Bastin, M. (1984). Tumorigenic activity of polyoma virus and SV40 DNAs in newborn rodents. *Virology* **135,** 53–64.

Boussif, O., Lezoualc'h, F., Zanta, M. A., Mergny, M. D., Scherman, D., Demeneix, B., and Behr, J. P. (1995). A versatile vector for gene and oligonucleotide transfer into cells in culture and *in vivo:* Polyethylenimine. *Proc. Natl. Acad. Sci. USA* **92,** 7297–7301.

Brigham, K. L., Meyrick, B., Christman, B., Magnuson, M., King, G., and Berry, L. C., Jr. (1989). *In vivo* transfection of murine lungs with a functioning prokaryotic gene using a liposome vehicle. *Am. J. Med. Sci.* **298,** 278–281.

Dubensky, T. W., Cambell, B. A., and Villarreal, L. P. (1984). Direct transfection of viral plasmid DNA into the liver or spleen of mice. *Proc. Natl. Acad. Sci. USA* **81,** 7529–7533.

Felgner, P. L., and Rhodes, G. (1991). Gene Therapeutics. *Nature* **349,** 351–352.

Felgner, P. L., and Ringold, G. M. (1989). Cationic liposome-mediated transfection. *Nature* **337,** 387–388.

Felgner, P. L., Barenholz, Y., Behr, J. P., Cheng, S. H., Cullis, P., Huang, L., Jessee, J. A., Seymour, L., Szoka, F., Thierry, A. R., Wagner, E., and Wu, G. (1997). Nomenclature for synthetic gene delivery systems. *Human Gene Ther.* **8,** 511–512.

Felgner, P. L., Gadek, T. R., Holm, M., Roman, R., Chan, H. W., Wenz, M., Northrop, J. P., Ringold, G. M., and Danielsen, M. (1987). Lipofection: A highly efficient, lipid mediated DNA-transfection procedure. *Proc. Natil. Acad. Sci. USA* **84,** 7413–7417.

Felgner, P. L., Zaugg, R. H., and Norman, J. A. (1995). Synthetic recombinant DNA delivery for cancer therapeutics. *Cancer Gene Ther.* **2,** 61–65.

Fraley, R., Subramani, S., Berg, P., and Papahadjopoulos, D. (1980). Introduction of liposome-encapsulated SV40 DNA into cells. *J. Biol. Chem.* **255,** 10431–10435.

Friedmann, T., and Roblin, R. (1972). Gene therapy for human genetic disease? *Science* **175,** 949–955.

Fung, Y. K., Crittenden, L. B., Fadly, A. M., and Kung, H. J. (1983). Tumor induction by direct injection of cloned v-src DNA into chickens. *Proc. Natl. Acad. Sci. USA* **80,** 353–357.

Gould-Fogerite, S., Mazurkiewicz, J. E., Raska, K., Jr., Voelkerding, K., Lehman, J. M., and Mannino, R. J. (1989). Chimerasome-mediated gene transfer *in vitro* and *in vivo. Gene* **84,** 429–438.

Graham, F. L., and Van Der Eb, A. J. (1973). A new technique for the assay of infectivity of human adenovirus 5 DNA. *Virology* **52,** 456–467.

Hakim, I., Levy, S., and Levy, R. (1986). A nine-amino acid peptide from IL-1beta augments antitumor immune responses induced by protein and DNA vaccines. *J. Immunol.* **157,** 5503–5511.

Holland, J. J., McLaren, L. C., and Syverton, J. T. (1959). The Mammalian cell-virus relationship. III. Poliovirus production by non-primate cells exposed to poliovirus ribonucleic acid. *Proc. Soc. Exp. Biol. Med.* **100,** 843–845.

Holt, C. E., Garlick, N., and Cornel, E. (1990). Lipofection of cDNAs in the embryonic vertebrate central nervous system. *Neuron* **4,** 203–214.

Hwang, L. S., and Gilboa, E. (1984). Expression of genes introduced into cells by retroviral infection is more efficient than that of genes introduced into cells by DNA transfection. *J. Virol.* **50,** 417–424.

Israel, M. A., Chan, H. W., Martin, M. A., and Rowe, W. P. (1979). Molecular cloning of polyoma virus DNA in *Escherichia coli:* Oncogenicity testing in hamsters. *Science* **205,** 1140–1142.

Johnston, S. A., Anziano, P. Q., Shark, K., Sanford, J. S., and Butow, R. A. (1988). Mitochondrial transformation in yeast by bombardment with microprojectiles. *Science* **240,** 1538–1541.

Kaneda, Y., Iwai, K., and Uchida, T. (1989). Increased expression of DNA cointroduced with nuclear protein in adult rat liver. *Science* **243,** 375–378.

Mann, R., Mulligan, R. C., and Baltimore, D. (1983). Construction of a retrovirus packaging mutant and its use to produce helper-free defective retrovirus. *Cell* **33,** 153–159.

Minson, A. C., Wildy, P., Buchan, A., and Darby, G. (1978). Introduction of the *herpes simplex* virus thymidine kinase gene into mouse cells using virus DNA or transformed cell DNA. *Cell* **13,** 581–587.

Mulligan, R. C., Howard, B. H., and Berg, P. (1979). Synthesis of rabbit beta-globin in cultured monkey kidney cells following infection with a SV40 beta-globin recombinant genome. *Nature* **277,** 108–114.

Nabel, E. G., Plautz, G., Boyce, F. M., Stanley, J. C., and Nabel, G. J. (1989). Recombinant gene expression *in vivo* within endothelial cells of the arterial wall. *Science* **244,** 1342–1344.

Nabel, G. J., and Felgner, P. L. (1993). Direct gene transfer for immunotherapy and immunization. *Trends Biotechnol.* **11,** 211–215.

Nabel, G. J., Yang, Z. Y., Nabel, E. G., Bishop, K., Marquet, M., Felgner, P. L., Gordon, D., Chang, A. E. (1995). Direct gene transfer for treatment of human cancer. *Ann. N. Y. Acad. Sci.* **772,** 227–231.

Nicolau, C., Le Pape, A., Soriano, P., Fargette, F., and Juhel, M. F. (1983). *In vivo* expression of rat insulin after intravenous administration of the liposome-entrapped gene for rat insulin I. *Proc. Natl. Acad. Sci. USA* **80,** 1068–1072.

Seeger, C., Ganem, D., and Varmus, H. E. (1984). The cloned genome of ground squirrel hepatitis virus is infectious in the animal. *Proc. Natl. Acad. Sci. USA* **81,** 5849–5852.

Smull, C. E., and Ludwig, E. H. (1962). Enhancement of the plaque-forming capacity of poliovirus ribonucleic acid with basic proteins. *J. Bacteriol.* **84,** 1035–1040.

Southern, P. J., and Berg, P. (1982). Transformation of mammalian cells to antibiotic resistance with a bacterial gene under control of the SV40 early region promoter. *J. Mol. Appl. Gen.* **1,** 327–341.

Tang, M. X., Redemann, C. T., and Szoka, F. C., Jr. (1996). *In vitro* gene delivery by degraded polyamidoamine dendrimers. *Bioconjugate Chem.* **7,** 703–714.

Tripathy, S. K., Svensson, E. C., Black, H. B., Goldwasser, E., Margalith, M., Hobart, P. M., Leiden, J. M. (1996). Long-term expression of erythropoietin in the systemic circulation of mice after intramuscular injection of a plasmid DNA vector. *Proc. Natl. Acad. Sci. USA* **93,** 10876–10880.

Ulmer, J. B., Donnelly, J. J., Parker, S. E., Rhodes, G. H., Felgner, P. L., Dwarki, V. J., Gromkowski, S. H., Deck, R. R., DeWitt, C. M., Friedman, A., *et al.* (1993). Heterologous protection against influenza by injection of DNA encoding a viral protein. *Science* **259,** 1745–1749.

Vaheri, A., and Pagano, J. S. (1965). Infectious poliovirus RNA: A sensitive method of assay. *Virology* **27,** 434–436.

Wagner, E., Zenke, M., Cotten, M., Beug, H., and Birnstiel, M. L. (1990). Transferrin-polycation conjugates as carriers for DNA uptake into cells. *Proc. Natl. Acad. Sci. USA* **87,** 3410–3414.

Wang, C. Y., and Huang, L. (1987). pH-sensitive immunoliposomes mediate target-cell-specific delivery and controlled expression of a foreign gene in mouse. *Proc. Natl. Acad. Sci. USA* **84,** 7851–7855.

Wolff, J. A., Malone, R. W., Williams, P., Chong, W., Acsadi, G., Jani, G., and Felgner, P. L. (1990). Direct gene transfer into mouse muscle *in vivo*. *Science* **247,** 1465–1468.

Wu, C. H., Wilson, J. M., and Wu, G. Y. (1989). Targeting genes: Delivery and persistent expression of a foreign gene driven by mammalian regulatory elements *in vivo*. *J. Biol. Chem.* **264,** 16985–16987.

Wu, G. Y., and Wu, C. H. (1987). Receptor mediated *in vitro* gene transformation by a soluble DNA carrier system. *J. Biol. Chem.* **262,** 4429–4432.

Yang, N. S., Burkholder, J., Roberts, B., Martinell, B., and McCabe, D. (1990). *In vivo* and *in vitro* gene transfer to mammalian somatic cells by particle bombardment. *Proc Natl Acad Sci USA* **87,** 9568–9572.

CHAPTER 3

Cationic Lipid-Mediated Gene Delivery to the Airways

John Marshall, Nelson S. Yew, Simon J. Eastman, Canwen Jiang, Ronald K. Scheule, and Seng H. Cheng
Genzyme Corporation, Framingham, Massachusetts

Cationic lipid-mediated delivery of therapeutic genes to the airways represents an attractive modality for treatment of a variety of inherited and acquired pulmonary diseases. Since the first description of their potential for gene transfer, much progress has been made in the development of improved cationic lipid structures and formulations with enhanced gene transfection activity. Furthermore, as our understanding of the mechanisms by which these agents effect gene transfer continues to improve, novel strategies to overcome the barriers limiting this vector system are being

Nonviral Vectors for Gene Therapy

developed. However, despite significant improvements, it is apparent that the efficiency of gene transfer with the present formulations of cationic lipid/pDNA complexes, particularly to the airway epithelium, is still relatively low. Nevertheless, because cationic lipids present a different safety profile than viral vectors, they are currently under active investigation for the treatment of several human indications including cystic fibrosis (CF). Initial human clinical studies in CF using either intranasal instillation or aerosolization to the lung have shown variable low levels of transgene expression and partial functional correction of the electrophysiological defects associated with CF. Although instillation of the complexes into the nasal epithelium of CF subjects was without any obvious clinical consequences, aerosolization to the lungs of CF subjects was associated with some minimal toxicity. While these findings are encouraging, they indicate that improvement in the efficiency of gene transfer and in the safety profile of the cationic lipids needs to be attained before they can be used for the long-term treatment of chronic diseases such as CF.

I. INTRODUCTION

Gene therapy refers to the transmission of DNA encoding a therapeutic gene of interest into appropriate target cells or organs with consequent expression of the transgene. Over the past several years, this approach has been increasingly considered as a modality for the treatment of a variety of genetic and acquired diseases (Crystal et al., 1994; Caplen et al., 1995; Sobol and Scanlon, 1995; Barranger et al., 1997). However, the success of gene therapy is predicated on the development of gene transfer vectors that are safe and efficacious. In this regard, several different gene transduction vectors that are based on either viral or synthetic self-assembling systems are concurrently being assessed. Of the nonviral vectors being considered, cationic lipids have attracted a significant amount of interest primarily because of their simplicity, purportedly low degree of toxicity, ability to package and deliver large transgenes, and relative ease of production. Additionally, because this system is nonproteinaceous, it is essentially devoid of many of the host immune-related issues that have plagued viral-based vectors (Yang et al., 1995; Kaplan et al., 1996). This review addresses current efforts at developing this particular nonviral gene delivery vector system for the treatment of lung diseases. An overview of the barriers currently limiting cationic lipid-mediated gene delivery to this organ and how these impediments might be surmounted for the treatment of CF are discussed.

II. THE LUNG AS A TARGET
FOR GENE TRANSFECTION

Several practical considerations make the lung an attractive target for gene therapy interventions. First, unlike most organs, the lung can be accessed directly and therefore is conducive to direct genetic modification. Delivery of gene transfer vectors by aerosolization or intratracheal instillation provides access to the conducting airways and lung parenchyma. This facile access to the target cells precludes the necessity to develop more complex delivery systems, for example those that incorporate targeting moieties for facilitated delivery to the diseased cells. Second, luminal delivery minimizes the access of the transfecting agent to other organs that may be unaffected by disease. Importantly, this restriction also limits potential access to the gonads, thereby providing added protection against germ-line alteration. Third, the lung, by virtue of being highly vascularized, can also be genetically modified via the systemic route. This method of administration, unlike the luminal route which results in transfection of the airway epithelia, delivers the genes predominantly to the endothelial cells (Canonico et al., 1994). In this regard, different subsets of the lung cells can potentially be modified depending on the disease manifestations.

Interest in the lung as a target for gene therapy is also borne of recent advances in our understanding of the molecular basis of some of the more common genetic and acquired diseases affecting the respiratory system. Examples of such ailments include CF (Cheng et al., 1990; Rich et al., 1990), α_1-antitrypsin deficiency (Crystal et al., 1989; Rosenfeld et al., 1991; Jaffe et al., 1992), lung carcinoma (Chen et al., 1990; Martin and Lemoine, 1996), and malignant mesothelioma (Hwang et al., 1995). CF and α_1-antitrypsin deficiency represent the two most prevalent fatal inherited disorders of Caucasians, and lung cancer is the leading cause of cancer-related mortality among men and women in the United States. Since no effective therapies for any of these afflictions are presently available, much of the recent efforts in gene therapy for respiratory diseases have been directed at these particular genetic and acquired disorders. Other pulmonary disorders that may also be conducive to treatment by this modality include asthma, pulmonary hypertension, and adult respiratory distress syndrome. Since a detailed discussion on the progress of treatment for each of these disorders is beyond the scope of this review, readers are referred to several recent reviews on these topics (Curiel et al., 1996; Canonico, 1997; Johnson and Boucher, 1997). By way of example we will highlight the cumulative experience to date at developing cationic lipids for gene therapy of CF.

III. CYSTIC FIBROSIS

Cystic fibrosis (CF) is a common lethal autosomal recessive disorder that affects approximately 1 in 2500 Caucasian births. The gene responsible for CF was isolated by positional cloning in 1989 (Rommens et al., 1989; Riordan et al., 1989) and was shown to encode a membrane-associated glycoprotein referred to as the cystic fibrosis transmembrane conductance regulator (CFTR). Several investigators have established that CFTR is a cAMP-mediated chloride (Cl^-) channel (Anderson et al., 1991; Kartner et al., 1991; Bear et al., 1992). It is also reportedly capable of regulating other ion channels such as the outwardly rectifying Cl^- channel (Egan *et al.*, 1992) and the sodium (Na^+) channel (Stutts *et al.*, 1995), perhaps through its interactions with the PDZ family of proteins (Short *et al.*, 1998; Hall *et al.*, 1998). Moreover, it is purportedly also indirectly involved in other cellular functions including sulfation of glycoconjugates (Cheng *et al.*, 1989), regulation of plasma membrane recycling (Bradbury *et al.*, 1992), acidification of intracellular organelles (Barasch *et al.*, 1992), and interactions with *Pseudomonas aeruginosa* and *Salmonella typhi* (Pier *et al.*, 1997, 1998).

Over 400 mutations in the CFTR gene have been identified, the most common of which is a 3-base-pair deletion leading to loss of a phenylalanine residue at position 508 (ΔF508-CFTR) (Welsh and Smith, 1993; Welsh *et al.*, 1995). The consequences of these mutations are varied but all lead to the generation of CFTR variants whose cAMP-stimulated Cl^- channel activity is diminished or absent from the cell surface. This loss of activity, it is proposed, results in the inability of CF-affected epithelia to regulate normal electrolyte transport and fluid secretion. These aberrations in turn are thought to be responsible for the observed dehydration of the airway epithelial lining fluid, reduction in mucociliary clearance, and loss of the innate antibacterial activity of the lung. If this is correct, then delivery of a normal copy of the CFTR cDNA to the affected airway epithelial cells should reverse the various manifestations associated with the disease. Indeed gene complementation studies using a variety of gene delivery systems have indicated that transduction of CF cells *in vitro* with a wild-type CFTR-encoding gene can result in restoration of both the defective ion and fluid transport properties (for review see Collins, 1992). Although CF is a generalized disease of the secretory epithelia, the principal clinical manifestations are realized in the airway epithelia. Chronic infection and inflammation followed by subsequent damage to the lung tissue account for most of the morbidity and mortality associated with CF. Thus, present attempts to alleviate the disease by gene therapy have focused primarily on delivering a copy of the wild-type CF gene to the affected cells of this organ.

A. GENE THERAPY CONSIDERATIONS FOR CF

Several different gene delivery vector systems are currently being evaluated for use in CF gene therapy. These can be broadly divided into two classes: (1) those that are viral-based such as recombinant retrovirus, adenovirus, and adeno-associated virus and (2) those that are composed of synthetic self-assembling systems such as cationic lipids and molecular conjugates. The virtues and limitations for each of these systems are varied but center around their ability to (1) efficiently transduce the slowly dividing airway epithelial cells *in vivo*, (2) confer long-term expression of CFTR, and (3) be safe and compatible with chronic repeated administration to the lung (with the possible exception of retrovirus and adeno-associated virus).

1. Efficiency of Gene Transduction

To be effective for CF, gene therapy vectors necessarily need to be efficient at delivering the CFTR gene to the airway epithelium. Since these cells normally have a low rate of proliferation (Leigh *et al.*, 1995), these gene transfer vectors also need to harbor the ability to facilitate translocation of the transgene into the nucleus in the absence of cell division. Based on several gene transfer and cell-mixing studies *in vitro* and *in vivo*, it has been proposed that correction of approximately 5 to 10% of the affected surface epithelial cells may be sufficient to restore the defective Cl^- conductance observed in CF cells (Johnson *et al.*, 1992; Dorin *et al.*, 1996). If this low percentage of cells needing correction also applies to CF lungs *in vivo*, then the low efficiency of transduction required should be within the capability of most of the different gene transfer vector systems currently being developed. However, such a low level of correction is unlikely to reverse the hyperabsorption of Na^+ from the lumen for which it has been suggested that all cells of the epithelium may necessarily need to be modified (Stutts *et al.*, 1995). Currently, it is still unclear whether complete normalization of Na^+ hyperabsorption is important for reverting the pathogenesis of airway disease. If full correction of this defect is required, then gene delivery vectors that are both indiscriminate with respect to cell type and very efficient need to be developed to restore normal airway function.

2. Transfection of Relevant Target Cells

Based on immunolocalization studies, endogenous CFTR is normally present at low levels in the airway epithelial cells, with the highest levels located in the serous and duct cells of the submucosal glands (Engelhardt *et al.*, 1992). Within the conducting airway epithelial cells, higher expression is observed in the distal than in the proximal airways. However, despite these observations, it still remains unclear which cells represent the appropriate *in vivo* targets for CF gene therapy. The arguments center around whether CF lung disease results from defects in the columnar epithelial cells located within the lumen of the small airways or in the serous cells of the submucosal glands. Although both viral and nonviral gene transfer vector systems are capable of transfecting the conducting airway epithelial cells in preclinical studies, neither is able to access the submucosal cells efficiently. Hence if gene transfer of CFTR to the submucosal glands is a necessary prerequisite, then current vectors are unlikely to affect the disease pathophysiology in the cartilaginous airways. However, it has been proposed that a high level of CFTR gene expression in the surface epithelium may suffice to ameliorate distal airway disease and thereby impede the clinical course of CF lung disease. Ultimately, the resolution of these concerns can probably only be realized from clinical studies with CF subjects.

3. Safety

Since CF is a chronic disease, vectors that facilitate persistent expression of CFTR in the appropriate cells are desirable to provide a sustained therapeutic effect. With the possible exception of retroviral and adeno-associated viral vectors that are able to integrate into the host cell genome, the other gene delivery systems, at least in their present configurations, are unable to facilitate long-term expression of the transgene. While this may argue that the former vectors may be more appropriate for use in this indication, this argument is weakened somewhat by our lack of understanding of the identity of the progenitor cells in the airway epithelium. Access to these cells, which are likely located basolaterally, may also be challenging. Additionally, the finite risk of insertional mutagenesis, particularly with the use of viral vectors, also must be considered. In consequence, effective therapy for CF, particularly with the use of cationic lipids, will likely require repeated administration of the gene delivery vectors over the lifetime of the patients. In this regard, some consideration of the acute and cumulative toxicity of the vector system must be factored into the evaluation process. Preclinical studies have indicated that administration of adenoviral vectors or cationic lipids into the lung are associated with inflammation (Goldman et al., 1995; Scheule et al., 1997). Thus the ability to readminister will depend in part on the longevity of expression of CFTR provided by the vector and the time for resolution of the inflammatory response. Finally, the host immune response to viral or peptide antigens (Kaplan et al., 1996) is also an additional major impediment to the readministration of viral-based systems.

4. Practical Considerations

An effective treatment of CF lung disease will likely require uniform, low-level transfection of the entire lung. To achieve this, nebulization of the gene delivery vector represents the most simple and direct approach. In principle, the anatomic deposition of the transfecting agent can be controlled by regulating the size of the aerosol droplets so that delivery can be directed to the most affected areas. However, delivery by aerosolization presents several challenges including (1) the lability of the vector to the shearing forces associated with nebulization, (2) the uniformity of the aerosolized material, (3) the inefficiency of delivery by this route, and (4) the need to generate stable and concentrated formulations of the vector system so that delivery of therapeutic quantities can be accomplished within a reasonable time. Alternatively, dry powder inhalation aerosols can be employed (Adjei and Gupta, 1997). Although these aerosols have been used effectively in the delivery of small proteins, as yet there have been no reports of their applicability at delivering gene transfer vectors.

An ideal gene delivery vector system for treatment of CF would therefore be one that is (1) capable of mediating efficient transfection of the target cells (submucosal gland and columnar epithelial cells), (2) able to effect long-term expression of CFTR, (3) innocuous to the lung, (4) conducive to repeat administration for protracted treatment of this chronic disease, and (5) compatible with nebulization.

IV. CATIONIC LIPIDS FOR CF GENE THERAPY

Cationic lipids generally refer to a class of lipids that by virtue of their positive charge(s) are capable of interacting with and condensing the oppositely charged DNA (Behr, 1986). Since the first description by Felgner and colleagues (Felgner *et al.*, 1987) of the potential of such a lipid for gene transfection, several hundred new cationic lipids of different structure types have been described. Although early-generation cationic lipids were used primarily for *in vitro* transfection of cells in culture (Felgner *et al.*, 1987; Behr *et al.*, 1989; Gao and Huang, 1991), subsequent studies have indicated that they are also capable of mediating gene transduction *in vivo* (Brigham *et al.*, 1989; Yoshimura *et al.*, 1992; Zhu *et al.*, 1993; Hyde *et al.*, 1994). These observations coupled with cationic lipids' (1) ease of production, (2) reportedly low order of toxicity, (3) lack of immunogenicity, and (4) ability to deliver large genes, continue to make them an increasingly popular choice for gene transfer applications.

The cationic lipid gene delivery system invariably comprises three components: a cationic lipid, a neutral colipid, which is most often dioleoylphosphatidyl-ethanolamine (DOPE), and the plasmid DNA (pDNA) that encodes the transgene of interest (Fig. 1). Assembly of the vector system is normally achieved by combining the liposome mixture with the pDNA to generate so-called cationic lipid/pDNA complexes or lipoplexes (Felgner *et al.*, 1997). Addition of pDNA to the liposome results in condensation and compaction of the liposomes and pDNA to generate small (200–500 nm) but heterogenous complexes (Fig. 1) (Sternberg *et al.*, 1994; Eastman *et al.*, 1996). Although more homogeneous populations can be achieved—for example, by extrusion—because it has not been determined which fraction of this heterogeneous population is responsible for the observed transfection activity, it is unclear whether homogenization will provide additional benefit. However, recent studies by Templeton *et al.* (1997) have indicated that extrusion of the complexes can lead to enhanced transfection activity, at least when tested by systemic administration.

To facilitate transfection of the airway epithelial cells, these cationic lipid/pDNA complexes will need to harbor properties that allow them to traverse several biological barriers. Thus in addition to being deposited on the appropriate target cells in the airways, these complexes also need to (1) be able to interact with the cell membranes to facilitate uptake into the intracellular compartment, (2) inherently be able to effect subsequent release of the pDNA, for example from endosomes into the cytoplasm, (3) ideally contain pDNA with the appropriate elements to enhance translocation of the pDNA from the cytoplasmic milieu into the nuclear compart-

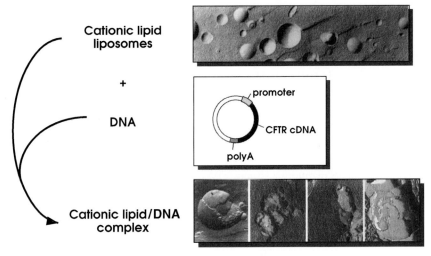

Figure 1   Formation of cationic lipid/pDNA complexes. Shown are electron micrographs of GL-67/ DOPE liposomes prior to and following addition of pDNA. Representative structures of the cationic lipid/pDNA complexes are indicated to highlight the heterogeneous nature of the resultant complexes.

ment, and (4) contain pDNA that harbor the necessary regulatory elements for high-level and persistent expression of the transgene product in these cells. Following is a discussion of some of the factors that have been identified that may aid in the development of these properties.

A. STRUCTURE–ACTIVITY RELATIONSHIPS

Typically, cationic lipids are composed of a positively charged headgroup, a hydrophobic lipid anchor, and a linker that bridges these two components together (Fig. 2). All three components appear to have a role in determining the transfection activity of the cationic lipid. For example, the chemical and structural details of the headgroup likely dictate its interaction with the pDNA and the ability to bind and thereby facilitate uptake by the cells; the linker bond determines the stability and biodegradability of the lipid; the length or spacing of this linker may regulate the accessibility of the protonatable amines on the headgroup with the phosphate groups on the nucleic acid backbone; and the hydrophobic lipid anchor determines the order of lipid packing and perhaps membrane fluidity.

Many new cationic lipids have been synthesized and their respective transfection activities in a wide variety of cell types have been reported (Remy *et al.*, 1994; Felgner *et al.*, 1994; Guy-Caffey *et al.*, 1995; Boussif *et al.*, 1995; Solodin *et al.*, 1995; Lee *et al.*, 1996; Walker *et al.*, 1996; Wheeler *et al.*, 1996; Gold-

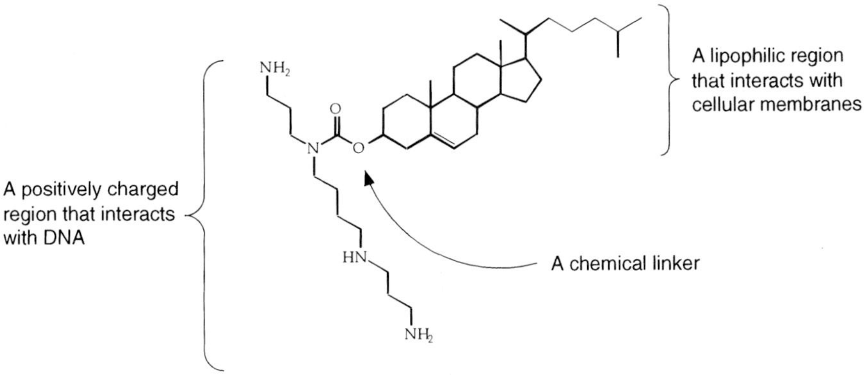

A lipophilic region that interacts with cellular membranes

A positively charged region that interacts with DNA

A chemical linker

Figure 2 Representative structure of a cationic lipid.

man et al., 1997; Oudrhiri et al., 1997; Paukku et al., 1997; Byk et al., 1998; Cooper et al., 1998; Huang et al., 1998; Wang et al., 1998). However, because these have been tested by different investigators using dissimilar cell lines and reporter genes, it is difficult to compare and contrast the relative activities of the different lipids generated thus far. Furthermore, results from several laboratories have illustrated that testing the activity of the lipids using cells in culture does not necessarily reflect their performance in vivo (Solodin et al., 1995; Lee et al., 1996; Gorman et al., 1997). Thus, it is uninformative to translate data generated using in vitro assays for in vivo applications. Furthermore, since a number of the cationic lipids are reportedly sensitive to the presence of serum (Fasbender et al., 1995; Lewis et al., 1996), comparing results of animal studies in which the complexes had been administered systemically with those in which the complexes had been instilled into the lung may not be instructional. Hence, despite systematic analysis of a large number of different cationic lipid compounds by several investigators, few structure–activity relationships have emerged. Notwithstanding these caveats, following is an analysis of cationic lipids that we and others have synthesized and their relative activities at least as tested in the lung.

1. Headgroup

A majority of the cationic lipids that have been described to date can broadly be divided into two classes: those that harbor cholesterol as the lipid anchor (Leventis and Silvius, 1990; Gao and Huang, 1991; Guy-Caffey et al., 1995; Lee et al., 1996) and those that utilize diacyl chains of varying lengths and extent of saturation of the hydrocarbon chains (Felgner et al., 1987; 1994; Behr et al., 1989; Hawley-Nelson et al., 1993; Wheeler et al., 1996). Irrespective of the anchor, it has been observed that in general the number of protonatable amines on the headgroup affects transfection activity, with multivalent headgroups more active than monovalent (Gao and Huang, 1995; Lee *et al.*, 1996). This may be related to the greater ability of multivalent headgroups to condense DNA into smaller and more compact lipid/pDNA complexes. However, there appears to be an upper limit on the number of useful protonatable amines on the headgroups. For example, progressively increasing the protonatable amines in the headgroup to four did not improve transfection activity beyond those with three (Lee *et al.*, 1996). It may be that too many charges render the cationic lipid more water soluble and hence give it a greater propensity to form micelles. This in turn may lead to the formation of complexes that are less stable and more toxic. Additionally, the inclusion of a large number of positive charges may result in an extremely strong interaction with pDNA that impedes the subsequent dissociation of pDNA and interaction with protein factors necessary for efficient transcription of the transgene. In the context of monovalent compounds, tertiary amino groups are generally more effective than those containing a quarternary amino group (Farhood *et al.*, 1992). In another study, Felgner *et al.* (1994) indicated that modification of the monovalent cationic lipid DOTMA by incorporation of a hydroxyethyl group also increased transfection activity (Fig. 3).

The orientation of the headgroup in relation to the backbone was also shown to be an important structure−activity consideration. Coupling of spermine or spermidine via a secondary amine to the lipid anchor such that they are presented as a T-shape (Fig. 3, GL-67) was ascertained to present a significantly more active configuration than coupling these same amines via a terminal amine, which generates the linear counterpart (Fig. 3, GL-62)(Lee *et al.*, 1996). The higher transfection

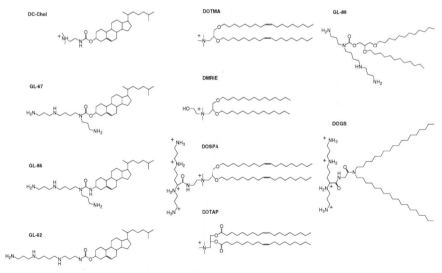

Figure 3   Structures of cationic lipids.

activity of the T-shaped cationic lipids in the lung was observed irrespective of whether the headgroups were coupled to cholesterol or diacyl chains as anchors. It has been proposed that perhaps the T-shaped configuration more closely resembles a ligand for a cell surface receptor, thereby facilitating attachment and entry into the target cells. This observation with the T-shaped compounds has since been independently confirmed by another group (Bischoff *et al.,* 1997). However, whether these findings in the lung will also be apparent following systemic administration has not been fully investigated.

2. Lipid Anchor

The importance of the lipid anchor selection is best illustrated by the observation that although T-shaped headgroups were more effective than their linear counterparts, those linked to a cholesterol anchor were invariably more active than those linked to diacyl backbones (Lee *et al.,* 1996). Hence, GL-67, a cationic lipid with a T-shaped spermine headgroup linked to a cholesterol anchor, was more active than its analog GL-89, whose headgroup was attached to dilauryl chains (Fig. 3). Use of a dihydrocholesterol moiety instead of cholesterol as the lipid anchor also reduced transfection activity. It is not readily apparent how these changes in the lipid anchor affect the transfection activity of the cationic lipids. A possible influence may be on membrane fluidity and the subsequent effects on lipid mixing within the bilayers. However, that a subtle change in structure (e.g., from cholesterol to dihydrocholesterol) can result in significant loss of activity argues for the need to improve our understanding of the role of the lipid anchor in this process.

In considering diacyl lipids, it has been observed that the hydrocarbon chain length is important. Cationic lipids with shorter acyl chains are generally more active, with dilauryl chains perhaps being the most active (Felgner *et al.,* 1994; Bennett *et al.,* 1996; Lee *et al.,* 1996; Wheeler *et al.,* 1996). This rule appears to apply to both monovalent and multivalent cationic lipids. Other cationic lipid analogs of this class that have been described (e.g., DOTMA, DOTAP, DOSPA) harbor unsaturated aliphatic chains or asymmetric chains (Balasubramaniam *et al.,* 1996), but it has not been determined how these different structures may influence the performance of the lipid (Fig. 3).

3. Linker

The selection of the linker that bridges the hydrophobic anchor with the cationic headgroup is also an important determinant of gene transfer activity. The stability and, consequently, the biodegradability of the cationic lipid are governed in large part by the choice of the chemical structure used to bridge the polar and hydrophobic components. Several different chemical linkers have been used, including ether, amide, carbamate, amine, urea, ester, and even peptide bonds. While the selection of the ether linker such as found in the cationic lipids DOTMA and

DMRIE (Felgner *et al.*, 1987, 1994) may provide good chemical stability, it is unlikely to be easily metabolized in the cell (Fig. 3). The use of cationic lipids with a poor biodegradability profile presents serious safety concerns and is therefore limiting, particularly in disease indications where chronic readministration of the lipid is required. Importantly, cationic lipids with more biodegradable bonds such as esters or tetraesters are also reportedly less cytotoxic (Silvius and Leventis, 1990; Aberle *et al.*, 1998). However, incorporation of linkers that are chemically unstable can also be constraining for practical reasons. For example, the half-life of a cationic lipid that utilized an ester bond was estimated to be approximately 24 h at 4°C (Farhood *et al.*, 1992). Furthermore, the use of less stable linkers can also affect the efficacy of the cationic lipid. Substitution of the carbamate linker in GL-67 (Fig. 3) with the relatively less stable urea (Fig. 3, GL-86) or amide linkers reduced transfection activity *in vitro* and *in vivo* (Lee *et al.*, 1996). If chemical instability was a factor, the observed loss in activity may have been due to premature dissolution of the cationic lipid/pDNA complex. Ultimately, linkers will be selected based on a balance between their biodegradabilities in the cell and their relative chemical stabilities. However, no intracellular degradation studies have yet been conducted for any cationic lipid. Such studies warrant attention, particularly as the design of future generations of cationic lipids are contemplated.

4. Counterion

Most cationic lipids are generally synthesized and used in their salt form. For example, DOTMA, DC-chol, and DOTAP are generated as chloride salts and DMRIE and DDAB as bromide salts. To date, there has not been a systematic analysis to compare the relative influence of all the different ions that have been reported. Therefore, it is unclear how the selection of different counterions may affect the performance of the cationic lipid. However a study by Bennett *et al.* (1996) indicated that counterions with a highly delocalized anionic charge performed more effectively. In Bennett *et al.*'s analysis, bisulfate > trifluoromethylsulfonate > iodide > bromide > chloride > acetate or sulfate at facilitating greater transfection activity *in vitro*. Whether this analysis will also hold for *in vivo* applications is unknown. A recent survey of a different series of cationic lipids (e.g., GL-67) indicated that free bases were more effective than their respective acetate and chloride counterparts, at least when tested in the lung (Lee *et al.*, 1996). Together, these studies suggest that counterion selection is a more important variable than was realized previously. Thus, more analysis and care should be exercised in their selection.

B. FORMULATION CONSIDERATIONS

In addition to the cationic lipid structure, many other factors are known to influence transfection activity *in vivo*. Such variables include the stoichiometry of the various components of the complex, the rate and order of mixing of these components, their respective concentrations, and the selection of excipients used in the formation of the complexes. Depending on the parameters used, different-sized cationic lipid/pDNA complexes with various net charges (zeta potential) realized at the surfaces of these complexes can be attained (Cheng *et al.*, 1998). These variables can result in the generation of complexes with significantly different transfection activities. However, despite studies by different investigators, our present understanding for the basis of this observed variability remains limited. Thus, much of the effort to identify the optimal formulations for use in any particular setting has been strictly empirical.

1. Neutral Colipid Selection

For maximal transfection activity, most cationic lipid formulations require the inclusion of a significant mole fraction of a neutral colipid (Felgner *et al.*, 1994). The most commonly used neutral colipids are DOPE and cholesterol. Both lipids contain a relatively small headgroup in relation to their hydrophobic domains, and in consequence exhibit the tendency to adopt nonbilayer or inverted hexagonal phases. Because such structures are often observed in regions where membranes fuse with each other (Litzinger and Huang, 1992), it has been proposed that these neutral colipids may act to destabilize membrane bilayers, thereby facilitating delivery of cationic lipid/pDNA complexes into the cells. In support of this hypothesis is the demonstration that substitution of DOPE with a bilayer-forming analog DOPC (dioleoylphosphatidylcholine) severely reduced transfection activity (Duzgunes *et al.*, 1989; Felgner *et al.*, 1994; Zhou and Huang, 1994). A recent survey of a number of different neutral colipids indicated that they also may have additional roles in enhancing gene delivery besides promoting cellular entry (Fasbender *et al.*, 1997). Depending on the selection of the neutral colipid, some facilitated escape of the complexes from endosomes, while others promoted the separation of the pDNA from the cationic lipid. In support of this proposal, they showed that greater transfection activity could be attained using formulations containing a mixture of neutral colipids with different properties than with those containing the individual colipids alone (Fasbender *et al.*, 1997).

Increasing the extent of acyl chain saturation on the neutral colipid was associated with progressively less activity when compared to the diunsaturated DOPE (Farhood *et al.*, 1995; Felgner *et al.*, 1994; Fasbender *et al.*, 1997). Presumably, this was related to the tighter packing of the phospholipid bilayer as the extent of acyl chain saturation was increased (Lichtenberg *et al.*, 1981). Decreasing the saturation of the neutral colipid might be predicted to reduce transfection activity if its primary role was to destabilize the membrane bilayers. However, while decreasing the extent of saturation might improve transfection activity, highly unsaturated neutral colipids are also likely to be more chemically unstable, which might limit their use.

Although both DOPE and cholesterol are widely used as colipids, there has not been a systematic comparison of their relative activities when formulated with different families of cationic lipids. There is a tendency to utilize DOPE as the neutral lipid with cationic lipids that are based on cholesterol (e.g., DC-chol, GL-67) and to use cholesterol as the neutral lipid with cationic lipids that harbor diacyl chains as the hydrophobic anchor (e.g., DOTAP, DOTIM). It is unclear whether there is any advantage to the use of either neutral lipid, except that perhaps cholesterol may be more biodegradable. However, recent studies by different investigators have suggested that use of cholesterol as a neutral lipid may confer an advantage, particularly for systemic delivery (Liu *et al.*, 1997; Templeton *et al.*, 1997; Crook *et al.*, 1998).

Finally, it should be noted that although most cationic lipids strictly require a neutral colipid for efficient gene transduction, some cationic lipids exhibit reasonable transfection activity in the absence of a neutral colipid. Examples of such cationic lipids include DOTAP (Porteous *et al.*, 1997) and DOGS (Behr *et al.*, 1989) (Fig. 3). This observation suggests that the mechanisms by which the various cationic lipids mediate gene delivery are different and are likely dependent on the chemical properties of the cationic lipid.

2. Stoichiometry of the Components

It is generally accepted that the relative proportions and concentrations of each of the components that make up the cationic lipid/pDNA complex need to be optimized for maximal activity. The ratio of cationic lipid to pDNA determines the surface charge and to some extent, the particle size of the complexes, two important variables known to influence transfection activity. In addition to defining the cationic lipid to neutral lipid and cationic lipid to pDNA ratios, other variables include the order in which the components are mixed and the presence or absence of excipients such as salt or serum. Protocols (*in vitro*) have been developed that titrate both the absolute concentrations of cationic lipid and pDNA as well as their ratio to determine the optimal set of conditions for transfection (Felgner *et al.*, 1994; Lee *et al.*, 1996; Cheng *et al.*, 1998). Provided that an appropriate cell model system that closely reflects the *in vivo* target is used, data generated using such *in vitro* screens can be useful for establishing the initial starting values for *in vivo* applications (Lee *et al.*, 1998). However, because the cationic lipids that have been described to date are structurally diverse, it is likely that they will all behave differently under the varied conditions and therefore optimization will require an empirical survey of the known variables. Furthermore, since the physiological milieu of specific organ targets for *in vivo* gene transfer are different, the optimal formulation will need to be redefined depending on the particular indication for which it is intended. For example, an optimized ratio of cationic lipid to pDNA for gene delivery to the lung is unlikely to be similar to that for systemic delivery. Additionally, a formulation that is optimized for instillation into the lung will necessarily need to be different from that for aerosolization. For the latter application, the ratio of cationic lipid to pDNA has to be such that maximal protection is afforded to the pDNA during nebulization (Eastman *et al.*, 1997).

3. Physical Characteristics of the Complexes

It has been proposed that positively charged cationic lipid/pDNA complexes, by virtue of their ability to interact with the oppositely charged cell membranes, have a greater tendency to be internalized by the cells with resultant higher expression of the transgene (Felgner *et al.*, 1987; Behr *et al.*, 1989; Fasbender *et al.*, 1995; Matsui *et al.*, 1997). However, other studies have indicated that cells are equally capable of being transfected by negatively charged complexes (Philip *et al.*, 1994; Lee *et al.*, 1996; Wheeler *et al.*, 1996). These discrepancies may reflect the different environments in which the lipids were studied. For example, for intravenous delivery, less positively charged formulations that limit their interactions with serum proteins may be more desirable. However, studies have indicated that an excess of cationic liposomes in the complexes can provide greater transfection activity (Li and Huang, 1997; Liu *et al.*, 1997). Negatively charged complexes were determined to be more active when instilled into the lung. Complexes at or close to charge neutrality are invariably unstable and exhibit a tendency to aggregate.

It should be noted that the net charge of the complex in most of these studies was calculated based on the stoichiometry of the cationic lipid and pDNA. However, such calculations were based on an assumption about the charge state(s) of the protonatable amines on the headgroups of the lipids, which was usually not characterized. In addition, there was a reliance on the assumption that all the cationic groups and anionic phosphates on the pDNA were equally available for interaction with each other, which may not be the case. Thus, much of what has been reported to date pertains to the theoretical net charge rather than to the absolute charge on the surface of these complexes. A parameter that is often measured to reflect the surface charge of the complex is the zeta potential. We have observed that although there is a correlation between the net charge balance and zeta potential, the zeta potential measured is significantly less positive than would be predicted based on the simple arithmetic of a cationic–anionic interaction.

The size of the complexes that are formed also appears to be an important variable. Studies using the cationic lipid DMRIE indicated that larger vesicles generated by simple vortexing were more active than smaller equivalents generated by sonication (Felgner *et al.*, 1994). However, the opposite was shown for DC-chol, where smaller vesicles were reportedly superior (Gao and Huang, 1995). Again, the basis for these observed differences among the different cationic lipids is not well understood. Of note, however, is the observation that freshly prepared cationic lipid/pDNA complexes tend to be heterogenous and it is unclear which subfraction of this heterogenous population is the most active fraction. Studies have shown that at least a proportion of the cationic lipid/pDNA complexes are internalized into cells by endocytosis (Zhou and Huang, 1994; Zabner *et al.*, 1995). Because endocytosis of large macromolecules of greater than 200 nm in size is relatively inefficient, it is rational to propose that smaller rather than larger complexes should be more efficacious at effecting gene transfer. However, our own experience with GL-67 and other cationic lipid formulations indicates that despite having a broad range of different-size complexes with a mean of around 450 nm, high-level transfection of airway cells *in vivo* could be attained. This indication argued that either endocytosis was not the only route by which the complexes were internalized or the fraction that was endocytosed was not the major one that led to productive transfection of the cell. Studies to further define the active fraction of the cationic lipid/pDNA complex and its mechanism of cellular entry need to be pursued to resolve these issues.

One of the principal limitations to the effective use of a number of cationic lipids is their inability to be formulated at high concentrations. Most cationic lipid/pDNA complexes formed at high concentrations tend to aggregate and precipitate out of suspension. This deficiency has obvious implications for their effective use, especially in *in vivo* settings. However, this restriction can be rectified in part through formulation development. For example, it has been shown that with DC-chol formulations, precipitation of the complexes could be minimized by raising the pH during complex formation (Caplen *et al.*, 1995). We have also observed that at least in the case of GL-67 and related compounds, inclusion of the lipid excipient DMPE-PEG$_{5000}$ to GL-67/DOPE/pDNA formulations could facilitate the generation of stable and highly concentrated formulations of the complexes (Eastman *et al.*, 1997). Inclusion of this bilayer-stabilizing lipid permitted the formation of complexes that harbor pDNA concentrations of greater than 4 mg/ml. This in turn permitted the development of an aerosol formulation for delivery to CF patients (Alton *et al.*, 1998; Eastman *et al.*, 1998).

V. EFFICIENCY OF CATIONIC LIPIDS AT MEDIATING GENE TRANSDUCTION TO THE LUNG

A large number of preclinical *in vivo* studies have been performed with cationic lipid/pDNA complexes in the lung using either instillation (Yoshimura *et al.*, 1992; Hyde *et al.*, 1993; Logan *et al.*, 1995; Lee *et al.*, 1996; Wheeler *et al.*, 1996; Gorman *et al.*, 1997; Jiang *et al.*, 1998) or nebulization (Stribling *et al.*, 1992; Alton *et al.*, 1993; Canonico *et al.*, 1994; Schwarz *et al.*, 1996; Eastman *et al.*, 1997; McDonald *et al.*, 1998). The results of these studies indicated that a variety of different cationic lipids and formulations are capable of mediating gene transfer to the airway epithelial cells. Although expression could be observed in animals treated with pDNA alone, complexing with cationic lipids invariably resulted in manyfold higher expression (Lee *et al.*, 1996). In some instances, these cationic lipid-mediated levels were sufficiently high to partially correct some of the electrophysiological defects shown associated with CF (Alton *et al.*, 1993; Hyde *et al.*, 1993; Logan *et al.*, 1995; Jiang *et al.*, 1998). Recent screens of hundreds of different compounds have yielded novel cationic lipids that are reportedly several hundred- to over a thousand-fold more active than free pDNA alone in the lung (Lee *et al.*, 1996; Wheeler *et al.*, 1996; Gorman *et al.*, 1997). In terms of expression levels as well as their ability to correct the CF electrophysiological defects, these lipids have been shown to be capable of performing at levels comparable to that attained with 10 to 20 moi of an adenovirus vector (Lee *et al.*, 1996).

Instillation of the complexes into the lungs of small animals resulted predominantly in transfection of the alveolar and terminal bronchial cells (Wheeler *et al.*, 1996; Scheule *et al.*, 1997; Cheng and Scheule, 1998), although other reports indicated that airway epithelial cells could also be efficiently transfected (Alton *et al.*, 1993; Hyde *et al.*, 1993; Logan *et al.*, 1995). However, aerosolization of the complexes consistently resulted in higher level transfection of the airway epithelium in mice (Stribling *et al.*, Canonico *et al.*, 1994, 1992; Eastman *et al.*, 1997) and monkeys (McDonald *et al.*, 1998). It is possible that this preferential transfection of the conducting airways following aerosolization may be due to more even deposition of the complexes throughout the lung than could be attained by instillation, which presumably primarily deposits the complexes in the parenchyma. Transfection efficiencies ranging from 5 to 50% of the airway cells judged using reporter genes have been reported following aerosolization. If correct this would argue that the present formulations of cationic lipids are sufficiently efficacious to reverse the Cl^--associated pathophysiology observed in CF lungs (Boucher, 1996). However, caution should be used with these reported frequencies of transfection since staining of reporter gene products reportedly can be artifactual.

VI. LIMITATIONS OF PRESENT CATIONIC LIPID FORMULATIONS

It is clear that significant advances have been made in the development of improved cationic lipids with higher gene transfection activity. However, despite this progress, it is apparent that present formulations of complexes are still relatively ineffectual, particularly when tested in human clinical studies (see the following and also Chapter 15 by Griesenbach, Geddes, and Alton). Next is a discussion of some of the barriers that still limit cationic lipid-mediated gene transfer.

A. Efficiency at Transfecting Differentiated Airway Epithelium

Although proliferating airway cells can be transfected at high efficiency with the improved cationic lipid formulations, the ability of these formulations to transduce polarized and differentiated airway epithelia is relatively low (Fasbender *et al.*, 1995; Zabner *et al.*, 1995; Matsui *et al.*, 1997; Jiang *et al.*, 1998). Studies designed to identify the rate-limiting factors indicated that binding and internalization of the complexes by well-differentiated airway epithelial cells were inefficient (Matsui *et al.*, 1997). This inability to be internalized by the differentiated airway epithelial cells was shown to be due to the low level of phagocytotic activity associated with these cells. Since the majority of the cationic lipid/pDNA complexes, as previously indicated, are greater than 200 nm in size, the loss of phagocytotic pathways will predictably reduce the extent of uptake by these differentiated airway cells. Thus, strategies to reduce the size of the complexes such that they can be internalized by pinocytosis may enhance transfection. However, this approach assumes that pinocytosis represents a significant route by which the complexes are internalized.

Yet another barrier to the successful transfection of these cells is the relative inefficiency by which the pDNA is translocated to the nuclear compartment (Zabner *et al.*, 1995; Jiang *et al.*, 1998). Since nuclear entry of pDNA is facilitated by the dissolution of the nuclear membrane during cell division, the problem of pDNA entry in well-differentiated airway cells is exacerbated because these cells divide relatively slowly (Leigh *et al.*, 1995). Attempts to enhance nuclear uptake by covalently attaching peptide fragments containing a nuclear localization signal to the pDNA have met with mixed results. Finally, it is apparent that for efficient transcription to occur, dissolution of the cationic lipid/pDNA complex needs to occur either prior to or following nuclear entry (Zabner *et al.*, 1995). This was borne from experiments showing that direct injection of pDNA into the nucleus resulted in greater levels of expression than injection of an equivalent amount of pDNA complexed to cationic lipid.

B. Toxicity of the Cationic Lipid/pDNA Complexes

Although early reports referred to cationic lipids as "safe," it is evident from recent studies that most elicit an inflammatory response when introduced into lungs of small animals (Logan *et al.*, 1995; Scheule *et al.*, 1996). This inflammation was found to be dose dependent and was characterized by cellular infiltrates, predominantly of polymorphonuclear leukocytes that resolved over a few weeks (Fig. 4). Morphologic changes, primarily at the alveolae and terminal bronchioles, such as necrosis with sloughing and denudation of underlying basement membranes, were

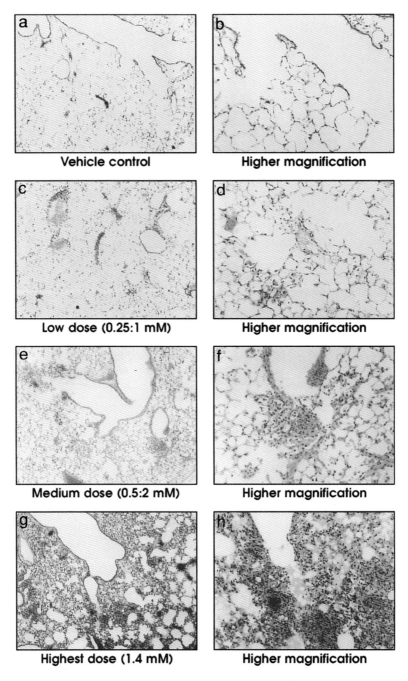

Figure 4 Dose-dependent inflammation following intranasal instillation of 100 μl of GL-67:pDNA complexes in water into lungs of BALB/c mice. The animals were sacrificed 2 days post-instillation and sections stained (h and e).

also evident. Concomitant with the increase in cellular influx were highly elevated levels of several proinflammatory cytokines that peaked at Day 1 and that resolved to normal limits by Day 3 (Scheule *et al.*, 1997). Histopathological analysis of lungs treated with the individual components of the cationic lipid/pDNA complex suggested that the cationic lipid itself was responsible for the majority of the observed changes. However, recent studies demonstrated that pDNA is also inflammatory in the lung and was causative of most of the observed increase in cytokine levels in the bronchoalveolar lavage fluids (Schwartz *et al.*, 1997; Yew *et al.*, 1999). This inflammation appeared to be mediated by the presence of unmethylated CpG dinucleotide sequences in the bacterially derived pDNA (Krieg *et al.*, 1995; Klinman *et al.*, 1996).

The acute toxicity associated with the cationic lipid/pDNA complexes was dramatically reduced when delivery was by aerosol rather than by instillation (Canonico *et al.*, 1994; Eastman *et al.*, 1997; McDonald *et al.*, 1998). This suggests that the toxicity observed in the instillation experiments may be due, at least in part, to "pooling" of the instilled bolus in the distal lung. In addition, there is also concern for the potential of chronic toxicity, particularly from long-term, repeated applications such as for CF. The resultant accumulation of large quantities of unmetabolized lipids over a protracted period in the transfected cells may lead to lysosomal storage disease-like abnormalities. The design of cationic lipids that are more biodegradable may help alleviate this potential problem.

C. LEVEL AND PERSISTENCE OF TRANSGENE EXPRESSION

One approach to minimize the toxicity associated with delivery of cationic lipid/pDNA complexes to the lung is to extend the persistence of expression of the therapeutic transgene, thereby reducing the frequency of repeat administration. Expression in the lung using the present generation of plasmid vectors appears to be transient, with maximal expression attained by Day 2 to 4, but dropping dramatically such that by Day 14, only approximately 10 to 20% of the peak activity remained (Lee *et al.*, 1996; Wheeler *et al.*, 1996; Yew *et al.*, 1997b). The transient nature of the expression may be related to the observation that the delivered pDNA was normally nonintegrated and remained in an episomal form. However, whether this transience was due to actual degradation of the transcriptionally active nuclear pDNA or rather to it being rendered transcriptionally inactive, perhaps by the presence of the cytokines, is not known (Harms and Splitter, 1995; Yew *et al.*, 1997b). Exceptions to these results come from studies, albeit using different transcriptional units, showing high-level expression in the lung for up to 4 weeks postinstillation (Yoshimura *et al.*, 1992; Zhu *et al.*, 1993). It is now evident that by careful manipulation of the transcriptional unit—for example, by incorporation of the adenovirus E4 coding sequence—that vectors capable of providing improved longevity of expression in the lung could be attained (Yew *et al.,* 1997a). Alternatively, use of viral genes that confer on a plasmid an ability to replicate episomally could also extend expression for several months or at least for the life of the transfected cell (Thierry *et al.,* 1995; Cooper *et al.,* 1997). However, the use of viral oncogenes for this purpose in a clinical context might be problematic. Yet another approach entails the incorporation of elements that facilitate integration of the pDNA into the genome. Examples include the use of transposons, retroviral integrase, and the terminal repeats of adeno-associated virus in conjunction with the Rep protein (Shoji-Tanaka *et al.,* 1994; Ivics *et al.,* 1997; Surosky *et al.,* 1997).

The fact that cationic lipids are unable to confer specificity to the cell types that may be transfected in the lung raises concerns regarding the ectopic expression of CFTR. Furthermore, because endogenous CFTR in the lung is normally only present at low levels, the effects of overexpression also must be considered. A study using transgenic mice indicated that overexpression of CFTR (under the control of an SP-C promoter) in the pulmonary epithelia had no adverse consequences (Whitsett *et al.,* 1992). In addition, incrementally increasing the expression of CFTR in a CF cell line did not cause an increase in chloride secretion above a certain level, suggesting that cellular processes may limit ion transport (Rosenfeld *et al.,* 1994). However, in subsequent studies where CFTR was overexpressed in two different fibroblast cell lines, abnormal growth rates and altered electrophysiological properties were observed (Stutts *et al.,* 1993; Schiavi *et al.,* 1996). Since fibroblasts lack distinct apical and basolateral domains, it is possible that the observed adverse events may be restricted to cells of nonepithelial origin. Despite this reservation, it is probably prudent to continue studies to develop an increased understanding of the importance of appropriate spatial and temporal expression of CFTR.

D. REPEAT ADMINISTRATION

Since delivery of pDNA does not normally result in stable integration and expression of the transgene, it is likely that repeated administration of the complex will be necessary for chronic diseases like CF over the lifetime of the patient. Thus, it is critical that the gene delivery vector system is compatible for use in a repeat-dosing regime that continually maintains therapeutic levels of the gene product. As previously indicated, the time period between readministrations will be dictated in part by the longevity of expression attained by the pDNA. Studies in lungs of rodents have indicated that, provided there is no immune response to the transgene product, repeated administration of cationic lipid/pDNA complexes is efficacious (Lee *et al.,* 1996; Wheeler *et al.,* 1996). However, the efficacy of repeat administration is dependent on the dose used and the time interval between administrations. It appears that the inflammation elicited by the complexes may affect the efficacy of repeat administration (Lee *et al.,* 1996; Marshall, Scheule, and Cheng, unpublished data). Hence a second administration that results in efficacious, high-level expression could be attained using a shorter time interval in animals treated with lower doses than in animals treated with higher doses. However, efficacy following a repeat administration could be achieved in animals receiving a high dose provided the time between successive administrations was spaced sufficiently to allow for the inflammatory response to subside. Beyond the influence of inflammation, it is not clearly understood what the basis is for this refractory period.

VII. CLINICAL STUDIES

Clinical trials to assess the safety and efficacy of cationic lipid/pDNA complexes in the airway epithelia of CF patients have been initiated in the United States and Europe (Table 1). To minimize the risk to CF patients, early trials were conducted in the nasal epithelium of CF patients (Caplen *et al.,* 1995; Gill *et al.,* 1997; Porteous *et al.,* 1997; Zabner *et al.,* 1997; Chapter 15 by Griesenbach, Geddes, and Alton). In addition to safety aspects, the nasal epithelium also offered other advantages, including ease of access for delivery of the complexes and subsequent evaluation of vector expression such as by RT-PCR and by functional analysis using nasal transepithelial potential difference measurements. However, a possible disadvantage pertains to the uncertainty of the relevance of the cells in the nasal epithelium as a target relative to the real target cells in the epithelia of the conducting airways. Given the lack of toxicity observed in these nasal clinical trials, studies have now progressed to delivering the complexes to the entire lung by aerosol, initially in normal volunteers (Chadwick *et al.,* 1997) and subsequently in CF subjects (Table 1).

Results from several intranasal instillation clinical studies using different cationic lipid formulations have been reported (Caplen *et al.,* 1995; Gill *et al.,* 1997; Porteous *et al.,* 1997; Zabner *et al.,* 1997; Alton *et al.,* 1998). In general, the results of these trials indicated that although delivery of cationic lipid/pDNA complexes to the nasal epithelia were associated with minimal toxicity, the efficacies attained were variable and marginal at best. Evidence of gene transfer was detected using RT-PCR, which in some instances was still measurable, albeit at very low levels, after 4 weeks. Using measurements of nasal epithelial potential differences, there was some suggestion of correction of the electrophysiological defects toward normal in a proportion of the patients, but there was no obvious dose response. Moreover, in those instances where partial functional correction was attained, the effects were transient. This measurement was, however, complicated by the large variations in potential difference observed both within and between patients, which made this measure of CFTR gene expression particularly difficult to interpret. Interestingly, a study that incorporated the use of pDNA alone as a control showed that pDNA was as effective as the complex at mediating gene transfer to the nasal

Table I

CF Gene Therapy Studies Using Cationic Lipids

Principal investigators	Cationic lipid formulation	Description of study
Geddes, Alton	DC-chol/DOPE	Delivery of cationic lipid/pCFTR of liposomes alone to nasal epithelia of CF subjects
Sorscher, Logan	DMRIE/DOPE	Delivery of cationic lipid/ppCFTR to nasal epithelia of CF subjects
Welsh, Zabner	GL-67	Delivery of liposomes alone to nasal epithelia of healthy volunteers; cationic lipid/pCFTR and pCFTR alone to nasal epithelia of CF subjects
Hyde, Gill, Higgins	DC-chol/DOPE	Delivery of cationic lipid/pCFTR or liposomes alone to nasal epithelia of CF subjects
Porteous, Innes	DOTAP	Delivery of cationic lipid/pCFTR or liposomes alone to nasal epithelia of CF subjects
Geddes, Alton	GL-67/DOPE/ DMPE-PEG$_{5000}$	Aerosol delivery of liposomes alone to lungs of healthy volunteers
Geddes, Alton	GL-67/DOPE/ DMPE-PEG$_{5000}$	Aerosol delivery of cationic lipid/cCFTR to nose and lungs of CF subjects
Sorscher	GL-67/DOPE/ DMPE-PEG$_{5000}$	Aerosol delivery of cationic lipid/pCFTR to lungs of CF subjects
Noone, Knowles	EDMPC/cholesterol	Delivery of cationic lipid/pCFTR to nasal epithelia of CF subjects

epithelium (Zabner *et al.*, 1997). These results are difficult to reconcile with pre-clinical studies showing that cationic lipid/pDNA complexes are several hundred-fold more active than pDNA alone when instilled into lungs of mice. There are many possible explanations for this discrepancy including the possibility that the epithelium of the nose may be more difficult to transfect than the conducting airways. In this regard a recent study in mice confirmed that although cationic lipid/pDNA complexes were more active than pDNA alone at transfecting the lung, administration of either to the nose did not exhibit a significantly marked difference in transfection activity (Jiang *et al.*, 1998). These results highlight the importance of model selection for assessing gene transfer and the inclusion of appropriate controls in the studies.

Given the concern regarding the nasal epithelium as an appropriate surrogate target for the lung, studies on delivering the complexes directly to the lung using an aerosol have also been initiated. A dose-ranging study using liposomes alone in normal healthy volunteers indicated that up to 229 mg of liposomes (GL-67/DOPE/DMPE-PEG$_{5000}$) could be aerosolized to individuals without any obvious adverse clinical indications (Chadwick *et al.*, 1997). Using this information, two aerosol studies—one in the United Kingdom and the other in the United States—were undertaken with CF subjects. Preliminary results of these studies were reported recently (Alton *et al.,* 1998, 1999; CF Annual Conference, Nashville, October 1997). Evidence of gene transfer by RT-PCR was detected and measurement of chloride secretion in the lower airways showed evidence of partial correction of the electrophysiological defect in the majority of the CF subjects. The extent of correction was estimated to be approximately 25% of that observed in non-CF individuals. However, some toxicity as evidenced by flu-like symptoms was observed in the group of CF subjects who received the complex but not in those who received liposomes alone. This response may be partly attributable to the inflammatory CpG motifs located on the bacterially derived pDNA as previously discussed.

VIII. SUMMARY

While the results of recent clinical studies in CF subjects are encouraging, it is clear that there are still many hurdles to overcome before cationic lipid-mediated gene transfer can be regarded as a viable therapy for CF. Significant improvements in (1) efficiency of gene transfer to the appropriate airway epithelial cells, (2) the safety profile of the cationic lipid complexes, (3) the persistence of expression, and (4) our understanding of the relationship between bioelectric measurements and clinical improvement need to be attained. Although these tasks are challenging, it is likely that as our basic understanding of the processes governing cationic lipid-mediated gene delivery continues to improve, more efficacious vector systems that are compatible with use in the treatment of these chronic diseases will emerge. When such vectors are identified we will move closer to attaining an effective treatment for CF.

REFERENCES

Aberle, A. M., Tablin, F., Zhu, J., Walker, N. J., Gruenert, D. C., and Nantz, M. H. (1998). A novel tetraester construct that reduces cationic lipid-associated cytotoxicity. Implications for the onset of cytotoxicity. *Biochemistry* **37,** 6533–6540.

Adjei, A. L., and Gupta, P. K. (1997). Dry-powder inhalation aerosols. *Lung Biol. Health Dis.* **107,** 625–665.

Alton, E. W. F. W., Geddes, D. M., Gill, D. R., Higgins, C. F., Hyde, S. C., Innes, J. A., and Porteous, D. J. (1998). Towards gene therapy for cystic fibrosis: A clinical progress report. *Gene Ther.* **5,** 291–292.

Alton, E. W. F. W., Middleton, P. G., Caplen, N. J., Smith, D. M., Munkonge, F. M., Jeffery, P. K., Geddes, D. M., Hart, S. L., Williamson, R., Fasold, K. I., Miller, A. D., Dickinson, P., Stevenson, B. J., McLachlan, G., Dorin, J. R., and Porteous, D. J. (1993). Non-invasive liposome-mediated gene delivery can correct the ion transport defect in cystic fibrosis mutant mice. *Nature Genet.* **5,** 135–142.

Alton, E. W. F. W., Stern, M., Farley, R., Jaffe, A., Chadwick, S. L., Phillips, J., Davis, J., Smith, S. N.,

Browning, J., Hodson, M. E., Durham, S. R., Li, D., Jeffery, P. K., Scallan, M., Balfour, R., Eastman, S. J., Cheng, S. H., Smith, A. E., Meeker, D., and Geddes, D. M. (1999). *Lancet* (in press).

Anderson M. P., Rich, D. P., Gregory, R. J., Smith, A. E., and Welsh, M. J. (1991). Generation of cAMP-activated chloride currents by expression of CFTR. *Science* **251**, 679–682.

Balasubramaniam, R. P., Bennet, M. J., Aberle, A. M., Malone, J. G., Nantz, M. H., and Malone, R. W. (1996). Structural and functional analysis of cationic transfection lipids: The hydrophobic domain. *Gene Ther.* **3**, 163–172.

Barasch, J., Kiss, B., Prince, A., Saiman, L., and Al-Awqati, Q. (1992). Acidification of intracellular organelles is defective in cystic fibrosis. *Nature* **352**, 70–73.

Barranger, J. A., Rice, E. O., Dunigan, J., Sansieri, C., Takiyama, N., Beeler, M., Lancia, J., Lucot, S., Scheirer-Fochler, S., Mohney, T., Swaney, W., Bahnson, A., and Ball, E. (1997). Gaucher's disease: Studies of gene transfer to haematopoietic cells. *Baillieres Clin. Haematol.* **10**, 765–778.

Bear, C. E., Li, C. H., Kartner, N., Bridges, R. J., Jensen, T. J., Ramjeesingh, M., and Riordan, J. R. (1992). Purification and functional reconstitution of the cystic fibrosis transmembrane conductance regulator (CFTR). *Cell* **68**, 809–818.

Behr, J. P. (1986). DNA strongly binds to micelles and vesicles containing lipopolyamines or lipointercalants. *Tetrahedron Lett.* **27**, 5861–5864.

Behr, J. P., Demeneix, B., Loeffler, J. P., and Mutul, J. P. (1989). Efficient gene transfer into mammalian primary endocrine cells with lipopolyamine-coated DNA. *Proc. Natl. Acad. Sci. USA* **86**, 6982–6989.

Bennett, M. J., Aberle, A. M., Balasubramanian, R. P., Grunert, D. C., and Malone, R. W. (1996). Considerations for the design of improved cationic amphiphiles based on transfection respects. *J. Liposome Res.* **6**, 545–566.

Bischoff, R., Cordier, Y., Perraud, F., Thioudellet, C., Braun, S., and Pavirani, A. (1997). Transfection of myoblasts in primary culture with isomeric cationic cholesterol derivatives. *Anal. Biochem.* **254**, 69–81.

Boucher, R. C. (1996). Current status of CF gene therapy. *Trends Genet.* **12**, 81–84.

Boussif, O., Lezoualch, F., Zanta, M. A., Mergny, M. D., Scherman, D., Demeneix, B., and Behr, J. P. (1995). A versatile vector for gene and oligonucleotide transfer into cells in culture and *in vivo*: Polyethylenimine. *Proc. Natl. Acad. Sci. USA* **92**, 7297–7301.

Bradbury, N. A., Jilling, T., Berta, G., Sorscher, E. J., Bridges, R. J., and Kirk, K. L. (1992). Regulation of plasma membrane recycling by CFTR. *Science* **256**, 530–532.

Brigham, K. L., Meyrick, B., Christman, B., Magnuson, M., King, G., and Berry, L. C. (1989). *In vivo* transfection of murine lungs with a functioning prokaryotic gene using a liposome vehicle. *Am. J. Med. Sci.* **298**, 278–281.

Byk, G., Dubertret, C., Escriou, V., Frederic, M., Jaslin, G., Rangara, R., Pitard, B., Crouzet, J., Wils, P., Schwartz, B., and Scherman, D. (1998). Synthesis, activity, and structure-activity relationship studies of novel cationic lipids for DNA transfer. *J. Med. Chem.* **41**, 229–235.

Canonico, A. E. (1997). Gene therapy for chronic inflammatory diseases of the lungs. *Lung Biol. Health Dis.* **104**, 285–307.

Canonico, A. E., Conary, J. T., Meyrick, B. O., and Brigham, K. L. (1994). Aerosol and intravenous transfection of human α_1-antitrypsin gene to the lungs of rabbits. *Am. J. Respir. Cell Mol. Biol.* **10**, 24–29.

Caplen, N. J., Alton, E. W. F. W., Middleton, P. G., Dorin, J. R., Stevenson, B. J., Gao, X., Durham, S. R., Jeffery, K., Hodson, M. E., Coutelle, C., Huang, L., Porteous, D. J., Williamson, R., and Geddes, D. M. (1995). Liposome-mediated CFTR gene transfer to the nasal epithelium of patients with cystic fibrosis. *Nature Med.* **1**, 39–46.

Caplen, N. J., Kinrade, E., Sorgi, F., Gao, X., Gruenert, D., Geddes, D., Coutelle, C., Huang, L., Alton, E. W. F. W., and Williamson, R. (1995). *In vitro* liposome-mediated DNA transfection of epithelial cell lines using the cationic liposome DC-Chol/DOPE. *Gene Ther.* **2**, 603–613.

Chadwick, S. L., Kingston, H. D., Stern, M., Cook, R. M., O'Connor, B. J., Lukason, M., Balfour, R. P., Rosenberg, M., Cheng, S. H., Smith, A. E., Meeker, D. P., Geddes, D. M., and Alton, E. W. (1997). Safety of a single aerosol administration of escalating doses of the cationic lipid GL67/DOPE/DMPE-PEG$_{5000}$ formulation to the lungs of normal volunteers. *Gene Ther.* **4,** 937–942.

Chen, P. L., Chen, Y., Bookstein, R., and Lee, W. H. (1990). Genetic mechanisms of tumor suppression by the human p53 gene. *Science* **250,** 1576–1580.

Cheng, P. W., Boat, T. F., Cranfield, K., Yankaskas, J. R., and Boucher, R. C. (1989). Increased sulfation of glycoconjugates by cultured nasal epithelial cells from patients with cystic fibrosis. *J. Clin. Invest.* **84,** 68–72.

Cheng, S. H., and Scheule, R. K. (1998). Airway delivery of cationic lipid:DNA complexes for cystic fibrosis. *Adv. Drug Del. Rev.* **30,** 173–184.

Cheng, S. H., Gregory, R. J., Marshall, J., Paul, S., Souza, D. W., White, G. A., O'Riordan, C. R., and Smith, A. E. (1990). Defective intracellular transport and processing of CFTR is the molecular basis of most cystic fibrosis. *Cell* **63,** 827–834.

Cheng, S. H., Marshall, J., Scheule, R. K., and Smith, A. E. (1998). Cationic lipid formulations for intracellular gene delivery of cystic fibrosis transmembrane conductance regulator to airway epithelia. *Methods Enzymol.* **292,** 697–717.

Collins, F. S. (1992). Cystic fibrosis: Molecular biology and therapeutic implications, *Science* **256,** 774–779.

Cooper, M. J., Lippa, M., Payne, J. M., Hatzivassiliou, G., Reifenberg, E., Fayazi, B., Perales, J. C., Morrison, L. J., Templeton, D., Pierarz, R. L., and Tan, J. (1997). Safety-modified episomal vectors for human gene therapy. *Proc. Natl. Acad. Sci. USA* **94,** 6450–6455.

Cooper, R. G., Etheridge, C. J., Stewart, L., Marshall, J., Rudginsky, S., Cheng, S. H., and Miller, A. D. (1998). Polyamine analogues of 3β-[N-(N′, N′-dimethylamino ethane)carbamoyl]-cholesterol (DC-Chol) as agents for gene delivery. *Chem. Eur. J.* **4,** 137–151.

Crook, K., Stevenson, B. J., Dubouchet, M., and Porteous, D. J. (1998). Inclusion of cholesterol in DOTAP transfection complexes increases the delivery of DNA to cells in vitro in the presence of serum. *Gene Ther.* **5,** 137–143.

Crystal, R. G., Brantly, M. L., Hubbard, R. C., Curiel, D. T., States, M. D., and Holmes, M. D. (1989). The alpha1-antitrypsin gene and its mutations. *Chest* **95,** 196–208.

Crystal, R. G., McElvaney, N. G., Rosenfeld, M. A., Chu, C. S., Hay, J. G., Jaffe, H. A., Eissa, N. T., and Danel, C. (1994). Administration of an adenovirus containing the human CFTR cDNA to the respiratory tract of individuals with cystic fibrosis. *Nature Genet.* **8,** 42–51.

Curiel, D. T., Pilewski, J. M., and Abelda, S. M. (1996). Gene therapy approaches for inherited and acquired lung diseases. *Am. J. Respir. Cell Mol. Biol.* **14,** 1–18.

Duzgunes, N., Goldstein, J. A., Friend, D. S., and Felgner, P. L. (1989). Fusion of liposomes containing a novel cationic lipid, N-[2,3 (dioleyoxy) propyl]-N,N,N-trimethylammonium: Induction by multivalent anions and asymmetric fusion with acidic phospholipid vesicles. *Biochemistry* **28,** 9179–9184.

Eastman, S. J., Lukason, M. J., Tousignant, J. D., Murray, H., Lane, M. D., St. George, J. A., Akita, G. Y., Cherry, M., Cheng, S. H., and Scheule, R. K. (1997). A concentrated and stable aerosol formulation of cationic lipid:DNA complexes giving high-level gene expression in mouse lung. *Hum. Gene Ther.* **8,** 765–773.

Eastman, S. J., Siegel, C., Tousignant, J., Smith, A. E., Cheng, S. H., and Scheule, R. K. (1996). Biophysical characterization of cationic lipid:DNA complexes. *Biochim. Biophys. Acta* **1325,** 41–62.

Eastman, S. J., Tousignant, J. D., Lukason, M. J., Chu, Q., Cheng, S. H., and Scheule, R. K. (1998). Aerosolization of cationic lipid/pDNA complexes—*in vitro* optimization of nebulizer parameters for human clinical studies. *Hum. Gene Ther.* **9,** 43–52.

Egan, M., Flotte, T., Afione, S., Solow, R., Zeitlin, P. L., Carter, B. J., and Guggino, W. B. (1992). Defective regulation of outwardly rectifying Cl⁻ channels by protein kinase A corrected by insertion of CFTR. *Nature* **358,** 581–584.

Engelhardt, J. F., Yankaskas, J. R., Ernst, S. A., Yang, Y., Marino, C. R., Boucher, R. C., Cohn, J. A., and Wilson, J. M. (1992). Submucosal glands are the predominant site of CFTR expression in the human bronchus. *Nature Genet.* **2**, 240–248.

Farhood, H., Bottega, R. Epand, R. M., and Huang, L. (1992). Effect of cationic cholesterol derivatives on gene transfer and protein kinase C. *Biochim. Biophys. Acta* **1111**, 239–246.

Farhood, H., Serbina, N., and Huang, L. (1995). The role of dioleoylphostidylethanolamine in cationic liposome mediated gene transfer. *Biochim. Biophys. Acta* **1235**, 289–295.

Fasbender, A., Marshall, J., Moninger, T. O., Grunst, T., Cheng, S. H., and Welsh, M. J. (1997). Effect of co-lipids in enhancing cationic lipid-mediated gene transfer *in vitro* and *in vivo*. *Gene Ther.* **4**, 716–725.

Fasbender, A. J., Zabner, J., and Welsh, M. J. (1995). Optimization of cationic lipid-mediated gene transfer to airway epithelia. *Am. J. Physiol.* **269**, L45-L51.

Felgner, J. H., Kumar, R., Sridhar, C. N., Wheeler, C. J., Tsai, Y. J., Border, R., Ramsey, P., Martin, M., and Felgner, P. L. (1994). Enhanced gene delivery and mechanism studies with a novel series of cationic lipid formulations. *J. Biol. Chem.* **269**, 2550–2561.

Felgner, P. L., Barenholz, Y., Behr, J. P., Cheng, S. H., Cullis, P., Huang, L., Jessee, J. A., Seymour, L., Szoka, F., Thierry, A. R., Wagner, E., and Wu, G. (1997). Nomenclature for synthetic gene delivery systems. *Hum. Gene Ther.* **8**, 511–512.

Felgner, P. L., Gadek, T. R., Holm, M., Roman, R., Chan, H. W., Wenz, M., Northrop, J. P., Ringold, G.M., and Danielsen, M. (1987). Lipofection: A highly efficient, lipid-mediated DNA-transfection procedure. *Proc. Natl. Acad. Sci. USA* **84**, 7413–7417.

Gao, X., and Huang, L. (1991). A novel cationic liposome reagent for efficient transfection of mammalian cells. *Biochem. Biophys. Res. Commun.* **179**, 280–285.

Gao, X., and Huang, L. (1995). Cationic liposome-mediated gene transfer. *Gene Ther.* **2**, 710–722.

Gill, D. R., Southern, K. W., Mofford, K. A., Seddon, T., Huang, L., Sorgi, F., Thomson, A., Mac-Vinish, L. J., Ratcliff, R., Bilton, D., Lane, D. J., Littlewood, J. M., Webb, A. K., Middleton, P. G., Colledge, W. H., Cuthbert, A. W., Evans, M. J., Higgins, C. F., and Hyde, S. C. (1997). A placebo-controlled study of liposome-mediated gene transfer to the nasal epithelium of patients with cystic fibrosis. *Gene Ther.* **4**, 199–209.

Goldman, C. K., Sorocaenu, L., Smith, N., Gillespie, G. Y., Shaw, W., Burgess, S., Bilbao, G., and Curiel, D. T. (1997). In vitro and in vivo gene delivery mediated by a synthetic polycationic amino polymer. *Nature Biotech.* **15**, 462–466.

Goldman, M. J., Litzky, L. A., Engelhardt, J. F., and Wilson, J. M. (1995). Transfer of the CFTR gene to the lung of nonhuman primates with E1-deleted, E2a-defective recombinant adenoviruses: A preclinical toxicology study. *Hum. Gene Ther.* **6**, 839–851.

Gorman, C. M., Aikawa, M., Fox, B., Lapuz, C., Michaud, B., Nguyen, H., Roche, E., Sawa, T., and Wiener-Kronish, J. P. (1997). Efficient *in vivo* delivery of DNA to pulmonary cells using the novel lipid EDMPC. *Gene Ther.* **4**, 983–992.

Guy-Caffey, J. K., Bodepudi, V., Bishop, J. S., Jayaraman, K., and Chaudhary, N. (1995). Novel polyaminolipids enhance the cellular uptake of oligonucleotides. *J. Biol. Chem.* **270**, 31391–31396.

Hall, R. A., Ostedgaard, L. S., Premont, R. T., Blitzer, J. T., Rahman, N., Welsh, M. J., and Lefkowitz, R. J. (1998). A V-terminal motif found in the b2-adrenergic receptor, P2Y1 receptor and cystic fibrosis transmembrane conductance regulator determines binding to the Na^+/H^+ exchanger regulatory factor family of PDZ proteins. *Proc. Natl. Acad. Sci. USA* **95**, 8496–8501.

Harms, J. S., and Splitter, G. A. (1995). Inteferon-γ inhibits transgene expression driven by SV40 or CMV promoters but augments expression driven by the mammalian MHC I promoter. *Hum. Gene Ther.* **6**, 1291–1297.

Hawley-Nelson, P., Ciccarone, V., Gebeyehu, G., and Jesse, J. (1993). Lipofectamine reagent: A new, higher efficiency polycationic liposome transfection agent. *Focus* **15**, 73–79.

Huang, C. Y., Uno, T., Murphy, J. E., Hamer, J. D., Escobedo, J. A., Cohen, F. E., Radhakrishnan, R., Dwarki, V., and Zuckerman, R. N. (1998). Lipitoids—novel cationic lipids for cellular delivery of plasmid DNA *in vitro*. *Chem. Biol.* **5,** 345–354.

Hwang, C. H., Smythe, W. R., Elshami, A. S., Kucharczuk, J. C., Amin, K. M., Williams, J. P., Litzky, L. A., Kaiser, L. R., and Albelda, S. M. (1995). Gene therapy using adenovirus carrying the herpes simplex-thymidine kinase gene to treat *in vivo* models of human malignant mesothelioma and lung cancer. *Am. J. Respir. Cell Mol. Biol.* **13,** 7–16.

Hyde, S. C., Gill, D. R., Higgins, C. F., Tresize, A. E. O., MacVinish, L. J., Cuthbert, A. W., Ratcliff, R., Evans, M. J., and Colledge, W. H. (1993). Correction of the ion transport defect in cystic fibrosis transgenic mice by gene therapy. *Nature* **362,** 250–255.

Ivics, Z., Hackett, P. B., Plasterk, R. H., and Izsvak, Z. (1997). Molecular reconstruction of Sleeping Beauty, a Tc1-like transposon from fish, and its transposition in human cells. *Cell* **91,** 501–510.

Jaffe, H. A., Danel, C., Longenecker, G., Metzger, M., Setoguchi, Y., Rosenfeld, M. A., Gant, T. W., Thorgeirsson, S. S., Stratford-Perricaudet, L. D., Perricaudet, M., Pavirani, A., Lecocq, J. P., and Crystal, R. G. (1992). Adenovirus-mediated *in vivo* gene transfer and expression in normal rat liver. *Nature Genet.* **1,** 372–378.

Jiang, C., O'Connor, S. P., Fang, S. L., Wang, K. X., Marshall, J., Williams, J. L., Wilburn, B., Echelard, Y., and Cheng, S. H. (1998). Efficiency of cationic lipid-mediated transfection of polarized and differentiated airway epithelial cells *in vitro* and *in vivo*. *Human Gene Ther.* **9,** 1531–1542.

Johnson, L. G., and Boucher, R. C. (1997). Gene therapy for lung diseases: molecular and clinical therapeutic issues. *Lung Biol. Health Disease* **107,** 515–553.

Kaplan, J. M., St. George, J. A., Pennington, S. E., Keyes, L. D., Johnson, R. P., Wadsworth, S. C., and Smith, A. E. (1996). Humoral and cellular immune responses of nonhuman primates to long-term repeated lung exposure to Ad2/CFTR-2. *Gene Ther.* **3,** 117–127.

Kartner, N., Hanrahan, J. W., Jensen, T. J., Naismith, A. L., Sun, S. Z., Ackerley, C. A., Reyes, E. F., Tsui, L. C., Rommens, J. M., Bear, C. E., and Riordan, J. R. (1991). Expression of the cystic fibrosis gene in non-epithelial invertebrate cells produce a regulated anion conductance. *Cell* **64,** 681–691.

Klinman, D. M., Yi, A. K., Beaucage, S. L., Conover, J., and Krieg, A. M. (1996). CpG motifs present in bacterial DNA rapidly induce lymphocytes to secrete interleukin-6, interleukin-12, and interferon-gamma. *Proc. Natl. Acad. Sci. USA* **83,** 2879–2883.

Krieg, A. M., Yi, A.K., Matson, S., Waldschmidt, T. J., Bishop, G. A., Teasdale, R., Koretzky, G. A., and Klinman, D. M. (1995). CpG motifs in bacterial DNA trigger B-cell activation. *Nature* **374,** 546–549.

Lee, E. R., Marshall, J., Siegel, C. S., Jiang, C., Yew, N. S., Nichols, M. R., Nietupski, J. B., Ziegler, R. J., Lane, M. B., Wang, K. X., Wan, N. C., Scheule, R. K., Harris, D. J., Smith, A. E., and Cheng, S. H. (1996). Detailed analysis of structures and formulations of cationic lipids for efficient gene transfer to the lung. *Human Gene Ther.* **7,** 1701–1717.

Leigh, M. W., Kylander, J. E., Yankaskas, J. R., and Boucher, R. C. (1995). Cell proliferation in bronchial epithelium and submucosal glands of cystic fibrosis patients. *Am. J. Respir. Cell Mol. Biol.* **12,** 605–612.

Leventis, R., and Silvius, J. R. (1990). Interactions of mammalian cells with lipid dispersions containing novel metabolizable cationic amphiphiles. *Biochim. Biophys. Acta* **1023,** 124–132.

Lewis, J. G., Lin, K. Y., Kothavale, A., Flanagan, W., Matteucci, M. D., DePrince, R. B., Mook, R., Hendren, R., and Wagner, R. W. (1996). A serum resistant cytofectin for cellular delivery of antisense oligonucleotides and plasmid DNA. *Proc. Natl. Acad. Sci. USA* **93,** 3176–3181.

Li, S., and Huang, L. (1997). *In vivo* gene transfer via intravenous administration of cationic lipid-protamine-DNA (LPD) complexes. *Gene Ther.* **4,** 891–900.

Lichtenberg, D., Freire, E., Schmidt, C. F., Barenholz, Y., Felgner, P. L., and Thompson, T. E. (1981). Effect of surface curvature on stability, thermodynamic behavior, and osmotic activity of dipalmitoylphosphatidylcholine single lamellar vesicles. *Biochemistry* **20,** 3462–3467.

Litzinger, D. C., and Huang, L. (1992). Phosphatidylethanolamine liposomes: Drug delivery, gene transfer and immunodiagnostic applications. *Biochim. Biophys. Acta* **1113**, 201–227.

Liu, Y., Mounkes, L. C., Liggitt, H. D., Brown, C. S., Solodin, I, Heath, T. D., and Debs, R. J. (1997). Factors influencing the efficiency of cationic liposome-mediated intravenous gene delivery. *Nature Biotechnol.* **15**, 167–173.

Logan, J. J., Bebok, Z., Walker, L. C., Peng, S., Felgner, P. L., Siegal, G. P., Frizzell, R. A., Dong, J., Howard, M., Matalon, S., Lindsey, J. R., DuVall, M., and Sorscher, E. J. (1995). Cationic lipids for reporter gene and CFTR transfer to rat pulmonary epithelium. *Gene Ther.* **2**, 38–49.

Martin, L. A., and Lemoine, N. R. (1996). Cancer gene therapy I: Genetic intervention strategies. *Gene therapy* In N. R. Lemoine and D. N. Cooper, (Eds.). (pp. 255–275). Oxford: Bios Scientific Publishers.

Matsui, H., Johnson, L. G., Randell, S. H., and Boucher, R. C. (1997). Loss of binding and entry of liposome-DNA complexes decreases transfection efficiency in differentiated airway epithelial cells. *J. Biol. Chem.* **272**, 1117–1126.

McDonald, R. J., Liggit, D. H., Roche, L., Nguyen, H. T., Pearlman, R., Raabe, O. G., Bussey, L. B., and Gorman, C. M. (1998). Aerosol delivery of lipid:DNA complexes to lungs of rhesus monkeys. *Pharm. Res.* **15**, 671–679.

Oudrhiri, N., Virgeron, J. P., Peuchmaur, M., Leclerc, T., Lehn, J. M., and Lehn, P. (1997). Gene transfer by guanidinium-cholesterol cationic lipids into airway epithelial cells *in vitro* and *in vivo. Proc. Natl. Acad. Sci. USA* **94**, 1651–1656.

Paukku, T., Lauraeus, S., Huhtaniemi, I., and Kinnunen, P. K. (1997). Novel cationic liposomes for DNA-transfection with high efficiency and low toxicity. *Chem. Phys. Lipids* **87**, 23–29.

Philip, R., Brunette, E., Kilinski, L., Murugesh, D., McNally, M. A., Ucar, K., Rosenblatt, J., Okarma, T. B., and Lebkowski, J. S. (1994). Efficient and sustained gene expression in primary T lymphocytes and primary and cultured tumor cells mediated by adeno-associated virus plasmid DNA complexed to cationic liposomes. *Mol. Cell. Biol.* **14**, 2411–2418.

Pier, G. B., Grout, M., and Zaidi, T. S. (1997). Cystic fibrosis transmembrane conductance regulator is an epithelial cell receptor for clearance of *Pseudomonas aeruginosa* from the lung. *Proc. Natl. Acad. Sci. USA* **94**, 12088–12093.

Pier, G. B., Grout, M., Zaidi, T., Meluleni, G., Mueschenborn, S. S., Banting, G., Ratcliff, R., Evans, M. J., and Colledge, W. H. (1998). *Salmonella typhi* uses CFTR to enter intestinal epithelial cells. *Nature* **393**, 79–82.

Porteous, D. J., Dorin, J. R., McLachlan, G., Davidson-Smith, H., Davidson, H., Stevenson, B. J., Carothers, A. D., Wallace, W. A., Moralee, S., Hoenes, C., Kallmeyer, G., Michaelis, U., Naujoks, K., Ho, L. P., Samways, J. M., Imrie, M., Greening, A. P., and Innes, J. A. (1997). Evidence for safety and efficacy of DOTAP cationic liposome mediated CFTR gene transfer to the nasal epithelium of patients with cystic fibrosis. *Gene Ther.* **4**, 210–218.

Remy, J. S., Sirlin, C., Vierling, P., and Behr, J. P. (1994). Gene transfer with a series of lipophilic DNA-binding molecules. *Bioconjugate Chem.* **5**, 647–654.

Rich, D. P., Anderson, M. P., Gregory, R. J., Cheng, S. H., Paul, S., Jefferson, D. M., McCann, J. D., Klinger, K. W., Smith, A. E., and Welsh, M. J. (1990). Expression of cystic fibrosis transmembrane conductance regulator corrects defective chloride channel regulation in cystic fibrosis airway epithelial cells. *Nature* **347**, 358–363.

Riordan, J. R., Rommens, J. M., Kerem, B. S., Alon, N., Rozmahel, R., Grzelczak, Z., Zielenski, J., Lok, S., Plavsic, N., Chou, J. L., Drumm, M. L., Iannuzzi, M. C., Collins, F. C., and Tsui, L. C. (1989). Identification of the cystic fibrosis gene: Cloning and characterization of complementary DNA. *Science* **245**, 1066–1073.

Rommens, J. M., Iannuzzi, M. C., Kerem, B. S., Drumm, M. L., Melmer, G., Dean, M., Rozmahel, R., Cole, J. L., Kennedy, D., Hidaka, N., Zsiga, M., Buchwald, M. J., Riordan, J. R., Tsui, L. C., and

Collins, F. C. (1989). Identification of the cystic fibrosis gene: Chromosome walking and jumping. *Science* **245**, 1059–1065.

Rosenfeld, M. A., Rosenfeld, S. J., Danel, C., Banks, T. C., and Crystal, R. G. (1994). Increasing expression of the normal human CFTR cDNA in cystic fibrosis epithelial cells results in a progressive increase in the level of CFTR protein expression, but a limit on the level of cAMP-stimulated chloride secretion. *Human Gene Ther.* **5**, 1121–1129.

Rosenfeld, M. A., Siegfried, W., Yoshimura, K., Yoneyama, K., Fukayama, M., Stier, L. E., Paako, P. K., Gilardi, P., Stratford-Perricaudet, L. D., Perricaudet, M., Jallat, S., Pavirani, A., Lecocq, J. P., and Crystal, R. G. (1991). Adenovirus-mediated transfer of a recombinant α_1-antitrypsin gene to the lung epithelium *in vivo*. *Science* **252**, 431–434.

Scheule, R. K., St. George, J. A., Bagley, R. G., Marshall, J., Kaplan, J. M., Akita, G. Y., Wang, K. X., Lee, E. R., Harris, D. J., Jiang, C., Yew, N. S., Smith, A. E., and Cheng, S. H. (1997). Basis of pulmonary toxicity associated with cationic lipid-mediated gene transfer to the mammalian lung. *Human Gene Ther.* **8**, 689–707.

Schiavi, S. C., Abelkader, N., Reber, S., Pennington, S., Narayana, R., McPherson, J. M., Smith, A. E., Hoppe, H., and Cheng, S. H. (1996). Biosynthetic and growth abnormalities are associated with high-level expression of CFTR in heterologous cells. *Am. J. Physiol.* **270**, C341-C351.

Schwartz, D. A., Quinn, T. J., Thorne, P. S., Sayeed, S., Yi, A. K., and Krieg, A. M. (1997). CpG motifs in bacterial DNA cause inflammation in the lower respiratory tract. *J. Clin. Invest.* **100**, 68–73.

Schwarz, L. A., Johnson, J. L., Black, M., Cheng, S. H., Hogan, M. E., and Waldrep, J. C. (1996). Delivery of DNA-cationic liposome complexes by small-particle aerosol. *Human Gene Ther.* **7**, 731–741.

Shoji-Tanaka, A., Mizuochi, T., and Komura, K. (1994). Gene transfer using purified retroviral integrase. *Biochem. Biophys. Res. Commun.* **203**, 1756–1764.

Short, D. B., Trotter, K. W., Reczek, D., Kreda, S. M., Bretscher, A., Boucher, R. C., Stutts, M. J., and Milgram, S. L. (1998). An apical PDZ protein anchors the cystic fibrosis transmembrane conductance regulator to the cytoskeleton. *J. Biol. Chem.* **273**, 19797–19801.

Sobol, R. E., and Scanlon, K. J. (1995). Clinical protocols. *Cancer Gene Ther.* **2**, 137–145.

Solodin, I., Brown, C. S., Bruno, M. S., Chow, C. Y., Jang, E., Debs, R. J., and Heath, T. D. (1995). A novel series of amphiphilic imidazolinium compounds for *in vitro* and *in vivo* gene delivery. *Biochemistry* **34**, 13537–13544.

Sternberg, N., Sorgi, F. L., and Huang, L. (1994). New structures in complex formation between DNA and cationic liposomes visualized by freeze-fracture electron microscopy. *FEBS Lett.* **356**, 361–366.

Stribling, R., Brunette, E., Liggitt, D., Gaensler, K., and Debs, R. (1992). Aerosol gene delivery *in vivo*. *Proc. Natl. Acad. Sci. USA* **89**, 11277–11281.

Stutts, M. J., Canessa, C. M., Olsen, J. C., Hamrick, M., Cohn, J. A., Rossier, B. C., and Boucher, R. C. (1995). CFTR as a cAMP-dependent regulator of sodium channels. *Science* **269**, 847–850.

Stutts, M. J., Gabriel, S. E., Olsen, J. C., Gatzy, J. T., O'Connell, T. L., Price, E. M., and Boucher, R. C. (1993). Functional consequences of heterologous expression of the cystic fibrosis transmembrane conductance regulator in fibroblasts. *J. Biol. Chem.* **268**, 20653–20658.

Surosky, R. T., Urabe, M., Godwin, S. G., McQuiston, S. A., Kurtzman, G. J., Ozawa, K., and Natsoulis, G. (1997). Adeno-associated virus Rep proteins target DNA sequences to a unique locus in the human genome. *J. Virol.* **71**, 7951–7959.

Templeton, N. S., Lasic, D. D., Frederik, P. M., Strey, H. H., Roberts, D. D., and Pavlakis, G. N. (1997). Improved DNA: Liposome complexes for increased systemic delivery and gene expression. *Nature Biotechnol.* **15**, 647–652.

Thierry, A. R., Lunardi-Iskandar, Y., Bryant, J. L., Rabinovich, P., Gallo, R. C., and Mahan, L. C. (1995). Systemic gene therapy: Biodistribution and long term expression of a transgene in mice. *Proc. Natl. Acad. Sci. USA* **92**, 9742–9746.

Walker, S., Sofia, M. J., Kakarla, R., Kogan, N. A., Wierichs, L., Longley, C. B., Bruker, K., Axelrod, H. R., Midha, S., Babu, S., and Kahne, D. (1996). Cationic amphiphiles: A promising class of transfection agents. *Proc. Natl. Acad. Sci. USA* **93,** 1585–1590.

Wang, J., Guo, X., Barron, L., and Szoka, F. C. (1998). Synthesis and characterization of long chain alkyl acyl carnitine esters. Potentially biodegradable cationic lipids for use in gene delivery. *J. Med. Chem.* **41,** 2207–2215.

Welsh, M. J., and Smith, A. E. (1993). Molecular mechanisms of CFTR chloride channel dysfunction in cystic fibrosis. *Cell* **73,** 1251–1254.

Welsh, M. J., Tsui, L. C., Boat, T. F., and Beaudet, A. L. (1995). Cystic fibrosis. In *The metabolic and molecular basis of inherited disease* (C.R. Scriver, A.L. Beaudet, W.S. Sly and D. Valle, (Eds.), Vol. III, pp. 3799–3876. New York: McGraw-Hill, Inc.

Wheeler, C. J., Felgner, P. L., Tsai, Y. J., Marshall, J., Sukhu, S. G., Doh, J., Hartikka, J., Nietupski, J., Manthorpe, M., Nichols, M., Plewe, M., Liang, X., Norman, J., Smith, A. E., and Cheng, S. H. (1996). A novel cationic lipid greatly enhances plasmid DNA delivery and expression in mouse lung. *Proc. Natl. Acad. Sci. USA* **93,** 11454–11459.

Whitsett, J. A., Dey, C. A., Stripp, B. R., Wikenheiser, K. A., Clark, J. C., Wert, S. E., Gregory, R. J., Smith, A. E., Wilson, J. M., and Engelhardt, J. (1992). Human cystic fibrosis transmembrane conductance regulator directed to respiratory epithelial cells of transgenic mice. *Nature Genet.* **2,** 13–20.

Yang, Y., Li, Q., Ertl, H. C., and Wilson, J. M. (1995). Cellular and humoral responses to viral antigens create barriers to lung-directed gene therapy with recombinant adenoviruses. *J. Virol.* **69,** 2004–2015.

Yew, N. S., Marshall, J., Przybylska, M. J., Ziegler, R. J., Rafter, P. W., Wysokenski, D. M., Armentano, D., Gregory, R. J., and Cheng, S. H. (1997a). Inclusion of the adenovirus E4 region in a plasmid vector harboring a CMV promoter confers more persistent expression in the lung (abstr). *Pediatr. Pulmonol.* **14** (Suppl.), 255.

Yew, N. S., Wang, K. X., Przybylska, M., Bagley, R. G., Stedman, M., Marshall, J., Scheule, R. K., and Cheng, S. H. (1999). Contribution of plasmid DNA to inflammation in the lung following administration of cationic lipid/pDNA complexes. *Human Gene Ther.* **10,** 223–234.

Yew, N. S., Wysokenski, D. M., Wang, K. X., Ziegler, R. J., Marshall, J., McNeilly, D., Cherry, M., Osburn, W., and Cheng, S. H. (1997b). Optimization of plasmid vectors for high-level expression in lung epithelial cells. *Human Gene Ther.* **8,** 575–584.

Yoshimura, K., Rosenfeld, M. A., Nakamura, H., Scherer, E. M., Pavirani, A., Lecocq, J. P., and Crystal, R. G. (1992). Expression of the human cystic fibrosis transmembrane conductance regulator gene in the mouse lung after *in vivo* intra-tracheal plasmid-mediated gene transfer. *Nucleic Acids Res.* **20,** 3233–3240.

Zabner, J., Cheng, S. H., Meeker, D., Launspach, J., Balfour, R., Perricone, M. A., Morris, J. E., Marshall, J., Fasbender, A., Smith, A. E., and Welsh, M. J. (1997). Comparison of DNA-lipid complexes and DNA alone for gene transfer to cystic fibrosis airway epithelia *in vivo. J. Clin. Invest.* **100,** 1529–1537.

Zabner, J., Fasbender, A. J., Moninger, T., Poellinger, K. A., and Welsh, M. J. (1995). Cellular and molecular barriers to gene transfer by a cationic lipid. *J. Biol. Chem.* **270,** 18997–19007.

Zhou, X., and Huang, L. (1994). DNA transfection mediated by cationic liposomes containing lipopolylysine: Characterization and mechanism of action. *Biochim. Biophys. Acta* **1189,** 195–203.

Zhu, N., Liggitt, D. Liu, Y., and Debs, R. (1993). Systemic gene expression after intravenous DNA delivery into adult mice. *Science* **261,** 209–211.

CHAPTER 4

Structure and Structure–Activity Relationships of Lipid-Based Gene Delivery Systems

Dan D. Lasic

Liposome Consultations, Newark, California

Recent years have witnessed substantial improvements in gene transfection and expression by using nonviral gene delivery carriers. Unfortunately, these developments are not paralleled by our understanding of molecular and colloidal mechanisms of gene transfection *in vivo*. While first ideas about the stability of DNA/lipid complexes in blood circulation, their pharmacokinetics, and biodistribution are starting to emerge, we still do not understand the entry of complexes into cells and nuclei. Also, structure–activity relationships are poorly understood. The situation is similar with the delivery of antisense oligonucleotides and ribozymes by cationic liposomes. This review attempts to discuss some of these challenges.

I. INTRODUCTION AND HISTORY

Gene therapy is a promising new modality in medical practice that is based on switching appropriate genes on or off. To turn on a gene, the DNA code, ligated into a suitable DNA plasmid, must be imported into the cell nucleus, while genes can be turned off by inhibiting transcription by specific binding of antisense oligonucleotides or degradation of messenger RNA by ribozymes.

Nonviral Vectors for Gene Therapy

DNA plasmids for gene therapy are synthetic constructs, typically produced in bacterial cultures. Methods of modern recombinant DNA technology can produce large quantities of pure plasmids containing powerful promoters and enhancers for gene expression. The problem in practice, however, is to bring these large macromolecules into appropriate cells *in vivo*. Naked DNA plasmids, which have contour lengths on the order of micrometers, are unlikely to cross undamaged cell membranes; therefore, several agents exist that can reduce the size and effective charge of DNA and improve transfection efficiency. As will be described, *in vivo* only genetically engineered viruses or nonviral DNA colloids can be used for effective gene transfer. Agents used to complex DNA and enhance its permeabilization into cells include multivalent cations, polycations, cationic polyelectrolytes, and, more recently, self-aggregating colloidal systems containing cationic lipids (Lasic, 1997).

The basic ideas of gene therapy were expressed in the 1960s. In 1961 it was found that the uptake of nucleic acids into cultured cells could be enhanced if RNA was complexed with protamine (Amos, 1961). Complexation with spermine, spermidine, and streptomycin protected RNA against enzymes but did not enhance the transfection. Several other articles in the next few years clearly established that complexation of DNA and RNA with agents such as methylated protein (basic protein), gelatin, polyornithine, DEAE-dextran, 125 mM $CaCl_2$, polylysine, polyarginine, and some other agents increased transfection and/or infectivity (Wolff and Lederberger, 1994).

Liposomes were first used for internalization of nucleic acids in the late 1970s (Dimitriadis, 1979; Fraley *et al.*, 1980). In the beginning these were mostly phosphatidylserine, large unilamellar vesicles prepared by the reverse phase evaporation method (Fraley and Papahadjopoulos, 1982). While the ideas of cationic lipids reacting with DNA may have originated in Tom Thompson's laboratory (Barenholz, private communication), cationic liposomes for DNA transfer were first described by Papahadjopoulos *et al.* in 1982 (Fraley and Papahadjopoulos, 1982). Stearylamine-doped liposomes, however, were found to be rather toxic and not more efficient than large negatively charged liposomes (Fraley and Papahadjopoulos, 1982). At the same time, several researchers in the Soviet Union studied DNA encapsulation into neutral and negatively charged liposomes with the aid of divalent counterions (Hoffmann *et al.*, 1978, Budker *et al.*, 1980). In the mid 1980s, Behr and colleagues studied interactions of DNA with cationic micelles and liposomes (Behr, 1986). It was only in 1987, however, that the potential of cationic liposomes for DNA transfection was fully realized (Felgner *et al.*, 1987). Some early approaches of liposomal gene delivery were described by Nicolau and Papahadjopoulos (1998).

Today, nonviral gene delivery *in vivo* is based mostly on DNA complexes with polycations, cationic liposomes, cationic polyelectrolytes, and combinations of these. Numerous papers describing gene transfection and gene expression by various delivery systems exist. Typically researchers study gene expression at fixed DNA concentration as a function of composition and concentration of the delivery system. To improve the transfection, researchers are working on improving the DNA constructs, synthesizing novel lipids and polymers, as well as on decorating the delivery systems with functional ligands. These include mostly fusogenic groups targeting ligands and polymer coatings. The latter can stabilize delivery systems and, upon the preprogrammed release of the polymer coat, induce destabilization of the DNA complex (Lasic, 1997).

Among the nonviral systems previously described, cationic liposomes seem to be the most widely used DNA delivery system. From the first reports on cationic liposomes, their interactions with DNA, and the use of such complexes for gene expression, numerous papers have described gene expression *in vitro* and *in vivo* as achieved by different cationic and neutral lipids, different charge ratios, and different transfection protocols. Hundreds of novel cationic lipids have been synthesized (Leventis and Silvius, 1990; Gao and Huang, 1992; Felgner *et al.,* 1994; Farhood *et al.,* 1992; Solodin *et al.,* 1995; Wheeler *et al.,* 1996; Lee *et al.,* 1996).

To understand any quantitative structure–activity relationships, detailed structure of the DNA/lipid complexes as well as the mechanism of transfection and gene expression should first be understood. Unfortunately, the structure of the complexes is still poorly understood while the mechanism of transfection is even more enigmatic. Most of the published data concentrate on optimizing cationic lipid and measuring transfection as a function of its mixing with colipid or the DNA/lipid charge ratio. We shall first review some of these data, then introduce phase behavior of DNA and lipids in order to introduce the structure of the complexes and gain some insight on possible molecular rearrangements during transfection.

II. STRUCTURE–ACTIVITY RELATIONSHIPS

Structure–activity relationships (SAR) for various DNA-lipid systems are a very complex issue. We shall define SAR on the molecular and the colloidal levels. The first describes the influence of lipid structure and stereochemistry on gene expression, while the second deals with the effects of morphology of DNA/lipid complexes on transfection.

The literature to date deals almost exclusively with the first paradigm. Before we discuss the information available we will begin with some general observations. Early work has clearly established that there is a great difference between *in vitro* and *in vivo* experiments. Furthermore, a great variation of transfection of various systems in different cell lines has been observed. Different cell lines behave very differently when subjected to the same series of cationic lipid/DNA complexes.

Many experiments have also shown that the presence of neutral lipid in the liposome bilayer increases transfection efficacy. In few systems, however, were pure cationic liposomes or micelles shown to be more efficient in transfection. One of the least disputed claims seems to be that dioleoylphosphatidylethanolamine is the optimal neutral lipid for most *in vitro* experiments (Felgner *et al.,* 1994; Farhood and Huang, 1996) while for *in vivo* transfection, cholesterol seems to be the most effective colipid (Zhu *et al.,* 1993; Lasic, 1997). The exact role of the neutral lipid is still not understood. The original hypothesis—that it helps to destabilize endosomal membrane upon endocytosis—is at odds with more recent findings that in gene expression experiments *in vivo,* cholesterol, which is known for its membrane stabilization effect, is by far a more efficient colipid. Whether this indicates that there are different mechanisms of cell entry in the two experiments is still not known. It may simply be that more stable complexes are a necessary condition for *in vivo* systemic gene delivery, while *in vitro,* especially in the absence of neutralizing plasma proteins, complex stability is not of great importance. Some cationic lipids, such as derivatized cholesterols or other sterols, or multivalent diacyl cationic lipids cannot form bilayers and lamellar phases and therefore the presence of colipids is necessary for the formation of liposomes. However, the effect of the nature and morphology of the cationic colloidal particle on its interaction with nucleic acids and transfection activity is still not understood. Some speculative explanations will be given in the section on SAR on the colloidal level.

In addition to stabilization or destabilization of liposome, DNA/lipid complex, or cell membrane, the presence of neutral lipid can have some other effects. These include the influence on charge density and elasticity of the bilayer, which may have crucial roles in interactions with DNA and preservation of the bilayered lipid phase, a state that might be more effective than micellar or monolayered organization of lipids, especially for *in vivo* delivery. With respect to *in vivo* application of the complexes, we shall refer mostly to systemic, chiefly intravenous administration. Local, parenteral, or topical delivery bypasses most of the challenges of the defenses of the immune system in the blood and therefore might resemble *in vitro* cell culture experiments more than systemic administration, and SAR rules might show some similarities with the *in vitro* optimization. Before we engage too deeply in such speculations, let us review what is known about structure–activity relationships.

Most of the researchers, at least in the period from the late 1980s to the mid-1990s, were searching for the optimal cationic lipid. Reports were constantly emerging of novel, more efficient, and less toxic lipids as most of the field was engaged in the search of the "magic" lipid. Furthermore, several groups were looking also for lipids (without any ligands!) that could target particular cells or tissues. The work involved mostly the synthesis of hundreds of novel cationic lipids and their testing in cell cultures. Table 1 shows some of the cationic lipids.

Some early studies examined the SAR of cationic lipids based on derivatized cholesterol (Leventis and Silvius, 1990; Farhood *et al.,* 1992). Later, most such investigations continued in a high throughput screening approach rather than in

Table 1

Abbreviations and Chemical Names of Some Cationic Lipids

Abbreviation*	Lipid	References
DODAB/C	Dioctadecyl diammonium bromide/chloride	Zhu *et al.*, 1993
DOTMA	N-[1-(2,3-dioleoyloxy)propyl]-N,N,N-trimethylammonium chloride	Felgner *et al.*, 1987
DOTAP	1,2 Dioleoyl 3-trimethyl ammonium propane	Boehringer Mannheim
DC-chol	3-β[N-(N',N'-Dimethylaminoethane carbamoyl cholesterol	Gao *et al.*, 1992
DOGS	Dioctadecyl amido glycil spermine	Behr, 1993
DOSPA	2,3, Dioleoyl–N [sperminecaroxamino)ethyl]-N,N-dimethyl–1-propanaminium	Behr, 1993
DMRIE	Dimyristoyl oxypropyl dimethyl hydroxyethyl ammonium bromide	Felgner *et al.*, 1994
SAINT-n	A series of dialkyl pyridinium–alkyl halides	Woude *et al.*, 1997
DOSPER	1,3-Di-oleoyloxy-2(6-carboxy spermyl)propylamide four acetate	Sigma
DPPES	Dipalmitoyl phosphatidylethanolaminyl spermine	Behr, 1993
DORIE	Dioleoyloxypropyl dimethyl hydroxyethyl ammonium bromide	Felgner *et al.*, 1994
GAP-DLRIE	$\pm$-N-(3-aminopropyl)-N,N-dimethyl bis(dodecyloxy)-1-propanamonium bromide	Wheeler *et al.*, 1996
DOTIM	1-[2-(oleoyloxy)-ethyl]-2-oleoyl-3-(2-hydroxyethyl) imidazolinium chloride	Solodin *et al.*, 1995

*Abbreviations for neutral lipids are chol for cholesterol and DOPE for dioleoylphosphatidylethanolamine.

rational, deductive research. A variety of data on different lipids in different mixtures with different colipids and different ratios with DNA in a variety of different cell lines exist, but no general conclusions have been extracted on how the nature of lipid, composition, and lipid/DNA ratio correlate to transfection. Researchers mostly concentrated on the geometry of cationic surfactants in their particular systems and some findings are shown next.

A. Structure–Activity Relationships: Molecular Level

It is convenient to divide the lipid molecule into the nonpolar part, the backbone or linker group, and the polar head, possibly with some decorations, and study each part independently.

The structure–activity transfection data are most consistent with respect to the nonpolar part. *In vitro* experiments have shown that dioleoyl (DO) and dimyristoyl (DM) chains are the most effective for transfection (Felgner *et al.*, 1994; Solodin *et al.*, 1995). This observation can be easily explained by the fact that for transfection, probably some membrane mixing between liposomal/complex membrane and cell or endosomal membrane must occur. Membranes must be in a fluid state for this mixing, because solid membranes cannot mix with other membranes. Although the mechanism of the complex entry into the cell is still not known and can vary from endocytosis, fusion, or poration of lipid membrane or combination of these, both hydrocarbon chains (DO and DM) provide liquid-like membranes at physiological temperatures.

Nonpolar and polar groups are often connected via a spacer group and/or molecular backbone. Obviously it is desirable to have polar and nonpolar groups attached via a biodegradable bond, such as an ester bond, to reduce toxicity. Furthermore, the linkage itself affects the geometry and stereochemistry of the attached polar head.

Investigations of DNA interactions and DNA condensation have revealed that the charge separation on polyions plays a crucial role. For instance, dialkylamines NH_3^+—$(CH_2)_n$—NH_3^+ with odd n were much more effective in DNA condensation than the ones with an even number of methylene groups. Similar effects are to be expected with cationic lipids, especially the multivalent ones. Not many equivalent studies have been performed with cationic lipids. For a DC-chol system, however, it has been shown that a spacer of $3-8$ atoms and carbamoyl linkage is optimal with respect to transfection efficiency and safety (Farhood *et al.*, 1992). In a large screening of almost 100 lipids, it was observed that spermine attached to cholesterol through secondary amine but not terminal primary amine gives rise to 100-fold improved transfection (Lee *et al.*, 1996). This pointed to a T-shaped motif as an important condition for an effective lipid. Such a configuration might improve binding to DNA and increase stability of the supporting bilayer. No such relations have been found for linkages that involve a carbon atom instead of nitrogen.

Polar heads consist of one or several positive charges and possibly some other active groups that can bind to DNA via H-bonds. Additionally, the presence of some bulkier hydrophobic groups on polar heads may increase binding to DNA, especially at the sites of defects that may occur at high curvatures of the polymer. Despite the fact that most of the polar heads are either quaternary ammonium salts or protonable amines, SAR of polar heads is the least understood. Only for DC-chol was it shown that ternary amines are more effective than quaternary and secondary and much more effective than primary. While quaternary amines $(RR'R''R'''N^+\text{-}X^-)$ are salts and are therefore charged and dissociated, tertiary $(RR'R''N)$, secondary $(R,R'NH)$, and primary (RNH_2) amines are characterized with a series of basic pK values, which affect the charge density and therefore binding and release of DNA. Because these groups are basic, their charge increases with lowering of pH, which must be reflected in the interactions. No study has addressed these questions yet.

Another unsolved issue is the nature of the anion in the cationic lipid. In contrast to anionic liposomes, where the counterions do not play an important role in the bilayer properties, anions have a much greater influence on cationic bilayers. This is due to their larger size, typically nonuniform distribution of their charge, and hydrophobic character. For instance, exchange of anions can cause phase transition from micelles to vesicles (Lasic, 1993). Also, entropic effects of anions are much more complex and theoretical treatments still rely on structure forming/breaking ions and Hoffmeister series, while, for instance, in the anionic bilayer/cation systems the thermodynamic behavior can be explained by the Poisson–Boltzmann equation using different radii of cations.

The optimal number of charges on the polar head group is not yet known. *In vitro* it seems that a higher positive/negative charge ratio, *r*, originating either from multivalency of cationic lipid or excess of univalent lipid neutralizes effects of plasma. Such formulations may be more toxic *in vivo*, where, as it seems now, mostly univalent lipids are used. *In vivo*, the optimal charge ratio (*r* between 2 and 4 – 6) seems to be independent of the number of charges on the polar head group.

Decoration of the polar head by attaching a group such as a hydroxy or amine or short alkyl may increase interactions with DNA (hydrogen bonding, electrostatics, specific interaction, and hydrophobic interaction). Reversibility of the DNA/lipid binding may be also important. In order to express, DNA must dissociate from cationic lipids (Xu and Szoka, 1996) and binding that is too tight may not be advantageous. For DNA release, therefore, weaker bonds might be favorable.

Because a positive charge is almost exclusively provided by quaternary ammonium ion or protonable amines, it is difficult to use pH sensitivity of the charge as a means to release DNA. By decreasing pH the charge increases and so does the DNA binding to the lipid (unless there is a competitive binding to an anionic lipid or polymer). pK values of the amines can be estimated from Hammett equations. Using Pallas software one gets a pK of DC-chol tertiary amine of 8.35, not far from an indirect experimental measurement reported by Zuidam and Barenholz (1997). Secondary amine can be protonated only at pH $= -3.8$. pK values are 10.45 for spermidine, 9.85 for the (equivalent) terminal primary amines, and 9.02 for the secondary amine. Similarly, the four protonable amines on spermine have pK values between 9.85 and 10.64, in line with the general fact that aliphatic amines are as basic as ammonia. This analysis shows that these lipids are almost completely protonated (charged) in media with physiological pH values.

Many studies have shown that the transfection of particular cell lines by DNA plasmids complexed to different cationic lipids can differ up to a few orders of magnitude. Nevertheless, no one lipid has been shown to perform well in many different cell lines. While several very potent novel lipids have been discovered, it is still not clear whether these lipids transfect better *in vivo* than the early ones such as DOTAP or DOTMA.

Definitively, however, many lipids that were proven not to be too effective were also discovered. Unfortunately, SAR relationships of these have never been thoroughly evaluated despite the fact that there is a great deal of information in such data. One such lipid family is cationic lipids based on phosphatidyl cholines with esterified phosphate charge. This might be due to a too-bulky polar head. In addition, from organic chemistry it is also well known that phospho-triesters are toxic.

We believe that SAR on the colloidal level is also crucial for effective gene expression *in vivo*. To be able to rationally design any gene carrier one must know its structure and mechanism of action.

It was discovered very early that the presence of plasma reduces or eliminates transfection. Several groups, however, observed that the neutralizing effect of plasma can be overcome by using a higher positive charge as it was understood that some multivalent cationic lipids are less affected by the plasma presence. The idea that a positive charge is needed to neutralize negatively charged plasma proteins was proven by Yang and Huang (1997), who showed that plasma can be inactivated by removal of negatively charged molecules, preincubation with polylysine, or by increasing the $+/-$ charge ratio.

It is likely that negative components of plasma simply aggregate and precipitate positively charged DNA/lipid particles. Therefore, one can expect that steric stabilization of complexes should increase expression of metastable DNA/lipid complexes. Indeed, addition of PEGylated detergent Tween 80 increased transfection of DOTMA-based cationic complexes in vitro (Liu *et al.,* 1996, 1997) as well as of PEGylated-lipid stabilized DOTAP and DODAB complexes with DNA *in vivo* (Hong *et al.,* 1997).

These results indicate that despite great variability in these systems, the transfection is still governed by basic colloidal laws such as colloidal interactions including electrostatics and steric repulsion. The well-known observation that lung surfactants reduce transfection upon pulmonary delivery of complexes can be understood by destabilizing activity of detergents, which are known to cause flocculation of lipid colloidal particles at lower concentrations and dissolution at higher concentrations.

The above facts could be qualitatively explained by the colloidal properties of these complex mixtures. To obtain more information on SAR *in vitro,* the transfection activity of various cationic lipids would need to be correlated also to their cytotoxicity and lytic activity. What is happening *in vivo,* however, is even more difficult to comprehend. Logical thinking implies that the complex must protect DNA in circulation and extracellular spaces, deliver it to and into cells, and release the DNA plasmid in the cytoplasm. The details of the processes are not known. Even the most basic question—How does the complex enter the cell?—has not yet been answered. Different groups of researchers seem to be polarized between endocytosis and direct entry into the cell. Both groups are presenting experimental proofs for their models and further work is required to address this issue.

Along similar lines one can try to understand a finding that surprised most researchers in the field—namely, it has been shown that a timely preinjection of empty cationic liposomes increases transfection of complexes or even naked DNA (Song and Liu, 1998). The following explanation seems to be the most logical: preinjected liposomes adsorb on the blood vessels (or become trapped after aggregation or coaggregation with some plasma components) and postinjected DNA interacts with adsorbed liposomes where capillaries are clogged. Such a mechanism prefers direct entry of DNA into the cell (membrane poration) rather than endocytosis.

Despite more than a decade of work, there is not much information on basic physico-chemical and biological properties of the cationic lipids. In addition to current work, several other factors should be measured, including binding constants of lipids to DNA, pK, solubility (critical micelle concentration), lytic activity (hemolysis), stability of the complexes upon dilution and in plasma as well as force/distance profiles and monolayer properties without and with DNA in the subphase. Such work will shed more light in the mechanism of transfection and allow formulation of more active complexes.

B. STRUCTURE–ACTIVITY RELATIONSHIPS: COLLOIDAL LEVEL

In contrast to studies of the SAR on the molecular level, almost nothing has been published on the SAR on the colloidal level, especially in systemic *in vivo* applications, where these properties may be most important for complex pharmacokinetics, biodistribution, DNA protection, and cell internalization. Due to the lack of the information available, we shall speculate on some results and present some logical analyses of the data.

Before we discuss these issues we shall briefly introduce some characteristics of DNA and lipids and their interactions, which dictate the colloidal properties of the complexes. With respect to the plasmid we believe that DNA compaction/condensation and its reversibility upon entry into the cytoplasm are extremely important. We believe that complex size distribution is very important for gene expression, especially *in vivo,* because it dictates its pharmacokinetics and biodistribution.

1. DNA Condensation

In solution, supercoiled DNA plasmids are random coils with radii of gyration in the micrometer range and coil volumes on the order of cubic micrometers (Bloomfield *et al.,* 1974; Watson *et al.,* 1983). By using appropriate condensing agents, the overall dimensions of these plasmids can be reduced 10,000 times and the outer diameter can be well below 100 nm. This random (worm-like) coil–globule transition is referred to as DNA condensation (Bloomfield, 1996). It was shown that this phase transition, which is common in viruses and cell nuclei (and allows, for instance, our genome with a length of ca 1 m to be packed into a ca~5 μm nucleus), can be performed *in vivo* upon addition of cations or polymers with charge $z > 2+$ (polyamines, lantanides, polylysine, basic proteins such as histones, self-assemblies of univalent cationic lipids, etc.).

Condensed DNA molecules can have different shapes. Apart from irregular, amorphous aggregates, the toroidal shape is the most frequently observed morphology. Depending on the size and electrostatic potential of counterions, the toroid can have as small an outer radius as 45 nm (inner 7.5 nm) in the case of condensation with polyamines. Normally, one turn contains ca 930 base pairs and the whole toroid around 60 kbp. It can be formed by one plasmid or several smaller ones (Bloomfield, 1996).

Another frequently formed structure is rod-like particles which typically occur when alcohols and $Co(NH_3)_6^{3+}$ are used. High curvature at DNA foldbacks is thought to be due to partial destabilization of double helix by alcohols. DNA condensation is a complex process that is kinetically and thermodynamically controlled. Energy contributions that favor DNA condensation are electrostatic attraction and bridging of condensing agents and correlation forces (counterion fluctuations), while entropy loss, bending elasticity (stiffness of DNA and tight curvature), electrostatic repulsion of helices, and hydration repulsion oppose the process (Bloomfield, 1996; Podgornik *et al.,* 1998).

Recent cryoelectron micrographs and SAXS measurements have revealed two-dimensional DNA condensation in the intercalated lamellar phase of DNA/lipid complexes (Lasic *et al.*, 1996, 1997; Rädler *et al.,* 1997). Similarly, planar condensation was observed upon adsorption of DNA on monolayers of cationic lipids (Fang and Yie, 1997). It is obvious that these morphologies are driven by interactions with lipids, which in some cases can dictate the symmetry of the mixed phase.

2. Lipid Polymorphism

Phase behavior of neutral and anionic lipids in aqueous mixtures is well known. No studies exist on liquid crystalline behavior of cationic lipids. Optical microscopy of concentrated phases in water show liquid crystalline behavior. The textures in the micrographs between crossed nicols reveal mostly lamellar phases, while occasionally inverted hexagonal phase texture can be observed. When these phases are diluted, colloidal particles can be observed. However, in many cases, aqueous mixtures gel even at very low lipid concentration (<1%) due to electrostatic repulsion between the particulates. Screening this interaction by addition of salt or diluting cationic lipid with neutral colipid fluidizes the gel. Colloidal particles can be characterized by micellar, lamellar, or inverse hexagonal symmetry. In contrast to DOPE, some cationic lipids disperse easily into hexagonal colloidal particles.

Several diacyl cationic lipids, such as DOGS, have many charges and form micelles in aqueous solutions.

Diluted lamellar phases consist of a broad distribution of vesicles in the size range from <100 nm to several μm. Lipids disperse well in water due to high charge density. Almost no large multilamellar liposomes can be observed. Large unilamellar vesicles can be prepared by extrusion while for the preparation of SUV almost all methods used in liposome technology can be used.

In mixtures with oil, these surfactants can form stable emulsions (with or without addition of cosurfactants) and microemulsions.

3. DNA–Cationic Colloid Interactions

Micelles and liposomes as well as emulsion droplets containing positively charged lipids interact with DNA. Obviously, the nature of the colloidal particle and, in the case of liposomes, the morphology (size, lamellarity) of the liposome dictate the reaction and the structure of the complex formed. Although many researchers report that the size of the complex is independent of the size of the DNA plasmid or lipid particle, we believe that such observations are due to the lack of precision. Although there are typically from 5 to 20 plasmid in a complex (Lasic *et al.*, 1997), obviously, large plasmids can never form small particles. Similarly, in the case of noncomplete lipid restructuring, large liposomes may not give rise to small complexes. In many experiments, however, the very broad size distribution of the particles formed wipes out such differences.

In most cases small particles are advantageous for *in vivo* delivery. We believe that the use of small vesicles in complex preparation is preferred because they give rise to smaller complexes. It would be especially effective if complexes contained only one condensed plasmid. Due to the random nature of particle interactions even in dilute colloid dispersions, such condensates are likely only in carefully designed experiments based on specific nanotechnology matrices, such as surfaces with specific micropatterns (silicon wafer with etched wells in micrometer range) in reverse emulsions or perhaps in large liposomes.

With respect to the size of the DNA/lipid complex, there are several conflicting reports, including some that indicate increased gene expression in the lungs when larger liposomes were used (Liu *et al.*, 1997, Templeton *et al.*, 1997). However, one must be cautious when analyzing some older data. Gene expression might have been simply a consequence of clogging of capillaries (the lung capillary bed is the first one tail-vein-injected complexes encounter in mice) by large and possibly aggregated complexes. Therefore, the transfection could have been a consequence of the reversible local damage and may have occurred also in the infiltrated cells.

Similarly, pulmonary expression upon intratracheal instillation might have been a simple consequence of flooding the lung and transient transfection of infiltrating macrophages. Anecdotal evidence that expression was the highest in animals that barely survived supports such conclusions and urges better design of experiments as well as controls.

Similarly, when assessing therapeutic activity of formulations, thorough controls must be in place to ensure that the results are due to the expressed gene. For instance, intratumor injection of complexes can easily decrease the tumor size due to the (cyto)toxicity of cationic lipids. Obviously, empty liposomes and irrelevant plasmid must be included as controls, in addition to the measurements of m-RNA of the therapeutic gene and tumor size.

4. Structure of the Complexes

In contrast to extensive studies of gene expression, investigations on the physico-chemical characteristics of the complexes are only starting to emerge.

Early hypothetical models of aggregates of unchanged liposomes bridged by partially adsorbed DNA chains (Felgner and Ringold, 1989) were never confirmed experimentally. First analyses of the structure of DNA/lipid complexes were shown only in the mid 1990s (Behr, 1993; Minsky et al., 1996). Typically electron micrographs showed aggregates of spherical structures with some attached fibers (Sternberg et al., 1994; Gustaffson et al., 1995). Electron microscopic observation of various DNA/lipid complexes in the literature reveals a variety of smaller or larger, less or more compact or dense complexes and their aggregates, with a possibility of attached fibrillar structures (Sternberg, 1996, 1997).

Fibers were shown to be DNA helices surrounded by a lipid tubule. Other studies, such as on measurements of the size, zeta potential, lipid mixing, or density of complexes, did not reveal much useful data regarding the structure and dynamics of the particle.

Size varies according to the preparation procedure and method of analysis. Zeta potentials are roughly in the expected range. If they do not deviate much from the expected values they may tell us of some problems in the complex formation (see kinetic effects following). They are not very useful for structural predictions. They can be used, however, to predict the stability of the complexes following the Derjaguin Landau Verwey Overbeek (DLVO) model. Lipid mixing seems to be always inconclusive; some, but not all lipid mixing occurs in all the published studies of transfection-effective and -ineffective complexes.

According to the DLVO model, colloids are stable if the repulsive electrostatic forces are larger than ubiquitous van der Waals attractive forces. Cationic lipid/ DNA complexes represent a curious paradox, because in the beginning of the interaction, electrostatic forces are attractive (positive lipid surfaces and negative phosphate groups on DNA). However, when the complexes are formed, if mixed properly they have a relatively uniform distribution of surface charge. In the case of excess of lipid, the complexes typically have positive surface charge which can be measured via zeta potential measurements. Because electrostatic forces are shielded by electrolytes, addition of salt can cause aggregation and precipitation of colloidal particles (phase separation) that can be calculated by the DLVO theory.

Small-angle x-ray scattering (SAXS) experiments of these complexes paralleled with cryoelectron microscopy revealed periodic structure (Lasic *et al.*, 1996). Two periodicities were apparent from cryoelectron micrographs and confirmed by small-angle x-ray scattering. The large periodicity of typically 5.5–6.5 nm was attributed to planar arrays of lipid bilayers with sandwiched DNA, while the small and less pronounced periodicity of 3–4 nm was attributed to the separation between the helices within the bilayered sandwich (Lasic *et al.*, 1996, 1997; Raedler *et al.*, 1997). To date, this is the only study in which fine structure has been confirmed by at least two independent techniques. This model was supported by atomic force microscopy observations of two-dimensional (short range order) DNA condensation on deposited cationic bilayers (Fang and Yie, 1997).

Analytical ultracentrifugation also revealed heterogeneous particles with a wide variation of the values of sedimentation coefficient. Typically, cationic complexes have shown a continuous population between very dense particles (sedimentation coefficient, $S > 1000$ Sv) to particles with the same S as cationic liposomes, meaning that in the cases of excess positive charge some liposomes did not react. Analogously, in the case of anionic complexes the trailing band was characterized by the same sedimentation coefficient as for naked DNA (Lasic, 1997).

The inverse hexagonal phase in which DNA helices are contained within the hydrophilic cylinders was also proposed (Felgner *et al.*, 1996). While we have never observed hexagonal arrangement (unless in the case of excess of DOPE), a hexagonal structure was recently reported (Koltover *et al.*, 1998). It was also reported that hexagonal structure is a necessary condition for high transfection activity, based on the interaction of the complexes with model anionic liposomes as viewed by fluorescence microscopy. While such extrapolation may have some value for *in vitro* cases, it is probably irrelevant for *in vivo* gene expression, where the most important condition for efficient transfection seems to be the stability of the complex. It is known that in several cases, structures with sharp edges and other "frozen fusion intermediates" (such as cochleates) show enhanced gene transfection *in vitro*. Such observations point to the direct entry of DNA through the cell membrane rather than endocytosis.

While the work to date mostly describes structure of the complexes, studies to elucidate the fundamental forces responsible for these interactions and their consequences to the observed morphologies have started to emerge (Campbell *et al.*, 1998). Adsorption of DNA on the cationic bilayers was shown to induce lipid phase separation. Surface force apparatus measurements at close approach of DNA and cationic lipid monolayers have shown that the structure is a result of strong electrostatic forces, steric repulsion, and van der Waals and hydrophobic attraction, which is due to defects in the lipid layers (Campbell *et al.,* 1998).

For site-specific DNA (or antisense oligonucleotide or ribozyme) delivery, it is very important (especially for systemic administration) to minimize nonspecific attractive electrostatic (and other) interactions. Steric stabilization seems to be the most convenient way to convey to the DNA/liposome particles long circulating times and subsequently the ability of small complexes to accumulate in specific tissues characterized by the leaky vascular system (such as sites of inflammation, infection, many tumors) upon intravenous administration. However, sterically stabilized cationic liposomes typically do not interact with DNA because steric repulsion exceeds electrostatic attraction between the two colloidal reagents. One can use higher solubility of PEGylated lipids (charged or uncharged) to insert them into the bilayer of the coated complexes upon incubation of preformed DNA/cationic lipid complexes and PEG-lipid micelles, analogous to the insertion of PEG-lipids from micelles into conventional liposomes (Uster *et al.,* 1996).

This brief review of the structures of liposome/DNA complexes shows many different morphologies and few conclusions that can be generalized. This should not be surprising. First, different techniques can unravel and report different parts or different specificities of the complexes. For instance, SAXS can see only periodic structure and its confirmation with optical microscopy is impossible because the latter can never distinguish the fraction of the sample that gives rise to the SAXS pattern. Cryoelectron microscopy, which has at least 100-fold better resolution, shows that all the sample is not characterized by ordered structure. Obviously, due to topological reasons, such ordered arrays must have disordered structure at the edges. Similarly, other microscopic methods such as freeze–fracture microscopy and negative stain show such heterogeneous structures, despite perennial problems with sample preparation and artifacts of it.

The above facts point to the specific viewpoints of different experimental techniques. In the case of DNA/lipid complexes, however, it seems that the observed heterogeneity with respect to size, shape (electron microscopy, light scattering), zeta potential, density, and sedimentation coefficient (ultracentrifuge) seems to be of natural origin, as will be described next.

In contrast to the majority of the researchers, who claim that the reaction between DNA and cationic liposomes is spontaneous, we believe that it is kinetically controlled. Therefore, the formation and morphology of the complexes is not determined solely by the absolute concentrations and charge ratios, ionic strength, temperature, and similar thermodynamic parameters but also by kinetic factors. A simple experiment of slow and fast mixing of the reactants reveals that the process is also kinetically controlled. For instance, slow mixing of liposomes and DNA at a charge ratio $+/- = 2$ results in precipitation for lipid concentrations above 0.5 mM for most of the lipids. Quick mixing, however, can result in colloidally stable suspension of much smaller particles at concentrations up to 5 mM.

If the reaction were thermodynamically controlled, it would occur spontaneously and would therefore always result in the same product characteristics, including extremely narrow size distribution of the colloidal particles, which would be independent of kinetic conditions (mixing rate, sequence of operations).

At high $-/+$ ratios it seems that mostly fibrillar structures are formed, while at $+/-$ ratios from 1 to 5 or 10 predominantly lamellar structures can be observed. We must be aware, however, that all samples are heterogeneous with respect to size and shape distribution. In reality, due to local conditions in random locations where the two reactants are being mixed and are interacting, probably the whole spectrum of structures between the two extrema occur. Therefore, many experiments simply yield one or several morphologies among many random structures that are possible in the range between the two limiting structures at thermodynamic equilibrium: coated DNA fibers at large excess DNA and intercalated lamellar phases at excess of lipid.

Thorough mixing and work at lower concentrations probably ensure that local concentrations in the complexes do not deviate much from the bulk and that structures have similar charge ratios. This in turn gives particles with tighter size distribution very similar surface charge densities, which significantly enhances their stability according to the DLVO model. The absence of fibers in well-condensed samples also seems to be correlated with increased test tube stability of the complexes.

In vivo data show that transfection is more effective when using liposomes compared to micelles and emulsions (Pinnaduwage *et al.*, 1989; Liu *et al.*, 1996). This may be due to better coating ability of DNA by bilayer-forming lipids as opposed to surface adsorption and subsequent DNA protection. In comparison to polymers, dendrimers, and similar agents, liposomes again may provide better protection of DNA because of the ability of lipid bilayers to coat surfaces. Additionally, as self-assembled vehicles they can be dispersed and are eliminated easier and faster than large polymers upon DNA release.

We were interested in the SAR on the colloid level. We wanted to find the correlation between physico-chemical properties of the complex and its transfection potency *in vivo*. To study the difference between various lipids, we generated phase diagrams of complex properties in the phase space of DNA and lipid concentration (Lasic, 1997; Lasic *et al.*, 1999). In such a phase diagram one can plot various characteristics such as complex size, turbidity, zeta potential, transfection activity (gene expression) for various DNA and lipid concentrations and study their correlations. The first observation was that some lipids, lipid mixtures, or liposome preparations were not able to colloidally suspend high enough concentrations of DNA. Gene expression should be proportional to the dose of DNA (Templeton *et al.*, 1997), which eliminates some lipids and complex preparation procedures (such as use of electrolytes as tonicifiers of the medium). This also explained the inability of commercially available transfection kits to achieve high transfection level *in vivo* because the concentration of these liposomes is simply too low to allow dispersing sufficient concentration of DNA into colloidal suspension.

Such phase diagram studies could clearly differentiate between various lipids, lipid compositions, and liposome sizes and explain why some systems were not effective *in vivo* (Lasic *et al.,* 1999). However, in many cases the transfection activity of complexes of similar size, turbidity, zeta potential, DNA concentration, and charge ratio seemed not to be correlated to physico-chemical characteristics of the complexes. The result of these studies was that a high concentration of colloidally suspended DNA in a small, compact particle is a necessary but not sufficient condition for effective gene expression upon intravenous administration.

We were interested in whether we could find any other difference between complexes that differ in *in vivo* activity but not in their physico-chemical properties, such as lipid and DNA concentrations, size distribution, turbidity, zeta potential, and density. In the next step we investigated the stability of these complexes. While most of the complexes were stable upon dilution with water, saline, or isotonic sucrose solution, we noticed that they exhibited markedly different stability in plasma. Surprisingly, the complexes that were the least stable in plasma were the ones that yielded the highest expression of marker genes in mice lungs upon tail-vein administration. This led us to the conclusion that the majority of the DNA probably enters the extravascular space at sites of damage such as clogged capillaries. At relatively low lipid doses such treatments are in general not fatal but results correlate with the anecdotal evidence that sicker animals typically show larger transfection.

This observation seems to be similar to the *in vitro* experiments in which it was seen that increasing lipid concentration increases transfection and also toxicity, so that the optimal dose depends on the balance of gene expression and cytotoxicity. This is, again, more consistent with direct entry of complexes into the cells (lipid-mediated poration) as opposed to endocytosis.

The lesson we have learned is that one needs positively charged complexes at the lowest possible dose of cationic lipid to minimize toxicity. One can control complex size and charge ratio and try to maximize DNA concentration. Complexes should be small to increase their stability in the test tube and decrease toxicity. Such complexes should contain reversibly condensed DNA that should be well protected.

Additional improvements may rely on novel designs, such as attachments of various ligands and other functional groups on the complexes. Indeed, preliminary experiments with asialofetuin-labeled complexes have shown sevenfold improved expression in the liver (Templeton *et al.*, 1997). However, we must be aware that nonspecific interactions of such complexes still predominate and that for such improvements they should be shielded, perhaps by coating of the DNA/lipid complex by (polyethylene) glycol chains, analogous to treatment of sterically stabilized liposomes (Lasic, 1997; Hong *et al.*, 1997).

While not much is known with respect to the exact mechanism of transfection, we can speculate that import of the released DNA in the nucleus might be the greatest barrier in achieving high levels of gene expression. For effective release of DNA in the cytoplasm, we have already mentioned that liposomes might be a better choice than polymers because of their self-disassociation in the media with lipid bilayers. Along these lines, relatively high values of the critical micelle concentration (cmc) of cationic lipids might be noteworthy; for instance, DOTAP has cmc of $7·10^{-5}$ M (Zuidam and Barenholz, 1997). With respect to the subsequent nuclear entry, it is well known that mitotic cells yield much higher gene expression and often two sister cells both show transfection. This can be explained by the fact that plasmids enter the nucleus mostly during cell division. To enhance nuclear entry, karyophilic peptide signal sequences should be attached to the complex, as will be briefly discussed next.

5. Nuclear Import and Export

In many cases large numbers of plasmids were delivered into the cytoplasm, as observed by fluorescence microscopy, and yet very low or no gene expression occurred. In many cases this was due to the condensed state of DNA. Another important factor is, however, the traffic of plasmid into the cell nucleus through the double membrane of the nuclear envelope (Weiss, 1998).

There is very dense traffic across nuclear membrane, similar to the traffic across the cell membrane. Every minute several million RNA and protein molecules are transported into the nucleus from cytoplasm and vice versa. It is believed that inbound and outbound traffic is regulated by transport factors, which are called importins and exportins. A large protein structure embedded in the nuclear membrane, called nuclear pore complex (NPC), is responsible for all nucleocytoplasmic transport. The channel with diameter of 9 nm allows permeation of smaller molecules, while up to three times larger particles or molecules can be transported with the help of appropriate nuclear transport signals. The first nuclear localization signals (NLS) were identified in SV40 large-antigen NLS (PKKKRKV) and bipartite nucleoplasmin NLS (KRPAAIKKAGQAKKKK). Such sequences are recognized by importins in the cytoplasm and the complex formed docks at the cytoplasmic side of the NPC, and the complex is then in an energy-dependent reaction translocated into the nucleus.

The entry of the DNA plasmids from the DNA/lipid complexes in the cytoplasm into the nucleus is not yet understood. It is likely that it is driven by importins in the previously described processes and therefore to some part of the plasmid these factors must bind. Obviously, for binding to the importins and entry in the nucleus, plasmid must be in a decondensed or uncompacted state. However, some results indicate that in some cases of polymer/DNA complexes, complexed plasmid can enter the nucleus (Pollard *et al.,* 1998; E. Wagner, personal communication). Understanding the whole mechanism in detail would allow us to design more efficient gene delivery systems, perhaps by attaching some of the NLS to the plasmid.

We have described many factors that are necessary conditions for gene

Table 2

Mechanism of Transfection: Steps in the Transfection Process for Systemic Administration

Step in transfection	Pros and cons*
Complex formation–preparation (thermo-dynamics and kinetics of complex formation)	Small, dense, condensed DNA. Maximize DNA conc; avoid electrolytes, excess lipid
Complex stability	**$r = 2-4$, no fibers (quick mixing?)
Complex stability *in vivo* (plasma, blood)	Small, dense complexes, possible stealth coating; avoid large, badly compacted aggregates
Complex pharmacokinetics, biodistribution	Reduce nonspecific and increase specific interactions (stealth coating, ligands)
Adsorption on the cells	Excess positive charge; possible targeting ligands to endocytotic receptors
Cell internalization	Size range $100-400$ nm? Fusogenic ligands, endosome disruptors?
DNA release in the cytoplasm	Lipid dependent (high cmc?); endosome disrupting agents? DNA decondensation
Nuclear entry	NLS sequences attached?
Duration of expression	Attachment to genomic DNA, self-replicating plasmids?

*Some comments on possible improvements, as seen from the liposome or DNA carrier perspective. Obviously, many improvements will come also from the DNA plasmid construction.

**r = positive/negative charge.

expression to occur. However, even after the gene is delivered into the nucleus, there are many factors that determine transfection. These include the current proliferation status of the cell, its polarization, and differentiation. Table 2 lists the steps in the transfection process with comments on the transfection activity.

III. CONCLUSION AND FUTURE PROSPECTS

With the development of DNA/plasmid constructs on one side and delivery systems on the other, gene expression *in vitro* and *in vivo* improved approximately 1,000-fold in the last decade (Felgner, 1996). For most applications, however, the levels achieved and short duration of protein synthesis are still not adequate. We believe that with improved understanding of the mechanism of transfection and SAR, improved systems can be designed. On the carrier side these particles must first be made refractory to nonspecific interactions in the body, such as via sterically stabilizing polymer coatings (Lasic, 1997). Then they may include targeting and fusogenic groups and, possibly, different condensing and encapsulating agents

(Wagner *et al.*, 1992; Sorgi *et al.*, 1997; Hong et al., 1997). Additionally, several proteins, such as HMG (high mobility group) non-histone chromosomal proteins, transcription factors, and protein nuclear localization factors can be used for activation of chromatin and to enhance import and transcription of transgene (Boulikas, 1998). On the plasmid side, some novel useful sequences from the viruses and nuclear targeting codons may improve the active part of the complex, and synthetic transfection vectors may someday match the natural ones.

REFERENCES

Amos, H. (1961). Protamine enhancement of RNA uptake by cultured chick cells. *Biochem. Biophys. Res. Comm.* **5,** 1–4.

Behr, J. P. (1986). DNA strongly binds to micelles and vesicles containing lipopolyamines or lipointercalants. *Tetrahedron Lett.* **27,** 5861–5864.

Behr, J. P. (1993). Synthetic gene transfer vectors. *Acc. Chem. Res.* **26,** 274–278.

Bloomfield, V. A. (1996). DNA condensation. *Curr. Opin. Struct. Biol.* **6,** 334–341.

Bloomfield, V. A., Crothers, D. M., and Tinoco, I. (1974). *Physical chemistry of nucleic acids.* New York: Harper & Row.

Boulikas, T. (Ed.). (1998). *Gene Therapy & Molecular Biology. From basic mechanisms to clinical applications.* Palo Alto, CA: Gene Therapy Press.

Budker, V. G., Godovikov, A. A., Naumova, L. P., and Slepneva, I. A., (1980). Interaction of polynucleotides with natural and model membranes. *Nucleic Acids Res.* **8,** 2499–2515.

Campbell, S. E., Park, C. K., Lasic, D. D., Israelachvili, J. N. (1998). Structure and forces in transfection related surfactant systems. *Mat. Res. Soc. Symp. Proc.* **489,** 19–24.

Dimitriadis, D. (1979). Entrapment of plasmid DNA in liposomes. *Nucleic Acids Res.* **6,** 2697–2705.

Fang, Y., and Yie, Y. (1997). Two dimensional DNA condensation on cationic lipid membrane, *J. Phys. Chem. B* **101,** 441–449.

Farhood, H., Bottega, R., Epand, R. M., and Huang, L. (1992). Effect of activity and the ultrastructure of DNA-cytofectin complexes. *Biochim. Biophys. Acta* **1280,** 1–11.

Farhood, H., Bottega, R., Epand, R. M., and Huang, L. (1992). Effect of cholesterol derivatives on gene transfer and protein kinase C activity. *Biochim. Biophys. Acta* **1111,** 239–246.

Farhood, H., and Huang, L. (1996). Delivery of DNA, RNA, and proteins by cationic liposomes. In Lasic, D. D., and Barenholz, Y., (Eds.), *Nonmedical applications of liposomes,* Vol. IV, *From gene delivery to diagnostics and ecology* (pp. 31–42). Boca Raton, FL: CRC Press.

Felgner, P. L., and Ringold, G. M. (1989). Cationic liposome mediated transfection. *Nature,* **337,** 387–388.

Felgner, P. L. (1996). Improvement in cationic liposomes for *in vivo* gene delivery. *Human Gene Ther.* **7,** 1791–1793.

Felgner, P., Tsai, Y., and Felgner, J. (1996). Advances in the design and application of cytofectin formulations. In Lasic, D. D., and Barenholz, Y. (Eds.), *Nonmedical applications of liposomes,* Vol. IV, *From gene delivery to diagnostics and ecology* (pp. 43–56). Boca Raton, FL: CRC Press.

Felgner, J. H., Kumar, R., Sridhar, R., Wheeler, C., Tsai, Y. J., Border, R., Ramsay, P., Martin, M., Felgner, P. L. (1994). Enhanced gene delivery and mechanism studies with novel series of cationic lipid formulations. *J. Biol. Chem.* **269,** 2550–2561.

Felgner, P. L., Gadek, T. R., Holm, M., Roman, R., Chan, H. S., Wenz, M., Northrop, J. P., Ringold, M., and Danielsen, H. (1987). Lipofection: A highly efficient lipid-mediated DNA transfection procedure. *Proc. Natl. Acad. Sci. USA* **84,** 7413–7417.

Fraley, P., and Papahadjopoulos, D. (1982). Liposomes: The development of a new carrier system for introducing nucleic acids into plant and animal cells. *Curr. Top. Microbiol. Immunol.* **96,** 171–187.

Fraley, P, Subramani, S., Berg, P., and Papahadjopoulos, D. (1980). Introduction of liposome encapsulated sv-40 DNA into cells. *J. Biol. Chem.* **255,** 10431–10435.

Gao, X., and Huang, L. (1996). Potentiation of cationic liposome-mediated gene delivery by polycations. *Biochemistry* **35,** 1027–1036.

Gustaffson, J., Almgrem, M., Karlsson G., and Arvidson, G. (1995). Complexes between cationic liposomes and DNA visualized by cryo EM. *Biochim. Biophys. Acta* **1235,** 305–317.

Hoffman, R. M., Margolis, L. B., and Bergelson, L. D., (1978). Binding and entrapment of high molecular weight DNA by lecithin liposomes. *FEBS Lett.* **93,** 365–368,

Hong, K., Zheng, W., Baker, A., and Papahadjopoulos, D. (1987). Stabilization of cationic liposome-DNA complexes by polyamines and PEG-lipid conjugates for efficient *in vivo* gene therapy. *FEBS Lett.* **400,** 233–237.

Koltover, I., Salditt, T., Radler, J., and Safynia, C. (1998). An inverted hexagonal phase of cationic liposomes-DNA complexes related to the DNA release and delivery. *Science,* **281,** 78–81

Lasic, D. D., (1993). *Liposomes: From physics to applications.* Amsterdam: Elsevier.

Lasic, D. D., Strey, H., Podgornik, R, and Frederik, P. M. (1997). Recent advances in medical applications of liposomes: Sterically stabilized and cationic liposomes. In *Book of Abstracts* (pp.61–65). Proceedings of Fourth European Symposium on Controlled Drug Delivery, 1996, Nordwijk ann Zee.

Lasic, D. D., Strey, H., Podgornik, R, and Frederik, P. M. (1997). The structure of DNA-liposome complexes. *J. Am. Chem. Soc.* **119,** 832–833.

Lasic, D. D. (1997). *Liposomes in gene delivery.* Boca Raton, FL: CRC Press.

Lasic, D. D. and Papahadjopoulos, D. (Eds.). (1998). *Nonmedical applications of liposomes.* Amsterdam: Elsevier.

Lasic, D. D. (1996). Liposomes in gene therapy. In Lasic, D. D. and Barenholz, Y., (Eds.), *Nonmedical applications of liposomes,* Vol. IV, *From gene delivery to diagnostics and ecology.* (pp.1–6). Boca Raton, FL: CRC Press.

Lasic, D. D., Podgornik, R., Frederik, P. M., Strey, H., and Yie, Y. (2000). In preparation.

Leventis, S., and Silvius, J. R. (1990). Interactions of mammalian cells with lipid dispersions containing novel metabolizable cationic amphiphiles. *Biochim. Biophys. Acta* **1023,** 124–132.

Lee, E. R., Marshall, J. M., Siegel, C. S., Jiang, C., Yew, N. S., Nichols, M. R., Nietupski, J. B., Ziegler, J. R., Lane, M. B., Wang, K. X., Wan, N. C., Scheule, R. K., Harris, D. J., Smith, A. E., and Cheng, S. H. (1996). Detailed analysis of structures and formulations of cationic lipids for efficient gene transfer to the lung. *Human Gene Ther.* **7,** 1701–1717.

Liu, F., Yang, J., Huang, L., and Liu, D. (1996). Effect of non-ionic surfactants on the formation of DNA/emulsion complexes and emulsion mediaed gene transfer. *Pharm. Res.* **13,** 1642–1648.

Liu, F., Qi, H., Huang, L., and Liu, D. (1997). Factors controlling the efficiency of cationic lipid-mediated transfection *in vivo* via iv administration. *Gene Ther.* **4,** 517–523.

Liu, Y., Mounkes, L. C., Liggit, H. D., and Debs, R. (1997). Factors influencing the efficiency of cationic liposome-mediated intravenous gene delivery. *Nature Biotech.* **15,** 167–173.

Minsky, A., Ghirlando, R., and Gerson, H. (1996). Structural features of DNA-cationic liposome complexes and their implication for transfection. In Lasic, D.D., and Barenholz, Y., (Eds.), *Nonmedical applications of liposomes,* Vol. IV, *From gene delivery to diagnostics and ecology* (pp.7–30). Boca Raton, FL: CRC Press.

Nicolau C., and Papahadjopoulos D. (Eds.). (1998). *Nonmedical applications of liposomes,* (pp. 347–352). Amsterdam: Elsevier.

Pinnaduwage, P., Schmitt, L., and Huang, L. (1989). Use of quaternary ammonium detergent in liposome mediated DNA transfection of mouse L-cells. *Biochim. Biophys. Acta* **985,** 33–37.

Podgornik, R., Strey, H., and Parsegian, A., (1998). Colloidal DNA. *Curr. Op. Coll. Inter. Sci.* **3,** 534–539.

Pollard, H., Remy, J.-S., Loussouarn, G., Demolombe, S., Behr, J. P., and Escande, D. (1998). Poly-ethyleneimine but not cationic lipids promotes transgene delivery to the nucleus of nondividing cells but at least some cationic lipid/DNA complexes cannot. *J. Biol. Chem.* **273,** 7507–7511.

Raedler, J., Koltover, I., Salditt, T., and Safinya, C. (1997). The structure of DNA-liposome complexes. *Science* **275,** 810–814.

Solodin, I., Brown, C., Bruno, M., Chow, C., Jang, E., Debs, R., and Heath, T. D. (1995). A novel series of amphiphilic imidazolinium compounds for *in vitro* and *in vivo* gene delivery. *Biochemistry* **34,** 13537–13544.

Song, Y. K., Liu, D. (1998). Free liposomes enhance transfection activity of DNA/lipid complexes by in vivo iv administration, *Biochim. Biophys. Acta* **1372,** 141–150.

Sorgi, F. L., Bhattacharaya, S., and Huang, L. (1997). Protamine sulfate enhances lipid mediated gene transfer. *Gene Ther.* **4,** 961–968.

Sternberg, B. (1996). Liposomes as model membranes for membrane structures and structural transformations. In Lasic, D. D., and Barenholz, Y., (Eds.), *Nonmedical applications of liposomes,* Vol. IV, *From gene delivery to diagnostics and ecology* (pp. 271–298). Boca Raton, FL: CRC Press.

Sternberg, B. (1998). Ultrastructural morphology of cationic liposome –DNA complexes in gene therapy. In Lasic, D. D., and Papahadjopoulos, D. (Eds.), *Nonmedical applications of liposomes,* (pp. 395–428). Amsterdam: Elsevier.

Sternberg, B., Sorgi, F., and Huanf, L. (1994). New structures in complex formation between DNA and cationic liposomes visualized by freeze fracture EM. *FEBS Lett.* **356,** 361–366.

Templeton, N. S., Lasic, D. D., Frederik, P. M., Strey, H. H., Roberts, D. D., and Pavlakis, G. (1997). Improved DNA:liposome complexes for increased systemic delivery and gene expression. *Nature Biotech.* **15,** 647–652.

Uster, P., Allen, T. M., Daniel, B, Mendez, C, Newman, M, and Zhu, G. (1996). Insertion of poly(ethylene) glycol derivatized phospholipid into pre-formed liposomes results in prolonged *in vivo* circulation time. *FEBS Lett.* **386,** 243–246.

Van der Woude, I., Wagenaar, A., Meekel, A., ter Beest, M. B. A., Ruiters, M., Engberts, J. B. F. N., and Hoekstra, D. (1997). Novel pyridinium surfactants for efficient, nontoxic *in vitro* gene delivery. *Proc. Natl. Acad. Sci. USA* **94,** 1160–1165.

Wagner, E., Cotton, M., Kirlaposs, H., Mechter, K., Curiel, D. T., and Birnsteil, M. L. (1992). Coupling of adenovirus to transferrin-polylysine/DNA complexes greatly enhances receptor-mediated gene delivery and expression of transfected genes. *Proc. Natl. Acad. Sci. USA* **89,** 7934–7938.

Watson, J. D., Toose, J., and Kurtz, D. T. (1983). *Recombinant DNA.* New York: Scientific American Books.

Weis, K. (1998). Importins and exportins: How to get in and out of the nucleus, *TIBS* **23,** 185–189.

Wheeler, C. J., Sukhu, L., Yang, G., Tsai, Y., Bustamante, C., Felgner, P., Norman J., and Manthorpe, M. (1996). Converting an alcohol to an amine in a cationic lipid dramatically alters the co-lipid requirement, cellular transfection activity and the ultrastructure of DNA-cytofectin complexes. *Biochim. Biophys. Acta* **1280,** 1–11.

Wolff, J. A., and Lederberg, J. (1994). A history of gene transfer and therapy. In Wolff, J. A. (Ed.). *Gene therapeutics.* Basel: Birkhauser.

Xu, Y., and Szoka F. (1996). Mechanism of DNA release from cationic liposome-DNA complexes used in cell transfection. *Biochemistry* **35,** 5616–5623.

Yang, J. P., and Huang, L. (1997). Overcoming the inhibitory effect of serum on lipofection by increasing the charge ratio of cationic liposome to DNA. *Gene Ther.* **4,** 950–967.

Zhu, N., Liggitt, D., Liu, Y., and Debs, R. (1993). Systemic gene expression after intravenous DNA delivery in adult mice. *Science* **261,** 209–211.

Zuidam, N., and Barenholz, Y. (1997). Electrostatic parameters of cationic liposomes commonly used in gene delivery, *Biochim. Biophys. Acta* **1329,** 211–222.

CHAPTER 5

Self-Assembled Structures of Lipid/DNA Nonviral Gene Delivery Systems from Synchrotron X-Ray Diffraction

Cyrus R. Safinya and Ilya Koltover
Materials Department, Physics Department, and Biochemistry and Molecular Biology Program, University of California, Santa Barbara, California

There is now a surge of activity in developing synthetic nonviral gene delivery systems for gene therapeutic applications because of their low toxicity, non-immunogenicity, and ease of production. Cationic liposome/DNA (CL/DNA) complexes have shown gene expression *in vivo* in targeted organs, and human clinical protocols are ongoing. Moreover, the single largest advantage of nonviral over viral methods for gene delivery is the potential of transferring extremely large pieces

Nonviral Vectors for Gene Therapy

of DNA into cells. This was clearly demonstrated when partial fractions of order 1 Mega base pairs of human artificial chromosome were recently transferred into cells using cationic liposomes as a vector, although extremely inefficiently. However, because the mechanism of action of CL/DNA complexes remains largely unknown, transfection efficiencies are at present very low and vary by up to a factor of 100 in different cell lines. The low transfection efficiencies with nonviral delivery methods are the result of poorly understood transfection-related mechanisms at the molecular and self-assembled levels.

In this chapter we review recent work aimed at elucidating the precise self-assembled structures of CL/DNA complexes by the quantitative techniques of synchrotron X-ray diffraction. Recently, we found that when DNA is mixed with cationic liposomes composed of DOPC/DOTAP, the result is a topological transition into condensed CL/DNA complexes with a multilamellar structure (L_α^C) composed of DNA monolayers sandwiched between cationic lipid bilayers. The existence of a completely different columnar inverted hexagonal H_{II}^C liquid-crystalline state in CL/DNA complexes was also demonstrated using synchrotron small-angle X-ray diffraction. The commonly used neutral helper lipid DOPE is found to induce the L_α^C to H_{II}^C structural transition by controlling the spontaneous curvature $C_o = 1/R_o$ of the lipid monolayer. Further, an entirely new class of helper molecules was introduced that controls the membrane bending rigidity κ and gives rise to a distinctly different pathway to the H_{II}^C phase. Significantly, optical microscopy has revealed that in contrast to the weakly transfectant L_α^C complexes that bind stably to anionic vesicles (models of cellular membranes), the more transfectant H_{II}^C complexes are unstable, rapidly fusing and releasing DNA upon adhering to anionic vesicles. The observations underscore the importance of self-assembled structure to "early-stage" gene delivery events and provide direct imaging support for either a mechanism of DNA escape from anionic endosomal vesicles or fusion of CL/DNA complexes with cell plasma membrane.

I. INTRODUCTION

Gene therapy depends on the successful transfer and expression of extracellular DNA to mammalian cells, with the aim of replacing a defective gene or adding a missing one (Friedmann, 1997; Felgner, 1997; Miller, 1998; Crystal, 1995; Zhu et al., 1993; Nabel et al., 1993; Mulligan, 1993; Felgner and Rhodes, 1991; Behr, 1994; Remy et al., 1994; Singhal and Huang, 1994; Lasic and Templeton, 1996; Marhsall, 1995). At present, viral-based vectors (retroviruses, adenoviruses, adeno-associated viruses) are the most common gene carriers used by researchers developing gene delivery systems because of their high efficiency of transfer and expression (Friedmann, 1997; Crystal, 1995; Mulligan, 1993). Each vector has advantages and disadvantages and it will be many years before the optimal vector is designed. Retrovirus vectors integrate into the host chromosomes, providing the prospect of lifelong cure. However, their disadvantages include the possibility that the crippled virus may potentially become infectious through recombination, the fact that the currently used vectors only integrate in dividing host cells, the limited size of the therapeutic gene that may be packed into the crippled viral gene, and the technical difficulty associated with production of high titer virus stocks (Friedmann, 1997; Crystal, 1995). Because adenovirus-based vectors do not integrate their genes into the chromosome but instead act transiently, repeated applications are necessary, which has resulted in undesirable immune responses in recent clinical trials. Adeno-associated viruses integrate into the host chromosomes but have a very small capsid size and are therefore limited in the size of the therapeutic gene that they may carry. Current viral vectors have a maximum gene-carrying capacity of 40k base pairs (Friedmann, 1997; Crystal, 1995).

At the same time there has been much recent activity for the development of synthetic nonviral gene delivery systems (Friedmann, 1997; Felgner, 1997; Miller, 1998; Zhu *et al.,* 1993; Nabel *et al.,* 1993; Felgner and Rhodes, 1991; Behr, 1994; Remy *et al.,* 1994; Singhal and Huang, 1994; Lasic and Templeton, 1996). The conventional nonviral transfer methodologies, which have transfection rates significantly lower than viral transfection rates, include anionic liposomes that encapsulate nucleic acid, calcium phosphate precipitation, and use of polycationic reagents (DEAE-dextran or polylysine). Some of the advantages of using nonviral vectors for gene delivery include the fact that plasmid DNA constructs used in deliveries are more readily prepared than viral constructs, they have no viral genes to cause disease, and that synthetic carriers are nonimmunogenic due to a lack of proteins.

The entire field of gene therapy based on synthetic nonviral delivery systems has recently undergone a renaissance since the initial paper by Felgner *et al.* (1987), which was soon to be followed by numerous other groups demonstrating gene expression *in vivo* in targeted organs (Zhu *et al.,* 1993; Singhal and Huang, 1994), and in human clinical trials (Nabel *et al.,* 1993). Felgner *et al.* discovered that cationic liposomes (CLs) (closed bilayer membrane shells of lipid molecules; see Fig. 1) when mixed with DNA to form (CL/DNA) complexes with an overall positive charge, enhance transfection (i.e., the transfer of plasmid into cells followed by expression).

They hypothesized that this was because CL/DNA complexes adsorbed more effectively to the anionic plasma membrane of mammalian cells via the electrostatic interactions. Compared to other nonviral delivery systems, cationic liposomes tend to mediate a higher level of transfection in the majority of cell lines studied to date (Lasic and Templeton, 1996). Using cationic liposomes, gene expression of chloramphenicol acetyltransferase (CAT) activity has been found in mice lungs and brain (Brigham *et al.,* 1989; Hazinski *et al.,* 1991; Ono *et al.,* 1990), and in brain tissue of frog embryos (Malone, 1989; Holt *et al.,* 1990). *In vivo* CAT expression of aerosol-administered plasmid DNA (pCIS-CAT) complexed with cationic liposomes in mouse lung has also been demonstrated (Stribling *et al.,* 1992).

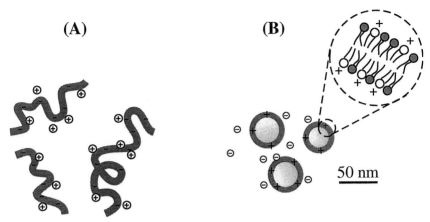

Figure 1   (A) Schematic of anionic DNA polyelectrolytes with cationic counterions condensed on the backbone due to Manning condensation. (B) Schematic of cationic liposomes (or vesicles), which consist of spherical membranes containing a bilayer of lipid molecules normally consisting of a mixture of cationic and neutral lipids. (Bar for (B) only.)

Without doubt, one of the principal and most exciting advantages of nonviral over viral methods for gene delivery is the potential of transferring and expressing (transfecting) large pieces of DNA into cells. The proof of this concept was clearly demonstrated when partial sections of first-generation human artificial chromosomes (HACs) of order 1 Mbp were transferred into cells with CLs, although extremely inefficiently (Harrington *et al.*, 1997; Roush, 1997). The future development of HAC vectors will be extremely important for gene therapy applications; because of their very large size capacity HACs would have the ability of delivering not only entire human genes (in many cases exceeding 100k bp) but also their regulatory sequences, which are needed for the spatial and temporal regulation of gene expression.

While the transfection rates and reproducibility in many cells have been found to be enhanced using CL/DNA complexes compared to other more traditional nonviral delivery systems, the mechanism of transfection via cationic liposomes remains largely unknown (Friedmann, 1997; Felgner, 1997; Miller, 1998; Crystal, 1995; Zhu *et al.*, 1993; Nabel *et al.*, 1993; Mulligan, 1993; Felgner and Rhodes, 1991; Behr, 1994; Remy *et al.*, 1994; Singhal and Huang, 1994; Lasic and Templeton, 1996; Marhsall, 1995). At present hundreds of plasmid DNA molecules are required for successful gene transfer and expression. The low transfection efficiencies with nonviral methods results from a general lack of knowledge regarding (1) the structures of CL/DNA complexes and (2) their interactions with cell membranes and events leading to release of DNA in the cytoplasm for delivery within the nucleus. We are now beginning to understand the precise nature of the self-assembled structures of CL/DNA complexes in different lipid-membrane systems (Raedler *et al.*, 1997; Spector and Schnur, 1997; Seachrist, 1997; Salditt *et al.*, 1997; Safinya *et al.*, 1998; Raedler *et al.*, 1998; Salditt *et al.*, 1998; Lasic *et al.*, 1997; Pitard *et al.*, 1997; Koltover *et al.*, 1998). The transfection efficiencies of nonviral delivery methods may be improved through insights into transfection-related mechanisms at the molecular and self-assembled levels.

Aside from the medical and biotechnological ramifications in gene therapy and gene and drug therapeutics, research on CL/DNA complexes should also shed light on other problems in biology. The development of efficient HAC vectors in the future, which will most likely occur once efficient synthetic nonviral delivery systems have been developed, is a long-range goal in studies designed to characterize chromosome structure and function. Furthermore, molecular biology studies would benefit substantially from the ability of transfecting hard-to-transfect cell lines with synthetic delivery systems; for example, in studies designed to characterize the structure of promoters of human genes in the appropriate cell lines.

DNA chains dissolved in solution are known to give rise to a rich variety of condensed and liquid crystalline phases. Studies show regular DNA condensed morphologies induced by multivalent cations (Bloomfield, 1991) and liquid crystalline (LC) phases at high concentrations of DNA both *in vitro* (Livolant and Leforestier, 1996) and *in vivo* in bacteria (Reich *et al.*, 1994). More recently there has been a flurry of experimental and theoretical work on DNA chains mixed with cationic liposomes. Early on oligo-lamellar structures had been reported in cryo-TEM studies by Gustafsson *et al.* (1995). Freeze−fracture electron microscopy study by Sternberg *et al.* had also observed isolated DNA chains coated with a lipid bilayer (Sternberg *et al.*, 1994). One of the self-assembled structures that forms spontaneously when DNA (a negative polyelectrolyte) is complexed with cationic liposomes containing a specific type of lipid is a multilayer assembly of DNA sandwiched between bilayer membranes shown schematically in Fig. 2 (Raedler *et al.*, 1997; Spector and Schnur, 1997; Seachrist, 1997; Salditt *et al.*, 1997; Safinya *et al.*, 1998; Raedler *et al.*, 1998; Salditt *et al.*, 1998; Lasic *et al.*, 1997; Pitard *et al.*, 1997).

The structure and thermodynamic stability of these CL/DNA complexes has also been the subject of much recent theoretical work (May and Ben-Shaul, 1997; Dan, 1998; Bruinsma, 1998; Bruinsma and Mashl, 1998; Harries *et al.*, 1998; O'Hern and Lubensky, 1998; Golubovic and Golubovic, 1998). Analytical and numerical studies of DNA−DNA interactions bound between membranes show the existence of a novel long-range repulsive electrostatic interaction (Bruinsma, 1998, Bruinsma and Mashl, 1998; Harries *et al.*, 1998). Theoretical work on CL/DNA complexes has also led to the realization of a variety of novel new phases of matter in DNA/lipid complexes (O'Hern and Lubensky, 1998; Golubovic and Golubovic, 1998). In particular, a novel new "sliding columnar phase," which remains to be discovered experimentally, is predicted where the positional coherence between DNA molecules in adjacent layers is lost without destroying orientational

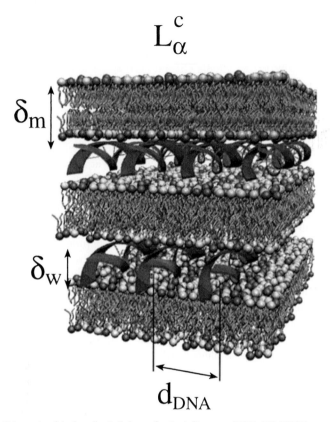

$$L_\alpha^c$$

δ_m

δ_w

d_{DNA}

Figure 2 Schematic of the lamellar L_α^c phase of cationic liposome/DNA (CL/DNA) complexes with alternating lipid bilayer and DNA monolayer. The DNA interaxial spacing is d_{DNA}. The interlayer spacing is $d = \delta_w + \delta_m$. [Redrawn from Raedler *et al.*, (1997).]

coherence of the chains from layer to layer. This new phase would be a remarkable new phase of matter if it exists, and shares many fascinating similarities with flux lattices in superconductors.

   In its own right from a biophysics perspective, it is important to explore the phase behavior of DNA attached to membranes in two dimensions (Fig. 2) as a tractable experimental and theoretical system for understanding DNA condensation. The mechanisms of DNA condensation *in vivo* (i.e., packing in a small space) are poorly understood (Lewin, 1997). DNA condensation and decondensation that happens, for example, during the cell cycle in eukaryotic cells involves different types of oppositely charged polyamines, peptides, and proteins (e.g., histones) where the nonspecific electrostatic interactions are clearly important. In bacteria that are the simplest cell types, it is thought that multivalent cationic polyamine molecules (spermine, spermidine) are responsible for DNA condensation in the three-dimensional space of the cell cytoplasm. DNA-membrane interactions might also provide clues for the relevant molecular forces in the packing of DNA in chromosomes and viral capsids.

II. SYNCHROTRON X-RAY DIFFRACTION STUDIES

A. THE LAMELLAR L_α^C PHASE OF CL/DNA COMPLEXES

We have recently carried out a combined *in situ* optical microscopy and synchrotron X-ray diffraction (XRD) study of CL/DNA complexes Raedler *et al.,*, 1997; Spector and Schnur, 1997; Seachrist, 1997; Salditt *et al.,* 1979; Safinya *et al.,* 1998; Raedler *et al.,* 1998; Salditt *et al.,* 1998) where the cationic liposomes consisted of mixtures of neutral (so called "helper lipid") DOPC (dioleoyl phosphatidyl cholin) and cationic DOTAP (dioleoyl trimethylammonium propane) (Fig. 3). High-resolution small-angle X-ray diffraction has revealed that the structure is different from the hypothesized "bead-on-string" structure, originally proposed by Felgner *et al.* for CL/DNA complexes in their seminal paper (Felgner and Rhodes, 1991; Felgner *et al.,* 1987), which pictures the DNA strand decorated with distinctly attached cationic liposomes. The addition of linear λ-phage DNA (48,502 bp, contour length = 16.5 μm) to binary mixtures of cationic liposomes (mean diameter of 70 nm) induces a topological transition from liposomes into collapsed condensates in the form of optically birefringent liquid crystalline (LC) globules with size on the order of 1 μm.

We show in Fig. 4(A) differential–interference–contrast (DIC) optical images of CL/DNA complexes for four lipid (L) to λ-DNA (D) ratios (L = DOTAP + DOPC (1/1)) (see Endnote 1). Similar images are observed with λ-DNA replaced by the pBR322 plasmid (4361 bp) DNA or DOPC replaced by DOPE. At low DNA concentrations (Fig. 4A, L/D = 50), in contrast to the pure liposome solution where no objects greater than 0.2 μm are seen, 1-μm large globules are observed. The globules coexist with excess liposomes. As more DNA is added, the globular condensates form larger chain-like structures (Fig. 4A, L/D = 10). At $L/D \approx 5$ the chain-like structures flocculate into large aggregates of distinct globules. For $L/D < 5$, the complex size was smaller and stable in time again (Fig. 4A, L/D = 2), and coexisted with excess DNA. Fluorescent microscopy of the DNA (labeled with YoYo) and the lipid (labeled with Texas Red-DHPE) also showed that the individual globules contain both lipid and DNA (see Fig. 15A and B). Polarized microscopy also shows that the distinct globules are birefringent indicative of their liquid crystalline nature.

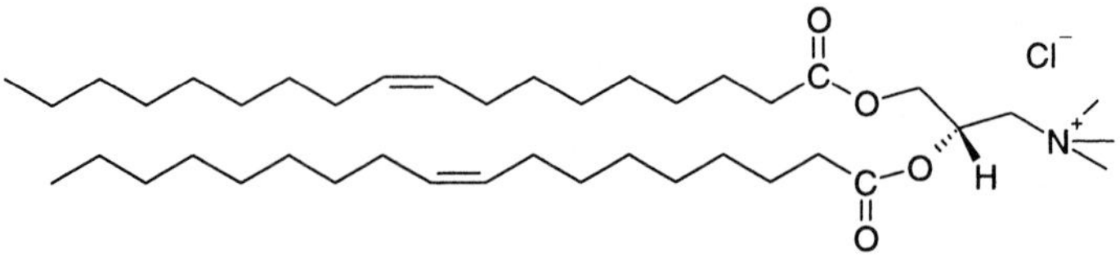

DOTAP (cationic lipid)

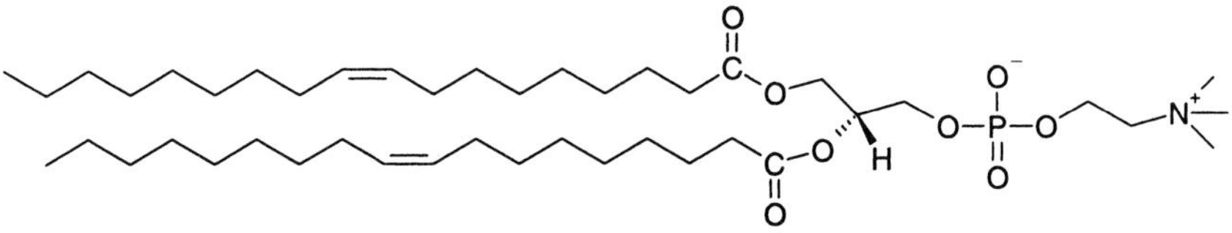

DOPC (neutral, helper lipid)

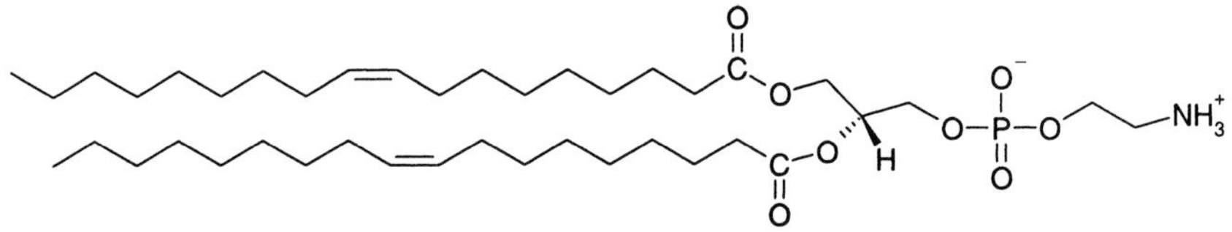

DOPE (neutral, helper lipid)

Figure 3   Top: The cationic lipid DOTAP (dioleoyl trimethylammonium propane). Middle: The neutral helper lipid DOPC (dioleoylphosphatidylcholine). Bottom: The neutral helper lipid DOPE (dioleoylphosphadtidylethanolamine). The neutral lipid is referred to as the "helper lipid."

The size dependence of the complexes as a function of L/D (Fig. 4B) was independently measured by dynamic light scattering (Microtrac UPA 150, Leeds and Northrup). The large error bars represent the broad polydispersity of the system. The size dependence of the aggregates can be understood in terms of a charge-stabilized colloidal suspension. The charge of the complexes was measured by their electrophoretic mobility in an external electric field. For $L/D > 5$ (Fig. 4A; $L/D = 50$ or 10) the complexes are positively charged, while for $L/D < 5$ (Fig. 4A; $L/D = 2$) the complexes are negatively charged. The charge reversal is in good agreement with the stoichiometrically expected charge balance of the components DOTAP and DNA at $L/D = 4.4$ (Wt./Wt.), where $L = $ DOTAP + DOPC in equal weights. Thus, the positively and negatively charged globules at $L/D = 50$ and $L/D = 2$, respectively, repel each other and remain separate, while as L/D approaches 5, the nearly neutral complexes collide and tend to stick due to van der Waals attraction.

The high resolution synchrotron small-angle X-ray scattering (SAXS) experi-

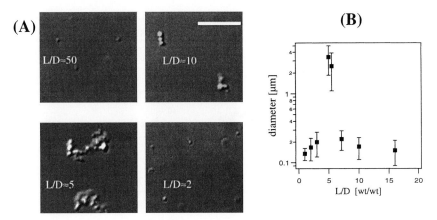

Figure 4 (A) High-resolution DIC optical images of CL/DNA complexes forming distinct condensed globules in mixtures of different lipid-to-DNA weight ratio (L/D). $L/D = 4.4$ is the isoelectric point, CL/DNA complexes are positively charged for $L/D = 50$ and 10, and negatively charged for $L/D = 2$. The positive (negative) regime contains excess lipid (DNA). (B) Average size of the lipid/DNA complexes measured by dynamic light scattering. Bar is 10 μm. [Adapted from Raedler *et al.*, (1997).]

ments carried out at the Stanford Synchrotron Radiation Laboratory revealed unexpected structures for mixtures of CLs and DNA (Raedler *et al.*, 1997; Salditt *et al.*, 1997; Seachrist, 1997; Spector and Schnur, 1997). SAXS data of dilute (Φ_w = the volume fraction of water = 98.6% $\pm$ 0.3%) DOPC/DOTAP (1/1)-λ-DNA mixtures as a function of L/D (L = DOPC + DOTAP) (Fig. 5A) are consistent with a complete topological rearrangement of liposomes and DNA into a multilayer structure with DNA intercalated between the bilayers (Fig. 2, denoted L_α^c). Two sharp peaks at $q = 0.0965 \pm 0.003$ and 0.193 ± 0.006 Å^{-1} correspond to the (00L) peaks of a layered structure with an interlayer spacing d ($= \delta_m + \delta_w$), which is in the range 65.1 $\pm$ 2 Å (Fig. 5B, open squares). The membrane thickness and water gap are denoted by δ_m and δ_w, respectively (Fig. 2). The middle broad peak q_{DNA} arises from DNA-DNA correlations and gives $d_{DNA} = 2\pi/q_{DNA}$ (Fig. 5B, solid circles). The multilamellar L_α^c structure with intercalated DNA is also observed in CL/DNA complexes containing supercoiled DNA both in water and also in Dulbecco's Modified Eagle's Medium used in transfection experiments (Lin *et al.*, in preparation). This novel multilamellar structure of the CL/DNA complexes is observed to protect DNA from being cut by restriction enzymes (Raedler *et al.*, 1998).

In the absence of DNA, membranes composed of mixtures of DOPC and the cationic lipid DOTAP (1/1) exhibit strong long-range interlayer electrostatic repulsions that overwhelm the van der Waals attraction (Roux and Safinya, 1988; Safinya 1989). In this case, as the volume fraction Φ_w of water is increased, the L_α phase swells and d is given by the simple geometric relation $d = \delta_m/(1 - \Phi_w)$. The

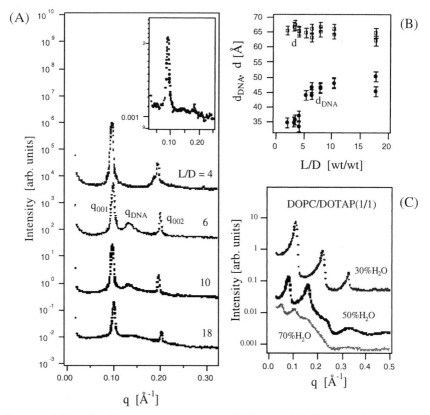

Figure 5 (A) A series of high-resolution synchrotron SAXS scans of CL/DNA complexes in excess water as a function of different lipid-to-DNA weight ratio (L/D). The Bragg-reflections at $q_{001} = 0.096$ Å^{-1} and $q_{002} = 0.192$ Å^{-1} result from the multilamellar L_α^c structure with intercalated monolayer DNA (see Fig. 2). The intermediate peak at q_{DNA} is due to the DNA-interaxial spacing d_{DNA} (Fig. 2). *Inset:* SAXS scan of an extremely dilute (lipid + DNA = 0.014% volume in water) λ-DNA-DOPC/DOTAP (1/1) complex at $L/D = 10$, which shows the same features as the more concentrated mixtures and confirms the multilamellar structure (with alternating lipid bilayer and DNA monolayers) of very dilute mixtures typically used in clinical gene therapy applications. (B) The spacings d and d_{DNA} as a function of L/D show that (i) d is nearly constant and (ii) there are two distinct regimes of DNA packing, one where the complexes are positive ($L/D > 5$, d_{DNA} 46 Å), and the other regime where the complexes are negative ($L/D < 5$, d_{DNA} 35 Å). (C) SAXS scans of the lamellar L_α phase of DOPC/DOTAP (cationic)–water mixtures done at lower X-ray resolution on a rotating anode X-ray generator. A dilution series of 30% ($d = 57.61$ Å), 50% ($d = 79.49$ Å), and 70% ($d = 123.13$ Å) H$_2$O by weight is shown. [Adapted from Raedler *et al.*, (1997).]

SAXS scans in Fig. 5(C) show this behavior with the (00L) peaks moving to lower q as Φ_w increases. From d ($= 2\pi/q_{00L}$) at a given Φ_w we obtain $\delta_m = 39 \pm 0.5$ Å for DOPC/DOTAP (1/1). Liposomes made of DOPC/DOTAP(1/1) with $\Phi_w \approx 98.5\%$ do not exhibit Bragg diffraction in the small wave-vector range covered in Fig. 5(A).

The DNA that condenses on the CLs strongly screens the electrostatic inter-action between lipid bilayers and leads to condensed multilayers. The average thick-ness of the water gap $\delta_w = d - \delta_m = 65.1$ Å $- 39$ Å $= 26.1$ Å ± 2.5 Å is just sufficient to accommodate one monolayer of B-DNA (diameter ≈ 20 Å) including a hydration shell (Podgornik *et al.*, 1994). We see in Fig. 5(B) that d is almost con-stant as expected, for a monolayer DNA intercalate (Fig. 2). In contrast, as L/D decreases from 18 to 2, d_{DNA} suddenly decreased from ≈ 44 Å in the positively charged regime just above $L/D = 5$ (near the stoichiometric charge neutral point $L/D = 4.4$) to ≈ 37 Å for the negatively charged regime (Fig. 5B). In these dis-tinctly different packing regimes, L_α^c complexes coexist with excess giant liposomes in the positive regime and with excess DNA in the negative regime. The excess DNA may in fact have a lipid bilayer coating as earlier freeze–fracture studies by Sternberg *et al.* had found ((Sternberg *et al.*, 1994).

The DNA/lipid condensation can be understood to occur as a result of re-lease of "bound" counterions in solution. DNA in solution (Fig. 6A) has a bare length between negative charges (phosphate groups) equal to $b_o = 1.7$ Å. This is substantially less than the Bjerrum length in water, $b_j = 7.1$ Å, which corresponds to the distance where the Coulomb energy between two unit charges is equal to the thermal energy $k_B T$. A nonlinear Poisson–Boltzmann analysis shows that counterions will condense on the DNA backbone until the Manning parameter $\xi = b_j/b'$ approaches 1 (Manning, 1969). (b' is the renormalized distance between negative charges after counterion condensation.) A similar analysis shows that coun-terions also condense near the surface of two-dimensional membranes (i.e., within the Gouy–Chapman layer) (Zimm and Le Bret, 1983). Through DNA/lipid con-densation the cationic lipid tends to fully neutralize the phosphate groups on the

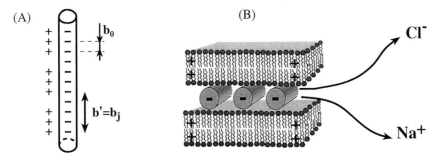

Figure 6 (A) Schematic of double-stranded DNA molecule with a bare distance between negative charges of $b_o = 1.7$ Å. From nonlinear Poisson–Boltzmann we know that positive counterions condense on DNA until the renormalized distance between the negative charges b' equals the Bjerrum length, which is $b_j = 7.1$ Å in water. (B) Schematic drawing showing that as DNA condenses onto the cationic membrane there is a simultaneous release of counterions and a gain in the entropy of solution when the previously condensed counterions (Na$^+$ on DNA and Cl$^-$ near the cationic liposome membrane) leave the immediate vicinity of DNA and the cationic membrane respectively.

DNA, in effect replacing and releasing the originally condensed counterions in solution. Thus, the driving force for higher order self-assembly is the release of counterions, which were one-dimensionally bound to DNA and two-dimensionally bound to cationic membranes, into solution (Fig. 6B).

The precise nature of the packing structure of λ-DNA within the lipid layers can be elucidated by conducting a lipid dilution experiment in the isoelectric point regime of the complex. In these experiments the total lipid (L = DOTAP + DOPC) is increased while the charge of the overall complex, given by the ratio of cationic DOTAP to DNA, is kept constant at DOTAP/DNA = 2.20. The SAXS scans in Fig. 7 (arrow points to the DNA peak) show that $d_{DNA} = 2\pi/q_{DNA}$ increases

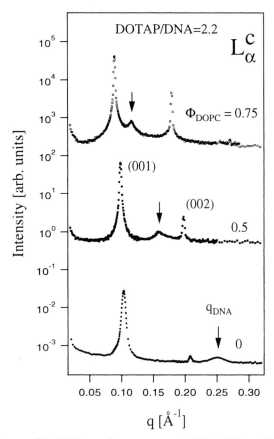

Figure 7   SAXS scans of CL/DNA complexes at constant DOTAP/DNA = 2.2 (at the isoelectric point) with increasing DOPC/DOTAP, which shows the DNA peak (arrow) moving toward smaller q as L/D (and Φ_{DOPC}) increases. L = DOTAP + DOPC, D = DNA. [Adapted from Raedler et al., (1997); Salditt et al., (1997).]

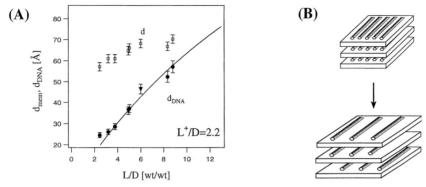

Figure 8   (A) The DNA interaxial distance d_{DNA} and the interlayer distance d in the L_α^c phase (Fig. 2) plotted as a function of lipid/DNA (L/D) (wt/wt) ratio at the isoelectric point of the complex DOTAP/DNA = 2.2. d_{DNA} is seen to expand from 24.5 Å to 57.1 Å. The solid line through the data is the prediction of a packing calculation where the DNA chains form a space-filling one-dimensional lattice. (B) Schematic drawing of DNA/membrane multilayers showing the increase in distance between DNA chains as the membrane charge density is decreased (i.e., as Φ_{DOPC} increases) at the isoelectric point.

with lipid dilution from 24.54 Å to 57.1 Å as Φ_{DOPC}, the weight fraction of DOPC in the DOPC/DOTAP cationic liposome mixtures, increases from 0 to 0.75 (or equivalently increasing L/D between 2.2 and 8.8). The most compressed interaxial spacing of 24.55 Å at $\Phi_{DOPC} = 0$ approaches the short-range repulsive hard-core interaction of the B-DNA rods containing a hydration layer (Pogornik *et al.*, 1994). Figure 8(A) plots d and d_{DNA} as a function L/D.

The observed behavior is depicted schematically in Fig. 8(B) showing that as we add neutral lipid (at the isoelectric point) and therefore expand the total cationic surface we expect the DNA chains to also expand and increase their interaxial spacing. The solid line in Fig. 8(A) is derived from the simple geometric packing relationship $d_{DNA} = (A_D/\delta_m)(\rho_D/\rho_L)(L/D)$, which equates the cationic charge density (due to the mixture DOTAP$^+$ and DOPC) with the anionic charge density (due to DNA$^-$). Here, $\rho_D = 1.7$ (g/cc) and $\rho_L = 1.07$ (g/cc) denote the densities of DNA and lipid respectively, δ_m the membrane thickness, and A_D the DNA area. $A_D = $ Wt(λ)/($\rho_D L(\lambda)$) = 186 Å^2, Wt(λ) = weight of λ-DNA = $31.5 * 10^6/(6.022 * 10^{23})$ g, and $L(\lambda)$ = contour length of λ-DNA = $48502 * 3.4$ Å.

The agreement between the packing relationship (solid line) with the data over the measured interaxial distance from 24.5 Å to 57.1 Å (Fig. 8A) is quite remarkable given the fact that there are no adjustable parameters. The variation in the interlayer spacing d ($= \delta_w + \delta_m$) (Fig. 8A, open squares) arises from the increase in the membrane bilayer thickness δ_m as L/D increases (each DOPC molecule is about 4 Å to 6 Å longer than a DOTAP molecule). The observation, of a *variation in the DNA interaxial distance* as a function of the lipid to DNA (L/D) ratio in multilayers (Fig. 8A), unambiguously demonstrates that x-ray diffraction directly probes the

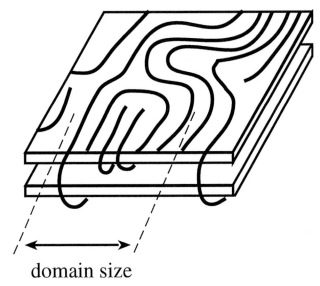

domain size

Figure 9 A schematic drawing of the local arrangement of DNA in the interior of the L_α^c cationic lipid DNA complex consistent with the X-ray diffraction data taking into account the broad width of the DNA peak due to the finite coherent domain size of the DNA chains adsorbed on lipid bilayers. The average domain size of the one-dimensional lattice of DNA chains derived from the width of the DNA peaks is about 10 unit cells.

DNA behavior in multilayer assemblies. From the linewidths of the DNA peaks (Fig. 7) the 1D lattice of DNA chains is found to consist of domains extending to near 10 neighboring chains (Salditt *et al.*, 1997, 1998). Thus, the CL/DNA complex is self-assembled into a new "hybrid" phase of matter, namely, a coupled two-dimensional smectic phase of DNA chains coupled to lipid bilayers of a three-dimensional smectic phase (Fig. 9). On larger length scales the lattice would melt into a 2D nematic phase of chains due to dislocations (O'Hern and Lubensky, 1998; Golubovic and Golubovic, 1998).

B. THE INVERTED HEXAGONAL $H_{II}{}^C$ PHASE OF CL/DNA COMPLEXES

A commonly used helper lipid in CL/DNA mixtures is DOPE (dioleoyl-phosphatidylethanolamine), shown in Fig. 3. Further, it is empirically known that transfection efficiency in mammalian cells is typically improved in CL/DNA complexes that contain DOPE instead of DOPC as the helper lipid (Felgner *et al.*, 1994; Farhood *et al.*, 1995; Hui *et al.*, 1996). Recent X-ray diffraction shows that DOPE-containing complexes may give rise to a completely different columnar inverted hexagonal $H_{II}{}^C$ liquid–crystalline structure (Fig. 11).

We show in Fig. 10 SAXS scans in positively charged CL/DNA complexes

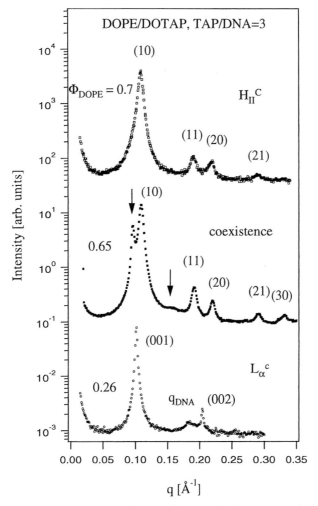

Figure 10 Synchrotron SAXS patterns of the lamellar $L_\alpha{}^C$ and columnar inverted hexagonal $H_{II}{}^C$ phases of positively charged CL/DNA complexes as a function of increasing weight fraction Φ_{DOPE}. At $\Phi_{DOPE} = 0.26$, the SAXS results from a single phase with the lamellar $L_\alpha{}^C$ structure shown in Fig. 2. At $\Phi_{PE} = 0.7$, the SAXS scan results from a single phase with the columnar inverted hexagonal $H_{II}{}^C$ structure shown in Fig. 11. At $\Phi_{DOPE} = 0.65$, the SAXS shows coexistence of the $L_\alpha{}^C$ (arrows) and $H_{II}{}^C$ phases. [Adapted from Koltover et al., (1998).]

for DOTAP/DNA (wt./wt.) $= 3$ as a function of increasing Φ_{DOPE} (weight fraction of DOPE) in the DOPE/DOTAP cationic liposome mixtures (Koltover *et al.*, 1998). We find that the internal structure of the complex changes completely with increasing DOPE/DOTAP ratios. SAXS data of complexes with $\Phi_{PE} = = 0.26$ and 0.70 clearly show the presence of two different structures. At $\Phi_{PE} = 0.26$, SAXS of the lamellar L_α^C complex shows sharp peaks at $q_{001} = 0.099$ Å^{-1} and $q_{002} = 0.198$ Å^{-1} resulting from the lamellar periodic structure ($d = 2\pi/q_{001} = 63.47$ Å) with DNA intercalated between cationic lipid analogous to the structure in DOPC/DOTAP–DNA complexes (Fig. 2).

For $0.7 < \Phi_{PE} < 0.85$, the peaks of the SAXS scans of the CL/DNA complexes are indexed perfectly on a two-dimensional (2D) hexagonal lattice with a unit cell spacing of $a = 4\pi/[(3)^{0.5}q_{10}] = 67.4$ Å for $\Phi_{PE} = 0.7$. Figure 10 at $\Phi_{PE} = 0.7$ shows the first four order Bragg peaks of this hexagonal structure at $q_{10} = 0.107$ Å^{-1}, $q_{11} = 0.185$ Å^{-1}, $q_{20} = 0.214$ Å^{-1}, and $q_{21} = 0.283$ Å^{-1}. The structure is consistent with a 2D columnar inverted hexagonal structure (Fig. 11), which we refer to as the H_{II}^C phase of CL/DNA complexes. The DNA molecules are surrounded by a lipid monolayer with the DNA/lipid inverted cylindrical micelles arranged on a hexagonal lattice. The structure resembles that of the inverted hexagonal H_{II} phase of pure DOPE in excess water (Seddon, 1989; Gruner, 1989), with the water space inside the lipid micelle filled by DNA. Assuming again an average lipid monolayer thickness of 20 Å, the diameter of micellar void in the H_{II}^C phase is close to 28 Å, again sufficient for a DNA molecule with approximately two hydration shells. For $\Phi_{PE} = 0.65$ the L_α^C and H_{II}^C structures coexist as shown in Fig. 10 (arrows point to the (001) and q_{DNA} peaks of the L_α^C phase) and are nearly epitaxially matched with $a \approx d$. For $\Phi_{PE} > 0.85$, the H_{II}^C phase coexists with the H_{II} phase of pure DOPE, which has peaks at $q_{10} = 0.0975$ Å^{-1}, $q_{11} = 0.169$ Å^{-1}, $q_{20} = 0.195$ Å^{-1} with a unit cell spacing of $a = 74.41$ Å.

C. THE ROLE OF THE NEUTRAL HELPER LIPID AND CATIONIC LIPID IN CONTROLLING THE INTERACTION FREE ENERGY IN DNA/CATIONIC LIPOSOME COMPLEXES: THE LAMELLAR AND HEXAGONAL PHASES OF CL/DNA COMPLEXES

To understand the stability of the lamellar and hexagonal phases we consider the interplay between the electrostatic and membrane elastic interactions in the CL/DNA complexes that is expected to determine the different structures. Recent theoretical work suggests that electrostatic interactions alone are expected to favor the inverted hexagonal H_{II}^C phase (Fig. 11) over the lamellar L_α^C, which minimizes the charge separation between the anionic groups on the DNA chain and the cationic lipids (May and Ben-Shaul, 1997; Dan, 1998). However, the electrostatic in-

$$H_{II}^C$$

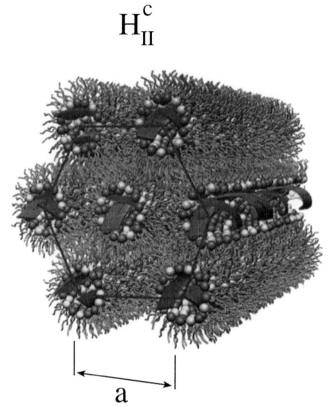

a

Figure 11 Schematic of the inverted hexagonal H_{II}^C phase (cylinders consisting of DNA coated with a lipid monolayer arranged on a hexagonal lattice) of cationic lipid/DNA (CL/DNA) complexes. [Adapted from Koltover *et al.*, (1998).]

teraction may be resisted by the membrane elastic cost (per unit area) (Seddon, 1989; Gruner, 1989; Israelachvili, 1978; Helfrich, 1978; Janiak *et al.,* 1979) of forming a cylindrical monolayer membrane around DNA:

$$F/A = 0.5 \ \kappa \ (1/R - 1/R_o)^2 \qquad (1)$$

Here, κ is the lipid monolayer bending rigidity, R is the actual radius, and R_o is the natural radius of curvature of the monolayer. Figure 12 shows schematically the possible "shapes" of many common lipids. For example, many lipids (e.g., phosphatidylcholine, phosphatidylserine, phosphotidylglycerol, cardiolipin) have a cylindrical shape, with the head group area $\approx$ the hydrophobic tail area, and tend to self-assemble into lamellar structures with a natural curvature $C_o = 1/R_o = 0$. Other

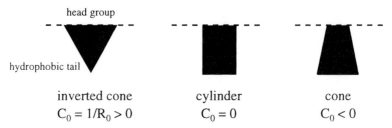

Figure 12   Three possible shapes of common lipid molecules as described in the text.

lipids (e.g., phosphotidylethanolamine) have a cone shape, with a smaller head group area than tail area, and give rise to a negative natural curvature $C_o < 0$. Alternatively, lipids with a larger head group than tail area have $C_o > 0$.

1. Control of CL/DNA Structure by Varying Shapes of Lipid Molecules: Helper Lipid and Cationic Lipid May Modify the Natural Radius of Curvature of Cationic Membranes

It is well appreciated (Seddon, 1989; Gruner, 1989; Israelachvili, 1978; Helfrich, 1978; Janiak *et al.*, 1979) that in many lipid systems the "shape" of the molecule that determines the natural curvature of the membrane $C_o = 1/R_o$ will also determine the actual curvature $C = 1/R$ which describes the structure of the lipid self-assembly (e.g., $C_o = 0 \rightarrow$ lamellar L_α; $C_o < 0 \rightarrow$ inverted hexagonal H_{II}; $C_o > 0 \rightarrow$ hexagonal H_I). This is particularly true if the bending rigidity of the membrane is large ($\kappa/k_B T >> 1$), because then a significant deviation of C from C_o would cost too much elastic energy. However, we see from Eq. (1) that if the bending cost is low with $\kappa \approx k_B T$, then C may deviate from C_o without costing much elastic energy, especially if another energy, is lowered in the process. Here we present experimental data pertaining to both situations: first, where the bending rigidity is large and the structure of the self-assembly is controlled by the "shape" of the molecule (described in the next paragraph), and second, where we lower κ enough that we find that C deviates from C_o due to the electrostatic energy favoring a different curvature and structure (described in Sect. C.2).

In the previous section we described recent synchrotron small-angle X-ray scattering (SAXS) studies in CL/DNA complexes using λ-DNA as a function of increasing Φ_{DOPE}, the weight fraction of DOPE in the DOPE/DOTAP cationic liposome mixtures (Koltover *et al.*, 1998). We found that the internal structure of the complex undergoes a structural transition from the L_α^C to the H_{II}^C. We can understand the L_α^C to H_{II}^C transition as a function of increasing Φ_{DOPE} (Fig. 10) by noting that in contrast to the helper lipid DOPC and the cationic lipid DOTAP (Fig. 3), which have a zero natural curvature ($C_o^{DOTAP,DOPC} = 1/R_o^{DOTAP,DOPC} = 0$),

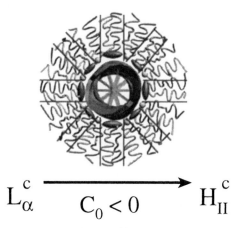

$$L_\alpha^c \xrightarrow{ C_0 < 0 } H_{II}^c$$

Figure 13   The H_{II}^c phase is expected to be the stable structure when the natural curvature (C_o) of the cationic lipid monolayer is driven negative by the addition of the helper lipid DOPE. This is shown schematically where the cationic lipid DOTAP is cylindrically shaped, while DOPE is cone-like with negative natural radius of curvature.

DOPE is cone shaped with $C_o^{DOPE} = 1/R_o^{DOPE} < 0$ (Fig. 12). Thus, the natural curvature of the monolayer mixture of DOTAP and DOPE is driven negative with $C_o = 1/R_o = \Phi_{DOPE} C_o^{DOPE}$. Hence, as a function of increasing Φ_{DOPE} we expect a transition from the L_α^c to the H_{II}^c phase (shown schematically in Fig. 13), which is observed experimentally and is now also expected to be favored by the elastic free energy. Thus, the helper lipid DOPE induces the L_α^c to H_{II}^c transition by controlling the spontaneous radius of curvature R_o of the lipid layers.

2. Control of CL/DNA Structure by Varying the Lipid Bilayer Bending Modulus through Modification of Helper Lipid: Electrostatic Energy Overwhelms the Membrane Elastic Energy as the Latter Is Reduced

The L_α^c to H_{II}^c transition was recently observed to occur along a different path by reducing the bending rigidity k of the lipid layer shown schematically in Fig. 14 (Koltover *et al.*, 1998). The reduction of κ along this pathway reduces the membrane elastic energy (determined by Eq. (1)), which otherwise prevents the formation of the H_{II}^c phase favored by electrostatics. In these experiments a new class of helper lipid molecules was used consisting of DOPC mixed with the membrane-soluble cosurfactant molecule hexanol (Koltover *et al.*, 1998). Although cosurfactant molecules, which typically consist of long-chain alcohols (e.g., pentanol, hexanol), are not able to stabilize an interface separating hydrophobic and hydrophilic regions, when mixed in with longer chain "true" surfactants they are

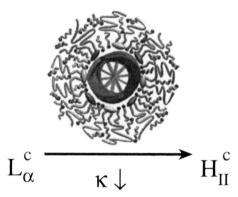

$$L_\alpha^c \xrightarrow[\kappa \downarrow]{} H_{II}^c$$

Figure 14   The H_{II}^c phase should be generally found when the membrane bending rigidity is lowered by thinning the membrane. One mechanism is to add cosurfactant molecules with short hydrophobic chains to the helper lipid as described.

known to lead to dramatic changes in interface elasticities. Simple compressional models of surfactant chains show that the bending rigidity of membranes κ scales with chain length l_n ($\propto \delta_m$, membrane thickness, n = number of carbons per chain) and the area per lipid chain A_L as $\kappa \propto l_n^3 / A_L^5$ (Szleifer *et al.*, 1988, 1990). The mixing of cosurfactants with lipids is expected to lead both to a thinner membrane and to a larger area per chain (Fig. 14) and result in a strong suppression of κ making the membrane highly flexible. Experimental studies have shown that the addition of cosurfactants like pentanol, hexanol, and heptanol to membranes of lamellar phases with a mole ratio of between two to four leads to a significant decrease of κ from about 20 $\kappa_B T$ to about 2–5 $k_B T$ (Safinya *et al.*, 1989; Safinya *et al.*, 1986).

D. The Relation between the Structure of CL/DNA Complexes and DNA Release from Complexes: Toward a Simple Model System for Studies of Transfection Efficiency

A major goal of research on CL/DNA complexes is to elucidate the key parameters resulting in the different CL/DNA complex structures and to establish the correlation between the different structures and transfection efficiency. It is known that transfection efficiency mediated by mixtures of cationic lipids and neutral helper lipids varies widely and unpredictably (Zhu *et al.*, 1993; Nabel *et al.*, 1993; Mulligan, 1993; Felgner and Rhodes, 1991; Behr, 1994; Remy *et al.*, 1994; Singhal and Huang, 1994; Lasic and Templeton, 1996). The choice of the helper lipid has been empirically established to be important (Felgner *et al.*, 1994; Farhood *et al.*, 1995; Hui *et al.*, 1996); for example, many papers report that transfection is believed to be significantly more efficient in mixtures of the cationic lipid DOTAP and the neutral helper lipid DOPE, then in mixtures of DOTAP and the similar helper lipid DOPC (Fig. 3). From X-ray diffraction work (Raedler *et al.*, 1997; Spector and Schnur, 1997; Seachrist, 1997; Salditt *et al.*, 1997; Safinya *et al.*, 1998; Raedler *et al.*, 1998; Salditt *et al.*, 1998) we know that DOTAP/DOPC/λ-DNA complexes form the multilamellar L_α^C structure (Fig. 2). Our work described previously shows that DOTAP/DOPE/λ-DNA containing complexes may also form the distinctly different self-assembled inverted hexagonal H_{II}^C structure.

The data represent one example of a correlation between the self-assembled structure of CL/DNA complexes and transfection efficiency for this particular concentration regime in DOTAP/DOPE and DOTAP/DOPC complexes: the empirically established more transfectant DOPE-containing complexes in mammalian cell culture exhibit the H_{II}^C structure rather than the L_α^C found in DOPC-containing complexes. What makes the H_{II}^C structure more transfectant than the L_α^C structure? Studies show that if complexes are physically microinjected into the nucleus, expression of reporter gene is strongly suppressed; thus, under normal entry circumstances plasmid DNA must be released from the complex in the cytoplasm before translocating into the nucleus (Zabner *et al.*, 1995). In particular, since DNA has to be released from the complex before expression can occur, it is clear that mechanisms of DNA release inside cells affect (and in most instances) increase transfection efficiency (Legendre and Szoha, 1992). As described next optical microscopy studies of interactions between CL/DNA complexes and giant anionic vesicles (i.e., models of cellular membranes, in particular, anionic endosomes and plasma membranes) have directly revealed such destabilizing events (Koltover *et al.*, 1998).

1. The Stability of CL/DNA Complexes: Fusion of Cationic Liposome–DNA Complexes with Oppositely Charged Giant Vesicles (Models of Cellular Membranes)

The importance of the precise self-assembled structures to biological function is underscored in optical imaging experiments described in this section, which show that interactions of CL/DNA complexes with anionic giant liposomes (model cell membranes) are structure dependent (e.g., H_{II}^C versus L_α^C). One of the simpler experimental designs that we have used to look for the effect of structure on the early stages of transfection (i.e., DNA release within cells) includes light microscopy studies of the interaction of CL/DNA complexes with giant anionic vesicles that are models of CL/DNA complexes interacting with cellular membranes and anionic endosomal vesicles (Koltover *et al.*, 1998). Current data from several laboratories (Zabner *et al.*, 1995; Wrobel and Collins, 1995; Legendre and Szolca, 1992; Lin *et al.*, in preparation) indicate that one of the main entry routes of complexes is endocytosis following attachment of the positive CL/DNA complexes to negatively charged cell surface proteoglycans (Mislich and Baldeschwieler, 1996). Thus, in many instances at the very early stages of cell transfection, an intact CL/DNA complex is captured inside an endosome that contains anionic lipids.

We have found that positively charged H_{II}^C and L_α^C complexes containing DOTAP interact very differently with giant vesicles (G-vesicles) even when both types of structures contained DOPE. Figure 15 (A and B) shows video-enhanced optical microscopy in differential-interference-contrast (DIC) and fluorescence configurations, for H_{II}^C (A, $\Phi_{DOPE} = 0.73$) and L_α^C (B, $\Phi_{DOPE} = 0.3$) complexes. Typical micrographs of positively charged (DOTAP/DNA = 4) complexes attached to the fluid membranes of G-vesicles show that the L_α^C complexes attach to the G-vesicles and remain stable (Fig. 15C); no fusion occurs between the complex and

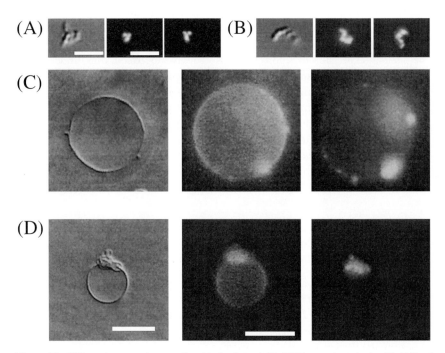

Figure 15 Video-microscopy images of positively charged CL/DNA complexes in the (A) H_{II}^C and (B) L_α^C phases, and interacting with negatively charged giant (G) vesicles (C and D). In all cases, complexes were viewed in (DIC) (left), lipid fluorescence (middle), and DNA fluorescence (right). The scale bar for DIC is 3 μm for (A) and (B) and 20 μm for (C) and (D), in fluorescence image is 6 μm for (A) and (B) and 20 μm for (C) and (D). In (C) the L_α^C complexes simply stick to the G-vesicle and remain stable for many hours, retaining their blob-like morphology. The blobs are localized in DIC as well as lipid and DNA fluorescence modes. In (D), the H_{II}^C complexes break up and spread immediately after attaching to G-vesicles, indicating a fusion process between the complex and the vesicle lipid bilayer and release of DNA. The loss of the compact structure of the complex is evident in both lipid and DNA fluorescence modes. [Adapted from Koltover *et al.*, (1998).]

the G-vesicle. L_α^C complexes containing DOPC show the same behavior. When H_{II}^C complexes attach to the G-vesicle, they rapidly fuse and lose their compact structure (Fig. 15D, left). The loss of the compact complex structure and the subsequent desorption of DNA molecules from membrane are seen in fluorescence (Fig. 15D right). This behavior is expected after fusion, which results in the mixing of cationic lipid (from the H_{II}^C complex) with anionic lipid (from the G-vesicle), effectively "turning off" the electrostatic interactions (which gave rise to the compact CL/DNA complexes) and releasing of DNA molecules inside the space between the lamellae and the G-vesicle bilayer. Because the geometry is the inverse of CL/DNA complexes inside anionic endosomal vesicles, we expect that upon fusion the inverse geometry will occur with DNA released and expelled outside the endosome within the cytoplasm. On longer time scales (a few hours) we observed lipid transfer between the L_α^C complexes and G-vesicles. Thus, the observation that DOPE/DOTAP L_α^C and DOPC/DOTAP L_α^C complexes do not fuse with G-vesicles is a kinetic rather than equilibrium effect. One may, in principle, be able to design L_α^C complexes with a lower kinetic barrier to fusion. Moreover, the behavior described in this review of complexes containing univalent cationic lipids may be different from that of multivalent cationic lipids. The data give evidence that H_{II}^C complexes may fuse with endosomal vesicles in cells and thus release their DNA cargo into the cytoplasm, resulting in higher transfection levels. Fluorescence microscopy studies in mouse fibroblast cell cultures, while showing more complex behavior overall, also show some similar features (Lin et al., in preparation).

The findings described in Section II unambiguously establish one particular correlation between the self-assembled structure of CL/DNA complexes and transfection efficiency: for a range of concentrations the empirically established transfectant complexes in mammalian cell cultures exhibit the H_{II}^C structure rather than the L_α^C. We caution that we believe that highly transfectant L_α^C complexes may be designed in certain concentration regimes (Lin et al., in preparation). Further, optical microscopy reveals a most likely origin for why different structures transfect cells with varying efficiency: in contrast to L_α^C complexes, H_{II}^C complexes are found to fuse and release DNA when in contact with anionic vesicles, which are cell free models of cellular organelle membranes, in particular, anionic endosomal vesicles. Thus, the data suggest a simple direct mechanism of DNA release into the cytoplasm from endosomal vesicles containing H_{II}^C complexes. This then paves the way for a fundamental understanding of the early-stage events following the endocytic uptake of CL/DNA complexes by mammalian cells in nonviral gene delivery applications.

III. FUTURE DIRECTIONS

Clearly much work remains before we have a complete understanding of the various possible self-assembled structures in CL/DNA complexes and an understanding, at the molecular and self-assembled levels, of all of the critical parameters that control the different structures. Even more work will be required to relate the structures to biological function, namely, the interactions of CL/DNA complexes with cellular components inside animal cells that lead to gene release and expression. The broad, long-term objective of our research is to develop a fundamental science base that will lead to the design and synthesis of optimal nonviral carriers of DNA for gene therapy and disease control. Simultaneously, a major long-term objective is to improve efficiency for delivering large pieces of DNA containing important human genes and their regulatory sequences ($> 100k$ base pairs), which at present can only be achieved with synthetic vectors. The structure–function data obtained from the research should allow us to begin the formidable task of a rational design of these self-assemblies for enhanced gene delivery applications from the ground up beginning with the chemical structure of the lipids and the correct compositions in mixtures including functional plasmid.

ACKNOWLEDGMENTS

First, we greatfully acknowledge the contributions of our collaborators Joachim Raedler, Tim Salditt, Alison Lin, Nelle Slack, Ayesha Ahmad, Cyril George, Charles Samuel, Uwe Schulze, and Hans-Werner Schmidt. It is a pleasure to acknowledge discussions with R. Bruinsma, A. Ben-Shaul, P. Pincus, W. Gelbart, T. Lubensky, and N. Dan. We would also like to acknowledge numerous enlightening discussions with Philip Felgner. Supported by the UC-Biotechnology Research and Education Program Grant No. 97–02, by NSF-DMR-9624091, PRF-31352-AC7, and Los Alamos-STB/UC:96–108. Partially supported by NIH-GM59288-01. The Materials Research Laboratory at Santa Barbara is supported by NSF-DMR-9632716. The X-ray experiments were carried out at the Stanford Synchrotron Radiation Laboratory supported by the U.S. Department of Energy.

REFERENCES

Behr, J. P. (1994). Gene transfers with synthetic cationic amphiphiles—prospects for gene therapy, *Bioconjugate Chem.* **5,** 382–389.

Bloomfield, V. A. (1991). Condensation of DNA by multivalent cations—considerations on mechanism, *Biopolymers* **31,** 1471–1481.

Brigham, K. L., Meyrick, B., Christman, B., Magnuson, M., King, G., and Berry, L. C. (1989). In vivo transfection of murine lungs with a functioning prokaryotic gene using a liposome vehicle, *Am. J. Med. Sci.* **298,** 278–281.

Bruinsma, R. (1998). Electrostatics of DNA cationic lipid complexes: isoelectric instability, *Eur. Phys. J. B* **4,** 75–88.

Bruinsma, R., and Mashl, J. (1998). Long-range electrostatic interaction in DNA cationic lipid complexes, *Europhys. Lett.,* **41,** (2), 165–170.

Crystal, R. G. (1995). Transfer of genes to humans—early lessons and obstacles to success", *Science* **270,** 404–410.

Dan, N. (1998). The structure of DNA complexes with cationic liposomes—cylindrical or flat bilayers?, *Biochim. Biophys. Acta* **1369,** 34–38.

Farhood, H., Serbina, N., and Huang, L. (1995). The role of dioleoyl phospatidylethanolamine in cationic liposome mediated gene therapy, *Biochim. Biophys. Acta* **1235,** 289–295.

Felgner, P. L. (1997). Nonviral strategies for gene therapy, *Sci. Am.* **276,** 102–106.

Felgner, P. L., Gadek, T. R., Holm, M., Roman, R., Chan, H.W., Wenz, M., Northrop, J. P., Ringold, G. M., and Danielsen, M. (1987). Lipofection: A highly efficient, lipid-mediated DNA-transfection procedure, *Proc. Nat. Acad. Sci. USA* **84,** 7413–7417.

Felgner, P. L., Rhodes, G. (1991). Gene therapeutics, *Nature* **349,** 351–352.

Felgner, J., *et al.,* (1994). Enhanced gene delivery and mechanism studies with a novel series of cationic lipid formulations, *J. Biol. Chem.* **269,** 2550–2561.

Friedmann, T. (1997). Overcoming obstacles to gene therapy, *Sci. Am.* **276,** 96–101.

Golubovic, L., and Golubovic, M. (1998). Fluctuations of quasi-two-dimensional smectics intercalated between membranes in multilamellar phases of DNA cationic lipid complexes, *Phys. Rev. Lett.* **80,** 4341–4344.

Gruner, S. M. (1989). Stability of lyotropic phases with curved interfaces, *J. Phys. Chem.* **93,** 7562–7570.

Gustafsson, J., Arvidson, G., Karlsson, G., and Almgren, M. (1995), Complexes between cationic liposomes and DNA visualized by cryo-TEM, *Biochim. Biophys. Acta* **1235,** 305–312.

Harries, D., May, S., Gelbart, W. M., and Ben-Shaul, A. (1998). Structure, stability, and thermodynamics of lamellar DNA-lipid complexes, *Biophys. J.* **75,** 159–173.

Harrington, J. J., Van Bokkelen, G., Mays, R. W., Gustashaw, K., and Williard, H. F. (1997). Formation of de novo centromeres and construction of first-generation human artificial microchromosomes, *Nature Gene.* **15,** 345–355.

Hazinski, T. A., Ladd, P. A., and Dematteo, C. A. (1991). Localization and induced expression of fusion genes in the rat lung, *Am. J. Respir. Cell Mol. Biol.* **4,** 206–209.

Helfrich, W. (1978). Steric interaction of fluid membranes in multilayer systems, *Zeitschift fur Naturforschung A* **33,** 305–315.

Holt, C. E., Garlick, N., and Cornel, E. (1990). Lipofection of CDNAS in the embryonic vertebrate central nervous system, *Neuron* **4,** 203–214.

Hui, S. W., Langner, M., Zhao, Y. L., Ross, P., *et al.* (1996). The role of helper lipids in cationic liposome-mediated gene transfer, *Biophys. J.* **71,** 590–599.

Israelachvili, J. N. (1992). *Intermolecular and surface forces* (2nd Ed.) London: Academic Press.

Janiak, M. J., Small, D. M., and Shipley, G. G. (1979). Temperature and compositional dependence of the structure of hydrated dimyristoyl lecithin, *J. Biol. Chem.* **254,** 6068–6078.

Koltover, I., Salditt, T., Raedler, J. O., and Safinya, C. R. (1998). An inverted hexagonal phase of DNA-cationic liposome complexes related to DNA release and delivery, *Science* **281,** 78–81.

Lasic, D., and Templeton, N.S. (1996). Liposomes in gene therapy, *Adv. Drug Del. Rev.* **20,** 221–266.

Lasic, D. D., Strey, H. H., Stuart, M. C. A., Podgornik, R., and Frederik, P. M. (1997). The structure of DNA-liposome complexes, *J. Am. Chem. Soc.* **119,** 832–833

Legendre, Y. J., and Szoka, F. C. (1992). Delivery of plasmid DNA into mammalian cell lines using PH-sensitive liposomes—comparison with cationic liposomes, *Pharm. Res.* **9,** 1235–1242.

Lewin, B. (1997). B: *Genes VI.* Oxford: Oxford University Press.

Lin, A., Slack, N., Ahmad, A., George, C., Samuel, C., and Safinya, C. R. Manuscript in preparation.

Livolant, F., and Leforestier, A. (1996). Condensed phases of DNA: structures and phase transitions, *Prog. Poly. Sci.* **21,** 1115–1164.

Malone, R. W. (1989). Expression of chloramphenicol acetyltransferase activity in brain tissue of frog embryos with cationic liposome vectors, *Focus* **11,** 4.

Manning, G. S. (1969). Limiting laws and counterion condensation in polyelectrolyte solutions. I. Colligative properties, *J. Chem. Phys.* **51,** 924–933.

Marshall, E. (1995). Gene therapys growing pains, *Science* **269,** 1050–1055.

May, S., and Ben-Shaul, A. (1997). DNA-lipid complexes: Stability of honeycomb-like and spaghetti-like structures, *Biophys. J.* **73,** 2427–2440.

Miller, A. D. (1998). Cationic liposomes for gene therapy, Angewandte Chemie (Intl. Ed.), *Reviews* **37,** 1768–1785.

Mislick, K. A. and Baldeschwieler, J. D. (1996). Evidence for the role of proteoglycans in cation mediated gene transfer, *Proc. Nat. Acad. Sci. USA* **93,** 12349–12354.

Mulligan, R. C. (1993). The basic science of gene therapy, *Science* **260,** 926–932.

Nabel, G. J., Nabel, E. G., Yang, Z. Y., Fox, B. A., Plautz, G. E., Gao, X., Huang. L., Shu, S., Gordon, D., and Chang, A. E. (1993). Direct gene transfer with DNA liposome complexes in melanoma-expression, biologic activity, and lack of toxicity in humans, *Proc. Nat. Acad. Sci. USA* **90,** 11307–11311.

O'Hern, C. S., and Lubensky, T. C. (1998). Sliding columnar phase of DNA lipid complexes, *Phys. Rev. Lett.* **80,** 4345–4348.

Ono, T., Fujino, Y., Tsuchiya, T., and Tsuda, M. (1990). Plasmid DNAS directly injected into mouse brain with lipofection can be incorporated and expressed by brain cells, *Neurosci. Lett.* **117,** 259–263.

Pitard, B., Aguerre, O., Airiau, M., Lachagés, A.-M., Boukhnikachvili, T., Byk, G., Dubertret, C., Herviou, C., Scherman, D., Mayaux, J.-F. and Crouzet, J. (1997). Virus sized self-assembled lamellar complexes between plasmid DNA and cationic micelles promote gene transfer, *Proc. Natl. Acad. Sci. USA* *94,* 14412–14417.

Podgornik, R., Rau, D. C., and Parsegian, V. A. (1994)., Parametrization of direct and soft steric-undulatory forces between DNA double helical polyelectrolytes in solutions of several different anions and cations, *Biophys. J.* **66,** 962–971.

Raedler, J. O., Koltover, I., Salditt, T., Jamieson, A., and Safinya, C. R. (1998). Structure and interfacial aspects of self-assembled cationic lipid-DNA gene carrier complexes, *Langmuir* **14,** 4272–4283.

Raedler, J. O., Koltover, I., Salditt, T., and Safinya, C. R. (1997). Structure of DNA-cationic liposome complexes: DNA intercalation in multi-lamellar membranes in distinct interhelical packing regimes, *Science* **275,** 810–814.

Reich, Z., Wachtel, E. J., and Minsky, A. (1994). Liquid crystalline-mesophases of plasmid DNA in bacteria, *Science* **264,** 1460–1463.

Remy, J. S., Sirlin, C., Vierling, P., and Behr, J. P. (1994). Gene transfer with a series of lipophilic DNA-binding molecules, *Bioconjugate Chem.* **5,** 647–654

Roush, W. (1997). Molecular biology—counterfeit chromosomes for humans, *Science* **276,** 38–39.

Roux, D., and Safinya, C. R. (1988). A synchroton x-ray study of competing undulation and electrostatic interlayer interactions in fluid multimembrane lyotropic phases, *J. Phys. France* **49,** 307–318.

Safinya, C. R., Koltover, I., and Raedler, J. O. (1998). DNA at membrane surfaces: An experimental overview, *Cur. Opin. Colloid Interface Sci.* **3 (1),** 69.

Safinya, C. R., Sirota, E. B., Roux, D. and Smith, G. S. (1989). Universality in interacting membranes: The effect of cosurfactants on the interfacial rigidity, *Phys. Rev. Lett.* **62,** 1134.

Safinya, C. R. Rigid and fluctuating surfaces: A series of synchrotron x-ray scattering studies of interacting stacked membranes. In Riste, R., and Sherrington, D. (Eds) (1989). *Phase transitions in soft condensed matter,* NATO ASI Series B **211,** 249–270.

Safinya, C. R., Roux, D., Smith G. S., Sinha S. K., Dimon P., Clark N. A., and Bellocq A. M. (1986). Steric interactions in a model membrane system: A synchrotron x-ray study. *Phys. Rev. Lett.* **57,** 2718–2721.

Salditt, T., Koltover, I., Radler, J. O., and Safinya, C. R. (1997). Two-dimensional smectic ordering of linear DNA chains in self-assembled DNA-cationic liposome mixture, *Phys. Rev. Lett.* **79,** 2582–2585.

Salditt, T., Koltover, I., Raedler, R. O., and Safinya, C. R. (1998). Self-assembled DNA-cationic lipid complexes: Two-dimensional smectic ordering, correlations, and interactions, *Phys. Rev. E* **58,** 889–904.

Seachrist, L. (1997). Researchers weigh advantages of using liposomes to deliver DNA to cells, *Bioworld Today* (The Daily Biotechnology Newspaper) **8,** (26), February 7, p.1.

Seddon, J. M. (1989). Structure of the inverted hexagonal phase and non-lamellar phase transitions of lipids, *Biochim. Biophys. Acta* **1031,** 1–69.

Singhal, A., and Huang, L. (1994); Wolff, J. A. (Ed.), In Gene therapeutics: Methods and applications of direct gene transfer.Boston: Birkhauser.

Spector, M. S. and Schnur, J. M. (1997). DNA ordering on a lipid membrane. In *Science* (Perspectives article), **275**, 791–792.

Sternberg, B., Sorgi, F. L., and Huang, L. (1994). New structures in complex formation between DNA and cationic liposomes visualized by freeze-fracture electron microscopy, *FEBS Lett.* **356**, 361–366.

Stribling, R., Brunette, E., Liggitt, D., Gaensler, K., and Debs, R. (1992). Aerosol gene delivery *in vivo*, *Proc. Natl. Acad. Sci. (USA)* **89**, 11277–11281.

Szleifer, I., Ben-Shaul, A., and Gelbart, W. M. (1990). Chain packing statistics and thermodynamics of amphilphile monolayers, *J. Phys. Chem.* **94**, 5081–5089

Szleifer, I., Kramer, D., Ben-Shaul, A., Roux, D., and Gelbart, W. M. (1988). *Phys. Rev. Lett.* **60**, 1966.

Wrobel, I., and Collins, D. (1995). Fusion of cationic liposomes with mammalian cells occurs after endocytosis, *Biochim. Biophys, Acta* **1235**, 296–304.

Zabner, J., Fasbender, A. J., Moninger,T., Poelinger, K. A., and Welsh, M. J. (1995). Cellular and molecular barriers to gene transfer by a cationic lipid, *J. Biol. Chem.* **270**, 18997–19007.

Zhu, N., Liggitt, D., Liu, Y., and Debs, R. (1993). Systemic gene expression after intravenous DNA delivery into adult mice, *Science* **261**, 209–211.

Zimm, B. H., and Le Bret, M. (1983). Counter-ion condensation and system dimensionality,*J. Biomol. Struc. Dyn.* **1**, 461–471.

ENDNOTE

A mixture of DOPC/DOTAP (1/1, wt/wt) was prepared in a 20 mg/ml chloroform stock solution. 500 ml was dried under nitrogen in a narrow glass beaker and desiccated under vacuum for 6 h. After addition of 2.5 ml Millipore water and 2 h incubation at $40°C$, the vesicle suspension was sonicated to clarity for 10 min. The resulting solution of liposomes, 25 mg/ml, was filtered through 0.2 μm Nucleopore filters. For optical measurements the concentration of SUV used was between 0.1mg/ml and 0.5 mg/ml. All lipids were purchased from Avanti Polar Lipids, Inc. (Alabaster, Alabama). For fluorescence work, sonicated DOPC-DOTAP(1/1) liposomes were prepared at 0.1mg/ml with 0.2 mol % DHPE-Texas Red fluorescence label. DNA stained by YOYO (Molecular Probes) was added under gentle mixing at different lipid to DNA ratios (L/D). For X-ray diffraction studies the DNA/lipid condensates were prepared from a 25 mg/ml liposome suspension and a 5mg/ml DNA solution. The solutions were filled in 2 mm diameter quartz capillaries with different ratios L/D respectively and mixed after flame sealing by gentle centrifugation up and down the capillary.

CHAPTER 6

Sites of Uptake and Expression of Cationic Liposome/DNA Complexes Injected Intravenously

John W. McLean, Gavin Thurston, and Donald M. McDonald
Cardiovascular Research Institute and Department of Anatomy, University of California, San Francisco, California

Cationic liposome/DNA complexes injected intravenously into mice are rapidly cleared from the circulation. Complexes initially coat the luminal surface of endothelial cells in many microvascular beds and are also associated with some intravascular macrophages. Thereafter, the complexes are internalized by endocytosis. The greatest uptake by endothelial cells is in the microvasculature of lung, lymph node, Peyer's patch, ovary, anterior pituitary, and adrenal medulla. Microvascular beds of the brain, kidney cortex, and pancreatic islets have little or no endothelial cell uptake. There is also avid uptake of the complexes by endothelial cells at sites of angiogenesis in tumors and chronic inflammation. The uptake of the complexes into endosomes and lysosomes in endothelial cells is consistent with a receptor-mediated uptake mechanism, whereas uptake in macrophages appears to involve phagocytosis. After injection of complexes, luciferase transgene expression measured by luminometry is several orders of magnitude greater in lungs than in lymph

Nonviral Vectors for Gene Therapy

nodes or ovaries despite the comparable amount of uptake by endothelial cells in these organs. The number of luciferase-expressing cells detected by immuno-histochemistry reflects the organ expression measured by luminometry. Of the cell types that take up liposome complexes, only endothelial cells unequivocally have luciferase immunoreactivity.

I. INTRODUCTION

Although cationic liposome-based DNA delivery systems are being used in clinical and preclinical gene therapy protocols, little is known about the distribution and cellular uptake of cationic liposomes *in vivo*. Our research has focused on the fate of systemically administered cationic liposomes and liposome/DNA complexes in mice (McLean *et al.*, 1997; Thurston *et al.*, 1998). We have been especially inter-ested in determining which cells within an animal are able to internalize liposomes and complexes after intravenous administration and to what extent uptake corre-lates with expression of plasmid reporter genes delivered by liposome complexes. Although delivery of DNA to a cell is a requirement for gene expression, it may not be sufficient, evidenced by the fact that there are marked differences in the efficacy of various lipid formulations in transfecting a variety of primary and transformed cells in culture. These *in vitro* experiments have shown that the amount of DNA internalized does not necessarily correlate with levels of gene expression, so corre-sponding differences might be expected to occur *in vivo* (Egilmez *et al.*, 1996; Fasbender *et al.*, 1997).

While the transfection of parenchymal cells in target organs would be the objective in many gene therapies, it has not been determined whether intravenously delivered cationic liposome/DNA complexes have access to cells outside the vas-culature. Based on the size of cationic liposome complexes, which are generally on the order of 50–500 nm, the endothelial barrier would normally prevent particles of this size from crossing, except in organs such as the spleen where the endothelium is discontinuous and complexes might reach cells outside the vasculature. Literature accounts of *in vivo* cationic liposome-mediated transfection have often reported to-tal gene expression by organ, without identifying the cell types involved, and usually only for a limited number of organs (Liu *et al.*, 1997; Song *et al.*, 1997; Templeton *et al.*, 1997). These are important issues that we have sought to clarify.

We have used several strategies to pursue our objectives. Primarily we have used fluorescent microscopy, both conventional and confocal, to track the fate of fluorescently labeled liposomes and DNA after intravenous delivery in mice. We have extended these studies by using electron microscopy to examine the intracel-lular fate of gold-labeled liposomes. In addition, we have used several different reporter genes to measure expression from plasmid DNA delivered to various cel-lular sites and have attempted to identify expressing cells by immunohistochemistry.

II. METHODOLOGY

Most of the experimental methodologies we use can be found in detail in our recent publications (McLean et al., 1997; Thurston et al., 1998), and are outlined here.

We have prepared liposomes with several different cationic lipids including DDAB (dimethyldioctadecyl ammonium bromide), DOTAP (1,2-dioleoyl-3-trimethylammonium-propane), and DOTIM (1-[2-(9(Z)-octadecenoyloxy)ethyl]-2-(8(Z)-heptadecenyl)-3-(2-hydroxylethyl)imidizolinium chloride, all formulated with cholesterol. Small unilamellar liposomes (SUVs) were prepared either by sonication or by extrusion. Cationic liposomes were complexed with plasmid DNA at high liposome to DNA ratios, which empirically gives the highest expression in vivo from reporter genes. Plasmid DNAs containing reporter genes such as luciferase and chloramphenicol acetyltransferase (CAT) cDNA sequences driven by the CMV immediate early promoter/enhancer have been used to examine the tissue distribution and amount of transgene expression.

Several different methods for labeling both liposomes and DNA were developed to enable us to follow their distribution in vivo. Initially we used the aldehyde-fixable red fluorescent carbocyanine dye CM-DiI, which was incorporated into liposomes postsynthetically. More recently we have incorporated phospholipid-linked fluorophores into liposomes during synthesis, most notably Texas Red 1,2-dipalmitoyl-sn-glycero-3-phosphoethanolamine. The incorporation of these lipid fluorophores at less than 1% of the total lipid does not cause any difference in the expression of reporter genes complexed to labeled liposomes. Texas Red-labeled, lysine-fixable dextran has also been encapsulated inside liposomes. Labeled DNA was prepared by the enzymatic addition of Cy3-labeled nucleotides to the termini of linearized plasmid DNA. All these different methodologies for fluorescently labeling liposomes or DNA appear to give identical results when liposomes or complexes are tracked in vivo (Thurston et al., 1998). For electron microscopy, the electron-dense lipid Nanogold 1,2-dipalmitoyl-sn-glycero-3-phosphoethanolamine was incorporated into the lipid bilayer of liposomes.

We have used several different strains of mice in our experiments, including inbred strains (C57BL/6, C3H, BALB/c) and outbred animals (ICR, CD-1) of both sexes and have noticed no significant differences in liposome distribution and uptake between them. Liposome distribution or transgene expression also does not appear to be affected by anesthesia, since femoral vein delivery of liposome complexes into anesthetized mice gave organ distribution and luciferase expression indistinguishable to that obtained by tail vein injection into conscious animals.

At various times after fluorescent liposome or complex injection, mice were anesthetized and fixed by vascular perfusion with paraformaldehyde-containing fixatives. To provide a three-dimensional, anatomical context for liposome uptake, fluorescein-labeled Lycopersicon esculentum lectin was perfused after the fixative

(Thurston *et al.*, 1996). This lectin uniformly stains the luminal surface of endothelial cells when perfused into the vasculature. After the mice were perfused, tissues were immediately removed and prepared for examination by fluorescent microscopy. Tissues such as skin, diaphragm, trachea, pituitary, bone marrow, and fat were mounted directly for observation. Otherwise tissues were embedded in agarose and sectioned by Vibratome at $50-200$ μm.

Tissues were examined with a Zeiss Axiophot epifluorescence microscope with Fluar objectives, and fluorescence filters specific for the fluorophores used to label the specimens. Images were recorded on transparency film and later digitally scanned. A Zeiss LSM410 inverted confocal laser scanning microscope equipped with a krypton-argon laser was used to examine selected tissues. CM-DiI, Texas Red, or Cy3 fluorescence was imaged with excitation at 568 nm and emission above 585 nm; fluorescein was imaged with excitation at 488 nm and emission from 515 to 540 nm. Sequences of images were recorded at 1 μm steps in the z-axis, compressed into a single plane, and combined with the corresponding projected digital image of the other fluorophore.

Transmission electron microscopy was performed on tissue samples from animals injected with liposomes incorporating Nanogold phospholipids. Animals were perfused with glutaraldhyde-containing fixatives before processing. Silver intensification was used to enhance the visibility of gold-labeled liposomes, and uranyl acetate and lead citrate were used to add contrast to cellular morphology (McDonald, 1988).

Expression from reporter gene plasmids was measured on tissue homogenates of animals perfused with phosphate buffered saline. Luciferase was measured by luminometry, and CAT was measured by the enzymatic transfer of butyryl groups from n-butyryl-CoA to [^{14}C]chloramphenicol. Immunohistochemistry was performed on formaldehyde-fixed permeabilized Vibratome or cryostat sections using polyclonal antibodies specific for the reporter protein, and alkaline phosphatase- or fluorophore-conjugated secondary antibodies.

III. INTERACTIONS OF LIPOSOMES AND LIPOSOME / DNA COMPLEXES WITH BLOOD

Most of the fluorescent liposomes used in our studies are generally too small to be individually resolved by conventional fluorescent microscopy but their size was estimated to be $60-100$ nm by dynamic laser light scattering. After complexing with DNA there is an approximately twofold increase in size of the particles of the formulations we have used.

Highly positively charged macromolecules such as liposomes or liposome complexes might be expected to interact with both the cellular and protein components of blood and these interactions may be very important in their uptake by cells. The bulk of the injected liposome dose appears to be rapidly cleared from the blood with a half-life in the order of minutes when injected into the tail vein of mice. Blood samples drawn from animals within the first few minutes after injection of labeled complexes show numerous bright fluorescent clusters that are much larger than the particles in the injected material. Many of these clusters are larger than 1 μm, but they are generally smaller than erythrocytes, and some of these clusters contain platelets as shown by staining with acridine yellow. However, these clusters do not grow any larger or block capillary beds, indicating that they result from relatively weak interactions. By 20 min, there is only a very small amount of fluorescent material left in blood, and this is generally associated with platelets and some leukocytes. Many monocytes have intense fluorescence, and some platelets and neutrophils have one or two punctate fluorescent dots.

Blood cell counts performed on animals after injection with DOTIM/ cholesterol complexes at 5 min, 4 h, and 24 h after injection revealed some surprising results (McLean et al., 1997). There was a biphasic reduction in platelets, which totaled 39% at 5 min, 7% at 4 h, and 30% at 24 h. There was also a biphasic drop in the leukocyte count: a 63% reduction at 5 min, 46% at 4 h, and 73% at 24 h. A differential leukocyte analysis revealed that the reductions were greatest in lymphocytes and neutrophils at 5 min, whereas at 24 h the reductions were greatest in eosinophils, lymphocytes, and monocytes. The thrombocytopenia and leukopenia seen with DOTIM/cholesterol complexes were also seen with the other cationic liposome formulations and appear to be a property of cationic liposomes, at least in the doses and formulations used in our studies.

The mechanisms responsible for the rapid fall in the number of circulating leukocytes and platelets after the injection of complexes is unknown. A rapid, transient thrombobocytopenia has previously been reported after the intravenous injection of charged liposomes followed by a rebound in platelet count by 1 h (Doerschuk et al., 1989; Reinish et al., 1988). There may be a direct interaction of liposome complexes with platelets and leukocytes leading to an upregulation of cellular receptors that mediate activation or adhesion. It is also possible that the liposome complexes have caused an upregulation of receptors on endothelial cells leading to leukocyte and platelet adhesion to vessel walls. These explanations are consistent with the appearance of fluorescently labeled adherent leukocytes on the luminal surface of venules in several organs, and adherent platelets in alveolar capillaries and high endothelial venules (HEV).

IV. LIPOSOME UPTAKE IN THE VASCULATURE

As liposome complexes disappear from the blood of mice in the first few minutes after intravenous injection, they coat the luminal surface of endothelial cells in many microvascular beds and are also associated with intravascular macrophages, especially in the lung and liver. Within the limits of fluorescence microscopy, liposomes and liposome complexes appear to be retained within the vasculature, except in the spleen, where they extravasate in the marginal zone of the white pulp. A comparison of many microvascular beds in a variety of organs revealed both organ- and vessel-specific differences in liposome uptake (Figs. 1–3).

The greatest amount of uptake is seen in endothelial cells of the microvasculature of the lung, lymph nodes (including Peyer's patches), ovary, anterior pituitary, and adrenal medulla. An intermediate amount of uptake is seen in endothelial cells of vessels in kidney medulla, thymus, uterus, acinar pancreas, adrenal cortex, adipose tissue, median eminence of hypothalamus, trachea, skeletal muscle, posterior pituitary, heart, and liver sinusoids. Sparse or no uptake is seen in endothelial cells of urinary bladder, pancreatic islets, kidney cortex, thyroid, spleen, intestinal smooth muscle, choroid plexus, and most of the brain (McLean et al., 1997). Capillaries and venules show much greater uptake than arterioles in those tissues in which uptake is present. The pattern of uptake of free cationic liposomes (not complexed with DNA) resembled that for corresponding liposome/DNA complexes when either the lipid or DNA was labeled. There was sometimes a difference in the magnitude of uptake between free liposomes and complexes, but the labeling pattern and rank order of tissue uptake was very similar or identical, and these slight differences may result from differences in particle size. Similarly, the pattern of uptake of complexes in which the DNA was labeled with Cy3 was identical to that observed when liposomes were fluorescently labeled. These findings indicate that at least in the high liposome to plasmid DNA formulations we have investigated, the DNA does not seem to play a major role in the specific uptake and tissue distribution of cationic liposome complexes. This simplifies the interpretations of our data since the liposome/DNA formulations we use also contain liposomes not associated with DNA, although all DNA appears to be associated with liposomes.

This biodistribution of cationic liposomes does not match any previously identified property of the microvasculature, or the uptake of any other systemically delivered exogenous macromolecule or particle reported in the literature. Since the mechanism of uptake has yet to be elucidated, we can only speculate that perhaps different vessels express in varying amounts cell surface binding protein(s) or receptor(s) that mediate the binding and internalization of cationic liposomes. The biodistribution of liposomes in the blood vessels of lymph nodes dramatically shows the difference in uptake by endothelial cells in different parts of the vasculature. The uptake by endothelial cells of HEV was among the highest found in any tissue, but there was a very striking transition between the uptake in these cells and the sparse uptake in the capillaries that drained into HEV. Another striking contrast is seen in the pituitary. The capillary endothelial cells of the posterior pituitary have little fluorescence, whereas those of the anterior pituitary have abundant fluorescence.

The organ- and vessel-specific uptake of liposomes by vascular endothelial cells might be expected given the heterogeneity of endothelial cells. This heterogeneity has been defined in terms of cellular structure, function, antigen expression,

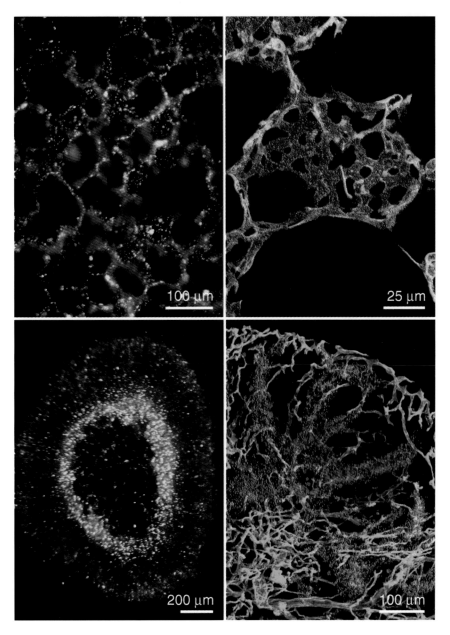

Figure 1 Micrographs showing uptake of fluorescently labeled DOTAP/cholesterol complexes 4 h after injection. Lungs (upper panels) show abundant punctate fluorescent labeling in alveolar capillaries. Adrenal glands (bottom left) have abundant labeling of microvessels in the medulla but not in the cortex. In cervical lymph nodes (bottom right), intense labeling is confined to high endothelial venules. Images in the right panels are from complexes containing Cy3-labeled DNA (golden yellow). Fluorescein lectin staining of vessels (green) is superimposed in some images.

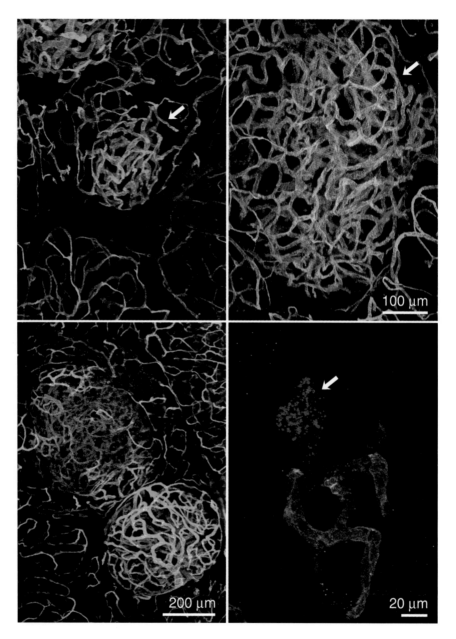

Figure 5 Confocal micrographs of angiogenic blood vessels in tumor islets 4 h after injection of fluorescently labeled DOTAP/cholesterol complexes. A normal islet (arrow, upper left panel) shows little uptake of fluorescent complexes compared to the vessels in a tumor islet (upper right). Hyperplastic islets (lower left) show uptake only in some vessels. A higher magnification view of a hyperplastic islet (lower right) shows focal sites of liposome uptake associated with a vessel protrusion (arrow) that ends blindly.

and response to stimuli, but the molecular bases for these differences have not been well characterized (Belloni *et al.*, 1992; Gerritsen, 1987; Gumkowski *et al.*, 1987). Vascular beds in specific organs are responsible for such diverse functions as solute and gas exchange and the binding and migration of lymphocytes. Given these widely diverse functions, the differential binding, internalization, and possible differences in transgene expression in different vascular beds should not be surprising.

When tissues are examined 4 h after injection, the overall distribution of fluorescence is similar to that observed at earlier times, but the fluorescence is now punctate. When examined by confocal microscopy, these dots appear to be beneath the luminal surface (where the fluorescent lectin binds) and inside endothelial cells. The fluorescent intensity of individual fluorescent dots at 4 h is greater than the clusters present at 5 min, but the overall amount of fluorescence associated with endothelial cells is about the same at these two time points. This intracellular translocation is in agreement with electron microscopic observations of lung capillaries, where at 20 min after injection, complexes could be seen in coated invaginations of the plasma membrane, in coated vesicles, and in endosomes or lysosomes of endothelial cells (Fig. 4).

At 24 h after injection of fluorescent liposome complexes the tissue-labeling pattern is generally similar to that at earlier time points, but with a somewhat decreased fluorescent intensity. However, some tissues such as mesenteric adipose tissue show a large decrease in microvasculature labeling, probably a result of the metabolism of the fluorophore.

Although the liver and spleen contain abundant fluorescence, most is associated with macrophages. In the liver the resident macrophages (Kupffer cells) contain most of the fluorescence, with only a small amount associated with the sinusoidal endothelial cells (Fig. 3). When 500 nm fluorescent polystyrene beads are coinjected with complexes, intracellular granules in Kupffer cells containing colocalized fluorescent complexes and polystyrene beads can be seen. At least some of the uptake by macrophages is probably by phagocytosis as this mechanism has been demonstrated previously for the clearance of liposomes from the circulation (Mahato *et al.*, 1995). Liposome fluorescence is not associated with hepatocytes, consistent with the complexes being too large to transverse the fenestrae of sinusoidal endothelial cells. Complexes that extravasate in the spleen impart an intense fluorescent to the marginal zone of the white pulp and to scattered regions of the red pulp (Fig. 3). These same regions are also labeled by extravasated fluorescent lectin, indicative of their inherent leakiness. The extravasated complexes are extracellular at 5 min, but by 4 h they are internalized by splenic cells in the region of extravasation, as evidenced by the appearance of punctate intracellular fluorescence. Many of the cells that internalize complexes also take up 500 nm fluorescent beads, again suggesting that they are macrophages. Only a small amount of labeling of endothelial cells is seen in splenic vessels at the perimeter of the white pulp.

Intravascular leukocytes containing fluorescent liposomes are found in several

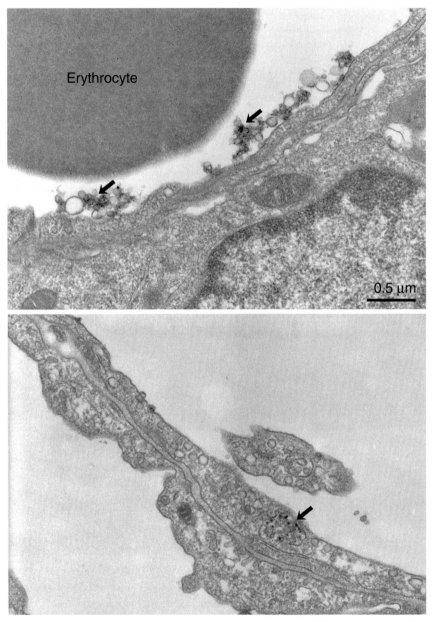

Figure 4 Electron micrographs of lung capillaries 5 and 20 min after intravenous injection of DOTAP/cholesterol liposomes labeled with an electron-dense Nanogold lipid. At 5 min (upper panel), clusters of liposomes (arrows) are attached to the endothelial surface. By 20 min (lower panel), clusters of liposomes can be seen inside alveolar endothelial cells in membrane-bounded vesicles.

organs 4 h after injection, and can be colabeled with 500 nm fluorescent beads injected at the same time as complexes. These labeled intravascular leukocytes are particularly abundant in venules of the urinary bladder and uterus. They are also prominent in the microvasculature of lung, lymph nodes, pancreas, and adipose tissue. The fluorescently labeled intravascular leukocytes are not present in most organs at 5 min after injection of complexes, and they are only seen at later times. We do not yet know at what point these cells take up fluorescent complexes or become associated with the vessel wall.

No evidence was found for the distribution of complexes being determined by the obstruction of the lumen of vessels, and vessel patency was confirmed by the uniform staining of perfused lectin in all the tissues that were examined. Electron microscopic analysis of lung capillaries at 5 min shows clusters of complexes on the luminal surface of the endothelium, but of a size that would not cause vessel blockage (Fig. 4). Even though the lung is the first microvascular bed through which all liposome complexes must pass when they are injected via either tail or femoral vein, a "first-pass" phenomenon cannot alone account for the large amount of uptake. HEV in lymph nodes throughout the body are at least as highly labeled as lung when uptake is considered per endothelial cell, and HEV are perfused with blood that has already passed through the lung microvasculature.

V. LIPOSOME UPTAKE IN ANGIOGENIC BLOOD VESSELS

The avid uptake of liposomes and liposome/DNA complexes in the blood vessels around large ovarian follicles prompted further examination. Since these are sites of physiologically normal angiogenesis, we decided to investigate the uptake of cationic liposomes in blood vessels in sites of pathological angiogenesis such as encountered in tumors and chronic inflammation.

We initially looked at tumors in a well-characterized transgenic mouse model (RIP-Tag2 C57BL/6) that develops pancreatic islet tumors (Thurston et al., 1998). The β cells of the pancreatic islets in these mice express the SV40 T antigen driven by the rat insulin gene promoter (Hanahan, 1985). In this animal model, various stages of tumor development are present concurrently in a single pancreas, enabling a side-by-side comparison of normal, hyperplastic, and tumor islets (Folkman et al., 1989). Normal islets have very little uptake of fluorescent liposomes or liposome/DNA complexes at 20 min or 4 h after intravenous injection. In contrast, the vasculature of RIP-Tag2 tumors has intense fluorescence at both time points, averaging 33 times more than in control vessels at 4 h (Fig. 5). In other organs of these transgenic animals the pattern of uptake is indistinguishable from corresponding wild-type mice. The fluorescent intensity in tumors correlates with their size, with the fluorescent liposomes being fairly uniformly distributed in the tumor vasculature. In hyperplastic islets, there are focal areas of liposomal fluorescence, many of which are associated with vessel protrusions that appear to end blindly. These protrusions may be vascular sprouts involved in the formation of new vessels.

When tumor sections are examined by confocal microscopy, the fluorescent liposomes and liposome complexes appear to be closely associated with the lectin-stained endothelium. No fluorescence is seen in the surrounding tissue, consistent with uptake by cells only in the vessel wall, and not by extravasation though a "leaky" endothelium. This was confirmed by electron microscopic examination following injection of gold-labeled cationic liposomes and complexes into tumor-bearing animals. After 20 min, liposomes are found mostly on the luminal surface of the endothelium and inside endothelial cells in multivesicular bodies or small vesicles. Some liposomes are also found on the abluminal side of the endothelium, but are confined to the basement membrane.

Chronic inflammation is another pathological situation associated with angiogenesis. Chronic airway inflammation of rodents infected with *Mycoplasma pulmonis* is associated with extensive vascular remodeling. Vessels in the tracheal mucosa of pathogen-free C3H/HeN mice have very little fluorescence at 20 min or 4 h after intravenous injection of cationic liposomes or complexes. However, mice infected with *M. pulmonis* for 4 weeks have conspicuous uptake of fluorescent liposomes or liposome/DNA complexes at these time points, averaging 15-fold more than mucosal vessels in control animals at 4 h after liposome injection (Thurston *et al.*, 1998). The pattern of cationic liposomal uptake appears very similar to that seen in the RIP-Tag tumors. At 20 min after injection, fluorescence liposomes coat the luminal surface of some vessels, followed by internalization and the appearance of punctate intracellular fluorescence at 4 h. These observations are supported by electronic microscopic observations of liposomes in endosomes and lysosomes in similarly treated animals.

The avid uptake by endothelial cells of angiogenic blood vessels appears to be restricted to cationic liposomes and cationic liposome/DNA complexes. The addition of DNA to cationic liposomes does not change the pattern of uptake, and this is not due to uncomplexed liposomes mediating the effect, since liposomes complexed with Cy3-labeled DNA behave identically. We also found no specific uptake of Texas Red-labeled anionic, neutral, or sterically stabilized neutral liposomes by blood vessels of RIP-Tag tumors, ovaries, or other organs at 20 min after intravenous injection.

VI. TISSUE-SPECIFIC EXPRESSION OF REPORTER GENES

Although the delivery of DNA to cells via liposomal complexes is necessary for transfection, other factors operate to determine the level of expression of reporter transgenes. We have used several different reporter genes, including chloramphenicol acetyltransferase and luciferase, linked to the CMV immediate-early enhancer/promoter. Transgenic mice are known to express transgenes linked to this promoter in a wide range of tissues, and it is the promoter of choice when pan-tissue expression is desired (Furth *et al.*, 1991; Schmidt *et al.*, 1990). For quantitation of gene expression we have relied primarily on the plasmids encoding luciferase. This reporter has several advantages, including the availability of a very sensitive luminometric assay and the absence of any background activity in mammalian tissues. In addition, the short half-life of luciferase in mammalian cells (Thompson *et al.*, 1991) provides an excellent indicator of the time course and duration of transcription and subsequent transgene synthesis. Although CAT has been widely used as a reporter for transfection, the longer half-life of this enzyme (Thompson *et al.*, 1991) and the high levels of CAT found in the plasma of transfected animals confound the interpretation of some expression data.

When cationic liposome luciferase plasmid complexes are injected via the tail vein, many tissues show detectable luciferase activity, but the principal site of expression is the lung (Fig. 7). In lungs, transgene expression per mg of tissue protein is at least 100 times greater than in any other organ examined at 24 h.

Although alveolar capillaries are sites of avid uptake of liposome complexes, tissues such as lymph nodes, adrenals, and ovaries also contain vessels that show comparable endothelial cell uptake. These latter tissues show much lower levels of expression, averaging four to five orders of magnitude lower than lung when expression is considered on a protein basis. Thus luciferase expression does not appear to be directly correlated with endothelial cell liposome complex uptake in these tissues. These differences in expression are evident when tissue luciferase expression

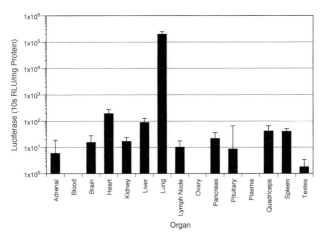

Figure 7   Luciferase expression in organs at 24 h after the intravenous injection of 60 μg of luciferase plasmid complexed to 1440 nmoles of DOTAP/cholesterol (1/1) liposomes (mean $\pm$ SD, $n = 6$). No detectable expression above background was seen in blood, ovary, or plasma.

is measured at either 5, 10 or 24 h after animals are injected, and together with the short half-life of luciferase is probably a reasonable indication of the differences in tissue transgene synthesis.

Although the binding and internalization of complexes look identical at a cellular level by microscopy in these tissues, the ability of intact DNA to dissociate from complexes and enter the nucleus where it is transcribed must be markedly different. These results are similar to that observed *in vitro*, where liposome transfections of cultured cell lines often show little correlation between DNA uptake and transgene expression (Coonrod *et al.*, 1997).

An examination of the time course of expression of luciferase in lung tissue shows detectable expression only 60 min after injection, with a rapid rise in levels reaching a maximum at 8–10 h (Fig. 8). Thereafter, lung luciferase levels fall rapidly to approximately 15% of peak levels by 24 h and to 0.4% by 48 h. These data indicate that plasmid DNA is able to reach the nucleus soon after cellular uptake, but transcription is only transient, probably for less than 10 h, after which the levels of tissue luciferase reflect translation of preexisting luciferase mRNA, with the subsequent decline due to the turnover of luciferase mRNA and protein. We do not know whether the transient nature of transgene transcription can be prolonged or what mediates this phenomenon. Plasmid DNA may become degraded or modified in the nucleus so that it is no longer transcriptionally active. The transient nature of expression is not confined to the lung, since other tissues also exhibit this phenomenon.

There is marked variability in luciferase expression among animals injected identically with luciferase complexes. A fivefold variation in expression is not uncommon, and there seems to be no consistent correlation between expression in different organs in the same animal. However, the uptake of fluorescent liposomes

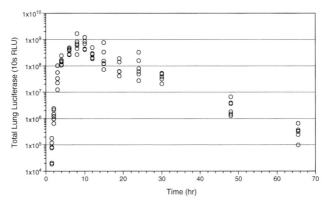

Figure 8   Time course of expression in lungs after the intravenous injection of 30 μg of luciferase plasmid complexed to 540 nmoles of DDAB/cholesterol (1/1) liposomes ($n = 6$ per time point).

or complexes does not show the same degree of variability. At present it is difficult to understand how littermates that appear to have essentially identical tissue amounts of uptake of complexes can have such markedly different levels of transgene expression.

The localization and identification of luciferase-expressing cells within organs was investigated by immunohistochemistry on tissue sections collected at 5–24 h after injection of complexes. Isolated areas of cellular transfection are seen in some sections, with the number of expressing cells per section correlating roughly with the amount of expression determined by luminometry for the corresponding organ from which the tissue was taken. Lung sections contain the largest number of expressing cells, with a range of bound antibody indicating a variability in the amount of luciferase expression between transfected cells (Fig. 6). Lung sections from identically treated animals also show an approximate fivefold variability in the number of expressing cells, consistent with the expression data obtained by luminometry.

As might be expected from the luminometry data, only a few scattered transfected cells are seen in liver, spleen, or heart sections, consistent with the greater than 100-fold less expression seen in these organs compared to lung. Other tissues routinely show no immunohistochemically positive cells, although this must be due to the detection sensitivity of the procedure in contrast to the much more sensitive luminometric assay. Because of the threshold of expression necessary for immunohistochemical detection, we cannot at present estimate the number of cells that may express the transgene. Although a very much larger number of alveolar capillary endothelial cells internalize liposome complexes than express immunohistochemically detectable transgene, we cannot be sure that a large number of cells have expression levels that fall below the detection threshold.

No reporter gene expression is seen in liver Kupffer cells, although these cells avidly internalize liposome complexes. Since these cells have specialized mechanisms for degrading phagocytosed material, the intracellular metabolism of liposome complexes might be different than in other cells.

We have also sought to identify by immunohistochemistry the cell type(s) in the lung that express transgenes from intravenously delivered liposome complexes. The colocalization of expressing cells with lectin binding on capillary walls is indicative of endothelial cells, and expressing cells do not bind antibodies for cell surface markers specific for macrophages or other leukocytes.

VII. IMPLICATIONS FOR THE USE OF CATIONIC LIPOSOMES FOR DNA DELIVERY

These studies have helped to refine our understanding of the cellular targets for intravenously administered cationic liposomes and liposome/DNA complexes.

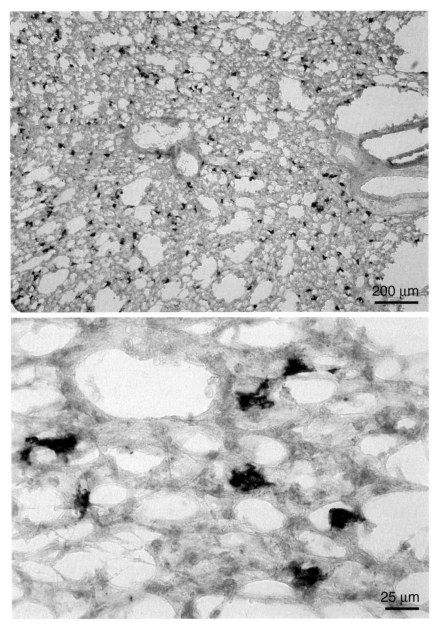

Figure 6 Luciferase expression in the lung microvasculature 10 h after injection of DOTAP/choles-
terol luciferase complexes. Immunohistochemistry for luciferase was performed on 20 μm cryostat sections of lung using an alkaline phosphatase-conjugated secondary antibody and BCIP/NBT substrate.
The upper panel is a low-magnification image showing uniform distribution of sites of expression
throughout the lung, and the lower panel is a higher-magnification image showing the polymorphic
nature of the cells expressing luciferase.

At least for the lipid and DNA formulations that we have investigated, the cells that have the ability to take up complexes from the circulation include a subpopulation of endothelial cells, intravascular leukocytes, macrophages, and extravascular cells in the spleen. These findings are consistent with the known size constraints for passage across that endothelial barrier. While these findings impose some limitations, they also offer some interesting possibilities for cationic liposome-mediated gene delivery. Of the cell types that take up liposome complexes, we have only been able to demonstrate transgene expression unequivocally by immunohistochemistry in endothelial cells. Since cationic liposomes are also able to complex with other polyanions such as oligonucleotides and ribozymes, these latter compounds might also be useful for gene therapy in those cells that do not appear to support transgene expression from the lipids and DNA formulations that we have used.

Potential applications of systemically delivered liposome complexes would therefore include altering endothelial cell and macrophage biology. Atherosclerosis, inflammation, immunity, or tissue remodeling are some of the candidate processes that may be amenable to modulation. Another obvious application would be the production of secreted proteins that act systemically, where the site of synthesis is irrelevant for their action.

Further studies are necessary to determine the cellular mechanisms that determine the levels of transgene expression from internalized liposome/DNA complexes. Several events need to be understood, such as how and where DNA and lipid dissociate once inside the cell, how the DNA translocates to the nucleus, and what determines the efficiency of these processes in different cell types. The transient nature of transgene transcription that we have observed with our formulations also needs further investigation if prolonged expression is desired. When we have a better grasp of these processes, we will have a better chance of modulating transgene expression *in vivo*, and perhaps be able to increase the amount and duration of expression comparable to levels being achieved with viral delivery systems, without the concomitant problems associated with those systems.

The novel finding of preferential uptake of cationic liposomes by endothelial cells in angiogenic blood vessels raises some exciting possibilities, particularly where inhibition of angiogenesis might halt the progression of a disease. The growth of tumors and the perpetuation of inflammation may well depend on continuing angiogenesis, and the inhibition of new vessel formation has been suggested as a powerful therapeutic strategy (Folkman and Ingber, 1992). The preferential uptake raises the possibility of using cationic liposomes to target not only DNA but also diagnostic or other therapeutic agents selectively to angiogenic blood vessels in tumors and sites of chronic inflammation.

We still do not understand how angiogenic endothelial cells bind and internalize cationic liposomes or whether this is a universal property of all angiogenic endothelial cells. It is known that these cells express distinctive cell surface markers, including oligosaccharides, adhesion molecules, and receptors, and the feasibility of using specific antibodies to target these cells has already been demonstrated (Sipkins *et al.*, 1998). Whether one or more of these molecules on the cell surface is responsible for the liposomal uptake remains an intriguing possibility. Angiogenic endothelial cells may upregulate the expression of those cell surface molecules that are responsible for the high uptake in other vascular beds, or they may express proteins that are not on the surface of normal endothelial cells.

Several considerations arise when considering the use of systemically delivered cationic liposome complexes for gene therapeutic applications in humans. Of primary concern is whether the amounts and duration of transgene expression are sufficient to elicit the required biological responses. The doses of DNA and lipid that we have used in mice would be unacceptably large if scaled according to human body mass. If the amount of transgene expression can be increased with the development of new cationic lipid formulations and vector systems, or strategies are found to increase the nuclear uptake and stabilization of internalized DNA, these problems should be surmountable. Another concern is whether transgene expression can be sufficiently restricted to desired cellular targets by the use of specific promoters, in conjunction with the known range of *in vivo* cellular targets for cationic liposomes.

We are still in the beginning stages of understanding the molecular mechanisms regulating the uptake and expression of DNA delivered by cationic liposomes. As our understanding increases through further research, new insights into strategies for the therapeutic utilization of these systems should ensue.

ACKNOWLEDGMENTS

The authors' research was funded in part by NIH Program Project Grants HL-24136 and HL-59157 from the NHLBI.

REFERENCES

Belloni, P. N., Carney, D. H., and Nicolson, G. L. (1992). Organ-derived microvessel endothelial cells exhibit differential responsiveness to thrombin and other growth factors. *Microvasc. Res.* **43**, 20–45.

Coonrod, A., Li, F. Q., and Horwitz, M. (1997). On the mechanism of DNA transfection: Efficient gene transfer without viruses. *Gene Ther.* **4**, 1313–1321.

Doerschuk, C. M., Gie, R. P., Bally, M. B., Cullis, P. R., and Reinish, L. W. (1989). Platelet distribution in rabbits following infusion of liposomes. *Thromb. Haemostasis* **61**, 392–396.

Egilmez, N. K., Iwanuma, Y., and Bankert, R. B. (1996). Evaluation and optimization of different cationic liposome formulations for *in vivo* gene transfer. *Biochem. Biophys. Res. Comm.* **221**, 169–173.

Fasbender, A., Zabner, J., Zeiher, B. G., and Welsh, M. J. (1997). A low rate of cell proliferation and reduced DNA uptake limit cationic lipid-mediated gene transfer to primary cultures of ciliated human airway epithelia. *Gene Ther.* **4**, 1173–1180.

Folkman, J., Watson, K., Ingber, D., and Hanahan, D. (1989). Induction of angiogenesis during the transition from hyperplasia to neoplasia. *Nature* **339**, 58–61.

Folkman, J., and Ingber, D. (1992). Inhibition of angiogenesis. *Semin. Cancer Biol.* **3**, 89–96.

Furth, P. A., Hennighausen, L., Baker, C., Beatty, B., and Woychick, R. (1991). The variability in activity of the universally expressed human cytomegalovirus immediate early gene 1 enhancer/promoter in transgenic mice. *Nucleic Acids Res.* **19**, 6205–6208.

Gerritsen, M. E. (1987). Functional heterogeneity of vascular endothelial cells. *Biochem. Pharmacol.* **36**, 2701–2711.

Gumkowski, F., Kaminska, G., Kaminski, M., Morrissey, L. W., and Auerbach, R. (1987). Heterogeneity of mouse vascular endothelium. *In vitro* studies of lymphatic, large blood vessel and microvascular endothelial cells. *Blood Vessels* **24**, 11–23.

Hanahan, D. (1985). Heritable formation of pancreatic beta-cell tumours in transgenic mice expressing recombinant insulin/simian virus 40 oncogenes. *Nature* **315**, 115–122.

Liu, F., Qi, H., Huang, L., and Liu, D. (1997). Factors controlling the efficiency of cationic lipid-mediated transfection *in vivo* via intravenous administration. *Gene Ther.* **4**, 517–523.

Mahato, R. I., Kawabata, K., Nomura, T., Takakura, Y., and Hashida, M. (1995). Physicochemical and pharmacokinetic characteristics of plasmid DNA/cationic liposome complexes. *J. Pharm. Sci.* **84**, 1267–1271.

McDonald, D. M. (1988). Neurogenic inflammation in the rat trachea. I. Changes in venules, leucocytes and epithelial cells. *J. Neurocytol.* **17**, 583–603.

McLean, J. W., Fox, E. A., Baluk, P., Bolton, P. B., Haskell, A., Pearlman, R., Thurston, G., Umemoto, E. Y., and McDonald, D. M. (1997). Organ-specific endothelial cell uptake of cationic liposome-DNA complexes in mice. *Am. J. Physiol.* **273**, H387–404.

Reinish, L. W., Bally, M. B., Loughrey, H. C., and Cullis, P. R. (1988). Interactions of liposomes and platelets. *Thromb. Haemostasis* **60**, 518–523.

Schmidt, E. V., Christoph, G., Zeller, R., and Leder, P. (1990). The cytomegalovirus enhancer: A pan-active control element in transgenic mice. *Mol. Cell. Biol.* **10**, 4406–4411.

Sipkins, D. A., Cheresh, D. A., Kazemi, M. R., Nevin, L. M., Bednarski, M. D., and Li, K. C. (1998). Detection of tumor angiogenesis *in vivo* by alphaVbeta3-targeted magnetic resonance imaging. *Nat. Med.* **4**, 623–626.

Song, Y. K., Liu, F., Chu, S., and Liu, D. (1997). Characterization of cationic liposome-mediated gene transfer *in vivo* by intravenous administration. *Human Gene Ther.* **8**, 1585–1594.

Templeton, N. S., Lasic, D. D., Frederik, P. M., Strey, H. H., Roberts, D. D., and Pavlakis, G. N. (1997). Improved DNA: liposome complexes for increased systemic delivery and gene expression. *Nature Biotechnol.* **15**, 647–652.

Thompson, J. F., Hayes, L. S., and Lloyd, D. B. (1991). Modulation of firefly luciferase stability and impact on studies of gene regulation. *Gene* **103**, 171–177.

Thurston, G., Baluk, P., Hirata, A., and McDonald, D. M. (1996). Permeability-related changes revealed at endothelial cell borders in inflamed venules by lectin binding. *Am. J. Physiol.* **271**, H2547–H2562.

Thurston, G., McLean, J. W., Rizen, M., Baluk, P., Haskell, A., Murphy, T. J., Hanahan, D., and McDonald, D. M. (1998). Cationic liposomes target angiogenic endothelial cells in tumors and chronic inflammation in mice. *J. Clin. Invest.* **101**, 1401–1413.

PART III

Other Vectors

CHAPTER 7

Nuclear Transport of Exogenous DNA

Magdolna G. Sebestyén* and Jon A. Wolff[†]

*Department of Pediatrics, University of Wisconsin–Madison, Madison,
Wisconsin

[†] Departments of Pediatrics and Medical Genetics, University of Wisconsin–
Madison, Madison, Wisconsin

Almost all the nonviral gene delivery protocols have to defeat a common
enemy: the nuclear envelope. The DNA carrying the therapeutic gene must reach
the nucleus to be transcribed. This nuclear import step has proven to be a major
barrier. In spite of its fundamental role in the advancement of gene therapy, the
nuclear transport of DNA has not won enough attention. To inspire more scientists
to devote their efforts to this field it is timely to review and discuss the data and

Nonviral Vectors for Gene Therapy

hypotheses accumulated during the last decade about the DNA nuclear transport itself and about various attempts to enhance it. This chapter also considers new gene delivery techniques developed by the evolution of viruses. Some of these fascinating viral strategies may one day be harnessed for the betterment of gene therapy protocols.

I. INTRODUCTION

The promise of gene therapy has ignited an immense interest in developing various gene delivery methods, each of which has its advantages and limitations. A common challenge for these gene transfer processes is that the therapeutic gene must cross several barriers before reaching the nucleoplasm, where transcription can finally take place using the host cell's transcription machinery. Recombinant DNA molecules constructed for gene therapy applications carry all the information required for gene expression in a mammalian cell. For effective gene delivery we should coach this DNA to enable it to jump all the hurdles and successfully cross the finish line in the "race" to the nucleus. To achieve this objective most of the protocols have a common trait: they wrap or complex the DNA molecules with either biological or chemical carriers to protect, guide, and help them during the race.

This book is dedicated to nonviral approaches: the use of different chemical carriers to form artificial microparticles of a great variety. Similarly to viral vectors, these particles are able to increase the efficiency of different steps (1) by targeting the DNA to the proper cell type and inducing its cellular uptake; (2) by changing some physical characteristics of the DNA (charge, structure, size, etc.); and (3) by protecting it from degradation.

In 1990, our studies revealed that injection of naked DNA into mouse skeletal muscle resulted in marker gene expression with efficiency comparable to transiently transfected fibroblasts (Wolff *et al.*, 1990). This discovery launched a new delivery method for certain applications, in which the DNA is not protected by any carrier, but rather runs the hurdles by itself. Recently we have achieved high levels of expression in skeletal muscle cells and hepatocytes by delivering naked DNA intravascularly under high pressure (Budker *et al.*, 1998; Budker *et al.*, 1996; Zhang *et al.*, 1997).

Naked DNA delivery raised the same kind of questions as any other direct gene delivery method: questions concerning cellular uptake, cytoplasmic and nuclear transport, the stability, safety, and immunogenicity of the vector (in this case the DNA itself), and, of course, efficacy. While the mechanism by which naked DNA traverses the cytoplasmic membrane is unknown, the past few years have seen a lot of progress in several laboratories working toward understanding the mechanism of DNA nuclear uptake. The goals of this chapter are (1) to review and discuss the data and hypotheses available today for the ultimate final step of any gene delivery method: the nuclear transport of exogenous DNA; and (2) to analyze some viral strategies that enable highly efficient nuclear delivery of viral genomes, making them desirable to mimic in order to dress up our naked DNA.

II. THE NUCLEAR TRANSPORT OF MACROMOLECULES

The study of the molecular mechanisms responsible for the highly organized nuclear import and export processes of a variety of molecules in eukaryotic cells has been an active field of cell biology in recent years. The elements and concepts of these transport pathways that are particularly relevant to DNA nuclear import are reviewed next.

In all eukaryotic cells the nucleus is confined within the double membrane layers of the nuclear envelope (NE), which is perforated by the nuclear pore complexes (NPC) to allow for molecular exchanges between the cytoplasmic and nuclear compartment. The NPC is a gigantic assembly of about 100 different polypeptides, many of which have been characterized (Pante and Aebi, 1996). Each component plays a well-defined role in either maintaining the structure or assisting the movement of different macromolecules and their carriers through the channel of the pore. The peripheral channels (with about 9–10 nm diameter) allow small solutes and macromolecules up to 50–60 kDa in size to freely diffuse in and out (except, when there is some kind of active retention on either side). Larger molecules need a nuclear localization signal (NLS) sequence in order to be actively transported through the central channel of the pore by the help of different transport factors. The number of characterized transport factors and distinct transport pathways that are specialized in the translocation of different cargo molecules grows steadily (Efthymiadis et al., 1998; Henderson and Percipalle, 1997; Pemberton et al., 1997; Pennisi, 1998). However, the pathway for which the most detailed information is available today is the so-called classical transport pathway of proteins carrying lysine- and arginine-rich (basic) NLSs. One of these typical signals is the SV40 large T antigen NLS (PKKKRKV (Kalderon et al., 1984)), which has been widely used both in fusion proteins and as a synthetic peptide conjugated to numerous different cargo molecules. Even colloidal gold particles as large as 15–25 nm diameter could be transported into the nuclei of HeLa cells when coated with SV40 NLS peptide-conjugated BSA, or with nucleoplasmin, an endogenous karyophilic protein (Dworetzky et al., 1988; Feldherr and Akin, 1990).

An in vitro nuclear transport system using digitonin permeabilized cells (Adam et al., 1992) has played an important role in elucidating the mechanisms of nuclear transport and has enabled the identification of several soluble factors such as the NLS receptor (karyopherin/importin α), p97 (karyopherin/importin β), p10

(NTF2), and the GTPase Ran (TC4). The NLS is recognized by the NLS receptor, which forms a complex with karyopherin-β. The latter enhances the affinity of the receptor for the cargo and docks the whole complex to the cytoplasmic filaments of the NPC. This first step of the transport process does not require ATP or GTP. However, in energy-depleted cells docking is not followed by the translocation of the cargo into the nucleus; thus, some steps of the transport are energy dependent. The source of energy, the steps actually coupled to NTP hydrolysis, and the proteins involved in it remain elusive. The small GTPase Ran has been shown to play a central role in almost all kinds of nuclear import and export pathways as a molecular switch (Goldfarb, 1997; Pennisi, 1998). Ran's activities are regulated by Ran-interacting proteins (RanBP1, RanBP2, RanGAP, and RCC1), the distribution of which is unequal on the two sides of the NE. As a result, these regulators maintain a steep concentration gradient of Ran/GTP and Ran/GDP on the two sides of the gate. The transport itself is a complex series of interactions between karyopherin-β (or its homologues), certain nucleoporins, p10, Ran/GDP, or Ran/GTP and its regulators. The molecular details of these interactions are still not fully understood, but the directionality of transport is thought to be due to the increased amount of Ran/GTP in the nucleus and the increased amount of Ran/GDP in the cytoplasm (Gorlich *et al.,* 1996; Izaurralde *et al.,* 1997). Since there is a limited number of each active component of this transport machinery in the cell, the transport can be saturated and blocked by a large excess of any of its players, crucial domains, signals, or binding sites. The details of the different nuclear transport pathways are summarized in numerous recent reviews (Gorlich and Mattaj, 1996; Jans and Hubner, 1996; Nigg, 1997; Ohno *et al.,* 1998; Pennisi, 1998).

III. THE NUCLEAR TRANSPORT OF EXOGENOUS DNA

A. TECHNICAL CHALLENGES

The pathway of DNA nuclear transport can be studied by determining the cellular location of foreign DNA under various conditions. A critical challenge is that cellular localization studies are correlative and not completely revealing about causative relationships. Given that the nuclear entry of only a couple of plasmid DNAs is sufficient for expression, it is formidable to determine the exact pathway that is responsible for expression.

Studying the nuclear transport of DNA was attempted decades ago, when very little was known about the molecular mechanisms responsible for any intracellular transport processes. Autoradiography was used to determine the subcellular location of exogenous radioactive DNA, not taking into account the possibility of potential binding to the surface of the nucleus rather than intranuclear accumulation

(Hill, 1961; Somlyai *et al.*, 1985). Even with isolation of nuclei by cell fractionation, it is possible that the radioactive DNA is only associated with the outside of nuclei rather than the inside (Wienhues *et al.*, 1987). Also, standard labeling techniques such as nick translation or multipriming do not in fact label the plasmid DNA (pDNA) but produce DNA of varying sizes and branching. The transport of this radioactive DNA product may be dissimilar to that of pDNA.

Using labeled and intact pDNA for histologic studies can also be problematic. For example, bromo-deoxyuridine (BrdU)-labeled DNA was used to follow its pathway after calcium phosphate-mediated transfection (Coonrod *et al.*, 1997). The BrdU signal was detected in the nucleus after being detected in endosomes and lysosomes. Given that the majority of the DNA is probably degraded in the lysosomes, the BrdU label becomes dissociated from the DNA and therefore the nuclear BrdU signal does not necessarily represent DNA nuclear uptake. This may explain why the BrdU nuclear labeling was lost after 8 h.

Today the technical advances brought about by the development of confocal microscopy make it possible to discern whether the DNA is within the nucleus. For most of the nuclear transport studies using confocal imaging became a must. Sensitivity did not suffer compared to when radioactive labeling was used, since fluorescent probes used today for the direct labeling of dsDNA or oligonucleotide hybridization probes can detect minuscule amount of target DNA (or RNA) in the nucleus (Femino *et al.*, 1998).

Another method is to use pDNA expression as an indirect indication of nuclear transport since expression can only occur if the pDNA enters the nucleus. However, pDNA expression can be affected by other factors that influence transcription and the nuclear location of transcription factors, mRNA nuclear export, and translation.

In summary, each of the above techniques has its limitations. If they are used together with an understanding of their pitfalls, then useful information concerning DNA nuclear transport can be learned.

B. DNA NUCLEAR TRANSPORT IS INEFFICIENT

The nuclear import step as a major barrier for gene expression has been recognized since the early 1980s. The direct delivery of DNA into the cytoplasm or nucleus of cells by microinjection is one of the most frequently used approaches in evaluating the importance of nucleocytoplasmic transport. With this approach the effects of other barriers (e.g., cellular uptake and endosomal release) can be eliminated. In 1980 Capecchi reported that 50–100% of thymidine kinase (TK)-deficient mouse fibroblast cells expressed TK after injecting plasmid DNA carrying the TK gene into their nuclei. Interestingly, cytoplasmic injection of over 1000 cells did not result in any detectable TK activity (Capecchi, 1980). Similar results were obtained by using human skin fibroblasts and a v-myc-expressing vector. Nuclear microinjection led to a high level of expression in approximately 30% of the cells, while cytoplasmic delivery initiated low-level expression only in about 2% (Mirzayans *et al.*, 1992). Rat embryo fibroblasts were also unable to express the β-galactosidase marker gene after cytoplasmic injection, although 60–70% of nuclear injections resulted in β-gal-positive cells (Thorburn and Alberts, 1993). Brinster and his co-workers used fertilized mouse eggs for the microinjection of linear and supercoiled plasmid DNAs of different sizes into either the egg cytoplasm or the male pronucleus. They concluded that integration of the DNA after cytoplasmic delivery is a very rare event (Brinster *et al.*, 1985), presumably due to inefficient nuclear uptake.

When the expression of the marker gene is driven by the T7 promoter and a source of cytoplasmic T7 polymerase is provided, transcription can take place in the cytoplasm (Deng *et al.*, 1991; Deng and Wolff, 1994; Elroy-Stein *et al.*, 1989; Gao and Huang, 1993). Studies using such T7 promoter-based plasmid expression vectors suggested that DNA nuclear transport is also rate limiting for cationic lipid-mediated transfection (Zabner *et al.*, 1995).

The importance of cell division for efficient gene transfer has been postulated in several studies. Quiescent or terminally differentiated cells are usually more difficult to transfect than dividing cells. Based on data previously mentioned, Zabner and his co-workers hypothesized that microinjected *Xenopus* oocytes showed absolutely no expression due to the lack of cell division, while the Cos cells used in parallel experiments expressed cytoplasmically delivered DNA because they were dividing (Zabner *et al.*, 1995). In a more recent publication, the same group addressed the importance of division in effective gene transfer into airway epithelia. They identified dividing cells by BrdU incorporation and concluded that BrdU-positive cells expressed the marker gene 10 times more frequently than BrdU-negative cells (Fasbender *et al.*, 1997). What these experimental data do not reflect is the mechanism by which a cell division can enhance expression. Putatively, when the nucleus is reorganized after the anaphase, the pDNA has a chance to sneak in. Another alternative is that the DNA enters the nucleus through the nuclear pore complexes even in a dividing cell, but the efficiency of this transport is increased because of cell cycle-dependent changes either in the structure of the NPC or the abundance or activity of factors mediating the transport. The cell cycle-dependent permeability of the NPC has been suggested by a study performed with NLS-coated colloidal gold particles of different sizes. Feldherr and Akin found that dividing cells could import larger particles: up to 230Å in diameter compared to 190Å in nondividing cells. Also, the relative import rate was sevenfold higher than in growth-arrested cells (Feldherr and Akin, 1990). Interestingly, when these growth-arrested (confluent) cells differentiated, their nuclear transport characteristics became similar again to those of the dividing cells. The authors proposed that physical changes in the pore size are responsible, possibly caused by alterations in the nuclear shape due to changes in the cytoskeletal and nuclear matrix structures; perhaps nuclear transport activities are affected by changes in overall energy levels in these dividing or differentiating cells (Feldherr and Akin, 1990). Further experiments will be needed to determine whether these changes have any relevance to the higher transfection efficiencies observed in actively dividing cells.

C. Plasmid DNA Is Able to Enter Postmitotic, Intact Nuclei

In 1995, our laboratory published the first evidence that cytoplasmically injected pDNA can enter postmitotic and thereby intact nuclei (Dowty et al., 1995). Expression levels of the β-galactosidase and luciferase marker genes increased with higher DNA concentrations and after microinjecting 10 mg/ml pSV2nLacZ DNA, 67% of the myotubes showed X-gal staining. Chilling the myotubes at 4 °C greatly reduced expression. Also, the well-known nuclear transport inhibitor WGA, which is thought to clog the channel by binding to N-acetyl glucosamine groups on nuclear pore proteins (Dabauvalle et al., 1988), inhibited expression. These data suggested transport through the NPC. Indeed, on electron microscopic images of myotubes microinjected with gold-labeled pDNA, gold particles could be observed within or near the nuclear pore complexes and inside nuclei (Dowty et al., 1995). In summary, these results indicate that pDNA has the ability to enter intact nuclei through the nuclear pore complex (Fig. 1).

D. Nuclear Transport of DNA in an *In Vitro* System

We used an *in vitro* nuclear transport system involving digitonin-permeabilized HeLa cells (Adam et al., 1992) and fluorescently labeled linear 1 kb dsDNA to further characterize the nuclear transport of DNA. The 1 kb DNA accumulated in the nuclei with a punctate staining pattern in the absence of any added cytosolic transport factors (Fig. 2J). While rabbit reticulocyte lysate (RRL), an exogenous source of soluble cytoplasmic proteins, is needed for nuclear uptake of NLS-containing proteins (Fig. 2A and 2B), the addition of RRL diminished DNA nuclear uptake (Fig. 2K). DNA nuclear transport was shown also to be mediated by the NPC, energy dependent and cold sensitive, and saturated by excess unlabeled DNA or excess RNA. Transport was unaffected by excess NLS-bearing proteins, suggesting that DNA may enter the nucleus by a distinct facilitated transport pathway. This model experiment also showed us that the nuclear uptake of DNA is size dependent

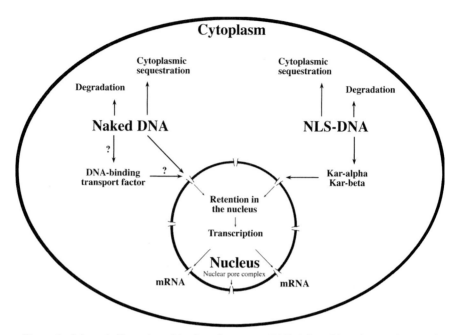

Figure 1   Schematic illustration of the fate of exogenous DNA delivered into the cytoplasm. Both naked and NLS-modified DNA can be degraded by nucleases and may be sequestered in the cytoplasm. These events greatly reduce the efficiency of gene delivery. Naked DNA has been shown to enter intact nuclei through the NPC (Dowty et al., 1995). In digitonin-permeabilized cells this transport occurs in the absence of known cytoplasmic transport factors, suggesting a direct, factor independent nuclear transport pathway for DNA (Hagstrom et al., 1997). The very low efficacy of nuclear uptake of plasmid DNA in intact cells also implies the lack of transport factors assisting DNA nuclear import, although the existence of proteins directly interacting with the DNA and the NPC cannot be excluded. In contrast, the NLS-conjugated DNA uses the same karyopherin-dependent pathway as karyophilic proteins do, resulting in immense nuclear accumulation in digitonin-permeabilized cells (Sebestyen et al., 1998).

and that linear dsDNA fragments greater than 2 kb were excluded from the nuclei of permeabilized cells (5 kb in Fig. 2G). This size constraint is consistent with the inefficient nuclear transport of the typical mammalian expression vector, which is above 5 kb. An important difference between this *in vitro* model and an intact cell is the presence of the unaltered, full cytoplasmic content of the latter. This must make some difference, even though digitonin has not been shown to disrupt the cytoskeleton at the concentration used for these studies (Fiskum et al., 1980). Microinjected intact HeLa cells do not accumulate dsDNA (1 kb or larger) in their nuclei to a microscopically detectable level (Fig. 2L and 2I, respectively), presumably due to some inhibitory effect presented by the intact cytoplasm (Fig. 1) (Hagstrom et al., 1997).

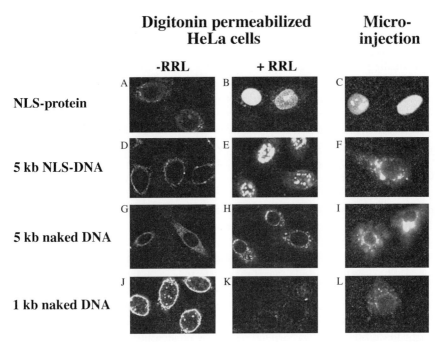

Figure 2   Summary of transport pathways in digitonin-treated and microinjected HeLa cells. NLS-bearing proteins (A–C) accumulate in the nucleus only in the presence of cytosolic transport factors provided either by RRL (B) or by the intact cytoplasm (C). The nuclear transport of large naked plasmid DNA remains undetectable under any of these conditions (G–I). The same DNA shows profound nuclear staining in the *in vitro* assay after being conjugated to NLS-peptides (E), although it remains mostly in the cytoplasm in microinjected cells (F). A short linear DNA fragment is able to enter the nuclei of digitonin-treated cells in the absence of cytosolic extract (−RRL; J), but its transport is inhibited in the presence of cytosol (K–L).

E. CYTOPLASMIC INTERACTIONS

The cytoplasm of mammalian cells itself may form a barrier for the entering DNA molecules. Its aqueous phase contains dissolved solutes and macromolecules at high concentrations, resulting in a slightly (approximately 4–5 times) decreased diffusion rate of even small molecules (Seksek *et al.*, 1997). Larger molecules like a pDNA must face other obstacles: the sieving effect of a crowded network of cytoskeletal filaments and the deflecting effects of frequent collisions with organelles and other large macromolecules (Luby-Phelps and Weisiger, 1996).

The highly organized architecture of the whole mammalian cell implies that molecules (even small ones), particles, or organelles are rarely allowed to move around freely, even if they are not compartmentalized. There are specific and well-regulated motor mechanisms for the directional movement of different cell components, usually along different cytoskeletal filaments. Besides the long-known protein-sorting phenomenon, mechanisms have been described also for the targeting of nucleic acids. mRNA molecules can be localized to specific subcellular regions to facilitate the sorting of their protein product. They are transported by specific mRNA-binding proteins that recognize a signal (usually in the 3' untranslated region of the message) and are able to move along the cytoskeleton to reach their destination (Hesketh, 1996; Hovland et al., 1996). At the present time there is no mechanism known to specifically move naked DNA within the cytoplasm. As we will elaborate later, some viral particles use cytoskeleton-related motor mechanisms to reach the close vicinity of the nuclear pore complexes (NPC), but in those cases the nucleic acid itself is presumably covered by core or capsid proteins, thus it is not available for direct interactions with cellular components.

A naked pDNA molecule of approximately 6–8 kb in size has a net negative charge around 10^4, and a radius of gyration of 100–120 nm (Fishman and Patterson, 1996). It is conceivable that it makes electrostatic interactions with the cell's polycations, which may result in complete charge neutralization and condensation. Theoretically, condensation of the DNA might help its cytoplasmic and/or nuclear transport (e.g., interaction with soluble, cytoplasmic histone proteins), but it also might inactivate the DNA, depending on the condensing agent. The DNA may also be actively sequestered in the cytoplasm by specific DNA-binding proteins (Fig. 1). Our observations after microinjection of labeled pDNA into intact myofibers and HeLa cells (Fig. 2I and 2L) suggest the existence of some kind of sequestration (Dowty et al., 1995; Hagstrom et al., 1997; Sebestyen et al., 1998), the nature of which has not yet been identified. In general, no specific molecular interactions have been pinpointed so far to either help or restrict the cytoplasmic transport of naked pDNA. Vimentin-containing intermediate filaments have been shown to bind oligonucleotides with varying affinity (depending on the nucleotide composition) (Hartig et al., 1997). However, the staining pattern of microinjected DNA sequestered in the cytoplasm does not resemble the staining pattern of the antibody-labeled vimentin network or of any other cytoskeletal filament network (Figs. 2I and 2L). It is also interesting that there are some muscle-specific proteins residing in the sarcoplasmic reticulum that bind double-stranded DNA (dsDNA) with high affinity (Hagstrom et al., 1996). Potentially, these may also play a role in sequestering DNA in muscle fibers.

IV. ATTEMPTS TO INCREASE THE EFFICIENCY OF NUCLEAR UPTAKE

It is well established that the attachment of NLSs to proteins not normally karyophilic enables their efficient nuclear targeting. This has inspired similar efforts to increase DNA nuclear transport by providing a NLS to the DNA. This has been attempted by noncovalently or covalently complexing the NLS-containing molecule with the DNA. Viral-derived sequences that can bind to NLS-containing proteins have been used as well.

A. Noncovalent Complexes

Kaneda *et al.* found that the high-mobility group-1 (HMG-1) protein increased the efficiency of complexes containing DNA-loaded liposomes, RBC membranes, and inactivated hemagglutinating virus of Japan (HVJ) (Kaneda *et al.,* 1989). After injection of these complexes into the portal vein of adult rats, complexes containing HMG-1 enabled an increased amount of nuclear-associated pDNA on Southern blot analysis and foreign gene expression in liver cells as compared to complexes containing bovine serum albumin. In Ltk cells in culture, HVJ liposome complexes containing HMG-1 protein (compared to complexes with IgG or no protein) increased the amount of nuclear-associated pDNA 6 h after addition of the complexes. Since no differences were noted by 24 h, the HMG-1 protein only increased the rate at which pDNA became nuclear associated but not the final amount of nuclear-associated pDNA. Nuclear localization was done using radioactive probes and *in situ* hybridization, but this technique cannot distinguish between exogenous DNA on the surface of the nucleus and DNA within the nucleus. The effect on nuclear association was only twofold, however, which is often within experimental error for most gene transfer systems. It would also have been of interest if a protein that was cationic but without a NLS was used for control purposes. Based on our experience with histone protein (see following), the effect of the HMG-1 protein may not be on the nuclear step of pDNA transport but rather on prior steps.

Collas and Alestrom have proposed that the noncovalent complexation of SV40 large T antigen NLS peptide with pDNA increases the nuclear targeting of the pDNA in sea urchin pronuclei and zebrafish embryos (Collas and Alestrom, 1996; Collas and Alestrom, 1997; Collas *et al.,* 1996). When $10-10^6$ molecules of a luciferase expression pDNA were injected into zebrafish embryo at the 1−2 cell stage, the addition of NLS peptide (100/1 and 1000/1 molar ratio of peptide to DNA) increased the percentage of embryos expressing luciferase. Enhanced expression only occurred when 10^3 or 10^4 DNA molecules were injected and the addition of the NLS peptide did not increase the percentage of positive embryo above the percentage achieved using 10^6 molecules without NLS peptide. PCR analysis also indicated that the NLS peptide enhanced nuclear uptake (Collas and Alestrom, 1996; Collas and Alestrom, 1997; Collas *et al.,* 1996). Although a control peptide (the NLS peptide sequence in reverse order) had no effect on pDNA uptake or expression, it lacked a cysteine residue. Since the NLS T antigen peptide had a cysteine, it most likely formed dimers that would interact with the DNA differently than the monomeric control peptide. The zebrafish embryos at the 1–2 cell stage contain rapidly dividing cells. Consequently, the overall level of luciferase activity may not have been augmented by increased DNA transport through the NPCs of intact nuclei. The NLS peptide might have helped the DNA gain access to re-forming nuclei of these rapidly dividing cells. From this point of view, the use of NLS peptides (or other oligocations) for transgenic studies might still be useful, enhancing the chance that more cells take up the DNA during the first few cycles of division.

Collas and Alestrom also observed increased physical uptake of DNA complexes with the NLS peptide in sea urchin pronuclei. However, the use of ethidium bromide for DNA labeling is problematic since it is efficiently transferred to excess unlabeled DNA or RNA. Studies in our laboratory indicated that the nuclear uptake in digitonin-treated HeLa cells of pDNA containing a covalently attached fluorochrome was not enhanced by the noncovalent association with SV40 T antigen NLS peptides. Perhaps the different cell types and assay methods are responsible for the discrepancy between their results and ours.

We found that a histone H1-SV40 NLS fusion protein substantially enhanced expression when added to a variety of cationic lipids and neutral or negative liposomes. Native histone H1 had a similar large effect on expression (Budker et al., 1997; Fritz et al., 1996; Hagstrom et al., 1996). However, these histone proteins did not increase the expression of pDNAs microinjected into the cytoplasm. This suggested that the enhanced expression was not due to increased nuclear transport but rather to increased efficiency of pDNA transport prior to nuclear entry. In fact, a variety of cationic proteins and polymers such as polylysine, spermine, and protamine increase the expression of pDNA/lipid complexes (Gao and Huang, 1996; Li and Huang, 1997). Another example is the SV40 T antigen NLS peptide-increased efficiency of DOTAP cationic liposome-mediated gene expression by increasing the delivery of pDNA to the cell (Aronsohn and Hughes, 1998).

Polyethylenimine (PEI) has been reported to enhance nuclear transport. Pollard and co-workers calculated a 10-fold increase in the efficiency of nuclear uptake of PEI-condensed plasmid DNA compared to naked DNA, based on the number of DNA copies needed to be delivered into the cytoplasm to yield the same percent of positive cells. Under optimal conditions the actual numerical increase was three- to fourfold. The PEI condensed the DNA into tight, spherical particles. The authors propose that these compact structures are responsible for enhanced nuclear targeting, although they also consider other potential functions of PEI, such as protection of the DNA from degradation in the cytoplasm and/or decreasing the chance of cytoplasmic retention (Pollard et al., 1998).

Recently, it was discovered that anti-DNA monoclonal antibodies are translocated from the extracellular space all the way to the nucleus (Avrameas et al., 1998). Peptides derived from these antibodies were also able to transport heterologous molecules to the nucleus. In fact, a peptide containing anti-DNA antibody sequences and 19 lysines mediated the transfection of pDNA.

In summary, several cationic peptides and proteins with or without NLSs enhance the expression of pDNA in conjunction with a variety of delivery systems, but it remains to be shown that they have this effect by increasing the nuclear transport of DNA.

B. Viral *cis* Sequences

SV40 and Epstein-Barr virus (EBV) *cis* DNA sequences have been reported to enhance the nuclear uptake of pDNA (Dean, 1997; Langle-Rouault *et al.*, 1998). Dean found that pBR322 plasmid containing the SV40 origin of replication and early and late promoters was transported to the nuclear rim by 2 h and into the nucleus by 8 h in several cell lines not expressing the SV40 T antigen. The cellular location of the pDNA was directly and physically assessed after cytoplasmic microinjection using fluorescent *in situ* hybridization (FISH) but the effect of these SV40 sequences on expression was not reported. Given the strong nuclear signal of these microinjected plasmids, there should be a large effect on expression. Plasmid expression vectors containing these SV40 sequences have been widely used and particularly strong expression of SV40 ori-containing plasmids has not been noted in T antigen-negative cells. Nonetheless, another study showed that the 72-bp SV40 early promoter enhanced chloramphenicol acetyl transferase (CAT) expression of plasmids microinjected into the cytoplasm of T antigen-negative cells (Graessmann *et al.*, 1989). In our laboratory the complete, supercoiled SV40 genome (fluorescently stained with a covalent but gentle DNA-labeling reagent) was not efficiently transported into the nuclei of digitonin-treated HeLa cells. Other expression constructs carrying the putatively crucial regions of the SV40 genome did not accumulate in the nucleus either.

Analogously, plasmids carrying the Epstein-Barr virus (EBV) oriP sequences enhanced 100-fold the expression of CMV promoter/luciferase expression vectors that were transfected with Lipofectin or microinjected into cells constitutively expressing the Epstein-Barr nuclear antigen 1 (EBNA1) (Langle-Rouault *et al.*, 1998). The oriP+ and oriP− constructs microinjected into the nuclei of these cells also showed a considerable 17-fold difference, suggesting a major role of oriP in the enhancement of transcription. In contrast to the SV40 DNA study previously mentioned, FISH did not detect any of these EBV oriP-containing plasmids in the nucleus.

Presumably, these *cis* viral sequences bind to NLS-containing proteins that enhance nuclear transport in a piggyback fashion. For example, EBNA1 can interact with karyopherin-α even after being bound to oriP-containing DNA (Fischer *et al.*, 1997; Kim *et al.*, 1997). In this case the nuclear transport is facilitated by an inherent viral protein, and this is a common feature of viral infection.

In summary, further studies are required to clarify the ability of these viral sequences to enhance DNA nuclear uptake. *Cis* viral sequences are not widely used currently for augmenting foreign gene transfer into the nucleus.

C. VECTOR CHEMISTRY: THE COVALENT MODIFICATION OF PLASMID DNA (pDNA)

The inability of noncovalent NLS to enhance DNA nuclear transport (in our laboratory) could be due to the weakness of the association of DNA with the NLS. We postulated that a stronger association would enable the NLS to enhance DNA nuclear transport. To this end a new conjugation technique has been developed (Sebestyen *et al.*, 1998) using a synthetic crosslinker, called CPI, and synthetic peptides carrying either the wild-type SV40 T antigen NLS, or mutant versions of it. Attaching a few peptides to a pDNA did not seem to cause any change in our *in vitro* nuclear import assay (Fig. 3A). However, above a certain threshold value, the DNA appeared in the nuclei of the digitonin permeabilized cells, and the more peptide was added, the more intense the nuclear uptake became (Figs. 3B and 3C). The punctate staining pattern resembled what we saw with small, linear naked DNA (Figs. 2E and 2J) (Hagstrom *et al.*, 1997). Since the attachment of similar peptides without a functional NLS did not initiate nuclear uptake while still resulting in the same level of charge neutralization and gel-shift, the transport must have been caused by the functional NLSs. Indeed, transport could be inhibited by adding excess NLS-bearing proteins to the assay, also indicating that the modified

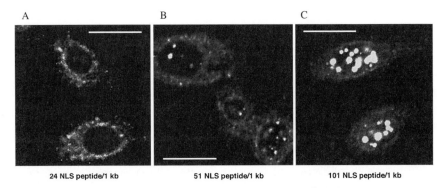

A	B	C
24 NLS peptide/1 kb	51 NLS peptide/1 kb	101 NLS peptide/1 kb

Figure 3 The intensity of nuclear transport of NLS-conjugated plasmid DNA is a function of the level of modification. Most of the DNA remains in the cytoplasm of digitonin–treated HeLa cells if fewer than 25 peptides are attached per 1 kb DNA (A), while 100 NLS per 1 kb drives almost all the DNA into the nucleus (C). Scale bars: 20 μm.

NLS DNA uses a different transport pathway than naked DNA (Fig. 1) (Sebestyen *et al.,* 1998).

The karyophilic NLS DNA has a major disadvantage: due to the covalent bonds on every 10th–20th nucleotide the DNA is completely inactive transcriptionally. To study the effect of the NLS conjugation on expression, hybrid molecules were created by ligating an unmodified expression cassette to a highly modified NLS-DNA fragment. Both the ligated hybrid and its unligated control resulted in a three- to fourfold increase in luciferase expression, compared to expression levels obtained with hybrids ligated to or mixed with unmodified fragments or mutant peptide conjugated fragments. We did not consider a three- to fourfold increase significant, even if numerical differences were statistically significant, since it did not reflect the increase in nuclear uptake observed visually in digitonin-permeabilized cells. Besides, in the unligated control the expression cassette had no covalent link to the NLS peptides; consequently that change must have been caused by an indirect effect of the NLS peptides, as discussed in Section IV.A. Since fluorescent plasmid DNA molecules highly modified with NLS peptides failed to accumulate in the nuclei of microinjected HeLa cells (Fig. 2F), the lack of a considerable increase in marker gene expression is not surprising. We assume that cytoplasmic sequestration abolishes the positive effect of the attached NLS peptides, and the efficiency of nuclear transport of the NLS-DNA in an intact cell is almost as low as that of naked DNA (Fig. 1 and Figs. 2F and 2I).

V. VIRAL STRATEGIES FOR DNA OR RNA NUCLEAR TRANSPORT

Viruses that replicate in the nucleus and are capable of infecting nondividing cells must deliver their genome into the host cell's nucleus at an early stage of infection. Recent progress in characterizing the molecular mechanisms of nuclear trafficking of macromolecules also triggered the unraveling of some fascinating strategies that viruses exploit to invade the nucleus of their host. Deciphering these strategies and understanding the differences between naked and viral DNA transport can give us new clues to manipulate the cytoplasmic and nuclear transport steps in many different nonviral gene delivery protocols. The following description of the early steps of the life cycle of certain animal and plant viruses is far from being complete. We tried to select examples from different virus families that represent the most efficient ways of intracellular and nuclear transport of their DNA or RNA genomes. Besides viral strategies, we also outline the fascinating maneuvers of a nonviral plant pathogen, *Agrobacterium tumefaciens,* since its transmissive pathogenic element, the Ti plasmid, greatly resembles the structure and function of viral nucleoprotein particles.

A. ANIMAL VIRUSES WITH DNA GENOME

1. Adenovirus

Adenoviruses form icosahedral virion particles of 70–100 nm in diameter, which consist of a protein shell (the capsid) surrounding the DNA-containing core. The dsDNA genome reaches the nucleus through a multistep process: attachment to cell surface receptors, endocytosis, entry into the cytosol by acidic pH triggered release, targeted intracellular transport to the nuclear pores and finally the "injection" of the genome into the nucleus. During these steps the viral particle weakens by gradually losing its structural elements (Greber *et al.*, 1994, 1993). The intact virion is assembled from nine different proteins and the DNA genome. The fibers (protein IV) and the penton base (III + IIIa) proteins initiate the uptake by endocytosis, and the penton base proteins are also thought to help the release of the particle from the acidified endosome. These two components are almost completely shed from the capsid by the time it gets released into the cytoplasm, about 15 min postinfection. Further dismantling occurs in the cytosol, where protein VI becomes degraded by a protease (Greber *et al.*, 1996), and soon protein VIII dissociates. At a later stage in the cytosol most of protein IX also detaches from the particle, leaving behind the DNA-containing core covered by a destabilized layer of the major building block of the capsid: the hexon (II) protein. Hexons remain attached and are thought to play a role in the directional intracellular transport of the core toward the nucleus, putatively by using the microtubular system (Dales and Chardonnet, 1973; Greber, 1996). It has been proposed that adenoviral particles move along microtubules toward the centrosomes. A viral protein (i.e., hexon protein) binds to the dynein/dynactin motor complex, a microtubule-associated molecular motor that moves organelles from the cell periphery toward the centrosome, which is often adjacent to the nucleus. Usually within 60 min postinfection, the particles accumulate around the nuclear envelope and on EM images they seem to interact with the nuclear pore complexes (Greber *et al.*, 1996). The final step is the release of the genome into the nucleus. Interestingly, for the final disassembly of the weakened capsid the participation of the NPC is also needed (Greber *et al.*, 1997), but the molecular details of this interaction have not yet been discerned. Finally, the linear dsDNA enters the nucleus together with multiple copies of protein VII and the covalently attached terminal protein. Possibly the transport itself is directed by the NLS of the terminal protein. Most of the hexon protein and other residual capsid proteins are left behind on the cytoplasmic side of the channel (Greber *et al.*, 1997).

There are two critical features of this life cycle that might be harnessed for nonviral gene delivery: (1) the targeting of the genome to the NPC in a protected, wrapped form, using microtubule-associated motor mechanisms, and (2) the presence of a covalently bound, NLS-carrying protein on the very end of the linear genome that putatively threads the linear DNA through the channel. It has been proposed that the adenoviral particles move along microtubules toward the centrosomes and then to the nucleus. If this viral–microtubule interaction does not require many viral contributors, the viral motif could be used to target nonviral vectors toward the nucleus. The role of protein VII in the actual transport step is unclear, although it might play a role in facilitating the process. Using adenoviral protein VII to form transfection competent DNA/protein complexes has been attempted (Wienhues *et al.*, 1987). Although the results of those experiments were dubious, they still show how long the idea of using viral strategies has existed. It would also be very exciting to see whether the terminal protein of adenovirus chemically attached to a linear, nonviral dsDNA fragment would initiate its nuclear uptake, even in the absence of the other components of the viral delivery system.

2. Hepatitis B Virus

Hepatitis B virus (HBV) is the prototypic member of hepadnaviruses. The 42-nm virions are wrapped in an envelope containing the small, medium, and large surface proteins, presumably responsible for receptor binding. Inside the envelope is the 28 to 32-nm icosahedral nucleocapsid core that is assembled from 180 subunits of the core protein, the 3.2 kb partially double-stranded DNA genome, and a reverse transcriptase/DNA polymerase protein, which is covalently attached to the genome by its terminal protein domain. Although the major steps of replication and virion assembly have been outlined (Nassal and Schaller, 1993, 1996), the molecular details of the hepatitis B life cycle remain enigmatic due to the lack of a susceptible *in vitro* model system. After receptor-mediated attachment to the surface of hepatocytes the virions enter the host cells by a largely uncharacterized step requiring proteolysis (Kann *et al.*, 1995). The core particles are released into the cytosol, but their fate there is also controversial. They are small enough to be transported into the nucleus through the NPCs as intact particles and to deliver the genome directly to the site of replication. Transport of recombinant core particles (CPs) through the NPC in yeast has been depicted by electron microscopic images (Yamaguchi *et al.*, 1994), while other investigators found that CPs cannot pass through the nuclear envelope (Guidotti *et al.*, 1994). Core particles accumulate in the nuclei of infected or transgenic mammalian cells (Guidotti *et al.*, 1994; Makarova *et al.*, 1994). However, since the core protein has an NLS, it can be transported into the nucleus after *de novo* synthesis, and CPs can be assembled within the nucleus. Thus, if intact CPs are ever migrating through NPCs, the direction of transport is thought to be outbound, delivering mature core particles from the nucleus into the cytoplasm in a late phase of infection rather than delivering the genome into the nucleus early after infection. Based on a recent study describing the nuclear uptake of the hepadnavirus genome in digitonin-permeabilized HuH-7 cells (Kann *et al.*, 1997), most of the incoming core particles reach the vicinity of the nucleus as intact particles. Once close to the NPC they disassemble and release the genome and the covalently DNA-bound polymerase. The putative NLS of the polymerase would then initiate the import of the partially double-stranded genome into the nucleus, where it is fully elongated and circularized.

This process is somewhat similar to the adenovirus life cycle previously described. In this case the destabilization of the core in the cytoplasm might be the result of phosphorylation of the core protein, the nature of which is unsettled. Protein kinase A and C and proline-directed cdc kinases have been postulated to play a role (Hild *et al.*, 1998; Kann and Gerlich, 1994; Kann *et al.*, 1995; Yu and Summers, 1994). Phosphorylation by PKC decreased the RNA/DNA binding capacity of the core protein (Kann and Gerlich, 1994); other protein kinases and phosphorylations probably weaken the interactions among the core proteins themselves. As a result the arginine-rich carboxy terminal domain originally bound to the RNA genome during core assembly becomes free, allowing for the dissociation of the core from the genome. The domain also contains an NLS motif, which may become exposed on the surface of the core particle. However, this NLS does not seem to be indispensable for nuclear transport, since purified polymerase/DNA complexes also accumulated in the nuclei of digitonin-permeabilized cells (Kann *et al.*, 1997). It cannot be excluded, however, that within an intact cell the core proteins do play a role in the directional transport of the particle. We would also like to emphasize that the fully de-proteinized HBV DNA remained completely in the cytoplasm of these digitonin-treated HuH7 cells (Kann *et al.*, 1997), which is in good correlation with our findings in digitonin-treated HeLa cells when using naked DNA of similar size (Hagstrom *et al.*, 1997; Sebestyen *et al.*, 1998). Our present knowledge is not sufficient to judge whether these steps of the hepatitis B life cycle would be the same in an intact hepatocyte after *de novo* infection; nevertheless, these experiments with the *in vitro* model system highlight some intriguing similarities between two distant groups of viruses.

3. Herpes Simplex Virus

Similar to adenoviruses and hepadnaviruses, herpes simplex virus (HSV) is able to infect nondividing cells by delivering its genome to the nucleus. This large and complex virus has (1) an outer envelope with characteristic spikes; (2) a capsid, containing a tightly packed core with the 152 kb linear, dsDNA genome; and (3) an amorphous tegument layer in between. The major stages of the life cycle have been well characterized (Roizman and Sears, 1993). Briefly, the first step is the attachment of the virion to heparan sulfate proteoglycans on the cell membrane, which triggers the subsequent fusion of the envelope with the plasma membrane. After entering the cytosol the capsid loses most of the tegument and within approximately an hour it becomes transported to the nuclear pore. As in the case of adenoviruses, it had been long postulated that microtubules mediate the intracellular transport of the HSV capsids, helping them move through long axons of peripherally infected neurons (Kristensson *et al.*, 1986; Penfold *et al.*, 1994; Topp *et al.*, 1994). Sodeik and her co-workers recently described the fascinating way these viruses take advantage of the host cell's molecular motors to reach their destination (Sodeik *et al.*, 1997). In HSV-infected Vero cells 15 min postinfection most of the capsids were still close to the plasma membrane, with their dense DNA core well visible. Two to four hours later the majority was located at the nucleus, more specifically at the NPCs, and most of the capsids looked empty, implying that they had released the DNA. It was also shown that the capsids were associated with microtubules and interacted with dynein. The dynein motor complex is a microtubule-dependent molecular motor that moves organelles from the cell periphery toward the centrosome (which is adjacent to the nucleus). By traveling along these tracks the capsids reached the centrosome, from where they were transported to the NPCs by another, yet unknown, mechanism (Sodeik *et al.*, 1997). The viral protein(s) responsible for the contact with dynein, the transport of the capsids from the centrosome to the NPCs, and details of the final step of releasing the DNA into the nucleus remain to be identified.

Two close relatives of HSV, Epstein–Barr virus and cytomegalovirus, have a mechanism for attachment, penetration, and uncoating similar to that of HSV. They both replicate in the nucleus; however, at present only little is known about the molecular details. A recent report claimed that the Epstein–Barr nuclear antigen 1 (EBNA1) is able to recognize the replication origin of the EBV even in another, exogenous pDNA. Increase of the nuclear uptake of pDNA is thought to be mediated by sequence-specific binding of the EBNA1 to the oriP and then initiating nuclear uptake by carrying a classical, karyopherin-α binding NLS sequence (Fischer *et al.*, 1997; Langle-Rouault *et al.*, 1998), These findings may have practical applications for gene delivery in the future.

4. Simian Virus 40

The simple, 45-nm icosahedral capsids of SV40 consist of only three proteins (VP1, VP2, and VP3), and contain the 5-kb circular, superhelical dsDNA genome in the form of a histone-condensed minichromosome. The efficiency of SV40 infection is very low, thus it may not provide a useful tool for the initiation of the nuclear uptake of exogenous DNA molecules, nonetheless it displays an interesting alternative pathway for the nuclear delivery of viral DNA. The virions enter the host cells by endocytosis, mediated by the attachment of the VP1 protein to cell surface receptors. Since intact viral particles were observed in the perinuclear cysternae on EM images (Griffith *et al.*, 1988), initially the nuclear entry of the virions was thought to happen by the fusion of late endosomes with the outer membrane of the nuclear envelope (Nishimura *et al.*, 1986). Later NLS sequences were identified on all three viral capsid proteins, and results from Harumi Kasamatsu's laboratory indicated a nuclear import process through the NPC, mediated by the capsid proteins. Their initial observations suggested the nuclear uptake of intact particles, followed by the dissociation of capsid proteins from the genome within the nucleoplasm (Clever *et al.,* 1991). This is an intriguing phenomenon, though: how can a 45- to 50-nm, fairly rigid icosahedral particle squeeze itself through the 25- to 30-nm central channel of the NPC (Davis, 1995; Pante and Aebi, 1996)? As Greber and Kasamatsu discussed later in their review, the process might involve some kind of conformational change or partial disassembly, also needed for exposing the originally hidden NLSs on the surface of the particle (Greber, 1996). If disassembly does happen, then the considerably smaller and probably more flexible histone condensed minichromosome would pass through the channel. VP3 has been implicated in accompanying it on this journey and providing it with NLSs (Nakanishi *et al.,* 1996)). The other capsid proteins, which also accumulate in the nucleus, probably enter independently, using their own NLSs. The histone proteins themselves did not initiate the nuclear targeting of the SV40 minichromosomes in the absence of viral capsid proteins (Nakanishi *et al.,* 1996), which is in agreement with our conclusions after using an SV40-NLS + histone-H1 fusion protein for the condensation of pDNA. Naked DNA resulted in a slightly higher level of expression than NLS-H1 condensed DNA did after microinjection of primary rat muscle cells (Fritz *et al.,* 1996). Nakanishi and his co-workers also found that the protein-free, naked SV40 DNA resulted in slower and low percentage of large T antigen expression in microinjected cells (less than 20% after 6 h), compared to expression after microinjecting intact SV40 particles (more than 40% after 4 h and about 80% after 6 h) (Nakanishi *et al.,* 1996). These data emphasize the importance of association with NLS-containing viral proteins. On the one hand, this observation is in accord with our unpublished data concerning the microscopically undetectable nuclear uptake of naked SV40 viral DNA (and as a matter of fact, of any kind of pDNA either containing SV40 sequences or not). On the other hand, Nakanishi's data are not completely inconsistent with David Dean's results either (Dean, 1997). Dean observed the nuclear accumulation of naked SV40 DNA 6−8 h after microinjection: the time at which Nakanishi's expression studies ended.

B. Animal Viruses with RNA Genome

1. Human Immunodeficiency Virus

Like all other retroviruses, HIV carries a positive strand RNA genome, packed into a protein core that is assembled mostly of Gag-derived proteins, Gag-recruited proteins, the reverse transcriptase (RT), and the integrase (IN). The core is covered by a lipid envelope, the major protein of which (gp120) attaches to the CD4 antigen on T lymphocytes and macrophages with very high affinity. Next, the envelope fuses into the plasma membrane with the assistance of other envelope and cell surface proteins and releases the nucleoprotein core. Once inside the cell, the core-associated (CA) protein detaches from the particle and the reverse transcriptase generates nascent linear cDNA. Both the RT and the cDNA remain associated with the tyrosine phosphorylated matrix (MA) protein, (an N-terminal cleavage product of Gag), with the integrase (IN) and with the Vpr protein (which is a Gag-recruited component of the core). This preintegration complex (PIC) must be transported into the nucleus for integration, gene expression, and replication. There have been numerous reports on which of these proteins is/are responsible for nuclear targeting and how they interact, but a final, coherent conception has not yet been reached. One element has been firmly established though: the transport is at least partially NLS dependent, using the classical importin/karyopherin pathway (Gulizia *et al.*, 1994). Essentially, two of the major candidates are the phosphorylated form of MA, which forms a tight complex with the other candidate, IN (Gallay *et al.*, 1995; Goldfarb, 1995). Both have an NLS, interacting with karyopherin α/β (Gallay *et al.*, 1997; Gallay *et al.*, 1996). The third candidate, Vpr, is also karyophilic, although it does not carry a classical NLS and uses a different pathway. Its nuclear targeting domain was mapped to a leucine/isoleucine-rich region of the molecule (Mahalingam *et al.*, 1997). Popov and his colleagues contemplated that Vpr is the key player in this complex transport process, by greatly increasing the affinity of karyopherin-α for NLS-bearing proteins. This ensures a sufficiently strong interaction between the huge nucleoprotein cargo displaying NLSs by its chaperones, MA and IN, and karyopherin-α (Popov *et al.*, 1998). Furthermore, Popov and his co-workers suggest that the enhancement of nuclear transport by Vpr also involves the regulation of binding to FxFG repeat containing nucleoporins (Popov *et al.*, 1998). This step of the nuclear uptake of the PICs is controversial though, since other investigators proposed that Vpr itself is a karyopherin-β-like transport factor, able to interact both with the PIC-associated karyopherin-α and with nucleoporins and to mediate the translocation through the NPC (Vodicka *et al.*, 1998). The ongoing intense research of this field in many more laboratories than we could mention here will explore and hopefully clarify the controversial role of MA and Vpr. Nevertheless, the reason why the above proposals are so exciting is that they assign distinct regulatory or executive function to Vpr that had not been described before in the host cell's own nuclear transport machinery. Thus HIV would not only exploit the host cell's transport system but also add to it.

Most members of the retroviruses are unable to infect quiescent cells: they need cell division and the disintegration of the nuclear envelope to reach the host cell's genome for integration. This property clearly limits their application as gene delivery vehicles. HIV, as briefly characterized previously, is one of the rare exceptions. However, a recent report about another retrovirus, the human foamy virus (HFV), described the nuclear uptake of the viral DNA in artificially growth-arrested (G_1/S) cells (Saib *et al.*, 1997). Since the nuclear import did not result in the expression of any viral gene product, this may not seem to be significant. However, the intracytoplasmic steps preceding the nuclear translocation of the HFV genome reiterate the importance of the cytoskeleton in targeting incoming virions toward the perinuclear region: the Gag antigen and the genome itself colocalized with the centrosome at 5–10 h postinfection, similar to herpes simplex virus previously described. The authors hypothesize that the Gag protein uses a microtubule-based motor mechanism to reach the centrosome, and during this movement it becomes processed by the viral protease to expose an NLS. The mature Gag protein can then initiate the nuclear uptake of the viral preintegration complex.

2. Influenza Virus A

Although influenza viruses have a negative strand RNA genome and carry their own RNA-dependent RNA polymerases in the virion, they still cannot transcribe their mRNAs in the cytoplasm. They need to "steal" $m^7GpppXm$-containing caps from host cell RNAs and they must use the splicing machinery located in the nucleus. Consequently, their genome must be transported into the nucleus. In influenza A the genome is fragmented into eight segments, each associated with viral nucleoproteins (NP) and polymerases (PA, PB1, and PB2), forming ribonucleoprotein complexes (vRNPs). These vRNP segments are held together by the viral matrix protein (M1) and are packed into an enveloped virion of 80–120 nm in size. The lipid envelope contains the integral membrane proteins neuraminidase (NA) and hemagglutinin (HA) and the M2 ion channel protein. NA and HA form the virus's characteristic spikes on the surface, and HA is responsible for binding to sialic acid residues on the cell surface. Attachment is followed by endocytosis into clathrin-coated vesicles. The acidic environment in late endosomes initiates a HA-mediated fusion between the virus envelope and the membrane of the vesicle. The fusion is followed by capsid disassembly, triggered by M2-mediated proton influx into the virion, presumably while it is still in the endosome. This acidification causes conformational changes in M1, the major structural component of the capsid, which dissociates from the RNP particles (Bui *et al.*, 1996). The vRNPs thus released into the cytosol quickly accumulate in the nucleus. *In vitro* model systems and the use of recombinant proteins helped to characterize some details of this transport machinery. Two central regulators of the import and export processes during the viral life cycle are M1 and NP. Both proteins are nucleophilic: M1 carries a classical basic NLS (which also functions as an RNA binding site (Eister *et al.*, 1997)) and NP has a nonconventional nuclear localization signal mapped close to its N terminus (Wang *et al.*, 1997). Since the deletion of the N-terminal NLS does not prevent its nuclear accumulation, NP is thought to also carry some other, yet unidentified, unusual NLSs (Neumann *et al.*, 1997; Stevens and Barclay, 1998; Wang *et al.*, 1997). Since pure viral RNA alone was not transported into the nuclei of digitonin-treated cells, while the addition of recombinant NP did initiate transport, NP seems to be responsible for the nuclear uptake of the vRNPs (O'Neill *et al.,* 1995). NP has been shown to interact with the karyopherin-α/β transport factors, and also needs the presence of Ran and p10 for viral RNA translocation, suggesting a classical, NLS-dependent nuclear transport pathway (O'Neill *et al.,* 1995; Wang *et al.,* 1997). M1 functions as a regulator: after infection it makes nuclear transport possible by releasing the vRNPs from the capsid due to the acid-induced conformational change. However, the freshly synthesized M1 proteins anchor the newly replicated RNPs in the cytoplasm to prevent their reimport before virion assembly can take place (Bui *et al.,* 1996).

NP is also known to be phosphorylated, and its phosphorylation may be responsible for the regulation of its nuclear import and export (Neumann *et al.,* 1997). During the viral life cycle freshly synthesized NP molecules accumulate in the nucleus to interact with the replicated new viral RNA segments and are exported into the cytoplasm as vRNP complexes. Although the phosphorylation of NP favors its cytoplasmic location (Neumann *et al.,* 1997), it is not clear how large a role the phosphorylated NP plays in the export of new vRNPs. Another protein, NEP (also called NS2), has been implicated as an adapter for the nuclear export step (O'Neill *et al.,* 1998).

C. Plant Pathogens

1. Plant Viruses

In general, plant viruses are much less well characterized than animal viruses, which reflects the importance of the latter as human and animal pathogens. There are many plant viruses, however, that replicate in the nucleus, and delineating the molecular machinery of nuclear delivery of these viral genomes would be of great importance. One recently identified example of a nuclear shuttle protein is the BR1 movement protein of the squash leaf curl virus. This virus has a ssDNA genome to be delivered to the nucleus for replication. Two NLSs have been identified in the N-terminal domain of BR1, which are similar to NLSs of yeast and animal origin, emphasizing the highly conserved nature of the nuclear transport machinery in probably all eukaryotes. BR1 is assumed to transport the DNA genome into the nucleus after infection and also out of the nucleus after replication (Sanderfoot *et al.,* 1996). Its C-terminal domain was shown to interact with the other viral movement protein BL1 (Sanderfoot *et al.,* 1996), which is responsible for redirecting the BR1-DNA complex out of the nucleus. In the cytoplasm BL1 interacts with tubular structures derived from the ER and presumably uses them for spreading the viral genome into uninfected neighboring cells (Ward *et al.,* 1997).

The nuclear transport of the NLS containing NIa protein of the tobacco etch virus has been observed (Carrington *et al.,* 1991; Li and Carrington, 1993), but the significance of its nuclear accumulation is debated, since the virus is thought to replicate in the cytoplasm. Even those viruses that replicate in the cytoplasm might give us valuable clues for strategies to move around in the host cell and pass freshly replicated virions into adjacent cells. However, intracellular movement is usually poorly characterized. More attention has been devoted to the characterization of movement proteins involved in cell-to-cell transport, which is crucial for the systemic spread of infection. Since this step seems to use specialized plant transport structures and mechanisms (like movement through plasmodesmata, the details of which have been summarized recently (Ghoshroy *et al.*, 1997; Lartey and Citovsky, 1997)), we do not wish to elaborate this subject. At this time we consider the group of plant viruses a potential treasure chest, the value of which cannot be appraised until after a closer look.

2. The T-DNA of *Agrobacterium tumefaciens*

Agrobacterium tumefaciens is a plant pathogen infecting wounded plant tissues and causing crown gall tumors. The bacterium carries a large Ti (tumor-inducing) plasmid. Wounded plant cells release phenolic molecules and sugars, which induce the expression of the *vir* genes, located on this Ti plasmid. Vir gene products initiate the production of the T–DNA: a single-stranded copy of an approximately 20-kb region of the Ti plasmid, which region is defined by two 25 bp direct repeats on the two sides. One of the enzymes participating in the initiation of the process, the VirD2 protein, remains covalently bound to the 5′ end of the generated T strand. This T–DNA–VirD2 complex is then exported into the plant cell, with the help of other membrane-associated Vir gene products (for review see Hooykaas and Beijersbergen, 1994; Zupan and Zambryski, 1997). The other most important player is VirE2: a single-stranded DNA binding protein. It is also transported into the host cell, where it binds to the T–DNA in a cooperative way (Sen *et al.*, 1989), fully covering it with about 600 molecules on the 20-kb fragment (Zupan *et al.*, 1996). This multimolecular complex is transported into the nucleus of the host cell, where it becomes integrated into the genome and initiates tumorigenesis. Similarly to viral mechanisms previously described, the nuclear uptake of the T–DNA complex is mediated by NLSs. VirD2 has one bipartite NLS close to its C terminus (Howard *et al.*, 1992) and VirE2 carries two similar NLSs in the middle of the protein (Citovsky *et al.*, 1992). These NLSs are homologous to each other and, notably, to the bipartite NLS of *Xenopus* nucleoplasmin and other NLSs of mammalian origin. Moreover, the VirD2 NLS is functional in *Xenopus* oocytes and in *Drosophila* embryos (Guralnick *et al.*, 1996), and a VirD2 binding receptor protein characterized from one of the natural plant hosts of *Agrobacterium* shows unmistakable homology to the yeast NLS receptor SRP1 and other mammalian karyopherin-α homologues (Ballas and Citovsky, 1997). The two NLSs of VirE2 are plant specific: they do not localize to nuclei of animal cells, unless one amino acid residue is repositioned within the bipartite NLS region (Guralnick *et al.*, 1996). Data obtained with VirD2 and the fact that changing a single residue in VirE2 made it acceptable for NLS receptors of animal origin suggest a highly conserved nuclear transport machinery in all eukaryotes. The modified version of the VirE2 protein induced the nuclear uptake of a 0.7-kb fluorescently labeled single-stranded DNA fragment in microinjected *Xenopus* oocytes, while the fragment alone remained sequestered in the cytosol (Guralnick *et al.*, 1996). Although VirE2 binds only single-stranded DNA and the transport of the 0.7−kb fragment does not necessarily mean a large expression cassette would also be transported, this experiment indicates the potentials of viral targeting proteins. One day components of this extremely efficient plant pathogen may be mixed and matched with elements of viral gene delivery systems to create an optimal, artificial DNA delivery protocol.

VI. CONCLUSIONS AND FUTURE PROSPECTS

The advent of using naked DNA as a gene delivery vehicle in 1990 foreshadowed a new era of safe and efficient yet simple and cheap *in vivo* gene therapy. The subsequent eight years of research improved the technique for certain applications and partly characterized its mode of action and limitations. The intravascular delivery of naked pDNA is one of the most efficient nonviral methods for *in vivo* expression. We anticipate an increasing interest in using naked DNA for clinical applications and at the same time expect to see a rapid development in harnessing viral strategies for DNA delivery in areas where naked DNA does not give satisfactory results. This will be accomplished in part by developing methods to increase the efficiency of DNA nuclear uptake. Given the remarkable efficiency of naked DNA expression, this may be easier to accomplish than commonly thought.

REFERENCES

Adam, S. A., Sterne-Marr, R., and Gerace, L. (1992). Nuclear protein import using digitonin-permeabilized cells. *Methods Enzymol.* **219,** 97−110.

Aronsohn, A. I., and Hughes, J. A. (1998). Nuclear localization signal peptides enhance cationic liposome-mediated gene therapy. *J. Drug Targeting* **5,** 163−169.

Avrameas, A., Ternynck, T., Nato, F., Buttin, G., and Avrameas, S. (1998). Polyreactive anti-DNA monoclonal antibodies and a derived peptide as vectors for the intracytoplasmic and intranuclear translocation of macromolecules. *Proc. Natl. Acad. Sci. USA* **95,** 5601−5606.

Ballas, N., and Citovsky, V. (1997). Nuclear localization signal binding protein from Arabidopsis mediates nuclear import of Agrobacterium VirD2 protein. *Proc. Natl. Acad. Sci. USA* **94,** 10723−10728.

Brinster, R. L., Chen, H. Y., Trumbauer, M. E., Yagle, M. K., and Palmiter, R. D. (1985). Factors affecting the efficiency of introducing foreign DNA into mice by microinjecting eggs. *Proc. Natl. Acad. Sci. USA* **82,** 4438−4442.

Budker, V., Hagstrom, J. E., Lapina, O., Eifrig, D., Fritz, J., and Wolff, J. A. (1997). Protein/amphipathic polyamine complexes enable highly efficient transfection with minimal toxicity. *Biotechniques* **23**, 139, 142–147.

Budker, V., Zhang, G., Danko, I., Williams, P., and Wolff, J. (1998). The efficient expression of intravascularly delivered DNA in rat muscle. *Gene Ther.* **5**, 272–276.

Budker, V., Zhang, G., Knechtle, S., and Wolff, J. A. (1996). Naked DNA delivered intraportally expresses efficiently in hepatocytes. *Gene Ther.* **3**, 593–598.

Bui, M., Whittaker, G., and Helenius, A. (1996). Effect of M1 protein and low pH on nuclear transport of influenza virus ribonucleoproteins. *J. Virol.* **70**, 8391–8401.

Capecchi, M. R. (1980). High efficiency transformation by direct microinjection of DNA into cultured mammalian cells. *Cell* **22**, 479–488.

Carrington, J. C., Freed, D. D., and Leinicke, A. J. (1991). Bipartite signal sequence mediates nuclear translocation of the plant potyviral NIa protein. *Plant Cell* **3**, 953–962.

Citovsky, V., Zupan, J., Warnick, D., and Zambryski, P. (1992). Nuclear localization of Agrobacterium VirE2 protein in plant cells. *Science* **256**, 1802–1805.

Clever, J., Yamada, M., and Kasamatsu, H. (1991). Import of Simian Virus 40 virions through nuclear pore complexes. *Proc. Natl. Acad. Sci. USA* **88**, 7333–7337.

Collas, P., and Alestrom, P. (1996). Nuclear localization signal of SV40 T antigen directs import of plasmid DNA into sea urchin male pronuclei in vitro. *Mol. Reprod. Dev.* **45**, 431–438.

Collas, P., and Alestrom, P. (1997). Nuclear localization signals: A driving force for nuclear transport of plasmid DNA in zebrafish. *Biochem. Cell. Biol.* **75**, 633–640.

Collas, P., Husebye, H., and Alestrom, P. (1996). The nuclear localization sequence of the SV40 T antigen promotes transgene uptake and expression in zebrafish embryo nuclei. *Transgenic Res.* **5**, 451–458.

Coonrod, A., Li, F. Q., and Horwitz, M. (1997). On the mechanism of DNA transfection: Efficient gene transfer without viruses. *Gene Ther.* **4**, 1313–1321.

Dabauvalle, M. C., Schulz, B., Scheer, U., and Peters, R. (1988). Inhibition of nuclear accumulation of karyophilic proteins in living cells by microinjection of the lectin wheat germ agglutinin. *Exp. Cell. Res.* **174**, 291–296.

Dales, S., and Chardonnet, Y. (1973). Early events in the interaction of adenoviruses with HeLa cells. IV. Association with microtubules and the nuclear pore complex during vectorial movement of the inoculum. *Virology* **56**, 465–483.

Davis, L. I. (1995). The nuclear pore complex. *Ann. Rev. Biochem.* **64**, 865–896.

Dean, D. A. (1997). Import of plasmid DNA into the nucleus is sequence specific. *Exp. Cell. Res.* **230**, 293–302.

Deng, H., Wang, C., Acsadi, G., and Wolff, J. A. (1991). High-efficiency protein synthesis from T7 RNA polymerase transcripts in 3T3 fibroblasts. *Gene* **109**, 193–201.

Deng, H., and Wolff, J. A. (1994). Self-amplifying expression from the T7 promoter in 3T3 mouse fibroblasts. *Gene* **143**, 245–249.

Dowty, M. E., Williams, P., Zhang, G., Hagstrom, J. E., and Wolff, J. A. (1995). Plasmid DNA entry into postmitotic nuclei of primary rat myotubes. *Proc. Natl. Acad. Sci. USA* **92**, 4572–4576.

Dworetzky, S. I., Lanford, R. E., and Feldherr, C. M. (1988). The effects of variations in the number and sequence of targeting signals on nuclear uptake. *J. Cell. Biol.* **107**, 1279–1287.

Efthymiadis, A., Briggs, L. J., and Jans, D. A. (1998). The HIV-1 Tat nuclear localization sequence confers novel nuclear import properties. *J. Biol. Chem.* **273**, 1623–1628.

Eister, C., Larsen, K., Gagnon, J., Ruigrok, R. W., and Baudin, F. (1997). Influenza virus M1 protein binds to RNA through its nuclear localization signal. *J. Gen. Virol.* **78**, 1589–1596.

Elroy-Stein, O., Fuerst, T. R., and Moss, B. (1989). Cap-independent translation of mRNA conferred by encephalomyocarditis virus 5′ sequence improves the performance of the vaccinia virus/bacteriophage T7 hybrid expression system. *Proc. Natl. Acad. Sci. USA* **86**, 6126–6130.

Fasbender, A., Zabner, J., Zeiher, B. G., and Welsh, M. J. (1997). A low rate of cell proliferation and reduced DNA uptake limit cationic lipid-mediated gene transfer to primary cultures of ciliated human airway epithelia. *Gene Therapy* **4**, 1173–1180.

Feldherr, C. M., and Akin, D. (1990). The permeability of the nuclear envelope in dividing and non-dividing cell cultures. *J. Cell. Biol.* **111**, 1–8.

Femino, A. M., Fay, F. S., Fogarty, K., and Singer, R. H. (1998). Visualization of a single transcript *in situ*. *Science* **280**, 585–590.

Fischer, N., Kremmer, E., Lautscham, G., Mueller-Lantzsch, N., and Grasser, F. A. (1997). Epstein-Barr virus nuclear antigen 1 forms a complex with the nuclear transporter karyopherin alpha2. *J. Biol. Chem.* **272**, 3999–4005.

Fishman, D. M., and Patterson, G. D. (1996). Light scattering studies of supercoiled and nicked DNA. *Biopolymers* **38**, 535–552.

Fiskum, G., Craig, S. W., Decker, G. L., and Lehninger, A. L. (1980). The cytoskeleton of digitonin-treated rat hepatocytes. *Proc. Natl. Acad. Sci. USA* **77**, 3430–3434.

Fritz, F. D., Herweijer, H., Zhang, G., and Wolff, J. A. (1996). Gene transfer into mammalian cells using histone-condensed plasmid DNA. *Human Gene Ther.* **7**, 1395–1404.

Gallay, P., Hope, T., Chin, D., and Trono, D. (1997). HIV-1 infection of nondividing cells through the recognition of integrase by the importin/karyopherin pathway. *Proc. Natl. Acad. Sci. USA* **94**, 9825–9830.

Gallay, P., Stitt, V., Mundy, C., Oettinger, M., and Trono, D. (1996). Role of the karyopherin pathway in human immunodeficiency virus type 1 nuclear import. *J. Virol.* **70**, 1027–1032.

Gallay, P., Swingler, S., Song, J., Bushman, F., and Trono, D. (1995). HIV nuclear import is governed by the phosphotyrosine-mediated binding of matrix to the core domain of integrase. *Cell* **83**, 569–576.

Gao, X., and Huang, L. (1993). Cytoplasmic expression of a reporter gene by co-delivery of T7 RNA polymerase and T7 promoter sequence with cationic liposomes. *Nucleic Acids Res.* **21**, 2867–2872.

Gao, X., and Huang, L. (1996). Potentiation of cationic liposome-mediated gene delivery by polycations. *Biochemistry* **35**, 1027–1036.

Ghoshroy, S., Lartey, R., Sheng, J. S., and Citovsky, V. (1997). Transport of proteins and nucleic acids through plasmodesmata. *Annu. Rev. Plant Physiol. Plant Mol. Biol.* **48**, 25–48.

Goldfarb, D. S. (1995). HIV-1 virology: Simply MArvelous nuclear transport. *Curr. Biol.* **5**, 570–573.

Goldfarb, D. S. (1997). Nuclear transport—whose finger is on the switch? *Science* **276**, 1814–1816.

Gorlich, D., and Mattaj, I. W. (1996). Nucleocytoplasmic transport. *Science* **271**, 1513–1518.

Gorlich, D., Pante, N., Kutay, U., Aebi, U., and Bischoff, F. R. (1996). Identification of different roles for RanGDP and RanGTP in nuclear protein import. *EMBO J.* **15**, 5584–5594.

Graessmann, M., Menne, J., Liebler, M., Graeber, I., and Graessmann, A. (1989). Helper activity for gene expression, a novel function of the SV40 enhancer. *Nucleic Acids Res.* **17**, 6603–6612.

Greber, U. F., Singh, I., and Helenius, A. (1994). Mechanisms of virus uncoating. *Trends Microbiol.* **2**, 52–56.

Greber, U. F., Suomalainen, M., Stidwill, R. P., Boucke, K., Ebersold, M. W., and Helenius, A. (1997). The role of the nuclear pore complex in adenovirus DNA entry. *EMBO J.* **16**, 5998–6007.

Greber, U. F., Webster, P., Weber, J., and Helenius, A. (1996). The role of the adenovirus protease in virus entry into cells. *EMBO J.* **15**, 1766–1777.

Greber, U. F., Willetts, M., Webster, P., and Helenius, A. (1993). Stepwise dismantling of adenovirus 2 during entry into cells. *Cell* **75**, 477–486.

Greber, U. F. a. K., H. (1996). Nuclear targeting of SV40 and adenovirus. *Trends Cell. Biol.* **6**, 189–195.

Griffith, G. R., Marriott, S. J., Rintoul, D. A., and Consigli, R. A. (1988). Early events in polyomavirus infection: Fusion of monopinocytotic vesicles containing virions with mouse kidney cell nuclei. *Virus Res.* **10**, 41–51.

Guidotti, L. G., Martinez, V., Loh, Y. T., Rogler, C. E., and Chisari, F. V. (1994). Hepatitis B virus nucleocapsid particles do not cross the hepatocyte nuclear membrane in transgenic mice. *J. Virol.* **68,** 5469–5475.

Gulizia, J., Dempsey, M. P., Sharova, N., Bukrinsky, M. I., Spitz, L., Goldfarb, D., and Stevenson, M. (1994). Reduced nuclear import of human immunodeficiency virus type 1 preintegration complexes in the presence of a prototypic nuclear targeting signal. *J. Virol.* **68,** 2021–2025.

Guralnick, B., Thomsen, G., and Citovsky, V. (1996). Transport of DNA into the nuclei of Xenopus oocytes by a modified VirE2 protein of Agrobacterium. *Plant Cell* **8,** 363–373.

Hagstrom, J. E., Ludtke, J. J., Bassik, M. C., Sebestyen, M. G., Adam, S. A., and Wolff, J. A. (1997). Nuclear import of DNA in digitonin-permeabilized cells. *J. Cell. Sci.* **110,** 2323–2331.

Hagstrom, J. E., Rybakova, I. N., Staeva, T., Wolff, J. A., and Ervasti, J. M. (1996). Nonnuclear DNA binding proteins in striated muscle. *Biochem. Mol. Med.* **58,** 113–121.

Hagstrom, J. E., Sebestyen, M. G., Budker, V., Ludtke, J. J., Fritz, J. D., and Wolff, J. A. (1996). Complexes of non-cationic liposomes and histone H1 mediate efficient transfection of DNA without encapsulation. *Biochim. Biophys. Acta* **1284,** 47–55.

Hartig, R., Huang, Y., Janetzko, A., Shoeman, R., Grub, S., and Traub, P. (1997). Binding of fluorescence- and gold-labeled oligodeoxyribonucleotides to cytoplasmic intermediate filaments in epithelial and fibroblast cells. *Exp. Cell. Res.* **233,** 169–186.

Henderson, B. R., and Percipalle, P. (1997). Interactions between HIV Rev and nuclear import and export factors: The Rev nuclear localization signal mediates specific binding to human importin-beta. *J. Mol. Biol.* **274,** 693–707.

Hesketh, J. E. (1996). mRNA targeting—signals in the 3'-untranslated sequences for sorting of some mRNAs. *Biochem. Soc. Trans.* **24,** 521–527.

Hild, M., Weber, O., and Schaller, H. (1998). Glucagon treatment interferes with an early step of duck hepatitis B virus infection. *J. Virol.* **72,** 2600–2606.

Hill, M. (1961). Uptake of deoxyribonucleic acid (DNA): A special property of the cell nucleus. *Nature* **189,** 916–917.

Hooykaas, P. J., and Beijersbergen, G.M. (1994). The virulence system of Agrobacterium tumefaciens. *Annu. Rev. Phytopath.* **32,** 157–179.

Hovland, R., Hesketh, J. E., and Pryme, I. F. (1996). The compartmentalization of protein synthesis—importance of cytoskeleton and role in mRNA targeting. *Int. J. Biochem. Cell. Biol.* **28,** 1089–1105.

Howard, E. A., Zupan, J. R., Citovsky, V., and Zambryski, P. C. (1992). The VirD2 protein of A. tumefaciens contains a C-terminal bipartite nuclear localization signal: Implications for nuclear uptake of DNA in plant cells. *Cell* **68,** 109–18.

Izaurralde, E., Kutay, U., von Kobbe, C., Mattaj, I. W., and Gorlich, D. (1997). The asymmetric distribution of the constituents of the Ran system is essential for transport into and out of the nucleus. *EMBO J.* **16,** 6535–6547.

Jans, D. A., and Hubner, S. (1996). Regulation of protein transport to the nucleus: Central role of phosphorylation. *Physiol. Rev.* **76,** 651–685.

Kalderon, D., Roberts, B. L., Richardson, W. D., and Smith, A. E. (1984). A short amino acid sequence able to specify nuclear location. *Cell* **39,** 499–509.

Kaneda, Y., Iwai, K., and Uchida, T. (1989). Increased expression of DNA cointroduced with nuclear protein in adult rat liver. *Science* **243,** 375–378.

Kann, M., Bischof, A., and Gerlich, W. H. (1997). *In vitro* model for the nuclear transport of the hepadnavirus genome. *J. Virol.* **71,** 1310–1316.

Kann, M., and Gerlich, W. H. (1994). Effect of core protein phosphorylation by protein kinase C on encapsidation of RNA within core particles of hepatitis B virus. *J. Virol.* **68,** 7993–8000.

Kann, M., Lu, X., and Gerlich, W. H. (1995). Recent studies on replication of hepatitis B virus. *J. Hepatol.* **22,** 9–13.

Kim, A. L., Maher, M., Hayman, J. B., Ozer, J., Zerby, D., Yates, J. L., and Lieberman, P. M. (1997). An imperfect correlation between DNA replication activity of Epstein–Barr virus nuclear antigen 1 (EBNA1) and binding to the nuclear import receptor, Rch1/importin alpha. *Virology* **239**, 340–51.

Kristensson, K., Lycke, E., Roytta, M., Svennerholm, B., and Vahlne, A. (1986). Neuritic transport of herpes simplex virus in rat sensory neurons in vitro. Effects of substances interacting with microtubular function and axonal flow. *J. Gen. Virol.* **67**, 2023–2028.

Langle-Rouault, F., Patzel, V., Benavente, A., Taillez, M., Silvestre, N., Bompard, A., Sczakiel, G., Jacobs, E., and Rittner, K. (1998). Up to 100-fold increase of apparent gene expression in the presence of Epstein-Barr virus oriP sequences and EBNA1—implications of the nuclear import of plasmids. *J. Virol.* **72**, 6181–6185.

Lartey, R., and Citovsky, V. (1997). Nucleic acid transport in plant-pathogen interactions. *Genet. Eng.* **19**, 201–214.

Li, S., and Huang, L. (1997). In vivo gene transfer via intravenous administration of cationic lipid-protamine-DNA (LPD) complexes. *Gene Ther.* **4**, 891–900.

Li, X. H., and Carrington, J. C. (1993). Nuclear transport of tobacco etch potyviral RNA-dependent RNA polymerase is highly sensitive to sequence alterations. *Virology* **193**, 951–958.

Luby-Phelps, K., and Weisiger, R. A. (1996). Role of cytoarchitecture in cytoplasmic transport. *Comp. Biochem. Physiol. B. Comp. Biochem.* **115**, 295–306.

Mahalingam, S., Ayyavoo, V., Patel, M., Kieber-Emmons, T., and Weiner, D. B. (1997). Nuclear import, virion incorporation and cell cycle arrest/differentiation are mediated by distinct functional domains of human immunodeficiency virus type 1 Vpr. *J. Virol.* **71**, 6339–6347.

Makarova, N., Pasquinelli, G., and Martinelli, G. N. (1994). Nuclear bodies associated with core particles of hepatitis B virus in a healthy carrier. *J. Submicrosc. Cytol. Pathol.* **26**, 569–575.

Mirzayans, R., Remy, A. A., and Malcolm, P. C. (1992). Differential expression and stability of foreign genes introduced into human fibroblasts by nuclear versus cytoplasmic microinjection. *Mutation Res.* **281**, 115–122.

Nakanishi, A., Clever, J., Yamada, M., Li, P. P., and Kasamatsu, H. (1996). Association with capsid proteins promotes nuclear targeting of Simian Virus 40 DNA. *Proc. Natl. Acad. Sci. USA* **93**, 96–100.

Nassal, M., and Schaller, H. (1993). Hepatitis B virus replication. *Trends Microbiol.* **1**, 221–228.

Nassal, M., and Schaller, H. (1996). Hepatitis B virus replication—an update. *J. of Viral Hepatitis* **3**, 217–226.

Neumann, G., Castrucci, M. R., and Kawaoka, Y. (1997). Nuclear import and export of influenza virus nucleoprotein. *J. Virol.* **71**, 9690–9700.

Nigg, E. (1997). Nucleocytoplasmic transport: Signals, mechanisms and regulation. *Nature* **386**, 779–787.

Nishimura, T., Kawai, N., Kawai, M., Notake, K., and Ichihara, I. (1986). Fusion of SV40-induced endocytotic vacuoles with the nuclear membrane. *Cell Struct. Funct.* **11**, 135–141.

O'Neill, R. E., Jaskunas, R., Blobel, G., Palese, P., and Moroianu, J. (1995). Nuclear import of influenza virus RNA can be mediated by viral nucleoprotein and transport factors required for protein import. *J. Biol. Chem.* **270**, 22701–22704.

O'Neill, R. E., Talon, J., and Palese, P. (1998). The influenza virus NEP (NS2 protein) mediates the nuclear export of viral ribonucleoproteins. *EMBO J.* **17**, 288–296.

Ohno, M., Fornerod, M., and Mattaj, I. W. (1998). Nucleocytoplasmic transport: The last 200 nanometers. *Cell* **92**, 327–336.

Pante, N., and Aebi, U. (1996). Molecular dissection of the nuclear pore complex. *Crit. Rev. Biochem. Mol. Biol.* **31**, 153–199.

Pemberton, L. F., Rosenblum, J. S., and Blobel, G. (1997). A distinct and parallel pathway for the nuclear import of an mRNA-binding protein. *J. Cell Biol.* **139**, 1645–1653.

Penfold, M. E., Armati, P., and Cunningham, A. L. (1994). Axonal transport of herpes simplex virions to epidermal cells: evidence for a specialized mode of virus transport and assembly. *Proc. Natl. Acad. Sci. USA* **91,** 6529–6533.

Pennisi, E. (1998). The nucleus's revolving door. *Science* **279,** 1129–1131.

Pollard, H., Remy, J. S., Loussouarn, G., Demolombe, S., Behr, J. P., and Escande, D. (1998). Polyethylenimine but not cationic lipids promotes transgene delivery to the nucleus in mammalian cells. *J. Biol. Chem.* **273,** 7507–7511.

Popov, S., Rexach, M., Ratner, L., Blobel, G., and Bukrinsky, M. (1998). Viral protein R regulates docking of the HIV-1 preintegration complex to the nuclear pore complex. *J. Biol. Chem.* **273,** 13347–13352.

Popov, S., Rexach, M., Zybarth, G., Reiling, N., Lee, M. A., Ratner, L., Lane, C. M., Moore, M. S., Blobel, G., and Bukrinsky, M. (1998). Viral protein R regulates nuclear import of the HIV-1 pre-integration complex. *EMBO J.* **17,** 909–917.

Roizman, B., and Sears, A. E. (1993). Herpes simplex viruses and their replication. In B. Roizman, R. J. Whitley and C. Lopez (Eds.), (pp. 11–68). *The human herpesviruses.* New York: Raven Press.

Saib, A., Puvion-Dutilleul, F., Schmid, M., Peries, J., and de The, H. (1997). Nuclear targeting of incoming human foamy virus Gag proteins involves a centriolar step. *J. Virol.* **71,** 1155–1161.

Sanderfoot, A. A., Ingham, D. J., and Lazarowitz, S. G. (1996). A viral movement protein as a nuclear shuttle. The geminivirus BR1 movement protein contains domains essential for interaction with BL1 and nuclear localization. *Plant Physiol.* **110,** 23–33.

Sebestyen, M. G., Ludtke, J. J., Bassik, M. C., Zhang, G., Budker, V., Lukhtanov, E. A., Hagstrom, J. E., and Wolff, J. A. (1998). DNA vector chemistry: The covalent attachment of signal peptides to plasmid DNA. *Nature Biotechnol.* **16,** 80–85.

Seksek, O., Biwersi, J., and Verkman, A. S. (1997). Translational diffusion of macromolecule-sized solutes in cytoplasm and nucleus. *J. Cell Biol.* **138,** 131–142.

Sen, P., Pazour, G. J., Anderson, D., and Das, A. (1989). Cooperative binding of Agrobacterium tumefaciens VirE2 protein to single-stranded DNA. *J. Bacteriol.* **171,** 2573–2580.

Sodeik, B., Ebersold, M. W., and Helenius, A. (1997). Microtubule-mediated transport of incoming herpes simplex virus 1 capsids to the nucleus. *J. Cell Biol.* **136,** 1007–1021.

Somlyai, G., Kondorosi, E., Kariko, K., and Duda, E. G. (1985). Liposome mediated DNA-transfer into mammalian cells. *Acta Biochim. Biophys. Acad. Sci. Hung.* **20,** 203–211.

Stevens, M. P., and Barclay, W. S. (1998). The N-terminal extension of the influenza B virus nucleoprotein is not required for nuclear accumulation or the expression and replication of a model RNA. *J. Virol.* **72,** 5307–5312.

Thorburn, A. M., and Alberts, A. S. (1993). Efficient expression of miniprep plasmid DNA after needle micro-injection into somatic cells. *Biotechniques* **14,** 356–358.

Topp, K. S., Meade, L. B., and LaVail, J. H. (1994). Microtubule polarity in the peripheral processes of trigeminal ganglion cells: Relevance for the retrograde transport of herpes simplex virus. *J. Neurosci.* **14,** 318–325.

Vodicka, M. A., Koepp, D. M., Silver, P. A., and Emerman, M. (1998). HIV-1 Vpr interacts with the nuclear transport pathway to promote macrophage infection. *Genes Dev.* **12,** 175–185.

Wang, P., Palese, P., and O'Neill, R. E. (1997). The NPI-1/NPI-3 (karyopherin alpha) binding site on the influenza A virus nucleoprotein NP is a nonconventional nuclear localization signal. *J. Virol.* **71,** 1850–1856.

Ward, B. M., Medville, R., Lazarowitz, S. G., and Turgeon, R. (1997). The geminivirus BL1 movement protein is associated with endoplasmic reticulum-derived tubules in developing phloem cells. *J. Virol.* **71,** 3726–3733.

Wienhues, U., Hosokawa, K., Hoveler, A., Siegmann, B., and Doerfler, W. (1987). A novel method for transfection and expression of reconstituted DNA-protein complexes in eukaryotic cells. *DNA* **6,** 81–89.

Wolff, J. A., Malone, R. W., Williams, P., Chong, W., Acsadi, G., Jani, A., and Felgner, P. L. (1990). Direct gene transfer into mouse muscle *in vivo. Science* **47,** 1465–1468.

Yamaguchi, M., Miyatsu, Y., Horikawa, Y., Sugahara, K., Mizokami, H., Kawase, M., and Tanaka, H. (1994). Dynamics of hepatitis B virus core antigen in a transformed yeast cell: Analysis with an inducible system. *J. Electron Microsc.* **43,** 386–393.

Yu, M., and Summers, J. (1994). Phosphorylation of the duck hepatitis B virus capsid protein associated with conformational changes in the C terminus. *J. Virol.* **68,** 2965–2969.

Zabner, J., Fasbender, A. J., Moninger, T., Poellinger, K. A., and Welsh, M. J. (1995). Cellular and molecular barriers to gene transfer by a cationic lipid. *J. Biol. Chem.* **270,** 18997-9007.

Zhang, G., Vargo, D., Budker, V., Armstrong, N., Knechtle, S., and Wolff, J. A. (1997). Expression of naked plasmid DNA injected into the afferent and efferent vessels of rodent and dog livers. *Human Gene Ther.* **8,** 1763–1772.

Zupan, J., and Zambryski, P. (1997). The Agrobacterium DNA transfer complex. *Crit. Rev. Plant Sci.* **16,** 279–295.

Zupan, J. R., Citovsky, V., and Zambryski, P. (1996). Agrobacterium VirE2 protein mediates nuclear uptake of single-stranded DNA in plant cells. *Proc. Natl. Acad. Sci. USA* **93,** 2392–2397.

CHAPTER 8

Particle-Mediated Gene Delivery: Applications to Canine and Other Larger Animal Systems

Ning-Sun Yang,* Gary S. Hogge,† and E. Gregory MacEwen†

*Comprehensive Cancer Center, University of Wisconsin Medical School, Madison, Wisconsin; and Institute of Bioagricultural Sciences, Academia Sinica, Taipei, Taiwan, Republic of China

† Department of Medical Sciences, School of Veterinary Sciences, University of Wisconsin, Madison, Wisconsin

I. INTRODUCTION

Particle-mediated gene delivery or the gene gun technology was initially designed as a nonviral method for transformation of plants and later was extended to mammalian gene transfer systems. More than a decade later (1987–1998), various researchers, including those in our own laboratories, have shown that the gene gun method can indeed be applied as a generic means for transfection of a wide variety of mammalian cells, under *in vivo, ex vivo,* or *in vitro* experimental conditions for all species examined to date (Yang, McCabe, and Swain, 1996). It has been demonstrated as effective in a variety of tissue and organ types, including skin, liver, muscle, spleen, pancreas, and other organs *in vivo* (Yang *et al.,* 1990; Chang *et al.,* 1993; Keller *et al.,* 1996). *Ex vivo* applications have included transfection of brain, mammary, leukocyte, and a variety of tumor tissue samples as primary cultures or explants (Burkholder *et al.*, 1993; Giao *et al.*, 1993; Thompson *et al.*, 1993). *In vitro*, several different mammalian cell lines have been transfected both in suspensions and as adherent populations (Yang *et al.*, 1990; Burkholder *et al.*, 1993; Thompson *et al.*, 1993).

The gene gun transfer method does not have the same requirements as many other methods of gene transfer. The direct insertion of genetic material into cells or tissue negates the need for specific cell surface receptors or ligands, biochemical conditions, structural components, endocytosis, or cell cycle status. Additionally, high copy numbers of DNA can be delivered into the targeted tissue or cells. A single particle may contain, on average, 5000 copies of a 5–10 Kb plasmid DNA (Yang *et al.*, 1997). The gene gun method allows transfer without size restrictions of DNA vectors or delivery of multiple transgenes (Albertini *et al.*, 1996). Stable transfections have been shown to be possible using selective media and antibiotic resistance genes (Yang *et al.*, 1997; Yang and Ziegelhoffer, 1994).

The actual gene delivery process is achieved by a physical force. With the Accell® (PowderJect Vaccines, Inc.) or the Helios™ (Bio-Rad) gene gun designs, a compressed shock wave of helium gas is created, accelerating DNA-coated gold or tungsten particles of varying sizes (1–3 μm) to high speed. Gold is the most commonly used substance for these particles due to its lack of cytotoxic or immunologic effects, its high density, and the commercial availability of quality controlled, defined-size elemental gold particles (Yang, McCabe, and Swain, 1996). This force is often of sufficient momentum to penetrate a target tissue, with the depth of penetration dependent on the pounds per square inch (psi) of helium pressure, size of particles, and the tissue or cells targeted. The accelerated particles actually have sufficient force to penetrate substantial physical barriers such as plant cell walls, cell membranes, and the stratum corneum of mammalian epidermis (Yang *et al.*, 1997). The shock wave is dissipated through a physics principle known as the Coanda curvature, which aids in reducing the shock wave, thereby maintaining viability of the targeted cells or tissue (Reba, 1966). Additionally, particle-mediated delivery has been adapted to other macromolecules such as RNA and proteins (Yang *et al.*, 1997).

For several years, it was difficult for many research laboratories to actively engage in gene gun experimentation. The lack of commercially available products made incorporation and application of the technique as part of routine research tools difficult. This shortage has now been overcome by the distribution of the Bio-Rad Helios™ gene gun and recent publications of standardized operational protocols (SOPs) and manufacturers' manuals (Yang, McCabe, and Swain, 1996; Yang *et al.*, 1997).

Using the gene gun technology, our laboratories have utilized the canine systems for a number of exploratory transgenic studies. We have demonstrated that it is readily applicable to the oral mucosal and dermal systems *in vivo* (Keller *et al.*, 1996) and to melanoma and soft-tissue sarcoma DNA tumor vaccine systems administered in an *ex vivo* vaccine fashion (Hogge *et al.*, 1998; Hogge *et al.*, 1999).

Recently published data have demonstrated that the canine systems appear to be ideally suited for many of the transfection techniques involving the gene gun. We have observed that canine tumor cells in the *ex vivo* tissue culture setting can be easily dissociated and manipulated from primary tumor sources, grow readily and effectively under standard culture conditions, and withstand direct bombardment both in suspension and as adherent cell layers (Hogge *et al.*, 1998a). *In vivo* transfection of canine tissue more closely approximates the biological conditions found in humans when compared to murine models due to the similar thickness of the canine skin and the overall body size of the animal. Additionally, protein expression levels in many of our primary cell culture samples, continuously growing proliferative cultures, and established cell lines appear to be substantially higher when compared to similar human conditions and cellular origin (Mahvi *et al.*, 1996). It is unclear to us at this time whether this is due to inherent canine cellular responses to promoters employed for various expression plasmids or is simply a matter of increased cell resistance to the physical transfection process compared to human samples. Nonetheless, due to these unique features and our experiences in the canine gene transfer systems, this chapter emphasizes and discusses our studies on the application of gene gun technology to the canine system and how these results offer both basic and clinical research applications to current and future particle-mediated gene delivery in humans.

Previous studies have successfully applied the gene gun technology to research into cancer gene therapy. These include both the *ex vivo* DNA cancer vaccine approach with cytokines, costimulatory molecules, or tumor associated antigens (Albertini *et al.*, 1996; Hogge *et al.*, 1998; Hogge *et al.*, 1999; Mahvi *et al.*, 1996; Tütting *et al.*, 1997) and the *in vivo* cytokine gene therapy approach (Keller *et al.*, 1996; Irvine *et al.*, 1996; Sun *et al.*, 1995; Rakhmilevich *et al.*, 1996; Conry *et al.*, 1996). Our recent studies in large animal models such as the dog can apparently offer many advantages to complement the murine model systems. The potential advantages of using these large animals as a "bridge system" between preclinical murine experiments and human clinical trials include spontaneously generated tumors, similar environmental influences, similar physiologic and biologic responses, and a body size approaching, if not exceeding, that of human beings. This chapter addresses our recent findings in these areas and provides data to support our belief in the dog as a new model for systematic "translational medicine" studies in cancer gene therapy strategies as both clinically relevant, experimentally feasible, and biologically significant.

II. *EX VIVO* GENE TRANSFER INTO CANINE TUMOR CELL EXPLANTS AND DERIVED PRIMARY CULTURES

We and others have shown that a variety of freshly isolated cell suspensions or tissue clumps (e.g., mammary gland organoids (Thompson *et al.*, 1993), splenocytes (Thompson *et al.*, 1993), prostate gland xenografts (Cheng *et al.*, 1996), fetal brain cells (Giao *et al.*, 1996), peripheral blood lymphocytes (Burkholder *et al.*, 1996; Rajogopalan *et al.*, 1995), and tumor cells from human clinical patient samples (Mahvi *et al.*, 1996)), can be effectively transfected *ex vivo* when placed into tissue culture vessels as fresh cell explant samples or as 1- to 4-day-old primary cell cultures (Fig. 1). In comparison, we have observed that various spontaneous tumor cell explants (melanoma and various soft-tissue sarcomas) derived from canine patients of our veterinary medical teaching hospital at the University of Wisconsin-Madison appeared to be exceptionally responsive to the *ex vivo* gene transfer procedure, yielding high-level transgene expression in short-term *ex vivo* extension culture systems. This may be attributed to the yield of highly viable, healthy, and proliferative epithelial cell populations and the high level of protein synthesis capacity of these cells following gene gun bombardment for gene transfer. Empirically, we have observed in a series of studies the following findings: (1) Primary culture tumor cell lines can be successfully grown and passaged multiple times following isolation of 10^7-10^8 cells from tumor-bearing dogs (Table 1). These numbers—from cells obtained either following initial tissue dissociation procedures or those generated from continuously growing, multiply passaged cells in culture—have proven sufficient to either begin tumor vaccine protocols without long-term cell culturing or create a large reserve of frozen stock tumor cells for future studies or re-vaccinations, respectively. (2) Tumor cell samples, as either freshly isolated cell explants or as frozen and thawed primary cell cultures, can express with good frequency, at the cellular level (>25%), high levels of transient transgenic proteins (as determined by flow cytometric analysis) such as granulocyte macrophage-colony stimulating factor (GM-CSF) to a mean level of 46 ng/10^6 cells/24 h *in vitro*, with a range of

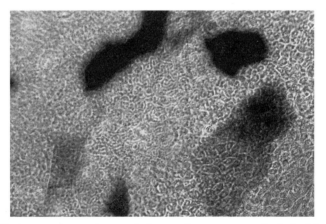

Figure 1   Canine mammary carcinoma cells from a primary tumor sample obtained from a clinical patient, dissociated, and transfected with β-gal using the gene gun. The cells in the staining areas are not all transfected, as there is some diffusion of the stain from those cells that were successfully transfected.

Table 1

In Vitro **Expression of hGM-CSF Protein in a Variety of Transfected Canine Primary Tumor Cells**

	Tumor type	Wt. (grams)	Cells/g tissue	*In vitro* expression
1	Mam Carc.	0.6	3.9×10^7	54
2	FSA	2.6	1.0×10^7	13
3	MEL	1.9	1.6×10^7	86
4	Mam Ad.	0.2	1.3×10^7	13
5	FSA	8.8	1.9×10^7	174
6	MEL	6.3	6.4×10^7	23
7	MEL	14.3	1.3×10^8	25
8	MEL	10.2	2.7×10^7	50
9	MEL	7.2	1.5×10^7	44
10	HSA	3.2	6.3×10^7	27
11	FSA	13.7	8.3×10^7	10
12	MEL	12.7	2.1×10^7	80
13	MEL	3.1	1.5×10^7	28
14	MEL	4.2	5.1×10^7	14
	Mean	6.4	4.0×10^7	46

FSA = Fibrosarcoma, MEL = Melanoma, HSA = Hemangiosarcoma, Mam. Carc. = Mammary carcinoma, Mam. Ad. = Mammary adenoma.

Tumor samples were surgically excised, made into single cell suspensions, transfected with hGM-CSF plasmid expression vector, allowed to incubate overnight, and expression of hGM-CSF quantified into ng/10^6 cells/24 h through ELISA. Cell yield and tumor sample weights are shown as well. [Adapted from Hogge *et al.* (1998).]

10–174 ng, as determined by ELISA (Hogge *et al.*, 1998). Postbombardment cell viability usually exceeds 80–90%. Additionally, when these cells are administered *in vivo* via intradermal injection, a mean transgenic expression level of 5.61 ng of GM-CSF protein/vaccine site/24 h (range of .10–14.2 ng) has been observed using a variety of tumor cell types in 16 different dog tumor cases (Table 2; Hogge *et al.*, 1998). These expression levels have been shown by various investigators to confer "tumor vaccine activity" and immunologic effects of GM-CSF as opposed to the hematopoietic activity observed with higher concentrations in mouse model systems or as used in the clinical setting of postchemotherapy hematopoietic recovery (Hill *et al.*, 1995; Arnberg *et al.*, 1998; Rosenthal *et al.*, 1996; Burns *et al.*, 1995). These expression levels are much higher than those obtained from primary human tumor cells transfected or injected using the gene gun technique (Mahvi *et al.*, 1996).

Table 2

In Vivo **Transgenic hGM–CSF Protein Expression**
at the Sites of Administration, 24 h Postvaccination
in ng / Vaccine Site

Case no.	Tumor type	24 h
1	FSA	6.14
2	MEL	5.80
3	MEL	10.74
4	MEL	5.03
5	HSA	0.34
6	FSA	0.38
7	MEL	9.25
8	MEL	5.49
9	MEL	11.74
10	OSA	0.21
11	MEL	1.93
12	FSA	11.16
13	MEL	14.15
14	MEL	2.05
15	MEL	5.32
16	OSA	0.10
Mean		5.61

FSA = Fibrosarcoma, MEL = Melanoma, HSA = Hemangiosarcoma, OSA = Osteosarcoma. [Adapted from Hogge *et al.*
(1998).]

The unique features of gene gun transfection of canine tumor cells include
(1) the convenience and simple mode of bombarding cells as single or aggregated
cells in a suspension of very small volumes (~20 μl) and (2) the very short turn-
around time for creation of such tumor cell vaccines—from time of obtaining the
surgical specimens until time of intradermal vaccine injection was usually less than
4 h (Hogge *et al.,* 1998). This was another attractive element of the gene gun,
making long-term cell culture with all of its potential complications such as tumor
antigen loss, use of culture facilities, infection, and contamination, a minimal concern issue. Finally, (3) levels of gamma irradiation required to induce blockage of
mitotic cell division do not inhibit protein synthesis, and in many cases appeared to
increase total transgenic protein expression. This was a finding we observed using
both human and murine established cell lines such as B16 (Mahvi *et al.,* 1996). The
lack of a need for mitotic cell division to ensure transgene expression using the gene
gun (differing substantially from the retroviral method for gene transfer, for example) further points to the practical application of the gene gun technology in
human clinical and animal model studies. Functional cellular protein synthesis machinery, which remains intact following radiation exposure, is all that is required for
tumor vaccine strategies using the particle-mediated gene transfer method. These

features allowed the approval of similar gene gun-mediated DNA cancer vaccine/ gene therapy protocols for clinical trials in both companion dogs and humans with spontaneous cancer. These studies, including the phase I clinical trial in dogs that helped lead to the trial in humans, are discussed next.

III. *EX VIVO* GENE TRANSFER TO NORMAL CANINES USING ESTABLISHED MELANOMA CELL LINES

Normal research dogs were entered into a pilot study to determine whether intradermal injection of an established canine melanoma tumor cell line, transfected using particle-mediated gene transfer with cDNA for hGM-CSF, would generate biologically relevant levels of protein, resulting in demonstrable, histological changes at sites of vaccination. We first determined optimal gene transfer conditions for two different canine cell lines using luciferase cDNA transfection utilizing a variety of cDNA loading rates and shock wave pressures. The more metastatic melanoma cell line (CML-6M, Lauren Wolfe, Auburn University) expressed significantly greater levels of luciferase than the nonmetastatic melanoma cell line (CML-1, L. Wolfe; Hogge *et al.*, 1999). These data enabled us to examine the frequency of tumor cells transfected using the gene gun and flow cytometric analysis of cytokine expression plasmid compared to mock transfected or control cells. Viability and integrity of the cells were generally in excess of 80% and transfection frequencies between 10 and 25% were observed in the test cell lines (Hogge *et al.*, 1999). The impact of irradiation on protein synthesis and the induction of cell death were examined using proliferation and clonogenic assays. The cells were exposed to various radiation doses (0–100 Gy) using a cesium 137 radiation source and assayed for growth or colony formation. We showed that the radiation dose described for most tumor cell vaccine protocols (100 Gy) can adequately induce reproductive cell death. Time course assays examining transgenic hGM-CSF protein expression for both irradiated and nonirradiated cells showed comparable levels for both cell lines (Hogge *et al.*, 1999). Transfection efficiency and expression was highest in the metastatic CML-6M line and based on the luciferase and flow cytometric analyses, we employed this cell line for the *ex vivo* tumor vaccine preclinical canine studies.

Ten million irradiated, hGM-CSF-transfected canine melanoma cells per intradermal vaccine site were administered. Twenty-four hours postinjection, induration, erythema, and pruritus at sites of hGM-CSF-transfected tumor cell injection were noted, while control site reactions were much less noticeable. All reactions were well tolerated in the dogs. The hGM-CSF-transfected sites eventually developed small eschars and maintained a continued mild inflammatory response (Hogge *et al.*, 1999).

Histopathologically, hGM-CSF-transfected cell injection sites demonstrated prominent neutrophilic infiltrate, lymphocytes, and macrophages. Disruption of collagen bundles and a diffuse inflammatory necrosis were also evident by Day 7. Control β-gal transfected or nontransfected vaccine samples gave only moderate eosinophilic infiltrate and minimal neutrophil and macrophage infiltration. No necrotic inflammation was apparent (Figs. 2A and B). Overall, the tumor vaccine

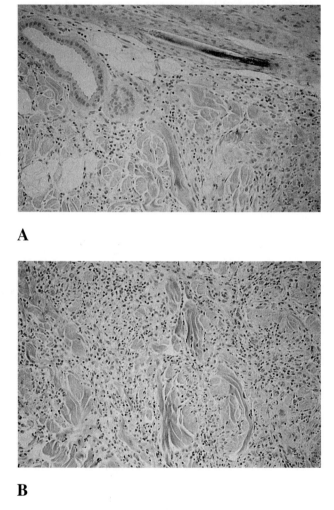

A

B

Figure 2 Representative histopathology from sites of tumor cell vaccination. 10^7 cells per site were injected intradermally and biopsied on D7. (A) β-gal-transfected, irradiated tumor cell control vaccine site biopsy, characterized by a mild inflammatory response with eosinophils and neutrophils present. (B) hGM-CSF-transfected tumor cell vaccination site biopsy, characterized by an intense inflammatory infiltrate of neutrophils, macrophages, and lymphocytes (250× magnification).

doses were well tolerated and all clinical assessments were within normal limits. ELISA demonstrated a range of 8.68–16.82 ng and a mean level of 12.58 ng of transgenic hGM-CSF protein/10^6 cells/24 h *in vivo* at the sites of injection. These levels decreased to a mean of 0.21 ng by Day 7 and 0.01 ng by Day 28, reflecting the transient transfection nature conferred by this particular application of particle-mediated gene transfer (Hogge *et al.*, 1999).

One of the most exciting results we obtained in this preclinical study was the obvious difference in the inflammatory responses between the hGM-CSF-transfected cancer vaccine sites and the control inoculation sites. No toxicity was observed in any animals using the dose injection level of 10^7 cells/site, which was well tolerated by the dogs. As a result, we initiated a phase I clinical trial using hGM-CSF-transfected, irradiated, autologous tumor cells in dogs with spontaneous melanoma and soft-tissue sarcomas.

IV. CLINICAL TRIAL OF *EX VIVO* TRANSFECTED AUTOLOGOUS CANCER VACCINES IN SPONTANEOUS TUMORS IN CANINES

Samples of spontaneous canine tumors were surgically obtained and enzymatically and mechanically dissociated to generate single cell suspensions for short-term cultivation and then transfected with reporter genes (β-gal or luciferase) or hGM-CSF. Tumor sample weights and cell numbers per gram of tissue were measured and supernatant expression levels for transgenic hGM-CSF in conditioned medium samples determined following 24 h of incubation (Table 1). We previously demonstrated the ability of gene gun transfection for production of transgenic cytokine proteins in primary tumor cell samples (Mahvi *et al.*, 1996). An interesting finding of our canine study was that tumors of a biologically less aggressive nature (e.g., fibrosarcoma) apparently express lower levels of protein than metastatic tumor samples (e.g., melanoma) (Hogge *et al.*, 1998). Given these results, we pursued a vaccine therapy strategy with autologous cytokine-gene-engineered tumor cell vaccines in client-owned companion dogs. Dogs admitted to the trial were vaccinated with multiple intradermal injection sites consisting of 10^7 irradiated, hGM-CSF- or reporter cDNA-transfected cells per site. The hGM-CSF sites revealed mild induration, erythema, and pruritus, all within acceptable limits, at the sites of injection 24 h postinjection. These responses were noticeably diminished at the reporter gene transfected sites (Figs. 3A and B). Both the hGM-CSF and the control vaccination site reactions were very similar to those observed in the normal dog study previously detailed.

As in the first study conducted by our group, prominent neutrophilic infiltrate, lymphocytes, and macrophages were visible on histopathologic evaluation of

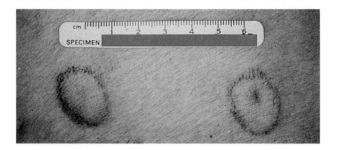

A

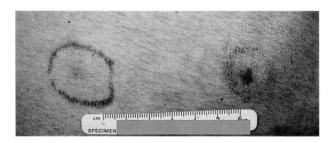

B

Figure 3 Photographs of gross inflammatory reactions at sites of tumor cell vaccine administration 24 h following injection. (A) 10^7 irradiated reporter cDNA transfected cells were administered per site and resulted in mild inflammation and small eschars. (B) hGM-CSF-transfected tumor cell vaccine sites revealed moderate induration, erythema, and pruritus, all within acceptable limits, at the sites of injection. Eschars were larger and persisted for longer periods of time than the control site reactions.

the hGM-CSF-transfected cell injection sites by 24 h postinjection. Disruption of collagen bundles, perivascular lymphoplasmacytic infiltrate, and diffuse inflammatory necrosis were also evident by Day 7. In contrast, control cDNA-transfected samples revealed moderate eosinophilic infiltrate with minimal neutrophil and macrophage infiltration, intact collagen bundles, and no necrotic inflammation. Histologically, the sites were identical to those demonstrated in Figs. 2A and B. Tumor cells were evident at the injection site for approximately 7 days post-vaccination (Fig. 4). Overall, tumor vaccine injections and transfected cell doses were well tolerated. All clinical assessments, hematological profiles, blood chemistry values, urinalysis, and so on, were within normal limits throughout the vaccination

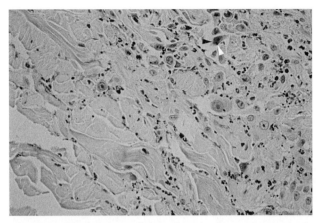

Figure 4 Representative histopathology of sites of vaccination to demonstrate intact, particle-mediated transfected tumor cells 24 h postinjection. Dark gray staining cells are tumor cells (black arrowhead). The $1-3$ μm dark, spherical objects are gold beads (white arrowhead) frequently associated with the tumor cells, although free beads can be located throughout the site of injection (400$\times$ magnification).

periods for all dogs entered and no hGM-CSF was detected in the serum samples at any time (Hogge *et al.*, 1998).

 In vivo transgenic hGM-CSF protein production levels at the sites of vaccination were assessed using ELISA and demonstrated a range of $0.10-14.15$ ng with a mean level of 5.61 ng of hGM-CSF/biopsy site/24 h at the sites of injection (Table 2). These expression levels decreased continually throughout the treatment periods. Samples obtained on Day 7 had a mean of .43 ng while those from Day 28 had a mean of .30 ng, reflecting the nature of transient transfection induced by the particle-mediated gene transfer technology and the destruction of tumor cells during the inflammatory responses at sites of vaccination (Hogge *et al.*, 1998).

 Sixteen animals with spontaneous tumor refractory to traditional therapy, often with advanced stage disease, were entered in the original clinical trial study. Objective evidence of tumor regression was observed in three animals (19%), with two dogs having measurable reduction at the primary tumor site or in metastatic regional lymph nodes as determined by three-dimensional measurement. Another dog had radiographic evidence of lung metastases, which had regressed following 3 months of therapy. This same animal later had a recurrence, which partially regressed in response to additional tumor vaccine administrations (Figs. 5A and B). After vaccination treatment, one dog with advanced melanoma had maintained a stable disease state for 125 days. Two animals died due to disease-related complications before completion of the tumor vaccine protocol. Two additional animals had

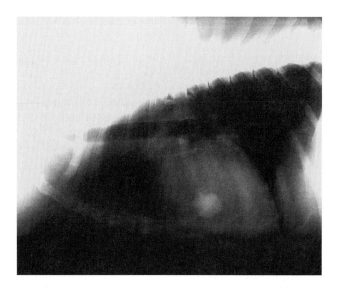

A

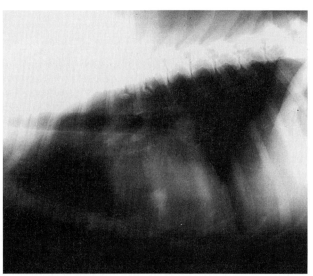

B

Figure 5 Lateral thoracic radiographs demonstrating a metastatic pulmonary lesion pretreatment (A). The lesion was histologically confirmed to be melanoma. (B) Partial response of the same pulmonary metastatic lesion following hGM-CSF tumor cell vaccine administration. Exposure time, body position, and magnification are the same between the two radiographs.

Table 3

Summary of *ex Vivo* Cancer Gene Therapy Clinical Trial in Dogs

Case #	Age (yrs)	Breed	Sex	Tumor	Response
1	3	Golden retriever	M	FSA	PD
2	11	Beagle cross	M	MEL	PD
3	8	Cocker	M	MEL	PR
4	9	Labrador	M	MEL	PD
5	11	German shepherd	F	HSA	PD
6	2	Labrador	F	FSA	PR
7	11	Newfoundland	M	MEL	CR
8	15	Cocker	F	MEL	SD
9	7	German shepherd	F	MEL	NED
10	9	Mastiff	M	OSA	PD
11	9	Golden retriever	M	MEL	PD
12	3	Terrier mix	F	FSA	NED
13	13	Golden retriever	M	MEL	PD
14	9	Labrador mix	M	MEL	PD
15	13	Airedale	M	MEL	PD
16	3	Labrador	M	OSA	PD

FSA = Fibrosarcoma, MEL = Melanoma, HSA = Hemangiosarcoma, OSA = Osteosarcoma, PD = Progressive disease, SD = Stable disease, PR = Partial response, CR = Complete response, NED = No Evidence of Disease. [Adapted from Hogge *et al.* (1998).]

no evidence of disease (NED) at the time of the conclusion of the first stage of the clinical trial (Table 3). No animals exhibited any signs of local or systemic toxicity. Results of this phase I clinical trial convinced our group of the utility of gene gun transfection methods and the clinical relevance of cytokine-transfected, locally delivered tumor cell vaccines. Our research group is currently conducting a phase II clinical trial in canine malignant melanoma.

V. *IN VIVO* GENE TRANSFER INTO SKIN AND ORAL MUCOSAL EPITHELIAL CELLS IN NORMAL DOGS

Several laboratories have shown that the gene gun method is an especially useful method in transfecting skin tissues or mucosal surfaces *in vivo,* not only for rodent systems (Larson *et al.,* 1998; Olsen *et al.,* 1997; Fynan *et al.,* 1993, 1995; Feltquate *et al.,* 1997; Eisenbraun *et al.,* 1993) but also for large animals (dogs, pigs, horses, and monkeys) (Keller *et al.,* 1996; Macklin *et al.,* 1998; Swain *et al.,* 1997; Lunn *et al.,* 1998; Fuller *et al.,* 1996). Using luciferase, β-galactosidase, or cytokine cDNA expression vectors as reporter or candidate therapeutic genes, we and others have shown that high levels of transgene expression can be routinely and readily obtained in a noninvasive, convenient, and versatile way. Epidermal cell layers can be effectively bombarded as a relatively uniform, well-organized cell or tissue sheet. Virtually all of the skin tissues tested, regardless of the animal species of origin, were capable of expressing similarly high levels of transgenic luciferase or β-gal activity when evaluated with a CMV-luc or β-gal cDNA expression vector. This indicates that the skin tissues of various animals can serve as excellent primary sites for a variety of transgenic studies, including gene therapy and DNA vaccination, as recently reviewed by Rakhmilevich and Yang (1997), Mahvi et al. (1997), and Yang et al. (Yang, Sun, and McCabe, 1996). We focus our next discussion on the dog oral mucosal epithelium as another primary target tissue site for in vivo gene transfer and transgenic expression studies.

As shown in Fig. 6, the dog oral mucosal epithelium contains multiple cell layers of varying thickness and, when gene gun transfected, express CMV β-gal or other reporter (e.g., luciferase) transgene activity at various cell layers (Keller et al., 1996). Expression of luciferase peaked at 1 day postbombardment and quickly decreased to baseline by Day 4. The return to baseline was not only due to the transient nature of the particle-mediated transfection process but also may reflect rapid epithelial renewal rates (Keller et al., 1996; Swain et al., 1997). Keller et al. also employed interleukin-2 (IL-2), IL-6, and GM-CSF to evaluate the levels and duration of expression, also finding a peak protein expression level at 24 h posttreatment and a return to baseline by 96 h. The experiments detailed here were also repeated in the skin of normal dogs, with similar findings (Keller et al., 1996).

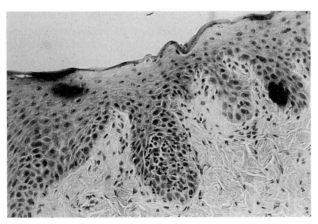

Figure 6 β-Gal expression in transfected canine oral mucosa 24 h after particle-mediated in vivo bombardment. The dark staining areas indicate successful transfection and protein expression within the area biopsied. Control transfected sites were negative for β-gal expression.

We have previously shown that by using the gene gun technique, various transcriptional promoters constructed to drive a luciferase reporter cDNA gene can be effectively evaluated for relative *in vivo* promoter strength in specific somatic tissues of rodent systems (Yang, Sun, and McCabe, 1996). In this approach, not only can different promoters be tested *in vivo* in the same tissue, the same promoter also can be tested in several different tissues or organs (e.g., skin vs muscle vs liver, etc.). It was further shown that inducible promoters, including the cAMP-inducible phosphoenolpyruvate carboxykinase promoter and the $ZnSO_4$-inducible metallo-thionine promoter, can be readily induced in these *in vivo* promoter assay experiments. As previously described, canine oral mucosal tissues and perhaps other epithelial surfaces (nasal, vaginal, etc.) can serve as excellent targets for the particle-mediated gene delivery system *in vivo*. We hence believe that a study of various *in vivo* promoter activity in canine oral mucosal epithelium at different anatomical sites may provide important and useful information on the capacity and limitation of promoter usage in canine somatic tissues.

These promoter applications and other transgene expression information generated may have useful implications for research into gene therapy and DNA vaccination using canine or other large animal models (see following). In pilot studies conducted in our laboratories, GM-CSF and IL-12 have been directly bombarded into oral tumor tissue (fibrosarcoma, melanoma, amelanotic melanoma) of companion dogs with spontaneous tumor. Histopathological results have shown an inflammatory response typically associated with the functional levels of protein expression for each of the transgenic cytokines, despite minimally detectable levels in biopsies of the bombarded tumor tissue (unpublished observations). We feel that the lack of a demonstrable tumor size reduction and poor expression may be due to a combination of rapid cellular turnover, the presence of necrotic tissue, the superficial penetration of the bead cDNA preparations employed, and the overall tumor burden. A potential application in the dog model with spontaneous tumors may be gene gun transfection of surgical margins postoperatively.

Similar to *ex vivo* gene transfer to tumor cell vaccines, gene gun-mediated *in vivo* gene transfer into skin tissue for transient production of potent, locally concentrated, physiologically relevant, transgenic cytokine proteins (e.g., IL-2, IL-6, IL-12, TNF-α, IFN-γ, and GM-CSF) has been demonstrated. This approach has been shown to eradicate or greatly inhibit the growth of established tumors in mouse tumor models (Sun *et al.,* 1995; Rakhmilevich *et al.,* 1996). In experimental mice, this potential cancer gene therapy schema utilizes three to five gene transfection treatments during the course of 1–2 weeks of therapy. Epidermal tissue overlying tumor sites or adjacent skin tissue sites were bombarded for gene transfer in a repetitive fashion, providing a short to mid-length (1–2 weeks) of high-level cytokine transgene expression. This schema resulted in local, regional and in some cases, systemic immunity against primary and metastatic tumors (Rakhmilevich *et al.,* 1996). These results suggest that transient cytokine transgene expressions, administered in a repetitive fashion, may effectively provide a useful gene therapy regime in appropriate clinical situations. These results were obtained in mouse models and have not been evaluated in the large animal tumor model systems. It has been shown that in many cases, data obtained from mouse tumor studies have unfortunately not been applicable to human cancer systems. We are of the opinion that a relevant large animal and its tumor system(s) can serve as a missing link to bridge this unfortunate gap in cancer therapy experimental systems.

We have obtained compelling data in *ex vivo* gene transfer derived cancer vaccines for treatments of dog oral melanoma and soft-tissue sarcoma. We hypothesize that a similar dog clinical trial study employing *in vivo* cytokine gene therapy using IL-12 or GM-CSF or other potent immune modulator cDNA expression vectors may generate useful transgene expression and immune response information, extending the mouse model data to human clinical trials. Due to the relatively large body size of most domestic dogs, it is possible that a larger area of mucosal or skin surface can be employed for multiple applications of gene transfer. The increased duration (e.g., 3−4 weeks) instead of 1−2 weeks commonly observed in murine studies can be evaluated for prolonged and increased transgenic expression of candidate therapeutic genes. Our laboratories are considering a systematic study with this approach.

VI. GENE GUN TECHNOLOGY AS A DELIVERY MECHANISM FOR DNA VACCINES

Nucleic acid vaccines are a method of inducing immunity, both humoral and cellular, by transfecting host cells with nonreplicating plasmid DNA or RNA that expresses immunogenic peptides or proteins. The advantages offered by nucleic acid vaccines include the lack of infectious nature of the vaccines, ease of mass production, cross-strain protection, generation of diverse immunologic responses, ease of administration, and the small total quantity of material required to induce a protective response (Jiang *et al.*, 1998). These factors make the gene gun an ideal delivery system.

Several laboratories have shown that gene gun transfection of skin tissue, ranging from mice (Larsen *et al.*, 1998; Olsen *et al.*, 1997) to several different large animals including pigs (Macklin *et al.*, 1998; Swain *et al.*, 1997), rhesus monkeys (Fuller *et al.*, 1996), and horses (Lunn *et al.*, 1998), can elicit humoral- as well as cellular-mediated immune responses, specifically cytotoxic T lymphocytes, resulting in immunization or vaccination against specific infectious pathogens or the related diseases (Fynan *et al.*, 1993, 1995; Zarozinski *et al.*, 1995). Both Th1 and Th2 T lymphocytes and the involved cytokines have been shown to become activated or expressed in these immune responses (Feltquate *et al.*, 1997). We hypothesize that these experimental studies can now be effectively evaluated in the canine oral mucosal experimental systems. We expect that valuable information on oral mucosal immunization, immune responses, and mucosal DNA vaccination can be obtained using the gene gun-mediated transgenic approach.

These various transgenic studies in pigs, horses, dogs, and rhesus monkeys strongly demonstrate that the gene gun technology can be effectively employed as a generic gene transfer tool for efficient transfection of animal skin tissue for gene delivery and expression applications.

A short-term burst of transient foreign gene expression in skin, muscle, or other tissues has been shown to be efficacious for eliciting both humoral- and cell-mediated immune responses against appropriate transgenic antigens. This can be achieved using gene gun delivery or intramuscular or intradermal injection of naked plasmid DNA carrying target cDNA gene expression vectors. Studies by a number of laboratories have shown that both DNA and RNA can be employed via particle-mediated bombardment to result in readily detectable production of antibodies against specific antigens synthesized transgenically *in vivo* in test animals (Feltquate *et al.*, 1997; Eisenbraun *et al.*, 1993; Swain *et al.*, 1997; Haynes *et al.*, 1996, 1994; Yang *et al.*, 1994). Particle-mediated gene transfer of a variety of growth factors, immunogenic components of infectious diseases, and such, have been and are currently being evaluated in monkeys, pigs, horses, and other large domestic animals.

Fuller *et al.* (1996) employed HIV and SIV expression constructs bombarded into the skin of rhesus monkeys and were able to elicit Env- and Gag-specific humoral responses. The authors found that 1 μg of DNA per dose was sufficient to induce immune responses in nonhuman primates using SIV gp160 and gp120 vectors driven by CMV-intron A promoters. Macklin *et al.* (1998) employed particle-mediated delivery of a DNA expression vector encoding the hemagglutinin (HA) of influenza virus to pig epidermis and mucosal surfaces. Humoral responses were generated that accelerated viral clearance and reduced the duration and extent of viral shed. The porcine influenza A virus model was shown to be relevant to humans both in disease progression and in the epidermis as a target for gene transfer. Swain *et al.* (1997) and Lunn *et al.* (1998), have demonstrated the effectiveness of human growth hormone (hGH) transfer to pigs and horses, respectively, using the gene gun method.

The DNA vaccination strategy using gene gun technology may have some advantages over other DNA immunization methods, including (1) the low dosage of DNA required for skin vaccination, (2) the rapid clearance or fate of plasmid DNA in targeted skin tissues, and (3) the noninvasive nature of gene delivery into the superficial 2–3 epidermal cell layers of skin tissue. Since different DNA vaccination techniques and different DNA dosages can apparently influence the balance of Th1 versus Th2 pathways (N-S. Yang *et al.* and M. Liu *et al.*, unpublished observations) the gene gun method may also be employed in combination with intramuscular injection of DNA or conventional protein antigens for vaccinations to bring about an optimal DNA vaccination effect.

VII. CONCLUSIONS

Gene gun technology is an emerging experimental tool that is yet to be fully realized for all of its potential in a variety of experimental and clinical applications in mammalian systems. Basic research in transgene biology, immunology, and tumor biology may all benefit from the gene gun approach. The implications for gene gun techniques in models that can effectively serve as "translational medicine research models" such as the dog appear to offer great potential. Spontaneous infectious diseases and neoplastic development, with emphasis on applications for both local and systemic control, may benefit substantially from the gene gun strategy. The growing field of DNA vaccines and cancer vaccines can utilize the gene gun and large animal models such as the canine to expand and integrate different gene delivery and transgene expression systems. For these basic and applied experimental techniques to be developed beyond animal models, the importance of an intermediate evaluation step cannot be overstated. Future applications in human clinical medicine (genetic diseases, biochemical deficiencies, infectious diseases, and neoplastic diseases) are dependent on data generated in these animal models and their related clinical trials. Safety, toxicity, dose schemes, and efficacy can be better evaluated in these translational models and complement the studies performed in murine or *in vitro* model systems. We believe that gene gun technology can offer an important option for these studies.

REFERENCES

Albertini, M. R., Emler, C. A., Schell, K., Tans, K. J., King, D. M., and Sheehy, M. J. (1996). Dual expression of human leukocyte antigen molecules and the B7-1 costimulatory molecule (CD80) on human melanoma cells after particle mediated gene transfer. *Cancer Gene Ther.* **3**, 192−201.

Arnberg, H., Letocha, H., Nou, F., Westlin, J.-F., and Nilsson, S. (1998). GM-CSF in chemotherapy-induced febrile neutropenia—a double blind study. *Anticancer Res.* **18**, 1255−1260.

Berns, A. J. M., Clift, S., Cohen, L. K., Donehower, R. C., Dranoff, G., Hauda, K. M., Jaffe, E. M., Lazenby, A. J., Levitsky, H. I., and Marshall, F. F. (1995). Phase I study of non-replicating autologous tumor cell injections using cells prepared with or without GM-CSF transduction in patients with metastatic renal cell carcinoma: Clinical protocol. *Human Gene Ther.* **6**, 347−368.

Burkholder, J. K., Decker, J., and Yang, N.-S. (1993). Rapid transgene expression in lymphocyte and macrophage primary cultures after particle bombardment-mediated gene transfer. *J. Immunol. Meth.* **165**, 149−156.

Cheng, L., Sun, J., Pretlow, T. G., Culp, J., and Yang, N.-S. (1996). CWR22 xenograft as an *ex vivo* human tumor model for prostate cancer gene therapy. *J. Nat. Cancer Inst.* **88**, 607−611.

Cheng, L., Ziegelhoffer, P. R., and Yang, N.-S. (1993). *In vivo* promoter activity and transgene expression in mammalian somatic tissues evaluated using particle bombardment. *Proc. Natl. Acad. Sci. USA* **90**, 4455−4459.

Conry, R. M., Widera, G., LoBuglio, A. F., Fuller, J. T., Moore, S. E., Barlow, D. L., Turner, J., Yang, N.-S., and Curiel, D. T.(1996). Selected strategies to augment polynucleotide immunization. *Gene Ther.* **3**, 67−74.

Eisenbraun, M. D., Fuller, D. H., and Haynes, J. R. (1993). Examination of parameters affecting the elicitation of humoral immune responses by particle bombardment mediated genetic immunization. *DNA Cell Biol.* **12,** 791–797.

Feltquate, D. M., Heaney, S., Webster, R. G., and Robinson, H. L. (1997). Different T helper cell types and antibody isotypes generated by saline and gene gun DNA immunizations. *J. Immunol.* **158,** 2278–2284.

Fuller, D. H., Murphy-Corb, M., Clements, J., Barnett, S., and Haynes, J. R. (1996). Induction of immunodeficiency virus-specific immune responses in rhesus monkeys following gene gun-mediated DNA vaccination. *J. Med. Primatol.* **25,** 236–241.

Fynan, E. F., Webster, R. G., Fuller, D. H., Haynes, J. R., Santoro, J. C., and Robinson, H. L. (1993). DNA vaccines: Protective immunizations by parental, mucosal, and gene-gun inoculations. *Proc. Natl. Acad. Sci. USA* **90,** 11478–11482.

Fynan, E. F., Webster, R. G., Fuller, D. H., Haynes, J. R., Santoro, J. C., and Robinson, H. L. (1995). DNA vaccines: A novel approach to immunization. *Int. J. Immunopharmacol.* **17,** 79–83.

Haynes, J. R., Fuller, D. H., Eisenbraun, M. D., Ford, M. J., and Pertmer, T. M. (1994). Accell® particle mediated DNA immunization elicits humoral, cytotoxic, and protective immune responses. *AIDS Res. Human Retroviruses* **10,** S43–S45.

Haynes, J. R., Fuller, D. H., McCabe, D., Swain, W. F., and Widera, G. (1996). Induction and characterization of humoral and cellular immune responses elicited via gene gun-mediated nucleic acid immunization. *Adv. Drug Del. Rev.* **21,** 3–18.

Hill, A. D. K., Naama, H. A., Calvano, S. E., and Daly, J. M. (1995). The effect of granulocyte macrophage-colony stimulating factor on myeloid cells and its clinical applications. *J. Leukocyte Biol.* **58,** 634–642.

Hogge, G. S., Burkholder, J. K., Culp, J., Albertini, M. R., Dubielzig, R., Keller, E., Yang, N.-S., and MacEwen, E. G. (1998). Development of hGM-CSF transfected tumor cell vaccines in spontaneous canine cancer. *Human Gene Ther.* **9,** 1851–1861.

Hogge, G. S., Burkholder, J. K., Culp, J., Albertini, M. R., Dubielzig, R., Yang, N.-S., and MacEwen, E. G. (1999). Preclinical development of hGM-CSF transfected melanoma cell vaccine using established canine cell lines and normal canines. *Cancer Gene Ther.*

Irvine, K. R., Rao, J. B., Rosenberg, S. A., and Restifo, N. P. (1996). Cytokine enhancement of DNA immunization leads to effective treatment of established pulmonary metastasis. *J. Immunol.* **156,** 229–245.

Jiang, W., Baker, H. J., Swango, L. J., Schorr, J., Self, M. J., and Smith, B. F. (1998). Nucleic acid immunization protects dogs against challenge with virulent canine parvovirus. *Vaccine* **16,** 601–607.

Jiao, S., Cheng, L., Wolff, J. A., and Yang, N.-S. (1993). Particle bombardment-mediated gene transfer and expression in rat brain tissues. *BioTechnology* **11,** 497–501.

Keller, E. T., Burkholder, J. K., Shi, F., Pugh, T. D., McCabe, D., Malter, J., MacEwen, E. G., Yang, N.-S., and Ershler, W. B. (1996). *In vivo* particle mediated cytokine gene transfer into canine oral mucosa and epidermis. *Cancer Gene Ther.* **3,** 186–191.

Larsen, D. L., Dybdahl-Sissoko, N., McGregor, M. W., Drape, R., Neumann, V., Swain, W. F., Lunn, D. P., and Olsen, C. W. (1998). Coadministration of DNA encoding interleukin-6 and hemagglutinin confers protection from influenza virus challenge in mice. *J. Virol.* **72,** 1704–1708.

Lunn, D. P., Olsen, C. W., Soboll, G., McGregor, M. W., Macklin, M. D., McCabe, D. E., and Swain, W. F. (1998). Development of practical DNA vaccination strategies for use in horses. In *Eighth international conference on equine infectious diseases,* Dubai, United Arab Emirates, March 23–26, 1998.

Macklin, M. D., McCabe, D., McGregor, M. W., Neumann, V., Meyer, T., Callan, R., Hinshaw, V. S., and Swain, W. F. (1998). Immunization of pigs with a particle-mediated DNA vaccine to influenza A virus protects against challenge with homologous virus. *J. Virol.* **72,** 1491–1496.

Mahvi, D. M., Burkholder, J. K., Turner, J., Culp, J., Malter, J. S., Sondel, P., and Yang, N.-S. (1996). Particle-mediated gene transfer of GM-CSF cDNA to tumor cells: Implications for a clinically relevant tumor vaccine. *Human Gene Ther.* **7,** 1535–1543.

Mahvi, D. M., Sheehy, M. J., and Yang, N.-S. (1997). DNA cancer vaccines: A gene gun approach. *Immunol. Cell Biol.* **75**, 456–460.

Olsen, C. W., McGregor, M. W., Dybdahl-Sissoko, N., Schram, B. R., Nelson, K. M., Lunn, D. P., Macklin, M. D., Swain, W. F., and Hinshaw, V. S. (1997). Immunogenicity and efficacy of baculovirus-expressed and DNA-based equine influenza virus hemagglutinin vaccines in mice. *Vaccine* **15**, 1149–1156.

Rajagopalan, L. E., Burkholder, J. K., Turner, J., Culp, J., Yang, N.-S., and Malter, J. S. (1995). Granulocyte macrophage-colony stimulating factor mRNA stabilization enhances transgenic expression in normal cells and tissues. *Blood* **86**, 2551–2558.

Rakhmilevich, A. L., Turner, J., Ford, M. J., McCabe, D., Sun, W. H., Sondel, P. M., Grota, K., and Yang, N.-S. (1996). Gene gun-mediated skin transfection with interleukin 12 gene results in regression of established primary and metastatic murine tumors. *Proc. Natl. Acad. Sci. USA* **93**, 6291–6296.

Rakhmilevich, A. L., and Yang, N.-S. (1997). Particle-mediated gene delivery system for cancer research. In M. Strauss and J. A. Barranger (Eds.), *Concepts in gene therapy* (pp. 109–120). New York: Walter de Gruyter.

Reba, I. (1966). Applications of the Coanda effect. *Sci.Am.* **214**, 84–92.

Rosenthal, F. M., Kulmburg, P., Früh, R., Pfeifer, C., Veelken, H., Mackensen, A., Köhler, G., Lindemann, A., and Mertelsmann, R. (1996). Systemic hematologic effects of granulocyte colony stimulating factor produced by irradiated gene-transfected fibroblasts. *Human Gene Ther.* **7**, 2147–2156.

Sun, W. H., Burkholder, J. K., Sun, J., Culp, J., Turner, J., Lu, X. G., Pugh, T. D., Ershler, W. B., and Yang, N.-S. (1995). *In vivo* cytokine gene transfer by gene gun suppresses tumor growth in mice. *Proc. Natl. Acad. Sci. USA* **92**, 2889–2893.

Swain, W. F., Macklin, M. D., Neumann, V., McCabe, D. E., Drape, R., Fuller, J. T., Widera, G., McGregor, M., Callan, R. J., and Hinshaw, V. (1997). Manipulation of immune responses via particle mediated polynucleotide vaccines. *Behring Inst. Mitteilungen* **98**, 73–78.

Thompson, T. A., Gould, M. N., Burkholder, J., and Yang, N.-S. (1993). Transient promoter activity in primary rat mammary epithelial cells evaluated using particle bombardment gene transfer. *In Vitro Cell. Dev. Biol.* **29A**, 165–170.

Tüting, T., Gambotto, A., Baar, J., David, I. D., Storkus, W. J., Zavodny, P. J., Narula, S., Tahara, H., Robbins, P. D., and Lotze, M. T. (1997). Interferon-α gene therapy for cancer: Retroviral transduction of fibroblasts and particle mediated transfection of tumor cells are both effective strategies for gene delivery in murine tumor models. *Gene Ther.* **4**, 1053–1060.

Yang, N.-S., Burkholder, J. K., McCabe, D. M., Neumann, V., and Fuller, D. H. (1997). Particle-mediated gene delivery *in vivo* and *in vitro*. In A. C. Boyle (Ed.), *Current protocols in human genetics* (Vol. 12.6, pp. 12.6.1–12.6.14). New York: Wiley-Liss.

Yang, N.-S., Burkholder, J., Roberts, B., Martinell, B., and McCabe, D. (1990). *In vivo* and *in vitro* gene transfer to mammalian somatic cells by particle bombardment. *Proc. Natl. Acad. Sci. USA* **87**, 9568–9572.

Yang, N.-S., McCabe, D. E., and Swain, W. F. (1996). Methods for particle mediated gene transfer into skin. In P. Robbins (Ed.), *Methods in molecular medicine, gene therapy protocols edition* (pp. 281–296). New York: Humana Press.

Yang, N.-S., Qiu, P., and Ziegelhoffer, P. (1994). Particle bombardment-mediated RNA delivery as an approach for gene therapy. *J. Cell Biochem.* **S18A**, 230.

Yang, N.-S., Sun, W. H., and McCabe, D. (1996). Developing particle-mediated gene-transfer technology for research into gene therapy of cancer. *Mol. Med. Today* November, 476–481.

Yang, N.-S., and Ziegelhoffer, P. R. (1994). The particle bombardment system for gene transfer. In N.-S. Yang and P. Christou (Eds.), *Particle bombardment technology for gene transfer* (pp. 117–142). New York: Oxford University Press.

Zarozinski, C. C., Fynan, E. F., Selin, L. K., Robinson, H. L., and Welsh, R. M. (1995). Protective CTL-dependent immunity and enhanced immunopathology in mice immunized by particle bombardment with DNA encoding an internal virion protein. *J. Immunol.* **154**, 4010–4017.

CHAPTER 9

Polyethylenimines: A Family of Potent Polymers for Nucleic Acid Delivery

Antoine Kichler, * **Jean-Paul Behr,**[†] **and Patrick Erbacher**[†]

* URA-CNRS 1923—Genethon III, Groupe de Vectorologie, F-91002, Evry Cedex 2, France
† Laboratoire de Chimie Génétique, UMR-7514 CNRS/Université Louis Pasteur, Faculté de Pharmacié, F-67401, Strasbourg-Illkirch, France

Ethylenimine polymers can be used as carriers to increase cellular uptake of the potent prodrug DNA. For cell targeting, ligands that are recognized by receptors present at the cell surface can easily be coupled to the cationic polymer. Packaging of DNA leads to compact particles, which are readily taken up by cells. Successful gene transfer requires subsequent protection of DNA from cellular nucleases, release of the DNA (complexes) from the endocytic vesicles, diffusion of the complexes through the cytosol, and nuclear import of the DNA. Complex disassembly must also occur at some stage. For *in vivo* applications, DNA complexes must in addition be able to penetrate the tissue of interest and the carrier/DNA particle should avoid activation of the complement system. Here we consider the structure and topology of polyethylenimines, their mode of action, and their *in vitro* and *in vivo* transfection potential.

I. INTRODUCTION

A better understanding of the cellular machinery has recently allowed the development of new drug candidates that are very efficient *in vitro*. However, their efficiency is often lowered at the final stage (*in vivo*) by a low bioavailability, a lack of selectivity, or even by an unexpected toxicity due to the high concentrations needed to obtain the desired effect. The therapeutic index of such bioactive molecules can be increased by their association with a carrier. The properties of these carriers, which were designed to transport "conventional" drugs, are also suitable and perhaps necessary for the delivery of what is potentially one of the most powerful drugs—namely, DNA. Beside natural carriers such as viruses, molecular carriers such as cationic lipids and cationic polymers were developed in order to enhance the delivery of DNA. Nonviral systems are essentially based on DNA compaction into nanometric particles by electrostatic interaction between the polyanion and the carrier. In such a condensed form, DNA is protected from degradation and provided the particles are positively charged, their binding to anionic cell-surface proteoglycans is an efficient means of entering adherent cultured cells (Fig. 1).

In order to develop nonviral systems for *in vivo* gene delivery by the systemic route, the candidate vector must meet several more stringent parameters (Fig. 1): (1) it must be able to condense DNA and to end up with neutral complexes, avoiding toxicity and unnecessary interactions with circulating cells and charged molecules (extracellular matrix, proteins, and especially complement proteins), which are widely distributed *in vivo*; (2) it must be able to form compact and stable complexes smaller than 100 nm and facilitate their diffusion within the blood flow, their extravasation through the fenestrated endothelium, and their size-dependent migration in tissue or interstitium to target cells; (3) it must be able to mediate cell-specific recognition and internalization and avoid nonspecific gene delivery; (4) it must have the ability to escape from the endosome and must facilitate exogenous gene entry and expression in the nucleus, even in nondividing cells. While existing vectors can fulfill some of these criteria, presently none can provide all of the necessary functions.

Among the synthetic transfection reagents developed recently, ethylenimine

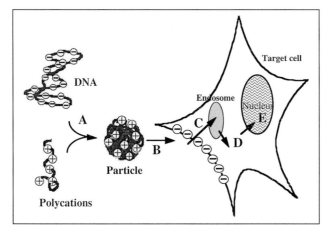

Figure 1   Cellular barriers for *in vitro* gene delivery. (A) Complex formation, (B) interaction of the complex with the target cell, (C) entry of the polyplex, (D) escape of the DNA from the endosome, and (E) nuclear transport of the DNA (followed by complex disassembly if the DNA carrier is still present).

polymers (PEIs) are among the most promising. This review focuses on various aspects of the use of these polymers.

II. SYNTHESIS, STRUCTURE, AND TOPOLOGY OF POLYETHYLENIMINES

Acid-catalyzed polymerization of aziridine produces polyethylenimines with a random branched topology (Dick and Ham, 1970; Fig. 2). These compounds thus contain ethylamine as the repeating unit, which gives them a high solubility in water. The fact that every third atom is an amino nitrogen makes them the organic macromolecules with the highest cationic charge density potential. Among the nitrogens present in branched PEIs, 25, 50, and 25% are primary, secondary, and tertiary amines, respectively. The ionization properties of PEI have been reported recently (Suh *et al.*, 1994). According to the protonation versus pH profile, only about 20% of the amino nitrogens are protonated under physiological conditions. Therefore, in contrast to polymers such as polylysine, PEI exhibits considerable buffer capacity over almost the entire pH range (Suh *et al.*, 1994; Tang and Szoka, 1997).

In addition to branched PEIs of various molecular weights, linear PEIs and several PEI derivatives such as ethoxylated PEI and PEI modified by epichlorohydrin are also available (Table 1; Boussif, 1996).

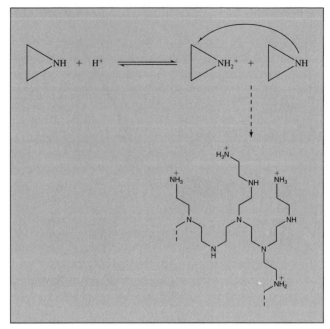

Figure 2 Synthesis of branched PEIs. [Dick and Ham (1970).]

Table 1

Transfection Efficiency of PEIs

PEI (average MW)	Topology	*In vitro* efficiency	*In vivo* efficiency	Selected references
800	Branched	−	Nd	Boussif, 1996
2,000	Branched	++++*	Nd	Boussif, 1996 Baker *et al.,* 1997
22,000	Linear	++++	++++	Boussif, 1996 Goula *et al.,* 1998a, b
25,000	Branched	++++	+++	Boussif, 1996 Abdallah *et al.,* 1996
50,000	Branched	++++	+	Boussif, 1996 Abdallah *et al.,* 1996
600,000–1,000,000 (800,000)	Branched	++++**	+**	Boussif *et al.,* 1996 Abdallah *et al.,* 1996
Ethoxylated PEI (50,000)	Branched	++++	Nd	Boussif, 1996
Epichlorohydrin modified PEI (20,000)	Branched	++++	Nd	Boussif, 1996

* Only in combination with replication-defective adenoviruses (Baker *et al.,* 1997).

** PEI 800 kDa without ligand (the efficiency can be increased by coupling a ligand such as transferrin to the carrier; Kircheis *et al.,* 1997 and E. Wagner, unpublished data).

III. DNA CONDENSATION

A. PEI/DNA COMPLEXES

Polycation-mediated gene delivery is based on the interaction of the polycation with the negative charges carried by the phosphate groups of DNA. DNA binding, as studied by means of gel retardation assays, was found to be a function of the cation/anion ratio. For the branched PEI 25 kDa, DNA is completely complexed at a molar ratio of PEI nitrogen atoms to DNA phosphate (N/P) above 2 (Zanta *et al.*, 1997). According to the pK profile of PEI, this N/P value gives a theoretical $+/-$ charge ratio of about 0.4. However, this calculation is based on the pK profile of PEI in the absence of DNA. Binding of nucleic acids to the polymer will shift the protonation profile of PEI. In fact, PEI 25 kDa leads to particles having a zeta potential (indicative of the particle's surface charge) ranging from strongly negative (around -50 mV at N/P $= 2$) to strongly positive (around 20 mV at N/P of 10) values, neutral particles being observed at N/P $= 3.5$ (Erbacher *et al.*, in press). Assuming a charge ratio of 1/1 at complex neutrality, the branched 25 kDa polymer has one-third of charged nitrogens at neutral pH, meaning that not only primary amines (25% content) are protonated but that 10% of the higher order amines are charged as well.

B. PREPARATION AND CHARACTERIZATION OF THE POLYPLEXES

Examination of PEI 25 kDa/DNA complexes using electron microscopy revealed a homogenous population of essentially toroidal particles with a size range of 40–60 nm (Tang and Szoka, 1997). Dunlap and co-workers (1997) found that PEI/DNA complexes made in 150 mM NaCl and imaged topographically by scanning atomic force microscopy in low salt (15 mM NaCl) have sizes ranging from 20 to 40 nm in diameter.

Although little is known about the size limitations for ligand binding and internalization of receptors, the size of the transfection particles is a fundamental factor not only for efficient internalization but also for *in vivo* diffusion. Since targeting of organs such as the liver after systemic administration of the polyplexes requires particles of small size (i.e., <100 nm) because the complexes must leave the vascular system, the size of the DNA complexes should be as small as possible.

The method of preparation of the complexes is important because DNA condensation by polycations is a function of the nature and concentration of all ions present. Boussif and co-workers (1995) therefore varied pH, volume, and sodium chloride concentration of the compaction medium and tested the transfection efficiency of all the resulting complexes. None of these modifications significantly increased the *in vitro* transfection efficiency of PEI. The order of adding reagents seemed to be important, however, since the dropwise addition of the polymer solution to the plasmid produced polyplexes that were 10 times more efficient *in vitro* than those obtained by adding the DNA to the polymer.

A recent study demonstrated that linear PEI 22 kDa/DNA complexes that are formulated in 5% glucose have a much increased *in vivo* transfection efficiency as well as an increased diffusibility of the particles in the tissues, compared to complexes prepared in 150 mM sodium chloride (Goula *et al.*, 1998a). Thus, one has to keep in mind that even if the *in vitro* transfection efficiencies are unchanged when the conditions of preparation of the polyplexes are varied, *in vitro* data can neither predict whether the polyplexes will have unchanged size and structure nor be predictive of *in vivo* efficacy.

IV. *IN VITRO* DNA DELIVERY

Polylysine is efficiently transported into mammalian cells by a process referred to as nonspecific adsorptive endocytosis (Leonetti *et al.*, 1990). The proposed mechanism involves binding of the polycation to the negatively charged plasma membrane, which is thought to stimulate pinocytosis (Duncan *et al.*, 1979). *In vitro* studies with both lipoplexes and polyplexes have shown that electrostatic interactions between the negatively charged cell membranes and the positively charged DNA complexes are enhanced by increasing the overall charge of the complexes, which in turn is achieved by increasing the ratio of carrier to DNA. Recent data gave evidence that the positively charged DNA complexes can enter cells via binding to membrane-associated sulfated proteoglycans (Labat-Moleur *et al.*, 1996; Mislick and Baldeschwieler, 1996). Electron microscopy has been used to follow these interactions *in vitro* (Zabner *et al.*, 1995; Labat-Moleur *et al.*, 1996).

The *in vitro* transfection efficiency of PEI 800 kDa has been compared to that of an effective cationic lipid, namely the lipopolyamine transfectam, on a large variety of cell lines and primary cultures (Boussif *et al.*, 1996). The results show efficiencies at least as high as those of the cationic lipid. PEIs of molecular weight of 2 kDa or less are not efficient in gene transfer even at high N/P ratios, although they are able to complex DNA (Boussif, 1996), unless endosomolytic agents such as replication-defective adenoviruses are added to the DNA complexes (Table 1).

V. RECEPTOR-MEDIATED GENE DELIVERY

In order to selectively transfect the cells of interest, a rational approach consists of coupling a ligand that is recognized by a receptor expressed at the cell surface to the carrier. For efficient delivery, receptors that mediate endocytosis of their ligand(s) and recycle to the plasma membrane are particularly attractive.

A. CHEMISTRY

Ligands can easily be coupled to the amino nitrogens of PEI. This can be achieved by using the following methods: (1) aldehydes such as reducing sugars or oxidized glycoproteins can be coupled to PEI by reductive amination (Zanta *et al.,* 1997); (2) compounds containing a carbodiimide-activated carboxyl group are coupled through amide bond formation; (3) isothiocyanate groups to form a thiourea derivative; and (4) the amino groups of PEI can be modified with bifunctional linkers containing a reactive ester such as SPDP (Carlsson *et al.,* 1978). Subsequent reaction with a thiol group leads to formation of a disulfide bond.

Purification of the conjugates is easily achieved by dialysis or gel filtration. Proteins such as transferrin are usually coupled to PEI in a 1/1 to 5/1 molar ratio (Kircheis *et al.,* 1997; transferrin-PEI mediated transfection is discussed elsewhere in this book), while 4–10% of the amino groups should be substituted with small molecules such as sugar residues (Zanta *et al.,* 1997) or short RGD-containing peptides (Erbacher *et al.,* 1999).

B. RESULTS AND COMMENTS

Receptor-mediated gene delivery was, until recently, essentially performed with the cationic polymer polylysine. A great variety of ligands such as asialoglycoproteins, transferrin, insulin, immunoglobulins, growth factors, or carbohydrates were coupled to polylysine in order to target selectively a cell of interest (for review see Kichler *et al.,* 1996). Similar targeting studies are now emerging with PEI as a DNA carrier. We will briefly present the data obtained with PEI-galactose and PEI-RGD.

1. PEI–Galactose

The "magic bullet" concept of Paul Ehrlich seems particularly suitable for cells that express carbohydrate-binding proteins (lectins) (Jones, 1994; Wadhwa and Rice, 1995). Hepatocytes are interesting target cells because the liver is a central organ of several genetic and acquired diseases amenable to gene therapy (Ledley, 1993). The asialoglycoprotein receptor of hepatocytes (ASGP-R or Gal/GalNAc receptor) has been investigated extensively (for review see Ashwell and Harford, 1982) and targeted for drug, oligonucleotide, or gene delivery (Wu and Wu, 1987; Plank *et al.,* 1992; Midoux *et al.,* 1993; Wadhwa and Rice, 1995).

Zanta and associates (1997) derivatized amino groups of the branched PEI of 25 kDa with simple galactose residues. A rough screening showed that substitution of about 5% of the total amino groups is optimal to target hepatocytes *in vitro*. These authors also showed that (1) in the human hepatocarcinoma cells HepG2 and the murine hepatocyte cell line BNL CL.2 an important part of the electroneutral PEI-galactose/DNA complex delivery is mediated through the ASGP-R; (2) the PEI-galactose/DNA particles, in contrast to PEI/DNA complexes, do not need to have a net positive charge to be efficient. This point could be particularly important for *in vivo* applications because neutral particles should not bind to anionic extracellular proteoglycans.

However, several problems remain unsolved, such as downsizing of the particles. Small DNA complexes are needed (1) to avoid phagocytic cells, especially Kupffer cells, which also possess membrane lectins that can recognize galactose residues, and (2) to cross the vascular endothelium fenestration (preliminary transmission electron microscopy pictures of PEI-galactose/DNA complexes showed rather large and polydisperse 100 to 400-nm particles). The problem of receptors that are present on more than one cell type can be partly overcome by the use of tissue-specific promoters (for review see Cooper, 1996). Nevertheless, the effects of introducing DNA into tissues other than the tissue of interest is a concern. Thus, despite the apparent flexibility in the design of targeted polyplexes, there are problems associated with their use.

2. PEI–RGD

Cell-surface integrins are naturally exploited for cell entry by certain viruses. Many integrins, including $\alpha 5\beta 1$ and most αV-containing proteins, recognize multiple Arg-Gly-Asp (RGD) peptide sequences present in cell adhesion, serum, or extracellular matrix proteins (Hynes, 1992). Short synthetic RGD-containing peptides can mimic the natural ligands, and integrin-mediated cell entry has been exploited to internalize DNA condensed with a RGD-containing peptide coupled to an oligolysine. This targeted gene delivery system is capable of transfecting epithelial and endothelial cells *in vitro* (Hart *et al.*, 1996; Harbottle *et al.*, 1998). Recently, thiol-derivatized PEI 25 kDa was conjugated to the integrin-binding peptide CYGGRGDTP via a disulfide bridge. In the presence of PEI-RGD, plasmid DNA was condensed into a rather homogeneous population of 30 to 100-nm toroidal particles (TEM observations) in 150 mM salt. Their surface charge was close to neutrality as a consequence of the shielding effect of the prominent zwitterionic peptide residues. Transfection efficiency of integrin-expressing epithelial (HeLa) and fibroblast (MRC5) cells was increased by 10- to 100-fold as compared to PEI even in serum (Fig. 3). This large enhancement factor was lost when aspartic acid was replaced by glutamic acid in the targeted peptide sequence (RGD/RGE), confirming the involvement of integrins in transfection. PEI-RGD/DNA complexes

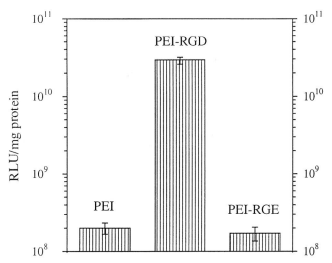

Figure 3 Integrin-mediated transfection. Transfection of HeLa cells with pCMVLuc plasmid (2 μg/well) complexed with PEI, PEI-CYGGRGDTP (PEI-RGD), or PEI-CYGGRGETP (PEI-RGE), at N/P ratio = 10. Luciferase activity was measured 24 h after transfection. Values are the mean ± s.d. of two independent experiments made in triplicate.

thus share constitutive properties with adenovirus such as size, a centrally protected DNA core, and early properties such as cell entry mediated by integrins and acid-triggered endosome escape. Recently, a RGD-containing peptide was shown to target preferentially a toxin to tumor blood vessels in a mouse model (Arap *et al.,* 1998). Using PEI-RGD, targeted gene delivery based on selective expression of $\alpha V\beta 3$ integrin in the tumor vasculature might become applicable to gene therapy for cancer.

VI. ENDOSOMAL RELEASE OF PEI / DNA COMPLEXES

Although large amounts of DNA complexes are taken up by most of the cells *in vitro* (up to half of the added material; Labat-Moleur *et al.,* 1996; Escriou *et al.,* 1998), the majority of the transfection particles seem to remain sequestered in endocytic vesicles. However, a small percentage of the plasmids can usually be delivered into the cytosol, as shown by experiments using a cytoplasmic expression system (Zabner *et al.,* 1995). In order to favor the transmembrane passage of the DNA into the cytosol, several helper molecules including replication-defective adenoviruses, chloroquine, glycerol, or fusogenic peptides are used, especially when polymers such as polylysine are used as the DNA carrier (for review see Kichler *et al.,* 1996). PEI, in contrast, does not need the addition of endosomolytic agents

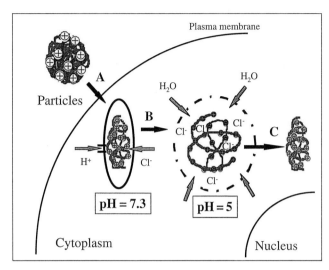

Figure 4   The proton sponge effect. After endocytosis of the cationic complexes (A), acidic endosome buffering (B) leads to increased osmotic pressure and finally to lysis (C).

to be efficient. A possible explanation for the high gene transfer efficiency of PEI is that the buffering of the endosomes through PEI provokes a massive proton accumulation followed by passive chloride influx. These events should cause osmotic swelling and subsequent endosome disruption ("proton sponge effect"; Boussif *et al.*, 1995), thus permitting the escape of endocytosed materials (Fig. 4). This hypothesis is based on the chemical structure of the polymer, that is, the pK of the amino groups; given the pH difference between the extracellular space and lysosomes, the estimated fraction of protonated nitrogens of PEI will increase from 20 to 45% (Suh *et al.*, 1994; Tang and Szoka, 1997).

We were interested to see whether the effects of an inhibitor of the proton-translocating ATPases could provide insights into the mechanism of escape of the PEI/DNA complexes from the endocytic vesicles. Therefore, we carried out transfection experiments in the absence and the presence of bafilomycin A1, an antibiotic isolated from *Streptomyces sp.*, which selectively inhibits vacuolar type H^+-ATPases (Bowman *et al.*, 1988; Yoshimori *et al.*, 1991). Our data show that in the presence of the drug (at 188 nM) the transfection of the human hepatocarcinoma cells HepG2 resulted in a decrease of the luciferase activity of about 100-fold compared to nontreated cells, regardless of the structure of the polymer (A. Kichler, F. Pulcini, and O. Danos, unpublished results). In contrast, the efficiency of the monocationic lipid DOTAP, whose protonation state does not significantly change in the pH range of 5−7.5, is not altered in the presence of bafilomycin A1. Thus, an important part of the efficiency of PEIs relies on their considerable buffer capacity, which results in an endosomolytic-like activity. However, this property is not sufficient to make a molecule a good carrier for gene delivery. Indeed, polyhistidine, which also possesses a buffer capacity at pH around 6.5, is unable to efficiently transfect cells *in vitro* (Boussif, 1996). It thus seems that other characteristics such as flexibility are necessary to make a polymer an efficient vehicle for DNA delivery (Tang *et al.*, 1996).

VII. NUCLEAR TRANSPORT OF THE DNA (COMPLEXES)

Once the DNA particle is in the cytosol it must still enter the nucleus. How this occurs is largely unknown, but the transport of the transfected DNA from the cytosol to the nucleus is certainly the most important limitation to successful gene transfer (Zabner *et al.*, 1995; Pollard *et al.*, 1998). The import of proteins into the nucleus occurs through the nuclear pore complexes, which allow diffusion of small molecules (<50 kDa) and can accommodate the active transport of particles of approximatively 25 nm in diameter (for review see Görlich, 1997; Nigg, 1997). Because nondividing cells are poorly transfectable by nonviral gene transfer systems and transfection particles generally have a diameter larger than 25 nm, it was postulated that nuclear uptake occurs preferentially in those cells that are entering mitosis, consequent to breakdown of the nuclear membrane. Moreover, very low levels of gene expression have been reported for LTK^- cells (Capecchi, 1980) and NIH 3T3 cells (Dowty *et al.*, 1995) after injection of plasmid DNA into the cytosol of these cells.

Recently, a nice study from Pollard and associates (1998) demonstrated that the PEI of 25 kDa, in contrast to cationic lipids, is able to promote gene delivery from the cytoplasm to the nucleus. Indeed, when 10,000 copies of naked DNA were injected into the cytoplasm of COS-7 cells, only 13% of the cells expressed the transgene, whereas the same efficiency could be obtained with 10 times fewer copies of the plasmid when DNA was associated with PEI. However, it must be stressed that the nuclear transport efficiency of PEI/DNA complexes markedly varies with the cell line; thus, the intracellular barriers to plasmid trafficking quantitatively vary with the cell type. For *in vivo* experiments, these data mean that at least 1000 copies of plasmid DNA must be present in the cytosol of each target cell. If we assume that 1 copy of 10^{4-5} will reach the cell of interest, enter the cell, and escape from the endosomes, 10^{7-8} copies of the gene are needed for each cell of interest!

Dissociation of PEI/DNA complexes to allow transcription seems not to be a limiting step since injection of these complexes into the nucleus produces comparable levels of gene expression to injection of uncoated DNA (Pollard *et al.*, 1998). This is in sharp contrast to other DNA carriers, especially cationic lipids when they are used at a $+/-$ charge ratio at which they are efficient (i.e., generally a $+/-$ charge ratio > 1) (Zabner *et al.*, 1995; Pollard *et al.*, 1998).

VIII. *IN VIVO* GENE DELIVERY

Whereas most of the cationic vectors are able to deliver plasmids to many cell lines cultured *in vitro*, only a few of them, including PEIs, were found to be capable of transfecting cells *in vivo*.

Stereotaxic intracerebral injections of PEI/DNA complexes in both adult and newborn mice provided levels of transfection equal to those found *in vitro* for the same amount of DNA applied to primary neuronal cells (Boussif *et al.*, 1995; Abdallah *et al.*, 1996). The best levels of expression were obtained with the linear 22 kDa molecule (Abdallah *et al.*, 1996; Goula *et al.*, 1998a). Immunostaining showed that both neurons and glia were transfected *in vivo*. The highest level of transfection in the adult brain was found with rather low amounts of PEI: ≤ 6 eq. nitrogens per DNA phosphate, which gives a theoretical $+/-$ charge ratio of 3 and a zeta potential of about 30 mV (P. E. and J. S. Remy, unpublished results). Intraventricular injection of complexes formulated in glucose showed the complexes with sizes ranging from 30 to 100 nm to be highly diffusible in the cerebrospinal fluid (Goula *et al.*, 1998a). Thus, it is possible that the use of PEI/DNA particles with a $+/-$ charge ratio close to 1 allows the complexes to diffuse widely through the tissue, whereas highly charged cationic complexes stick to substrates around the injection site.

In another *in vivo* study, the PEI of 22 kDa complexed to DNA in 5% glucose was delivered to adult mice through the tail vein (Goula *et al.*, 1998b). High levels of luciferase expression (10^7 RLU/mg protein, i.e., about 3 ng of luciferase/mg protein) were found in the lung when DNA was complexed with PEI at a N/P ratio of 4. In these experiments, PEI vectorization provided 4 orders of magnitude higher transgene expression over controls (animals injected with an equivalent amount of naked DNA). Lower levels of transfection were found in the heart, spleen, liver, and kidney. The expression was dose and time dependent in all tissues. It is noteworthy that experiments carried out with the branched PEI of 25 kDa showed this polymer to be highly toxic even when low N/P ratios were used.

For *in vivo* applications one has to keep in mind that activation of the complement system by the DNA complexes constitutes a potential barrier for efficient gene transfer. Indeed, opsonization by complement components is known to lead to the clearance by macrophages. Plank and associates (1996) demonstrated that the most important factor for complement activation is the number of cationic charges that are accessible to the complement proteins, meaning that electroneutral polyplexes will not, or only poorly, activate the complement system. It is possible, however, that other proteins such as the negatively charged serum albumin coat the injected positively charged complex so that the activation of the complement system will be prevented (Barron *et al.,* 1998). It remains to be shown whether such a coating still allows the specific delivery of the polyplexes. In any case, electroneutral transfection complexes presenting a ligand recognized by a specific receptor seem to be suitable because (1) such particles should not be prone to nonspecific interactions with negatively charged cell surfaces, (2) they activate the complement system only poorly, and (3) the electroneutrality will prevent the coating of the DNA complexes by serum proteins, thus allowing their targeting.

Finally, the immunogenicity of PEI is still an unknown factor.

IX. PEI: AN EFFICIENT CARRIER FOR OLIGONUCLEOTIDE DELIVERY

Boussif and co-workers (1995) efficiently delivered oligonucleotides (ON) *in vitro* into the nucleus of postmitotic neurons using PEI as a carrier. These results show that the complexes probably disassemble in the cytosol, where free ONs display nuclear tropism. Transfections with ONs were successful in every neuronal cell type tested: embryonic neurons such as motoneurons or Purkinje cells as well as postnatal neurons (Lambert *et al.,* 1996).

Moreover, the branched PEI of 25 kDa presenting galactose residues was recently shown to be able to deliver RNA/DNA hybrids to the liver after intravenous injection of the polyplexes in rats (Kren *et al.,* 1998). The delivered chimeric RNA/DNA ON was able to induce, in a dose-dependent manner, a site-specific nucleotide exchange in the single copy factor IX gene. (It was previously shown that chimeric RNA/DNA oligonucleotides are effective in introducing single nucleotide conversion in episomal and genomic DNA of cultured cells; see Kren *et al.,* 1997).

X. CONCLUSIONS AND PROSPECTS

In summary, PEI is an effective and versatile macromolecular carrier because of the following properties: (1) it associates tightly with the negatively charged DNA; (2) it prevents degradation of the DNA once the PEI/DNA complex has been formed; (3) the DNA complexes interact with the surface of mammalian cells and are rapidly internalized; (4) it contains a large number of amino groups to which ligands can be covalently coupled; and (5) it is available in a large range of molecular sizes.

A better understanding of the mechanism of the action of synthetic vectors will lead to the design of improved DNA carriers. Simultaneously, studies aimed at producing formulations that are stable, homogenous, and easy to prepare must be undertaken. Finally, more *in vivo* experiments must be done to identify more clearly the problems associated with the use of PEI/DNA complexes.

ACKNOWLEDGMENTS

We would like to thank the scientists from Genethon for critical reading of the manuscript.

REFERENCES

Abdallah, B., Hassan, A., Benoist, C., Goula, D., Behr, J. P., and Demeneix, B. A. (1996). A powerful nonviral vector for *in vivo* gene transfer into the adult mammalian brain: Polyethylenimine. *Human Gene Ther.* **7**, 1947–1954.

Arap, W., Pasqualini, R., and Ruoslahti, E. (1998). Cancer treatment by targeted drug delivery to tumor vasculature in a mouse model. *Science* **276**, 377–380.

Ashwell, G., and Harford, J. (1982). Carbohydrate-specific receptors of the liver. *Ann. Rev. Biochem.* **51**, 531–542.

Baker, A., Saltik, M., Lehrmann, H., Killisch, I., Mautner, V., Lamm, G., Christofori, G., and Cotten, M. (1997). Polyethylenimine (PEI) is a simple, inexpensive and effective reagent for condensing and linking plasmid DNA to adenovirus for gene delivery. *Gene Ther.* **4**, 773–782.

Barron, L. G., Meyer, K. B., and Szoka, F. C. (1998). Effects of complement depletion on the pharmacokinetics and gene delivery mediated by cationic lipid-DNA complexes. *Human Gene Ther.* **9**, 315–323.

Boussif, O., Lezoualc'h, F., Zanta, M. A., Mergny, M. D., Scherman, D., Demeneix, B., and Behr, J. P. (1995). A versatile vector for gene and oligonucleotide transfer into cells in culture and *in vivo*: Polyethylenimine. *Proc. Natl. Acad. Sci. USA* **92**, 7297–7301.

Boussif, O. (1996). Transfert de gènes médié par des polymères cationiques. Ph.D thesis, Louis Pasteur University, Strasbourg, France.

Boussif, O., Zanta, M. A., and Behr, J. P. (1996). Optimized galenics improve *in vitro* gene transfer with cationic molecules up to a thousand-fold. *Gene Ther.* **3**, 1074–1080.

Bowman, E. J., Siebers, A., and Altendorf, K. (1988). Bafilomycins: A class of inhibitors of membrane ATPases for microorganisms, animal cells, and plant cells. *Proc. Natl. Acad. Sci. USA* **85**, 7972–7976.

Capecchi, M. R. (1980). High efficiency transformation by direct microinjection of DNA into cultured mammalian cells. *Cell* **22**, 479–488.

Carlsson, J., Drevin, H., and Axen, R. (1978). Protein thiolation and reversible protein-protein conjugation. *N*-succinimidyl-3-(2-pyridyldithio)propionate, a new heterobifunctional reagent. *Biochem. J.* **173**, 723–737.

Cooper, M. J. (1996). Noninfectious gene transfer and expression systems for cancer gene therapy. *Semin. Oncol.* **23**, 172–187.

Dick, C. R., and Ham, G. E. (1970). Characterization of polyethylenimine. *J. Macromol. Sci. Chem.* **A4**, 1301–1314.

Dowty, M. E., Williams, P., Zhang, G., Hagstrom, J. E., and Wolff, J. A. (1995). Plasmid DNA entry into postmitotic nuclei of primary rat myotubes. *Proc. Natl. Acad. Sci. USA* **92**, 4572–4576.

Duncan, R., Pratten, M. K., and Lloyd, J. B. (1979). Mechanism of polycation stimulation of pinocytosis. *Biochim. Biophys. Acta* **587**, 463–475.

Dunlap, D. D., Maggi, A., Soria, M. R., and Monaco, L. (1997). Nanoscopic structure of DNA condensed for gene delivery. *Nucleic Acids Res.* **25**, 3095–3101.

Erbacher, P., Bettinger, T., Belguise, P., Zou, S., Behr, J-P., and Remy, J-S. (in press). Biophysical characteristics and transfection efficiency of new polyethylenimine conjugates, based on maltose, dextran, poly(ethylene glycol), and antibody. Implications for *in vivo* gene transfer. *J. Gene Med.*

Erbacher, P., Remy, J-S., and Behr, J-P. (1999). Gene transfer with synthetic virus-like particles via the integrin-mediated endocytosis pathway. *Gene Ther.* **6**, 138–145.

Escriou, V., Ciolina, C., Lacroix, F., Byk, G., Scherman, D., and Wils, P. (1998). Cationic lipid-mediated gene transfer: Effect of serum on cellular uptake and intracellular fate of lipopolyamine/DNA complexes. *Biochim. Biophys. Acta* **1368**, 276–288.

Görlich, D. (1997). Nuclear protein import. *Curr. Opin. Cell Biol.* **9**, 412–419.

Goula, D., Remy, J. S., Erbacher, P., Wasowicz, M., Levi, G., Abdallah, B., and Demeneix, B. A. (1998a). Size, diffusibility and transfection performance of linear PEI/DNA complexes in the mouse central nervous system. *Gene Ther.* **5**, 712–717.

Goula, D., Benoist, C., Mantero, S., Merlo, G., Levi, G., and Demeneix, B. A. (1998b). Polyethylenimine-based intravenous delivery of transgenes to mouse lung. *Gene Ther.* **5**, 1291–1295.

Hagstrom, J. E., Ludtke, J. J., Bassik, M. C., Sebestyen, M. G., Adam, S. A., and Wolff, J. A. (1997). Nuclear import of DNA in digitonin-permeabilized cells. *J. Cell Sci.* **110**, 2323–2331.

Harbottle, R. P., Cooper, R. G., Hart, S. L., Ladhoff, A., McKay, T., Knight, A. M., Wagner, E., Miller, A. D., and Coutelle, C. (1998). An RGD-oligolysine peptide: A prototype construct for integrin-mediated gene delivery. *Human Gene Ther.* **9**, 1037–1047.

Hart, S. L., Harbottle, R. P., Cooper, R., Miller, A., Williamson, R., and Coutelle, C. (1995). Gene delivery and expression mediated by an integrin-binding peptide. *Gene Ther.* **2**, 552–554.

Hynes, R. O. (1992). Integrins: versality, modulation, and signalling in cell adhesion. *Cell* **69**, 11–25.

Jones, M. N. (1994). Carbohydrate-mediated liposomal targeting and drug delivery. *Adv. Drug Del. Rev.* **13**, 215–250.

Kichler, A., Zauner, W., Morrison, C., and Wagner, E. (1996). Ligand-polylysine mediated gene transfer. In P. L. Felgner *et al.,* (Eds.), *Artificial self-assembling systems for gene delivery* (Ch. 12, pp. 120–128). Washington, DC: American Chemical Society.

Kircheis, R., Kichler, A., Wallner, G., Kursa, M., Ogris, M., Felzmann, T., Buchberger, M., and Wagner, E. (1997). Coupling of cell-binding ligands to polyethyleneimine for targeted gene delivery. *Gene Ther.* **4**, 409–418.

Kren, B. T., Cole-Strauss, A., Kmiec, E. B., and Steer, C. (1997). Targeted nucleotide exchange in the alkaline phosphatase gene of HuH-7 cells mediated by a chimeric RNA/DNA oligonucleotide. *Hepatology* **25**, 1462–1468.

Kren, B. T., Bandyopadhyay, P., and Steer, C. J. (1998). *In vivo* site-directed mutagenesis of the factor IX gene by chimeric RNA/DNA oligonucleotides. *Nature Med.* **4**, 285–290.

Labat-Moleur, F., Steffan, A. M., Brisson, C., Perron, H., Feugeas, O., Furstenberger, P., Oberling, F., Brambilla, E., and Behr, J. P. (1996). An electron microscopy study into the mechanism of gene transfer with lipopolyamines. *Gene Ther.* **3**, 1010–1017.

Lambert, R. C., Maulet, Y., Dupont, J. L., Mykita, S., Craig, P., Volsen, S., and Feltz, A. (1996). Polyethylenimine-mediated DNA transfection of peripheral and central neurons in primary culture: probing Ca2+ channel structure and function with antisense oligonucleotides. *Mol. Cell Neurosci.* **7**, 239–246.

Ledley, F. D. (1993). Hepatic gene therapy: Present and future. *Hepatology* **18**, 1263–1273.

Leonetti, J. P., Degols, G., and Lebleu, B. (1990). Biological activity of oligonucleotide-poly(L-lysine) conjugates: Mechanism of celluptake. *Bioconjugate Chem.* **1**, 149–153.

Midoux, P., Mendes, C., Legrand, A., Raimond, J., Mayer, R., Monsigny, M., and Roche, A. C. (1993). Specific gene transfer mediated by lactosylated poly-L-lysine into hepatoma cells. *Nucleic Acids Res.* **21**, 871–878.

Mislick, K. A., and Baldeschwieler, J. D. (1996). Evidence for the role of proteoglycans in cation-mediated gene transfer. *Proc. Natl. Acad. Sci. USA* **93**, 12349–12354.

Nigg, E. A. (1997). Nucleocytoplasmic transport: Signals, mechanisms and regulation. *Nature* **386**, 779–787.

Plank, C., Zatloukal, K., Cotten, M., and Wagner, E. (1992). Gene transfer into hepatocytes using asialoglycoprotein receptor mediated endocytosis of DNA complexed with an artificial tetra-antennary galactose ligand. *Bioconjugate Chem.* **3**, 533–539.

Plank, C., Mechtler, K., Szoka, F. C., and Wagner, E. (1996). Activation of the complement system by synthetic DNA complexes: A potential barrier for intravenous gene delivery. *Human Gene Ther.* **7**, 1437–1446.

Pollard, H., Remy, J. S., Loussouarn, G., Demolombe, S., Behr, J. P., and Escande, D. (1998). Polyethyl-enimine but not cationic lipids promotes transgene delivery to the nucleus in mammalian cells. *J. Biol. Chem.* **273**, 7507–7511

Remy, J. S., Kichler, A., Mordvinov, V., Schuber, F., and Behr., J. P. (1995). Targeted gene transfer into hepatoma cells with lipopolyamine-condensed DNA particles presenting galactose ligands: A stage toward artificial viruses. *Proc. Natl. Acad. Sci. USA* **92**, 1744–1748.

Suh, J., Paik, H. J., and Hwang, B. K. (1994) Ionization of polyethylenimine and polyallylamine at various pHs. *Bioorg. Chem.* **22**, 318–327.

Tang, M. X., Redemann, C. T., and Szoka, F. C. (1996). *In vitro* gene delivery by degraded polyami-doamine dendrimers. *Bioconjugate Chem.* **7**, 703–714.

Tang, M. X., and Szoka, F. C. (1997). The influence of polymer structure on the interactions of cationic polymers with DNA and morphology of the resulting complexes. *Gene Ther.* **4**, 823–832.

Wadhwa, M. S., and Rice, K. G. (1995). Receptor-mediated glycotargeting. *J. Drug Targeting* **3**, 111–127.

Wu, G. Y., and Wu, C. H. (1987). Receptor-mediated *in vitro* gene transformation by a soluble DNA carrier system. *J. Biol. Chem.* **262**, 4429–4432.

Yoshimori, T., Yamamoto, A., Moriyama, Y., Futai, M., and Tashiro, Y. (1991). Bafilomycin A1, a specific inhibitor of vacuolar-type H$^+$-ATPase, inhibits acidification and protein degradation in lysosomes of cultured cells. *J. Biol. Chem.* **266**, 17707–17712.

Zabner, J., Fasbender, A. J., Moninger, T., Poellinger, K. A., and Welsh, M. J. (1995). Cellular and molecular barriers to gene transfer by a cationic lipid. *J. Biol. Chem.* **270**, 18997–19007.

Zanta, M. A., Boussif, O., Adib, A., and Behr, J. P. (1997). *In vitro* gene delivery to hepatocytes with galactosylated polyethylenimine. *Bioconjugate Chem.* **8**, 839–844.

CHAPTER 10

Ligand–Polycation Conjugates for Receptor-Targeted Gene Transfer

Ernst Wagner

Institute of Biochemistry, Vienna University Biocenter, Vienna, Austria

Many cell surface proteins (receptors) recognize and bind specific extracellular molecules (ligands) with high affinity and in a specific mode. Ligands can be proteins, peptides, carbohydrates, vitamins, or antibodies. They have been incorporated into DNA complexes with the aim of achieving targeted gene delivery to specific cell types. The choice of a specific receptor/ligand pair also determines the intracellular uptake of the bound material: for example, ligands such as asialoglycoproteins or transferrin can trigger efficient internalization by receptor-mediated endocytosis. For binding ligands to DNA in a noncovalent mode, polycations like

Nonviral Vectors for Gene Therapy

polylysines, protamines, or polyethylenimines have been chemically coupled to the ligand. Examples of ligand–polycation conjugates and the involved conjugation chemistry are described. In most cases the polycation not only binds the ligand to the DNA but also packages it into discrete particles. The applied protocols for complex formation and subsequent modifications strongly influence the properties of the transfection particle. In addition to the DNA/carrier composition, the size, charge, solubility, and stability of the DNA complex determines the extracellular fate (body distribution, diffusion into target tissues) and intracellular fate (transfer into the nucleus). For direct *in vivo* administration of DNA complexes, hurdles such as interaction with blood and plasma and clearance have to be considered. Once the DNA particles are cell-surface bound, they face a series of cellular barriers such as the escape from endosomes or lysosomes into the cytoplasm or the entry into the nucleus. The perspectives of receptor-targeted complexes for *in vitro* and *in vivo* gene therapy are reviewed.

I. INTRODUCTION

A gene transfer vehicle must serve at least two major separate delivery functions: to deliver the gene from the application site into the appropriate tissue and to the surface of the target cells, and to transport the DNA into the nucleus of these cells. Targeting ligands have been incorporated into gene transfer systems for two major reasons: (1) to target specific cell types and (2) to enhance intracellular uptake after binding the target cell. Both viral and nonviral vectors have been modified into this direction. The concept of receptor-targeted gene transfer (RTGT) has been pioneered by the development of synthetic ligand-containing gene delivery systems over the last two decades: efforts have been made to link cell-binding domains either directly to DNA (Cheng *et al.*, 1983), to DNA/liposome systems, called *lipoplexes* (Remy *et al.*, 1995; Lee and Huang, 1996; Compagnon *et al.*, 1997), or to DNA complexed with polycationic molecules, called *polyplexes*. Viral vectors have their own endogenous cell-binding domains. To alter or broaden the target cell specificity, the surface of retroviral vectors (reviewed in Cosset and Russell, 1996) and adenoviral vectors (Wickham *et al.*, 1997; Watkins *et al.*, 1997) has been modified either chemically (e.g., lactosylation), biochemically (e.g., by antibody binding), or genetically with targeting molecules. In several (but not all) cases, cell culture experiments have demonstrated successful ligand-mediated transduction of cells that cannot be transduced by the unmodified, parental viral vector. This review focuses on receptor-targeted polyplexes, the current situation of this field, and its opportunities.

Besides its attractive aspects the targeting concept also carries critical points and pitfalls. For example, the presence of specific targeting ligands does not necessarily mean that the whole ligand-coated complex has the same high specificity; in many cases the DNA/polycation particle contains domains that mediate unspecific interactions. To keep the desired targeting specificity, unspecific domains need to be masked. Moreover, even if the DNA complex has the desired specificity, targeting can only be successful if the DNA complex has the chance to reach the target cell for binding the cell surface receptor. This is no obvious step, because due to physical restriction only a fraction of the applied material arrives at the target tissue. In addition, successful targeting is a prerequisite but is not sufficient for obtaining useful gene expression levels. Additional functions are required for efficient intracellular delivery of the gene into the nucleus of the cell.

Despite all these obstacles, the first prototype receptor-targeted gene transfer systems have already shown encouraging evidence for *in vivo* gene delivery and expression. Further refinements are necessary to make the systems more efficient and applicable for braoder clinical application.

II. TARGETING LIGANDS

A large variety of biological agents contain domains by which they are bound and internalized by cells through specific interactions with cell surface receptors. These receptor–ligand interactions are involved in biological processes such as entry of viruses, bacteria, or toxins; provision of nutrients (e.g., LDL, transferrin); cellular signaling through binding of growth factors and hormones (e.g., insulin, EGF, FGF, VEGF); or removal of modified molecules from the circulation (e.g., asialoglycoproteins). Targeting ligands can be natural or recombinant proteins, synthetic peptides, carbohydrates, vitamins, or antibodies that specifically recognize a cell surface receptor. They can be incorporated into DNA complexes with the aim of achieving targeted gene delivery to specific cell types. For the choice of the appropriate targeting ligand several aspects need to be considered, such as the biological function and distribution of the cellular receptor, the degree of tissue specificity dictated by the mode of administration, and therapeutic principle.

Regarding the receptor of choice, the target cells should contain sufficiently high receptor levels not only in cell culture, but also *in vivo*; this can be verified for example by immunohistochemical analysis of the target tissue. In the case of polarized cells such as epithelial or endothelial cells, an appropriate cellular receptor distribution is required; to be accessible for the ligand, the receptor must be located on the right side; for example, on polarized epithelial cells many receptors are primarily located at the basal cell membrane and would not be accessible for delivery of ligand–DNA complexes from the apical side.

Besides the targeting requirement, ligands also can promote receptor-mediated intracellular uptake of DNA complexes that would be relevant both for *in vitro* and *in vivo* applications. Therefore the biological function of the various ligand/receptor pairs (rate of endocytosis, intracellular trafficking/recycling to the cell surface) can be quite important. For example, natural shedding of receptors from the cell surface into the surrounding area should be low. Also there should be no local excess of natural ligand with similar or higher affinity that might compete. Ideally, the ligand of choice (within the context of the DNA complex) should have a high affinity for the target receptor.

The choice of receptor/ligand pairs with a high rate of endocytosis seems preferable. Some ligands are internalized very efficiently (e.g., hepatocyte-specific asialoglycoproteins, T-cell-specific anti-CD3 antibodies, transferrin), whereas some others are not. In regard to specificity, certain ligands are very specific for certain cells in the body (e.g., asialoglycoproteins for hepatocytes), whereas others are not (e.g., transferrin for iron supply to many cell types). Depending on the mode of administration (*ex vivo,* where no specificity is required; localized *in vivo* application such as into muscle, tumors, or topical application to skin of lung; systemic application) a lower degree of tissue specificity might be sufficient.

Example of ligands currently used for RTGT are shown in Table 1.

III. LIGAND–POLYCATION CONJUGATES AND DNA COMPLEXES

The simplest version of a RTGT system contains only three molecular elements: the DNA, which is complexed with a DNA-binding polycation, and the targeting ligand, which is covalently linked to the polycation (Fig. 1). In the majority of applications the DNA is a plasmid, but oligonucleotides or large DNA constructs such as BACs (bacterial artificial chromosomes) also can be used.

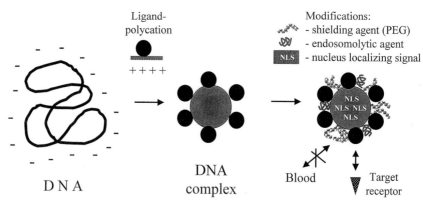

Figure 1 Receptor-targeted polyplexes. Gene transfer particles consisting of DNA, ligand/polycation conjugates, and—optionally—additional elements, such as shielding agents, endosome-destabilizing elements, or nuclear targeting functions.

Besides the basic elements of the DNA complex, additional elements can be incorporated that modify the properties of the complex, such as virus-like elements promoting intracellular release, nuclear targeting, or persistence of the introduced gene (Fig. 1). As these elements mimic viral delivery functions, the more complex systems are being termed "artificial viruses" (Wagner *et al.*, 1992; Plank *et al.*, 1992; Remy *et al.*, 1995).

A. Conjugates

For binding ligands to DNA in a noncovalent, reversible and nondamaging mode, polycations like polylysines, protamines, or polyethylenimines have been chemically coupled to the ligand. Examples of ligand–polycation conjugates are listed in Table 1.

Conjugates have been commonly synthesized by covalently coupling the ligand to the DNA-binding polycation. Generation of recombinant chimeric proteins that carry both ligand and (not polycationic) DNA-binding domains have also been described (Fominaya and Wels, 1996). Described examples for conjugation of proteins such as asialoglycoprotein (Wu and Wu, 1987), transferrin (Wagner *et al.*, 1990), or antibodies (Buschle *et al.*, 1995; Kircheis *et al.*, 1997) involve the modification of protein and polycation with bifunctional reagents such as SPDP [succinimidyl-3-(2-pyridyldithio) propionate]. In the first step, in two separate reactions a (more or less) defined number of lysine amino groups of either the protein or the polycation (polylysine, polyethylenimine) are modified by the activated (*N*-hydroxysuccinimidyl) esters of SPDP. Subsequent steps, including conversion of the introduced linker groups of the polycation into mercapto propionate groups (by reduction), followed by incubation with the modified protein, result in (reducible) disulfide bonds between the polycation and the protein. Alternative linker reagents such as agents containing maleimido groups can be used (Taxman *et al.*, 1993) resulting in a (nonreducible) thioether linkage. Such approaches are usually not specific, because the actual positions and numbers of site(s) of ligation between the protein and the polycation are unknown. In some cases conjugate synthesis can also be achieved in a more specific manner, for example, by *N*-terminal modification of the polycation polylysine (Trubetskoy *et al.*, 1992) or, in the case of glycoproteins, via ligation through the carbohydrate moiety (Wagner *et al.*, 1991a). Transferrin contains two carbohydrate chains that are attached by *N*-glycosylation to Asn-413 and Asn-611. The glycan chains have a biantennary structure, bearing two terminal sialic acid units. The glycosylation on the transferrin has no known influence on receptor binding or any other biological function (apart from the clearance of asialotransferrin from the plasma). Thus, these sites of transferrin are well suited for attachment of polylysine and other nucleic acid-binding elements. For coupling, the carbohydrate groups are activated by removal of the two terminal

Table 1

Ligand/Polycation Conjugates Currently Used for Receptor-Targeted Gene Transfer

Receptor	Ligand	Polycation used	Target cells tested	Endosomolytic agent	Selected references
ASGP receptor	asialoglycoproteins	polylysine	HepG2, HUH-7, primary hepatocytes	none, adenovirus, diphtheria toxin fragment	Wu, 1987, 1988, 1994; Chowdhury, 1996; Christiano, 1993; Fisher, 1994
ASGP receptor	synthetic galactosylated ligands	polylysine, oligolysine, bisacridine	HepG2, BNL Cl.2, HUH-7	none, chloroquine, peptide, adenovirus	Plank, 1992; Haensler, 1993; Midoux, 1993; Merwin, 1994; Wadhwa, 1995; Perales, 1994
Transferrin receptor	transferrin	polylysine, protamine, PEI, ethidium dimer	K562, J2E, F-MEL, HeLa, CFT 1, MRC-5, NIH 3T3, melanoma, neuroblastoma, fibroblasts, epithelial cells, endothelial cells	none, chloroquine, adenovirus, CELO virus, rhinovirus, peptides, glycerol	Cotten, 1990; Wagner, 1990; Zenke, 1990; Taxman, 1993; Kircheis, 1997
Insulin receptor	insulin	cation-modified albumin, polylysine	PLC/PRF/5 (hepatoma)	none	Huckett, 1990; Rosenkranz, 1992
FGF2-R	basic FGF	polylysine	Cos-1, 3T3, BHK, B16	none, chloroquine	Sosnowski, 1996
Folate receptor	folate	polylysine	KB, HeLa, Caco-2, SW620, SKOV	chloroquine, adenovirus	Mislick, 1995; Gottschalk, 1994
Carbohydrates	lectins	polylysine	airway cells, muscle cells	none, adenovirus, CELO virus	Batra, 1994; Cotten, 1993; Yin, 1994

Receptor	Ligand	Polymer	Cell line	Endosomolytic agent	Reference
Integrin	RGD peptide	oligolysine	Caco-2	chloroquine, PEI, adenovirus	Harbottle, 1998; Hart, 1995
Mannose receptor	synthetic ligands, glycosylated/mannosylated	polylysine	primary macrophages	none, chloroquine	Ferkol, 1996; Erbacher, 1996
Unknown	malarial circumsporozoite protein	polylysine	HepG2, primary hepatocytes, NIH 3T3, K562, HeLa, CHO	adenovirus	Ding, 1995
Unknown	surfactant proteins A and B	polylysine	H441 (pulmonary adenocarcinoma)	none, adenovirus	Baatz, 1994; Ross, 1995
PIG-R	anti-secretory component	polylysine	HT 29, tracheal epithelial cells	none	Ferkol, 1995
Tn carbohydrate	anti-Tn	polylysien	Jurkat	chloroquine, adenovirus	Thurnher, 1994
CD3	anti-CD 3	polylysine, PEI	Jurkat, H9, CCRF CEM, PBL, CIK	chloroquine, peptide, adenovirus	Buschle, 1995; Finke, 1998; Ebert, 1997; Kircheis, 1997
CD5	anti-CD 5	polylysine		adenovirus	Merwin, 1995
CD117	steel factor anti CD117	polylysine	MBO2, M07e, TF1, HEL	adenovirus	Schwarzenberger, 1996; Zauner, 1998
EGF-R	EGF anti-EGF	polylysine	NA. A549	none, adenovirus	Chen, 1994; Christiano, 1996
Her2	anti HER2	polylysine			Foster, 1997
Thrombomodulin	anti-thrombomodulin	polylysine	lung endothelial cells	none, liposomes	Trubetzkoy, 1992
Neuroblastoma	antibody ChCE7	polylysine			Coll, 1997
Surface immunoglobulin	anti-IgG, anti-idiotype	polylysine	B-LCLs, B-cell lymphoma	none, adenovirus	Curiel, 1994; Schachtschabel, 1996
FcR	IgG	polylysine	B-LCLs, alveolar macrophages	none, adenovirus	Curiel, 1994; Rojanasakul, 1994

exocyclic carbon atoms of the sialic acids by periodate oxidation, resulting in the formation of aldehyde groups at the end of the carbohydrate chains. Coupling to the amino groups of polylysine or polyethylenimine is performed by reductive amination with sodium cyanoborohydride through an aldimine intermediate.

Conjugates are purified from starting materials by cation exchange chromatography (unmodified ligand elutes first at low salt concentration, polycations and polycation conjugates elute at high salt), gel permeation chromatography (sizing columns), preparative gel electrophoresis (for polycations), in combination with gel filtration, dialysis or ultrafiltration steps. Characterization of conjugates (McKee *et al.*, 1994) includes determination of ligand/polycation ratios by UV and ninhydrin assay, assays specific for the ligands (e.g., sugar), SDS electrophoresis (detection of free ligand), or acid urea gel electrophoresis (conjugate charge/size).

B. DNA COMPLEXES

In most cases the polycation also packages the DNA into discrete particles. The type of polycationic carrier, DNA/carrier charge ratio, content of ligand, and applied protocol for complex formation strongly influence the properties of the transfection particle. Biophysical characteristics include the size and condensation of DNA particles (as determined by electron microscopy, laser light scattering, atomic force microscopy, centrifugation techniques, or filtration assays), solubility (recovery of DNA by UV or fluorescence assays), charge (as evaluated by zeta potential measurement and agarose gel electroporesis), and stability of DNA (protection against degradation by nucleases).

A series of different protocols for DNA complex formation have been used:

(1) Wu and Wu (1987) have formed complexes by mixing DNA and asialoorosomucoid (ASOR)-polylysine solutions at high salt (2M), followed by gradual reduction of the salt concentration to 150 mM by dialysis, resulting in a thermodynamically controlled complex formation. Charge ratios of well below 1 were used. The authors did not report the sizes of these complexes; however, the complex solution (containing particles of approximately 50 nm and larger) was filtered through a 0.45-μm membrane, giving an upper size limit to the DNA complex.

(2) Ferkol and colleagues (Ferkol *et al.*, 1995, 1996; Perales *et al.*, 1994) added polylysine conjugates slowly, in several small portions, to a vortexing solution of DNA in approximately 0.5–0.9M sodium chloride until a charge ratio of polylysine/DNA of approximately 0.7 is reached. This has been reported to generate monomeric DNA complexes with sizes of approximately 15–30 nm, which form aggregates. Aggregation is reverted by subsequent addition of salt. The exact salt concentration for mixing as well as the concentration of salt necessary to dissolve the complex aggregates depends on the length of polylysine as well as on the ligand used to modify the polylysine.

(3) We described (Wagner *et al.*, 1991b) the formation of kinetically controlled complexes by flash mixing of dilute solutions of DNA ($\geq$50 μg/ml) and polylysine conjugates in physiological buffer. Charge ratios of polylysine/DNA from smaller than 1/2 to larger than 2/1 have been applied. At a ratio of electroneutrality or higher, dense structures of 50–150 nm in size and donut-like or rod-like shape are formed. The size of the DNA complexes is independent of the size of DNA used (from 0.7 to 48 kb). In the case of DNA/polylysine particles, due to their hydrophobic nature at electroneutrality, aggregation occurs at concentration above 20 μg DNA/ml. Complexes containing transferrin-conjugated polylysine have increased solubility (up to 200 μg DNA/ml) compared to those using unmodified polylysine.

Besides the mixing protocol, factors such as the polymer and buffer play an important role. Wolfert and Seymour (1996) have shown that the molecular weight of polylysine has a profound effect on the size and size distribution of the resulting DNA/polylysine particles. For a charge ratio of 12, the size of the DNA complexes decreased with decreasing molecular weight of polylysine.

In the case of polyethylenimine (PEI) and PEI conjugates, charge ratio and ionic strength of the buffer determine the size of DNA complexes. Mixing of DNA/PEI complexes in 150 mM saline at molar ratios of N/P (PEI nitrogen/DNA phosphate) above 6 results in particle sizes of approximately 50 nm, while at N/P below 6 rapid aggregation to particle sizes of several 100 nm occurs. Aggregation can be avoided by complex formation at low ionic strength ($\geq$25 mM salt buffers) at $\geq$50 μg/ml DNA, generating particles with an average diameter of approximately 40–50 nm (Ogris *et al.*, 1998a).

Incorporation of additional elements can modify the properties of the complex. For example, a check must be made on whether noncovalent ionic binding of negatively charged endosomolytic peptides (see sect. IV.B) to positively charged DNA complexes results in aggregation of particles. Salt- or plasma-induced aggregation of DNA/ligand-PEI complexes can be prevented by first coating the DNA complexes with polyethylenglycol (PEG) through covalent coupling to PEI (Ogris *et al.*, 1998b). Complexes have reduced surface charge (zeta potential close to zero) and do not bind plasma proteins (see sect. IV.A). Because of the stabilizing effect of PEG, complexes formed at 50 μg/ml can be ultraconcentrated to a 1 mg/ml DNA concentration.

IV. CELLULAR BARRIERS
AND *IN VITRO* APPLICATIONS

A. CELL BINDING AND UPTAKE

Ideally, the ligand-coated DNA complex is recognized through specific interaction with the corresponding cell surface receptors only. However, it must be kept in mind that DNA particle properties such as positive charge can promote unspecific interaction with the cell surface (Fig. 2). We have previously shown that minor changes in the DNA/polylysine conjugate ratio of the complex (resulting in positively charged complexes) can convert a ligand-specific gene transfer into a completely unspecific process (Zauner *et al.*, 1998). Therefore, proper *in vitro* tests are required to evaluate receptor-dependent uptake of the ligand-containing complexes. Testing includes binding studies of the DNA complex to the target receptor; binding and uptake studies at 4°C (no endocytosis) and 37°(endocytosis) using for example fluorescence-labeled DNA complexes; a check on whether gene expression is reduced in competition experiments with free ligand or with DNA complexes lacking the ligand, whether it is low in cells without receptor or whether transfection efficiency is enhanced after upregulation of the cellular receptor. Several examples have demonstrated that incorporation of ligands to the carrier enhances cellular binding and internalization of DNA complexes with no or only

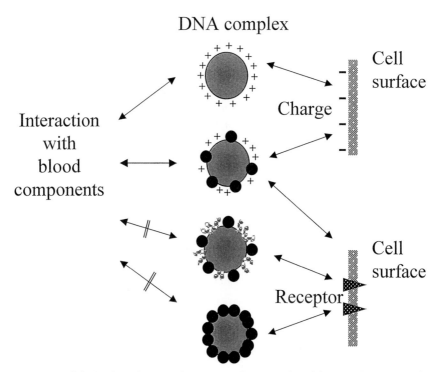

Figure 2 Cellular binding of DNA/polycation complexes. *Top:* ligand-free complexes. *Bottom:* ligand-coated complexes. Binding to cell surface can be unspecific (charge-mediated) or through ligand–receptor interaction. Surface properties of the complexes (charge density, degree of coating with ligand and/or shielding agents like PEG) can influence cell binding.

slight negative charge, thereby avoiding problems of positively charged complexes such as cytotoxicity or undesired interactions with other cells and proteins.

DNA particle size is another parameter that can influence binding and uptake. For example, large (several hundered nm) transferrin–PEI/DNA particles formed in physiological salt solutions showed an up to more than 100-fold higher activity in transfections in comparison to 40 nm small complexes formed in water (Ogris *et al.*, 1998a). Among several possible explanations (see below), limited transport and binding of the small particles to the cell surfaces is one possible reason for this effect. Larger particles can sediment onto the cultured cells, whereas small particles stay in solution and the Brownian molecular motion limits the contact with cells. FACS analysis of cells incubated with fluorescently labeled DNA complexes showed a 10-fold higher association between cells and the large complexes compared to the small complexes. Application of the small particles in more concentrated form and over extended periods of time improves transfection activity (Ogris *et al.*, 1998a).

B. Vesicular Escape

In the majority of cases, ligand-coated DNA complexes are internalized into intracellular vesicles such as endosomes, lysosomes, or phagocytic vesicles. Escape from these intracellular vesicles into the cytoplasm presents a major bottleneck to many types of DNA complexes. Entrapment in lysosomal or phagocytic vesicles is thought to be associated with degradation of the complexes in these compartments. The fate of the delivered DNA strongly depends on the complex composition and the chosen polycationic carrier.

For example, transferrin–polylysine DNA complexes have been found to be efficiently internalized into receptor-positive cells, but in most instances accumulation in intracellular vesicles prevents gene expression (Cotten *et al.*, 1990). Only positively charged complexes with excess of polylysine show some modest levels of gene expression in some cell lines. The use of the enzymatically stable poly(D)lysine, with the unnatural (D)lysine amino acid monomer, which should exclude degradation of the DNA in lysosomes, does not enhance transfection efficiency (Plank *et al.*, 1994).

Other polycationic carriers, such as polyethylenimine (Boussif *et al.*, 1995), dendrimer (Haensler and Szoka, 1993; Tang and Szoka, 1997), or several cationic lipids such as lipospermine (Behr *et al.*, 1989) can promote escape from intracellular vesicles to some degree. The high transfection efficiency of polyethylenimine (PEI) and ligand–PEI conjugates (Kircheis *et al.*, 1997; Zanta *et al.*, 1997) is thought to be based on a "proton sponge" effect. PEI within the DNA complex is only partially protonated at physiological pH (only one of two or three amino nitrogens). Upon intracellular delivery of the DNA particle, the natural acidification within endosome/lysosome triggers protonation of the complex-bound PEI, inducing osmotic swelling and destabilization of the endosomal/lysosomal vesicle. However, it must be kept in mind that in addition to the cationic carrier the actual formulation can have a large influence on gene transfer efficiency (Boussif et al., 1996). While with large PEI/DNA the endosomal escape is no major bottleneck, inefficient endosomal release contributes to the lower efficacy of the small PEI/DNA particles (Ogris et al., 1998a).

Several approaches have been developed to enhance escape of DNA complexes into the cytoplasm. The addition of the lysosomotropic agent chloroquine to the transfection medium has been shown to increase efficiency of polylysine-based transfection complexes, probably by interfering with lysosomal degradation and enhancing the release of the DNA into the cytoplasm. The positive effect of chloroquine is especially strong in K562 cells (Cotten et al., 1990) which can be explained by the unusually low pH in K562 early endosomes, resulting in accumulation of large amounts of the weak base in the endocytic vesicles, swelling, and subsequent destabilization of the endosomes. Chloroquine can also enhance efficacy of small (but not large!) PEI-based DNA complexes (Ogris et al., 1998a). However, the use of chloroquine is limited because of its significant cytotoxicity and the small number of cell lines that respond to chloroquine.

Incubation of cells with positively charged transferrin–polylysine/DNA complexes in the presence of 1–1.5 M glycerol results in a strongly enhanced transfection efficiency in a series of cell lines and primary fibroblasts (Zauner et al., 1996, 1997). However, several other cell lines, including suspension cells such as K562, were refractory to this technique.

The inclusion of endosomolytic agents (see Table 1) has been shown to dramatically enhance ligand–polylysine based gene transfer. Addition of replication-defective adenovirus or rhinovirus in the transfection medium augments the levels of transferrin-mediated gene transfer up to more than 1000-fold in cell lines that express high levels of both virus and transferrin receptor (Curiel et al., 1991; Cotten et al., 1992; Zauner et al., 1995). To broaden the applicability of these findings, endosomolytic agents were directly attached or incorporated into the DNA complexes, either by chemical or enzymatic linkage to polylysine (Christiano et al., 1993; Wu et al., 1994; Fisher and Wilson, 1994; Wagner et al., 1992a,b), by biotinylation (Wagner et al., 1992; Zauner et al., 1995), which allows subsequent binding to streptavidin-polylysine, by an antibody bridge (Curiel et al., 1992), or simply by ionic interaction to the polycation (Plank et al., 1994; Mechtler and Wagner, 1997; Baker and Cotten, 1997). Beside whole virus particles (inactivated human or chicken adenovirus, rhinovirus) also proteins such as adenovirus penton proteins (Fender et al., 1997), bacterial cytolysines (Gottschalk et al., 1995), or the transmembrane domain of diphtheria toxin (Fisher and Wilson, 1997) and synthetic peptides with virus-derived (influenza, rhinovirus) sequences or artificial sequences (Gottschalk et al., 1996; Haensler and Szoka, 1993; Mechtler and Wagner, 1997) have been used.

C. NUCLEAR ENTRY AND PERSISTENCE

Entry of DNA into the nucleus is another limiting step to gene transfer. This has been demonstrated for liposomal systems and lipopeptide DNA complexes (Zabner *et al.*, 1995; Wilke *et al.*, 1996). In standard cell cultures with growing cells, nuclear membrane breakdown during cell division may allow the passive inclusion of DNA or its complexes into the nuclear compartment. For example, high transfection levels are achieved with DNA/lipofectamine complexes on growing human primary fibroblasts. Using the fibroblasts in a confluent nondividing stage, transfection efficiency dropped to very low levels (Zauner and Wagner, unpublished data). In sharp contrast, transferrin/polylysine complexes applied in the presence of glycerol (Zauner *et al.*, 1996) gave high expression levels on both growing and growth-arrested fibroblasts. This finding is consistent with recent reports that some polycations may facilitate the nuclear uptake of DNA complexes. Pollard *et al.* (1998) reported that microinjection of DNA/polyethylenimine (PEI) complexes into the cytoplasm of several cell lines led to a higher fraction of expressing cells than the injection of the same amount of naked plasmid DNA. The authors also demonstrated that in contrast to lipoplexes, injection of polyplexes (PEI, polylysine) into the nucleus leads to gene expression. Consistent findings were also made by Wolfert and Seymour (1998).

Therefore, it remains open to which extent the incorporation of additional biological nuclear targeting peptides or DNA-based recognition sequences for nuclear targeting into the gene transfer particle will be required. This must be checked case by case and depends on the application (e.g., postmitotic cells, growing tumors). DNA structure and sequence elements can also be critical for persistence of the delivered genetic material (Haase *et al.*, 1994; Harrington *et al.*, 1997).

D. *IN VITRO* APPLICATIONS

Gene transfer complexes, such as based on adenovirus-enhanced transferrin-mediated transfection (AVET), can efficiently deliver genes to a large proportion of primary cultured cells such as primary fibroblasts, endothelial cells, hepatocytes, myoblasts, or tumor cells. The highly efficient delivery has resulted in the development of *ex vivo* gene therapy approaches, that is, genetic modification of cells harvested from a patient followed by reimplantation into the patient. For example, gene-modified cancer cell vaccines (autologous = cells derived from the patient's own tumor; allogeneic = standardized tumor cell lines) have been designed for the treatment of malignant melanoma. DNA complexes are used to deliver immunostimulatory genes (such as interleukin-2) into melanoma cells *in vitro*. After irradiation (to block tumor cell growth) the transfected cells are applied into the skin to trigger an antitumor immune response. These protocols are currently being evaluated in clinical trials (Stingl *et al.*, 1996).

V. EXTRACELLULAR HURDLES AND *IN VIVO* APPLICATIONS

A. EXTRACELLULAR HURDLES

A series of nonviral vectors with high efficiency for gene transfer in cell culture are available, but efficient and targeted *in vivo* gene delivery remains a major challenge. Depending on the route of *in vivo* administration of gene transfer particles, several extracellular barriers must be overcome (Fig. 3). Blood components may coat the DNA complex and cause aggregation or degradation. The reticuloendothelial system presents a major sink where DNA particles are removed from circulation. Extravasation from blood vessels into surrounding tissues is a major barrier and largely depends on the presence of fenestrations in the vessel walls (e.g., liver, tumors). Within tissues, delivery is influenced by the extracellular matrix; small

systemic application	blood components RES vascular barrier	shielding agents (PEG) small size

⇩

local application	diffusion in organ extracellular matrix	small size inert surface

⇩

target cell / in vitro	receptor binding endocytosis vesicular escape nuclear transport transcription persistence	suitable ligand suitable carrier endosomolytic agent NLS signal suitable DNA sequences

Figure 3 Extracellular and intracellular barriers for gene delivery. The hurdles (dependent on the mode of application) that DNA particles have to overcome and some of the required functions are shown.

complexes are obviously required for both extravasation and a broader distribution within tissues.

Previous studies have demonstrated the inactivation of polylysine-based DNA complexes by blood components (Wagner *et al.,* 1994). One of the factors was identified as the alternative pathway of the complement system (Plank *et al.,* 1996). Upon incubation of DNA/PEI or DNA/transferrin–PEI complexes with human or murine citrate plasma, followed by biochemical analysis of the complex bound protein fractions, specific protein bands for IgM, fibrinogen, fibronectin, and complement C3 could be identified (Ogris *et al.,* 1999). Coating the DNA/PEI complexes with polyethylenglycol (see Sect. III.B) through covalent coupling to PEI strongly reduced plasma protein binding.

B. *In Vivo* Applications

Depending on the indication and therapeutic strategy, different routes of administration can be considered, ranging from direct injection into the target tissue (e.g., intratumoral application or genetic vaccination into the muscle or skin), broader regional delivery (e.g., topical delivery to the airways, administration to the peritoneal cavity, delivery into artherosclerotic vessels or into joints), to systemic delivery (e.g., targeting of specific organs or tumors through the bloodstream). The route of administration influences the desired properties of the gene transfer complexes. For example, large and/or positively charged DNA particles might be applicable in administrations where the target cells are directly accessible (e.g., local injection into tumor tissue). Upon systemic administration only small complexes might have a chance to leave the circulation and migrate into a specific target tissue.

Local injection of naked DNA or DNA complexes directly into subcutaneously growing tumors result in significant reporter gene expression, with DNA/transferrin–PEI complexes or adenovirus-linked DNA/transferrin–polylysine complexes being 10- to 100-fold more efficient than naked DNA (Kircheis *et al.,* 1999).

Systemic *in vivo* gene transfer using polylysine condensed DNA was reported for the first time by Wu and colleagues (1988, 1989, 1991). CAT reporter gene expression after i.v. injection of asialo-orosomucoid–polylysine/DNA complexes in rats was highest after 24 h. Activity then declined to zero within 96 h. However, when partial hepatectomy (two-thirds of the liver) was performed 30 minutes after injection of the complexes, gene expression was shown to persist for several weeks postinjection. There was no evidence of integration of the foreign gene. Using galactose-modified polylysine and a different protocol for complex formation (see Sect. III.B), Perales *et al.* (1994) reported prolonged expression of a reporter gene in hepatocytes without liver surgery. Systemic delivery of anti-pIgR Fc-polylysine–coated DNA resulted in *in vivo* expression in the lung (Ferkol *et al.,* 1995).

Intravenous application of Tf-PEI/DNA complexes through the tail vein into the tumor-bearing mice resulted in preferential gene expression in the tail and lung, but also serious toxicity and lethality, with clinical signs of acute lung embolism. DNA complexes coated with polyethylenglycol through covalent coupling to PEI (see Sect. III) showed prolonged circulation upon intravenous application (Ogris *et al.*, 1996b). The complexes were far less toxic; gene expression in tumor-bearing mice was almost exclusively found at the application site (in the tail) and at the tumor (Ogris *et al.*, 1998b).

VI. CONCLUSIONS

As a result of more than ten years of research, receptor-targeting has become an integral part in the development of nonviral gene transfer systems. Numerous examples of receptor–ligand pairs have already been tested, and knowledge about the requirements for targeting and intracellular delivery has strongly expanded. Several systems have been established that have high *in vitro* efficiency and also show encouraging signs of *in vivo* activity. In the current stage, increased understanding about the relationship between biophysical structure of DNA particles and their biological activity is essential for the development of optimized systems that are more broadly applicable for *in vivo* gene therapies.

ACKNOWLEDGMENTS

The contributions of Sylvia Brunner, Antoine Kichler, Ralf Kircheis, Karl Mechtler, Manfred Ogris, Chistian Plank, Susanne Schüller, and Wolfgang Zauner to the research from our lab are greatly appreciated. The work is supported by the grant S07405 from the Austrian Science Foundation and by Boehringer Ingelheim R&D Vienna, Austria.

REFERENCES

Baatz, J. E., Bruno, M. D., Ciraolo, P. J., Glasser, S. W., Stripp, B. R., Smyth, K. L., and Korfhagen, T. R. (1994). Utilization of modified surfactant-associated protein B for delivery of DNA to airway cells in culture. *Proc. Natl. Acad. Sci. USA* **91**, 2547–2551.

Baker, A., and Cotten, M. (1997). Delivery of bacterial artificial chromosomes into mammalian cells with psoralen-inactivated adenovirus carrier. *Nucleic Acids Res.* **25**, 1950–1956.

Batra, R. K., Wang-Johanning, F., Wagner, E., Garver, R. I., and Curiel, D. T. (1994). Receptor-mediated gene delivery employing lectin-binding specificity. *Gene Ther.* **1**, 255–260.

Behr, J-P., Demeneix, B., Loeffler, J-P., and Perez-Mutul, J. (1989). Efficient gene transfer into mammalian primary endocrine cells with lipopolyamine-coated DNA. *Proc. Natl. Acad. Sci. USA* **86**, 6982–6986.

Boussif, O., Lezoualc'h, F., Zanta, M. A., Mergny, M., Scherman, D., Demeneix, B. and Behr, J. P.

(1995). A novel, versatile vector for gene and oligonucleotide transfer into cells in culture and *in vivo:* polyethyleneimine. *Proc. Natl. Acad. Sci. USA* **92,** 7297–7301.

Boussif, O., Zanta, M. A., Behr, J. P. (1996). Optimized galenics improve *in vitro* gene transfer with cationic molecules up to a 1000-fold. *Gene Ther.* **3,** 1074–1080.

Buschle, M., Cotten, M., Kirlappos, H., Mechtler, K., Birnstiel, M. L., and Wagner, E. (1995). Receptor-mediated gene transfer into T-lymphocytes via binding of DNA/CD3 antibody particles to the CD3 T cell receptor complex. *Human Gene Ther.* **6,** 753–761.

Chen, J., Gamou, S., Takayanagi, A., and Shimizu, N. (1994). A novel gene delivery system using EGF receptor-mediated endocytosis. *FEBS Lett.* **338,** 167–169.

Cheng, S. Y., Merlino, G. T., Pastan, I. H. (1983). A method for coupling of proteins to DNA: Synthesis of alpha$_2$-macroglobulin-DNA conjugates. *Nucleic Acids Res.* **11,** 659–669.

Chowdhury, N. R., Wu, C. H., Wu, G. Y., Yerneni, P. C., Bommineni, V. R. and Chowdhury, J. R. (1993). Fate of DNA targeted to the liver by asialoglycoprotein receptor-mediated endocytosis *in vivo. J. Biol. Chem.* **268,** 11265–11271.

Chowdhury, N. R., Hays, R. M., Bommineni, V. R., Franki, N., Chowdhury, J. R., Wu, C. H., and Wu, G. Y. (1996). Microtubular disruption prolongs the expression of human bilirubin-uridinediphosphoglucuronate-glucuronsyltransferase-1 gene transferred into Gunn rat livers. *J. Biol. Chem.* **271,** 2341–2346.

Coll, J.-L., Wagner, E., Combaret, V., Mechtler, K., Amstutz, H., Iacono-DiCacito, I., Simon, N., and Favrot, M. C. (1997). *In vitro* targeting and specific transfection of human neuroblastoma cells by chCE7 antibody-mediated gene transfer. *Gene Ther.* **4,** 156–161.

Compagnon, B., Moradpour, D., Alford, D. R., Larsen, C. E., Stevenson, M. J., Wands, J. R., and Nicolau, C. (1997). Enhanced gene delivery and expression in human hepatocellular carcinoma cells by cationic immunoliposomes. *J. Liposome Res.* **7,** 127–141.

Cosset, F.-L., and Russell, S. J. (1996). Targeting retrovirus entry. *Gene Ther.* **3,** 946–956.

Cotten, M., Laengle-Rouault, F., Kirlappos, H., Wagner, E., Mechtler, K., Zenke, M., Beug, H., and Birnstiel, M. L. (1990). Transferrin-polycation-mediated introduction of DNA into human leukemic cells: Stimulation by agents that affect the survival of transfected DNA or modulate transferrin receptor levels. *Proc. Natl. Acad. Sci. USA* **87,** 4033–4037.

Cotten, M., Wagner, E., Zatloukal, K., Phillips, S., Curiel, D. T., and Birnstiel, M. L. (1992). High-efficiency receptor-mediated delivery of small and large (48kb) gene constructs using the endosome disruption activity of defective or chemically inactivated adenovirus particles. *Proc. Natl. Acad. Sci. USA* **89,** 6094–6098.

Cotten, M., Wagner, E., Zatloukal, K., and Birnstiel, M. L. (1993). Chicken adenovirus (CELO virus) particles augment receptor-mediated DNA delivery to mammalian cells and yield exceptional levels of stable transformants. *J. Virol.* **67,** 3777–3785.

Cristiano, R., Smith, L., and Woo, S. (1993). Hepatic gene therapy: Efficient gene delivery and expression in primary hepatocytes utilizing a conjugated adenovirus-DNA complex. *Proc. Natl. Acad. Sci. USA* **90,** 11548–11552.

Cristiano, R. J., and Roth, J. A. (1996). Epidermal growth factor mediated DNA delivery into lung cancer via the epidermal growth factor receptor. *Cancer Gene Ther.* **3,** 4–10.

Curiel, D. T., Agarwal, S., Wagner, E., and Cotten, M. (1991). Adenovirus enhancement of transferrin-polylysine-mediated gene delivery. *Proc. Natl. Acad. Sci. USA* **88,** 8850–8854.

Curiel, T. J., Cook, D. R., Bogedain, C., Jilg, W., Harrison, G. S., Cotten, M., Curiel, D. T., and Wagner, E. (1994). Foreign gene expression in Epstein-Barr virus transformed human B cells, *Virology* **198,** 577–585.

Ding, Z.-M., Cristiano, R. J., Roth, J. A., Takacs, B., and Kuo, M. T. (1995). Malarial circumsporozoite protein is a novel gene delivery vehicle to primary hepatocyta cultures and cultured cells. *J. Biol. Chem.* **270,** 3667–3676.

Ebert, O., Finke, S., Salahi, A., Herrmann, M., Trojaneck, B., Lefterova, P., Wagner, E., Kircheis, R.,

Huhn, D., Schriever, F., and Schmidt-Wolf, I. G. H. (1997). Lymphocyte apoptosis: Induction by gene transfer techniques. *Gene Ther.* **4**, 296–302.

Erbacher, P., Bousser, M.-T., Raimond, J., Monsigny, M., Midoux, P., and Roche, A. C. (1996). Gene transfer by DNA/glycosylated polylysine complexes into human blood monocyte-derived macrophages. *Human Gene Ther.* **7**, 721–729.

Fender, P., Ruigrok, R. W. H., Gout, E., Buffet, S., and Chroboczek, J. (1997). Adenovirus dodecahedron, a new vector for human gene transfer. *Nature Biotechnol.* **15**, 52–56.

Ferkol, T., Perales, J. C., Eckman, E., Kaetzel, C. S., Hanson, R. W., and Davis, P. (1995). Gene transfer into the airway epithelium of animals by targeting the polymeric immunoglobulin receptor. *J. Clin. Invest.* **95**, 493–502.

Ferkol, T., Perales, J. C., Mularo, F., and Hanson, R. W. (1996). Receptor-mediated gene transfer into macrophages. *Proc. Natl. Acad. Sci. USA* **93**, 101–105.

Finke, S., Trojanek, B., Lefterova, P., Csipai, M., Wagner, E., Kircheis, R., Neubauer, A., Huhn, D., Wittig, B., and Schmidt-Wolf, I. G. H. (1998). Increase of proliferation rate and enhancement of antitumor cytotoxicity of expanded human CD3+CD56+ immunological effector cells by receptor-mediated transfection with the IL-7 gene. *Gene Ther.* **5**, 31–39.

Fisher, K. J., and Wilson, J. M. (1994). Biochemical and funcsher, K. J. and Wilson, J. M. (1997). The transmembrane domain of diphtheria toxin improves molecular conjugate gene transfer. *Biochem. J.* **321**, 49–58.

Fominaya, J., and Wels, W. (1996). Target cell-specific DNA transfer mediated by a chimeric multidomain protein. Novel non-viral gene delivery system. *J. Biol. Chem.* **271**, 10560–10568.

Foster, B. J. and Kern, J. A. (1997). HER2-targeted gene transfer. *Human Gene Ther.* **8**, 719–727.

Gottschalk, S., Cristiano, R. J., Smith, L. C., and Woo, S. L. C. (1994). Folate receptor mediated DNA delivery into tumor cells: Potosomal disruption results in enhanced gene expression. *Gene Ther.* **1**, 185–191.

Gottschalk, S., Tweten, R. K., Smith, L. C., and Woo, S. L. C. (1995). Efficient gene delivery and expression in mammalian cells using DNA coupled with perfringolysin O. *Gene Ther.* **2**, 498–503.

Gottschalk, S., Sparrow, J. T., Hauer, J., Mims, M. P., Leland, F. E., Woo, S. L. C., and Smith, L. C. (1996). A novel DNA-peptide complex for efficient gene transfer and expression in mammalian cells. *Gene Ther.* **3**, 448–457.

Haase, S. B., Heinzel, S. S., and Calos, M. P. (1994). Transcription inhibits the replication of autonomously replicating plasmids in human cells. *Mol. Cell. Biol.* **14**, 2516–2524.

Haensler, J., and Szoka, F. C. (1993). Polyamidoamine cascade polymers mediate efficient transfection of cells in culture. *Bioconjugate Chem.* **4**, 372–379.

Haensler, J., and Szoka, F. C. (1993). Synthesis and characterization of a trigalactosylated bisacridine compound to target DNA to hepatocytes. *Bioconjugate Chem.* **4**, 85–93.

Harbottle, R. P., Cooper, R. G., Hart, S. L., Ladhoff, A., McKay, T., Knight, A. M., Wagner, E., Miller, A. D., and Coutelle, C. (1998). An RGD-oligolysine peptide: A prototype construct for integrin-mediated gene delivery. *Human Gene Ther.* **9**, 1037–1047.

Harrington, J. J., Bokkelen, G. V., Mays, R. W., Gustashaw, K., and Willard, H. F. (1997). Formation of *de novo* centromeres and construction of first-generation human artificial microchromosomes. *Nature Genet.* **15**, 345–355.

Hart, S. L., Harbottle, R. P., Cooper, R., Miller, A., Williamson, R., and Coutelle, C. (1995). Gene delivery and expression mediated by an integrin-binding peptide. *Gene Ther.* **2**, 552–554.

Huckett, B., Ariatti, M., and Hawtrey, A. O. (1990). Evidence for targeted gene transfer by receptor-mediated endocytosis: Stable expression following insulin-directed entry of *neo* into HepG2 cells, *Biochem. Pharmacol.* **40**, 253–263.

Kircheis, R., Kichler, A., Wallner, G., Kursa, M., Ogris, M., Felzmann, T., Buchberger, M., and Wagner, E. (1997). Coupling of cell-binding ligands to polyethylenimine for targeted delivery. *Gene Ther.* **4**, 409–418.

Kircheis, R., Schüller, S., Brunner, S., Ogris, M., Heider, K.-H., Zauner, W., and Wagner, E. (1999). Polycation-based DNA complexes for tumor-targeted gene delivery *in vivo. J.Gene Med.*, in press.

Lee, R. J., and Huang, L. (1996). Folate-targeted, anionic liposome-entrapped polylysine-condensed DNA for tumor cell-specific gene transfer. *J. Biol. Chem.* **271**, 8481–8487.

McKee, T. D., DeRome, M. E., Wu, G. Y., and Findeis, M. A. (1994). Preparation of asialoorosomu-coid-polylysine conjugates. *Bioconjugate Chem.* **5**, 306–311.

Mechtler, K., and Wagner, E. (1997). Gene transfer mediated by Influenza virus peptides: The role of peptide sequences. *New J. Chem.* **21**, 105–111.

Merwin, J. R., Carmichael, E. P., Noell, G. S., DeRome, M. E., Thomas, W. L., Robert, N., Spitalny, G., and Chiou, H. C. (1995). CD5-mediated specific delivery of DNA to T lymphocytes: Compartmentalization augmented by adenovirus. *J. Immunol. Methods* **186**, 257–266.

Merwin, J. R., Noell, G. S., Thomas, W. L., Chiou, H. C., DeRome, M. E., McKee, T. D., Spitalny, G. L., and Findeis, M. A. (1994). Targeted delivery of DNA using YEE(GalNAcAH)3, a synthetic glycopeptide ligand for the asialoglycoprotein receptor. *Bioconjugate Chem.* **5**, 612–620.

Midoux, P., Mendes, C., Legrand, A., Raimond, J., Mayer, R., Monsigny, M., and Roche, A. C. (1993). Specific gene transfer mediated by lactosylated poly-L-lysine into hepatoma cells. *Nucleic Acids Res.* **21**, 871–878.

Mislick, K. A., Baldeschwieler, J. D., Kayyem, J. F., and Meade, T. J. (1995). Transfection of folate-polylysine DNA complexes: Evidence for lysosomal delivery. *Bioconjugate Chem.* **6**, 512–515.

Ogris, M., Steinlein, P., Kursa, M., Mechtler, K., Kircheis, R., and Wagner, E. (1998a). The size of DNA/transferrin-PEI complexes is an important factor for gene expression in cultured cells. *Gene Ther.* **5**, 1425–1433.

Ogris, M., Brunner, S., Schueller, S., Kircheis, R., and Wagner, E. (1999). PEGylated DNA/Transferrin-PEI complexes show reduced plasma interaction, extended blood circulation and *in vivo* gene delivery to tumors. *Gene Ther.* **6**, in press.

Perales, J. C., Ferkol, T., Beegen, H., Ratnoff, O. D., and Hanson, R. W. (1994). Gene transfer *in vivo:* Sustained expression and regulation of genes introduced into the liver by receptor-targeted uptake. *Proc. Natl. Acad. Sci. USA* **91**, 4086–4090.

Plank, C., Zatloukal, K., Cotten, M., Mechtler, K., and Wagner, E. (1992). Gene transfer into hepatocytes using asialoglycoprotein receptor mediated endocytosis of DNA complexed with an artificial tetra-antennary galactose ligand. *Bioconjugate Chem.* **3**, 533–539.

Plank, C., Oberhauser, B., Mechtler, K., Koch, C., and Wagner, E. (1994). The influence of endosome-disruptive peptides on gene transfer using synthetic virus-like gene transfer systems. *J. Biol. Chem.* **269**, 12918–12924.

Plank, C., Mechtler, K., Szoka, F., and Wagner, E. (1996). Activation of the complement system by synthetic DNA complexes: A potential barrier for intravenous gene delivery. *Human Gene Ther.* **7**, 1437–1446.

Pollard, H., Remy, J.-S., Loussouarn, G., Demolombe, S., Behr, J.-P., and Escande, D. (1998). Poly-ethyleneimine but not cationic lipids promotes transgene delivery to the nucleus in mammalian cells. *J. Biol. Chem.* **273**, 7507–7511.

Remy, J. S., Kichler, A., Mordvinov, V., Schuber, F., and Behr, J. P. (1995). Targeted gene transfer into hepatoma cells with lipopolyamine-condensed DNA particles presenting galactose ligands; a stage toward artificial viruses. *Proc. Natl. Acad. Sci. USA* **92**, 1744–1748.

Rojanasakul, Y., Wang, L. Y., Malanga, C. J., Ma, J. K. H., and Liaw, J. (1994). Targeted gene delivery to alveolar macrophages via Fc receptor-mediated endocytosis. *Pharm. Res.* **11**, 1731–1736.

Rosenkranz, A. A., Yachmenev, S. V., Jans, D. A., Serebryakova, N. V., Murav'ev, V. I., Peters, R., and Sobolev, A. S. (1992). Receptor-mediated endocytosis and nuclear transport of a transfecting DNA construct. *Exp. Cell Res.* **199**, 323–329.

Ross, G. F., Morris, R. E., Ciraolo, G., Huelsman, K., Bruno, M., Whitsett, J. A., Baatz, J. E., and

Korfhagen, T. R. (1995). Surfactant protein A-polylysine conjugates for delivery of DNA to airway cells in culture. *Human Gene Ther.* **6**, 31–40.

Schachtschabel, U., Pavlinkova, G., Lou, D., and Köhler, H. (1996). Antibody-mediated gene delivery for B-cell lymphoma *in vitro. Cancer Gene Ther.* **3**, 365–372.

Shwarzenberger, P., Spence, S. E., Gooya, J. M., Michiel, D., Curiel, D. T., Ruscetti, F. W., and Keller, J. R. (1996). Targeted gene transfer to human hematopoietic progenitor cell lines through the c-kit receptor. *Blood* **87**, 472–478.

Sosnowski, B. A., Gonzalez, A. M., Chandler, L. A., Buechler, Y. J., Pierce, G. F., and Baird, A. (1996). Targeting DNA to cells with basic fibroblast growth factor (FGF2). *J. Biol. Chem.* **271**, 33647–33653.

Stingl, G., Wolff, K., Bröcker, E.-B., Mertelsmann, R., Wolff, Schreiber, S. *et al.* (1996). Phase I study to the immunotherapy of metastatic malignant melanoma by a cancer vaccine consisting of autologous cancer cells transfected with the human IL-2 gene. *Human Gene Ther.* **7**, 551–563.

Tang, M. X., and Szoka, F. C. (1997). The influence of polymer structure on the interactions of cationic polymers with DNA and morphology of the resulting complexes. *Gene Ther.* **4**, 823–832.

Taxman, D. J., Lee, E. S., and Wojchowski, D. M. (1993). Receptor-targeted transfection using stable maleimido-transferrin/thio-poly-L-lysine conjugates. *Anal. Biochem.* **213**, 97–103.

Thurnher, M., Wagner, E., Clausen, H., Mechtler, K., Rusconi, S., Dinter, A., Berger, E. G., Birnstiel, M. L., and Cotten, M. (1994). Carbohydrate receptor-mediated gene transfer to human T-leukemic cells. *Glycobiology* **4**, 429–435.

Trubetskoy, V. S., Torchilin, V. P., Kennel, S. J., and Huang, L. (1992b). Use of N-terminal modified poly(L-lysine)-antibody conjugate as a carrier for targeted gene delivery in mouse lung endothelial cells. *Bioconjugate Chem.* **3**, 323–327.

Trubetskoy, V. S., Torchilin, V. P., Kennel, S., and Huang, L. (1992b). Cationic liposomes enhance targeted delivery and expression of exogenous DNA mediated by N-terminal modified poly(L-lysine)-antibody conjugate in mouse lung endothelial cells. *Biochim. Biophys. Acta* **1131**, 311–313.

Wadhwa, M. S., Knoell, D. L., Young, A. P., and Rice, K. G. (1995). Targeted gene delivery with a low molecular weight glycopeptide carrier. *Bioconjugate Chem.* **6**, 283–291.

Wagner, E., Zenke, M., Cotten, M., Beug, H., and Birnstiel, M. L. (1990). Transferrin-polycation conjugates as carriers for DNA uptake into cells. *Proc. Natl. Acad. Sci. USA* **87**, 3410–3414.

Wagner, E., Cotten, M., Mechtler, K., Kirlappos, H., and Birnstiel, M. L. (1991a). DNA-binding transferrin conjugates as functional gene-delivery agents: Synthesis by linkage of polylysine or ethidium homodimer to the transferrin carbohydrate moiety. *Bioconjugate Chem.* **2**, 226–231.

Wagner, E., Cotten, M., Foisner, R., and Birnstiel, M. L. (1991b). Transferrin-polycation-DNA complexes: The effect of polycations on the structure of the complex and DNA delivery to cells. *Proc. Natl. Acad. Sci. USA* **88**, 4255–4259.

Wagner, E., Zatloukal, K., Cotten, M., Kirlappos, H., Mechtler, K., Curiel, D. T., and Birnstiel, M. L. (1992). Coupling of adenovirus to transferrin-polylysine/DNA complexes greatly enhances receptor-mediated gene delivery and expression of transfected genes. *Proc. Natl. Acad. Sci. USA* **89**, 6099–6103.

Wagner, E., Plank, C., Zatloukal, K., Cotten, M., and Birnstiel, M. L. (1992b). Influenza virus hemagglutinin HA-2 N-terminal fusogenic peptides augment gene transfer by transferrin-polylysine/DNA complexes: Towards a synthetic virus-like gene transfer vehicle. *Proc. Natl. Acad. Sci. USA* **89**, 7934–7938.

Wagner, E., Curiel, D., and Cotten, M. (1994). Delivery of drugs, proteins and genes into cells using transferrin as a ligand for receptor-mediated endocytosis. *Adv. Drug Del. Rev.* **14**, 113–136.

Watkins, S. J., Mesyanzhinov, V. V., Kurochkina, L. P., and Hawkins, R. E. (1997). The 'adenobody' approach to viral targeting: specific and enhanced adenoviral delivery. *Gene Ther.* **4**, 1004–1012.

Wickham, T. J., Tzeng, E., Shears, L. L., Roelvink, P. W., Li, Y., Lee, G. M., Brough, D. E.,

Lizonova, A., and Kovesdi, I. (1997). Increased *in vitro* and *in vivo* gene transfer by adenovirus vectors containing chimeric fiber proteins. *J. Virol.* **71**, 8221–8229.

Wilke, M., Fortunati, E., van den Broek, M., Hoogeveen, A. T., and Scholte, B. J. (1996). Efficacy of a peptide-based gene delivery system depends on mitotic activity. *Gene Ther.* **3**, 1133–1142.

Wolfert, M. A., and Seymour, L. W. (1996). Atomic force microscopic analysis of the influence of the molecular weight of poly-L-lysine on the size of polyelectrolyte complexes formed with DNA. *Gene Ther.* **3**, 269–273.

Wolfert, M. A., and Seymour, L. W. (1998). Chloroquine and amphipathic peptide helices show synergistic transfection *in vitro*. *Gene Ther.* **5**, 409–414.

Wu, G. Y., and Wu, C. H. (1987). Receptor-mediated *in vitro* gene transformation by a soluble DNA carrier system. *J. Biol. Chem.* **262**, 4429–4432.

Wu, G. Y., and Wu, C. H. (1988). Receptor-mediated gene delivery and expression *in vivo*. *J. Biol. Chem.* **263**, 14621–14624.

Wu, G. Y., Wilson, J. M., and Wu, C. H. (1989). Targeting genes: Delivery and persistent expression of a foreign gene driven by mammalian regulatory elements *in vivo*. *J. Biol. Chem.* **264**, 16985–16987.

Wu, G. Y., Wilson, J. M., Shalaby, F., Grossman, M., Shafritz, D. A., and Wu, C. H. (1991). Receptor-mediated gene delivery *in vivo*. Partial correction of genetic analbumenia in Nagase rats. *J. Biol. Chem.* **266**, 14338–14342.

Wu, G. Y., Zhan, P., Sze, L. L., Rosenberg, A. R., and Wu, C. H. (1994). Incorporation of adenovirus into a ligand-based DNA carrier system results in retention of original receptor specificity and enhances targeted gene expression. *J. Biol. Chem.* **269**, 11542–11546.

Yin, W., and Cheng, P. W. (1994). Lectin conjugate-directed gene transfer to airway epithelial cells. *Biochem. Biophys. Res. Commun.* **205**, 826–833.

Zabner, J., Fasbender, A. J., Moninger, T. Poellinger, K. A., and Welsh, M. J. (1995). Cellular and Molecular Barriers to Gene Transfer by a cationic Lipid. *J. Biol. Chem.* **270**, 18997–19007.

Zanta, M. A., Boussif, O., Adib, A., and Behr, J. P. (1997). *In vitro* gene delivery to hepatocytes with galactosylated polyethylenimine. *Bioconjugate Chem.* **8**, 841–844.

Zauner, W., Blaas, D., Küchler, E., and Wagner, E. (1995). Rhinovirus mediated endosomal release of transfection complexes. *J. Virol.* **69**, 1085–1092.

Zauner, W., Kichler, A., Schmidt, W., Sinski, A., and Wagner, E. (1996). Glycerol enhancement of ligand-polylysine/DNA transfection. *Biotechniques* **20**, 905–913.

Zauner, W. Kichler, A., Mechtler, K., Schmidt, W., and Wagner, E. (1997). Glycerol and polylysine synergize in their ability to rupture vesicular membranes and increase transferrin-polylysine mediated gene transfer. *Exp. Cell Res.* **232**, 137–145.

Zauner, W., Ogris, M., and Wagner, E. (1998). Polylysine-based transfection systems utilizing receptor-mediated delivery. *Adv. Drug Del. Rev.* **30**, 97–114.

Zenke, M., Steinlein, P., Wagner, E., Cotten, M., Beug, H., and Birnstiel, M. L. (1990). Receptor-mediated endocytosis of transferrin polycation conjugates: An efficient way to introduce DNA into hematopoietic cells. *Proc. Natl. Acad. Sci. USA* **87**, 3655–3659.

CHAPTER 11

The Perplexing Delivery Mechanism of Lipoplexes

Lee G. Barron and Francis C. Szoka, Jr.
School of Pharmacy, University of California, San Francisco,
California

We summarize the process of lipoplex-mediated *in vitro* gene transfer and describe recent advances in the understanding of *in vivo* gene transfer with an emphasis on systemic lipoplex administration. Lipoplexes are formed through electrostatis interactions between cationic lipids and DNA components. The electrostatic charge on the resulting particle depends on the starting charge ratio (cationic group/nucleotide phosphate) used to prepare the complex, the method of mixing, and the ionic strength of the suspending medium. The resulting particles associate

Nonviral Vectors for Gene Therapy

with the target cell membrane *in vitro* through electrostatic interactions and are subsequently endocytosed. A limited amount of DNA is released into the cytoplasm from the endosome and gains access to the nucleus. The mechanism of this process has not been completely elucidated but has been suggested to be due to displacement of the DNA from the complexes by anionic cellular lipids. Gene delivery *in vivo* involves interactions with the biophase prior to reaching the target cell, which complicates efforts to understand the mechanism of the delivery process. In the context of intravenous delivery, the principal site of transfection is in the lung. There are a number of possibilities to account for this high level of lung expression relative to expression in other organs: (1) the cationic lipid component may interact with the cell membrane directly, either by binding to it and initiating the endocytosis of DNA or by altering the biomembrane properties to facilitate DNA uptake; (2) plasma protein interactions with the lipoplex result in specific plasma proteins associating with the lipoplex. These adsorbed proteins may target the lipoplex to cell surface receptors in the lung endothelium leading to lung transfection. Recently, the intravenous injection of a preformed lipoplex in rodents was found to be unnecessary for lung transfection; rather the cationic lipid may be injected 5 minutes prior to the plasmid DNA and similar levels of gene expression to that obtained with the preformed lipoplex are observed in the lung. Thus the mechanism of lipoplex-mediated transfection remains perplexing; however, the potential of lipoplexes for gene delivery makes understanding the mechanism an alluring goal, since this knowledge may be applied in the development of greatly improved synthetic gene delivery vectors.

I. INTRODUCTION

Gene delivery to somatic cells is a clinical strategy for the treatment of genetic and acquired diseases through the alteration of the patient's genetic repertoire. Two distinctive technologies have arisen to achieve this goal: viral and nonviral.

The technology of nonviral gene delivery was significantly advanced by Felgner and colleagues, who conceived the use of cationic lipid-mediated gene delivery (Felgner *et al.*, 1987). This group formed a complex between N-(2,3-(dioleoyloxy)propyl)-N,N,N-trimethyl ammonium chloride (DOTMA) liposomes and plasmid DNA and observed that the cationic lipid/DNA complex or lipoplex (Felgner *et al.*, 1997) successfully transfected several cell lines with a reporter gene. Since then, a number of groups (Gao and Huang, 1995; Lee *et al.*, 1996) using a variety of cationic lipids have shown gene transfer in animal models after administration of lipoplexes. This has led to the use of several cationic lipid-based gene delivery protocols in clinical trials (Caplen *et al.*, 1995; Chadwick *et al.*, 1997; McLachlan *et al.*, 1996; Nabel *et al.*, 1994; Nabel *et al.*, 1993; Sorscher *et al.*, 1994; Stopeck *et al.*, 1997) (Table 1).

Table I

In Vivo **Gene Delivery Studies**

Cationic lipid formulation	Model	Route of administration	Tissues assayed	Reference
DOTMA/DOPE	Mouse	iv	Lung, liver, and kidney	Brigham *et al.*, 1989
DOTMA/DOPE	Mouse	iv	Various tissues	Zhu *et al.*, 1993
DOTMA/DOPE	Mouse	iv	Lung, liver, and kidney	Brigham *et al.*, 1993
DDAB/Chol	Mouse	iv	Many	Liu *et al.*, 1995
DOTMA/DOPE	Rabbit	iv and aerosol	Pulmonary endothelium and epithelium	Canonico *et al.*, 1994
DOTMA/DOPE and DMRIE/DOPE	Rat	it instillation	Pulmonary endothelium	Logan *et al.*, 1995
Imidazolinium-based lipids	Mouse	iv	Lung, heart, and liver	Solodin *et al.*, 1995
GAP-DLRIE/DOPE	Pig	ia	Arteries	Stephan *et al.*, 1996
DOTAP/Protamine sulfate	Mouse	iv/Intraportal	Heart, lung, liver, spleen, and kidney	Li and Huang, 1997
DOTMA/Tween 80	Mouse	iv	Heart, lung, liver, spleen, and kidney	Liu *et al.*, 1997b
DOTIM/Chol and DOTIM/DOPE	Mouse	iv	Heart, lung, liver, and spleen	Liu *et al.*, 1997a
DOTAP/Chol, DOTAP/DOPE, DOTMA/Chol and DOTMA/DOPE	Mouse	iv	Heart, lung, liver, spleen, and kidney	Song *et al.*, 1997
DOTIM/Chol	Mouse	iv	Various tissues	McClean *et al.*, 1997
DOTAP/Chol	Mouse	iv	Various tissues	Smyth-Templeton *et al.*, 1997
DOTAP/Chol	Mouse	iv	Heart, lung, liver, and spleen	Barron *et al.*, 1998

A. Cationic Lipid Structure

The prototypical cationic lipid is composed of a hydrophobic anchor connected by a linker to a positively charged headgroup (Table 2). The hydrophobic moiety generally consists of tandem aliphatic chains or a cholesterol ring. In spite

Table II

Structure–Function Relationship Between Cationic Lipids and Gene Transfer

	Types	Effects
Cationic lipid	• Number of positive charges (1, 2, 3, 4, 5)	• Charge density
	• Types of amino group (primary, secondary, tertiary, quaternary)	• Hydrogen bonding
	• Other helper groups (hydroxyl)	• Membrane fusion
Linker group	• Amide • Ester • Carbamate • Ether • Phosphate	• Length • Hydrophilic property • Flexibility • Hydrogen bonding • Sensitivity to cleavage
Hydrophobic region	• Saturated • Unsaturated • Fused ring	• Length • Transition temperature • Flexibility

of considerable research, attempts to correlate lipid structure with transfection activity have met with limited success. Akao and colleagues (Akao *et al.,* 1991) synthesized a series of double acyl chain ammonium amphiphiles and correlated the transfection efficiency of the resulting liposomes with their physicochemical properties. They found that lipids with phase transition temperatures (T_m) below 37°C were more efficient in *in vitro* gene delivery than those with a T_m above 37°C. Similarly, the Felgner group (Felgner *et al.,* 1994) found that as the length of the acyl chain associated with a given polar headgroup was decreased from 18 to 14 carbons, yielding a lower T_m, the *in vitro* transfection activity of the corresponding lipoplex increased. A similar pattern of *in vitro* transfection was observed by Solodin and colleagues with cationic amphiphiles composed of alkylated imidazolinium salts (Solodin *et al.,* 1995). The longer alkyl chain length elevates the T_m and bilayer stiffness of ensuing vesicles (Felgner *et al.,* 1994). Lee and co-workers (Lee *et al.,* 1996) also reported greater transfection activity *in vivo* upon decreasing chain length for a series of cationic lipids with a constant polar headgroup. Solodin and co-workers noted that while the saturated 14-carbon imidazolinium derivative was the most effective lipid for *in vitro* transfection, the optimal *in vivo* counterpart consisted of 18-carbon, *cis* unsaturated alkyl groups. Similarly, Wang and colleagues, (Wang *et al.,* 1998), in studying a series of carnitine ester-based lipids, noted that the optimal hydrophobic moiety for *in vitro* gene delivery was a 14-carbon alkyl acyl group, while the most efficient *in vivo* was an 18-carbon derivative with 1-*cis* unsaturation. The incorporation of a double bond has an effect similar to that of shortening acyl chain length in that it decreases the rigidity of the resulting bilayer.

The lack of correlation between *in vitro* and *in vivo* transfection results has been commented on before (Lasic *et al.*, 1997; Lee *et al.*, 1996; Solodin *et al.*, 1995). Balasubramaniam and colleagues identified a cell-type dependency for optimal T_m and transfection and suggested that cationic lipid structures would require optimization for a given cell type (Balasubramaniam *et al.*, 1996). This group also demonstrated that asymmetric hydrophobic domains transfected as well as or better than their symmetric counterparts. Gao and Huang and later Lee and co-workers (Gao and Huang, 1991; Lee *et al.*, 1996) showed that the complete alteration of the hydrophobic domain from that of paired acyl chains to a cholesterol ring structure increased transfection activity in some contexts.

The linker group that connects the hydrophobic anchor to the polar headgroup also influences transfection activity (Lee *et al.*, 1996). For the compounds studied, substitution of amide, amine, or urea linkages in place of the existing carbamate function decreased transfection activity *in vivo;* however, this effect may be dependent on the lipid context (Lee *et al.*, 1996). An important parameter associated with the linker group is its chemical stability and biodegradability. A lack of chemical stability can manifest itself in reduced transfection efficiency, as may be the case in the study by Lee and colleagues, where the substitution of less stable amide and urea linkers for carbamate yielded lower transfection activity. Reduced chemical stability may also limit the practical potential of a given formulation by decreasing shelf-life (Gao and Huang, 1995).

The positively charged headgroup of most cationic lipids consists of one or more amine groups subject to varying degrees of substitution. The functions of this domain in gene delivery include facilitating electrostatic interactions with negatively charged DNA to form the lipoplex as well as mediating interactions between the lipoplex and target cells. Typical amine group substitution is carried out via methylation, but Felgner and associates have shown that hydroxyethylation of the terminal amine of DOTMA yielded compounds that mediated greater *in vitro* transfection activity than the parent compound (Felgner *et al.*, 1994). These authors suggested that the role of the hydroxyethyl group is to preserve bilayer integrity through the maintenance of membrane hydration.

The ability of polyvalent headgroups to form more compact lipoplexes has been implicated in the observation that amphiphiles containing this motif frequently mediate higher degrees of gene delivery than their monovalent counterparts (Behr *et al.*, 1989; Remy *et al.*, 1994; Zhou *et al.*, 1991). Lee and associates have observed that cationic lipids with multiple protonatable amines, when coupled to their respective lipid anchors in a T-shaped structure, yielded greater transfection activity than the linear construct (Lee *et al.*, 1996). The authors suggested two roles for the T-shaped motif: first, it may promote better complexation with plasmid

DNA and second, it might resemble a ligand for a particular cell surface receptor, although no evidence was presented to support either of these conjectures.

B. Role of Helper Lipid

A typical cationic liposome formulation contains a neutral helper lipid, in addition to the cationic lipid. The most common helper lipid employed *in vitro* is dioleoylphosphatidylethanolamine (DOPE). This lipid is necessary for the establishment of a bilayered vesicle morphology with some cationic lipids (Gao and Huang, 1995). In addition, the fusogenic capacity of DOPE (Hui *et al.*, 1981) enhances lipoplex-mediated transfection efficiency when compared to DOPC (Farhood *et al.*, 1995; Wrobel and Collins, 1995) and several DOPE structural analogs (Zhou and Huang, 1994). It has been suggested that the augmentation of gene delivery is a reflection of increased endosomal DNA release into the cytoplasm as facilitated by DOPE-mediated endosomal membrane disruption (Zhou and Huang, 1994; Farhood *et al.*, 1995).

C. Lipoplex Formation

The initial step in the formation of cationic lipid/DNA complexes is mediated by spontaneous electrostatic interactions between the positively charged headgroup of the cationic lipid and the negatively charged phosphate backbone of the nucleotide. The lipids serve to condense DNA into chemically and physically diverse structures (Gershon *et al.*, 1993) characterized as multilamellar globules. The precise character of these aggregates depends both qualitatively and quantitatively on the lipid components of the liposome, the method of lipid preparation including ionic strength of the buffers employed, the method of mixing, and the elapsed time of incubation (Gershon *et al.*, 1993; Sternberg *et al.*, 1994; Yang and Huang, 1997).

Gershon and colleagues have demonstrated that as the lipid concentration increases relative to that of DNA, concomitant membrane fusion and DNA condensation rapidly occur (Gershon *et al.*, 1993). These phenomena yield a complex that protects the DNA from DNase I digestion and ethidium bromide intercalation although, as Eastman and co-workers state, the latter may simply reflect compaction rather than encapsulation (Eastman *et al.*, 1997a). Investigations using small-angle X-ray scattering (SAXS) (Rädler *et al.*, 1997; Lasic *et al.*, 1997) have revealed that liposomes, upon mixing with DNA, undergo significant rearrangements resulting in the formation of multilamellar structures with a periodicity of 6.5 nm. This measurement corresponds with a model of a hydrated DNA helix with a diameter of 2.5 nm, intercalated between a cationic bilayer 4 nm thick.

Analysis of lipoplex structure using electron microscopy revealed amorphous aggregates with diameters in the submicron range (Gershon *et al.*, 1993; Gustafsson *et al.*, 1995; Sternberg *et al.*, 1994). Gershon and colleagues, utilizing Kleinschmidt metal rotary shadowing electron microscopy, observed that DOTMA–DOPE lipoplexes consisted of vesicles arranged along DNA strands at low lipid to DNA ratios, but as the amount of lipid was increased, the complexes assumed rod-like structures in which the DNA was presumably entirely encapsulated.

Sternberg and co-workers (Sternberg *et al.*, 1994), through freeze–fracture electron microscopic imaging techniques, observed that both tube-like structures and globules formed upon mixing DC cholesterol lipids with DNA and coined the term "spaghetti and meatballs" to describe these moieties. Based on the diameter of the tubular structures (10 nm), it was suggested that they consisted of single DNA strands ensheathed in cationic lipid membranes. These tubules were found both freely suspended and associated with the "meatball" components of the preparations.

Gustafsson and colleagues (Gustafsson *et al.*, 1995) generated cryotransmission electron micrographs of cationic detergents and lipids combined with DNA exhibiting oligolamellar structures that appeared to entrap DNA upon its addition to cationic liposomes.

In summary, cationic lipid/DNA complex formation is characterized by rapid charge neutralization of DNA upon addition of lipid accompanied by thorough lipid mixing. The lipids and DNA arrange themselves into multilamellar formations consisting of DNA aligned in strata between lipid bilayers with a periodicity of 6.5 nm yielding the large globules that are apparent in electron micrographs. In some instances, the formation of tubular structures with a diameter of 10 nm occurs (Gershon *et al.*, 1993; Sternberg *et al.*, 1994). These tubules appear to form when DOPE is included in the lipid formulation and may be attributed to the ability of DOPE to accommodate structures of a high radius of curvature upon phase separation (Seddon, 1990). In support of this point, formulations prepared without DOPE tend not to yield lipoplexes with extensive tube-like structures (Eastman *et al.*, 1997a; Lasic *et al.*, 1997; Xu *et al.*, 1998). Although colloidal properties do not vary considerably between different lipid compositions, lipoplex aggregates tend to become larger and more polydisperse as the charge equivalency point is approached (Mahato *et al.*, 1995a; Rädler *et al.*, 1997; Xu *et al.*, 1998). The issue of lipoplex size is considered important in that the endocytotic process, whether receptor or fluid-phase-mediated, is limited to particles of approximately 200 nm or less, although larger particles may be phagocytosed by some cell types. This may be due to a cooperative endocytotic event that results in the internalization of large diameter lipoplexes that become associated, through electrostatic interactions, with multiple coated pits.

II. PATHWAY FOR INTERACTION
WITH CELLS IN CULTURE

Viruses have developed complex strategies to overcome numerous barriers to exogenous gene delivery and expression. These impediments include the initial interaction between the vector and the cell membrane, followed by internalization, endosomal release, transport into the nucleus, and finally, transcription and translation of the therapeutic gene. It is reasonable to assume that the mediation of high transfection levels requires nonviral vectors to overcome the same series of barriers to transgene delivery and expression (Fig. 1). Therefore, multiple functions would be needed to overcome these barriers. Currently employed cationic lipoplexes add one, or at most two, components to the DNA, so it is surprising that lipoplexes deliver genes as well as they do.

A. CELL ADHESION

The initial step for lipoplex-mediated transfection *in vitro* is thought to be the electrostatic interaction between the cationic lipid and the anionic plasma membrane (Felgner *et al.*, 1987). As mentioned previously, it is possible to manipulate

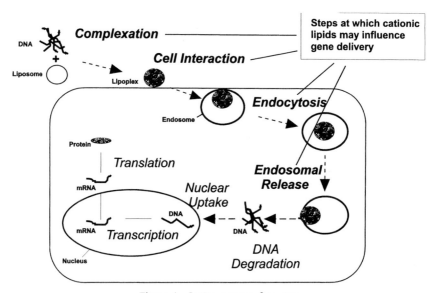

Figure 1   *In vitro* gene transfer events.

the colloidal properties of the lipoplex by varying the ratio of cationic lipid to anionic nucleotide such that a charge excess of lipid is included in the lipoplex to neutralize the DNA. This yields a complex with a net positive charge, thus facilitating its interaction with the plasma membrane. The precise nature of this coupling is not clear; however, it may include binding of the lipoplex to negatively charged membrane components such as sialylated glycoproteins or possibly coupling with sulfated membrane-associated proteoglycans (Mislick and Baldeschwieler, 1996; Labat-Moleur *et al.*, 1996).

B. ROUTE OF INTERNALIZATION

Following membrane association, the cellular uptake of cationic lipid/DNA complexes was originally thought to occur through fusion of the liposome and plasma membranes (Felgner *et al.*, 1987; Felgner and Ringold, 1989). This mechanism was based on the observation that fluorescently labeled DOPE distributed into the plasma and intracellular membranes of the target cells 4 h after exposure to the lipoplex (Felgner *et al.*, 1987). Moreover, it was demonstrated that cationic liposomes are capable of fusion with anionic liposomes (Felgner and Ringold, 1989; Düzgünes *et al.*, 1989) although fusion activity was attenuated by preincubation with DNA (Leventis and Silvius, 1990).

More recently, two lines of evidence suggest that the lipoplexes gain entry to the cytoplasm of target cells via endocytosis: (1) enhancement of transfection by lysosomotrophic agents such as chloroquine (Legendre and Szoka, 1992; Felgner *et al.*, 1994) and (2) morphologic studies that visually monitor intracellular lipoplex trafficking (Zhou and Huang, 1994; Zabner *et al.*, 1995; Friend *et al.*, 1996). Legendre and Szoka (1992) have demonstrated that lysosomotrophic agents enhanced DOTMA/DOPE mediated transfection in some cell types. Similarly, Felgner and colleagues (Felgner *et al.*, 1994), have shown that chloroquine influenced transfection mediated by both DORIE/DOPE and DMRIE/DOPE-based lipoplexes. The buffering capacity of chloroquine alters the trafficking of cationic lipid-based systems by preventing endosomal/lysosomal fusion; thus, the influence of this molecule on lipoplex-mediated transfection suggests the involvement of an endocytic uptake process.

Electron microscopy (EM) studies revealed electron-dense cationic lipopolylysine-containing liposomes in the endosomes of treated cells within 1 h of administration (Zhou and Huang, 1994). Similar *in vitro* studies by Zabner and colleagues (Zabner *et al.*, 1995), showed the exclusive localization of gold-labeled DNA complexed with DMRIE/DOPE liposomes in cytoplasmic vesicles or endosomes. Friend and co-workers (Friend *et al.*, 1996) found that the gold-labeled lipid component of the lipoplex associated with the clathrin coat assembly of the plasma

membrane upon *in vitro* administration. This interaction was followed by observation of the lipid label being taken up by endosomes. Unlike the gold-labeled DNA
experiments mentioned, these investigators noted extravesicular lipid label present
in the cytoplasm. This phenomenon was attributed to the release of endocytosed
material rather than the manifestation of a plasma membrane fusion event.

Although these data do not exclude the possibility that some lipoplex uptake
occurs via membrane fusion, they suggest that the principal mode of cell uptake
involves endocytosis.

C. Endosomal Release Mechanism

Zabner and colleagues have postulated that one of the salient factors limiting *in vitro* nonviral gene delivery is the release of free DNA from the endosome
(Zabner *et al.*, 1995). This conclusion was based on light microscopic visualization
of fluorescently labeled DNA and lipids, electron microscopic observation of gold-
labeled DNA, and quantitative analysis of cellular DNA uptake via dot blots
(Zabner *et al.*, 1995). Although DNA release from the endosome appears to be an
inefficient process, transgene expression is realized upon administration of lipoplexes to cells, demonstrating that a small amount of DNA gains access to the nucleus where it can be transcribed. Moreover, the DNA apparently enters the nucleus
as a free molecule, since nuclear or cytoplasmic microinjection of cationic lipid/
DNA complexes does not express the encoded protein (Zabner *et al.*, 1995; Pollard
et al., 1998). In addition, the fact that T7 promoter-containing plasmid DNA is
accessible to cytoplasmic T7 polymerase upon lipoplex delivery, suggests that DNA
is cytoplasmically released (Gao and Huang, 1993). Fluorescent confocal imaging
of cultured cells exposed to cationic lipid/oligonucleotide complexes shows cytoplasmic dissociation of lipid and ODN with subsequent nuclear localization of the
ODN while the lipid component of the complex remained in the cytoplasm
(Zelphati and Szoka, 1996).

A model has been presented that explicates endosome release of DNA (Xu and
Szoka, 1996; Zelphati and Szoka, 1997) and is consistent with the preceding observations: The paradigm describes the uptake of lipoplex by endocytosis followed by
destabilization of the endosomal membrane (Fig. 2). Destabilization results in the
flip-flop of anionic lipids, located principally in the cytoplasmic monolayer, and
their subsequent lateral diffusion into the complex where they form ion pairs with
cationic lipids. Charge neutralization of the cationic lipids results in the release of
DNA into the cytoplasm. Based on this mechanism, the cytoplasmic release of
DNA is clearly dependent on the presence of sufficient numbers of anionic lipids
to displace the DNA molecule. An approximation of the number of anionic lipids
present on an endosomal vesicle monolayer of 150 nm diameter was made based on

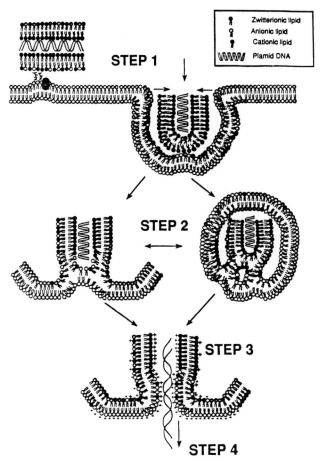

Figure 2   Mechanism of uptake and release of plasmid DNA from the lipoplex. Step 1: After electro-static interaction with the cell membrane, cationic liposome/DNA complexes are endocytosed. Step 2: In the early endosome, membrane destabilization results in anionic phospholipid flip-flop. Step 3: The anionic lipids diffuse into the complex and form a charge neutral ion pair with cationic lipids. Step 4: The DNA dis-associates from the complex and is released into the cytoplasm. Reprinted with permission from Xu and Szoka, *Biochemistry* **35**, pp. 3616–3623. Copyright 1996 American Chemical Society.

the assumption that the monolayer has a surface area of 1.76×10^4 nm^2. In addition, it was assumed that the monolayer consists of 70% phospholipids of which 25% are negatively charged and that each lipid has a surface area of 0.7 nm^2. Given these assumptions, the total number of anionic lipids available for the plasmid DNA molecule to interact with is approximately 4.4×10^3. The complete dissociation of a 4×10^3 base-pair plasmid DNA molecule would require ca 8,000 anionic residues; hence there may be an insufficient number of anionic lipids in the vesicle membrane of small endosomes to entirely release the plasmid DNA. This phenomenon may explain the lack of efficient cytoplasmic plasmid DNA release. However, the delivery of oligonucleotides appears to be more efficient (Zelphati and Szoka, 1996). The model can account for oligonucleotide release because the previously mentioned 4.4×10^3 interactions between anionic and cationic lipids would lead to the complete release of 220 20-mer oligonucleotides. This model has been described and rationalized in detail (Szoka *et al.*, 1996; Zelphati and Szoka, 1997).

D. Nuclear Uptake

A second potentially rate-limiting step in lipoplex-mediated transfection is at the point of plasmid nuclear uptake (Zabner *et al.*, 1995). As mentioned, the cytoplasmic dissociation of the DNA from the cationic lipid appears to occur and, moreover, is necessary for transgene expression since nuclear injection of lipoplex results in little or no transfection (Zabner *et al.*, 1995; Pollard *et al.*, 1998). A detailed account of cytoplasmic trafficking and nuclear transport of nucleic acids has been compiled by Meyer and colleagues (Meyer *et al.*, 1997).

E. Cell-Type Dependence of Transfection

In general, it has proved difficult to establish a firm correlation between cell type and transfectability. Indeed, Boussif and colleagues have found luciferase activity from various transfected-cell extracts to span five orders of magnitude (Boussif *et al.*, 1996). Commercial cationic lipid transfection reagents have been used to transfect numerous different cell lines but the literature from such companies is difficult to interpret in terms of the mechanism of action for a given cationic lipid. It does appear, however, that adherent cells are more amenable to lipoplex transfection than cells in suspension (Labat-Moleur *et al.*, 1996; Boussif *et al.*, 1996). Recently, however, Floch and colleagues have shown K562 cells, a nonadherent cell type, to be as transfectable as adherent cells (Floch *et al.*, 1998).

Transfection data associated with a given synthetic delivery system *in vitro* are not generally predictive of data generated *in vivo*. Efforts to improve this correlation have involved the addition of serum in the cell culture medium to more closely resemble the *in vivo* gene transfer context. Although some cationic lipid/DNA complexes demonstrate limited resistance to the deleterious effects of serum (Hofland *et al.*, 1996; Yang and Huang, 1998) the addition of serum to culture medium typically results in the reduction of transfection *in vitro* (Felgner *et al.*, 1987). The decrease in transfection has been attributed to a reduction in cell association due to charge neutralization of the complex (Yang and Huang, 1997) or compromise of the stability of the complex (Zelphati *et al.*, 1998). The effects of blood components will be discussed in further detail next.

F. *In Vitro* Summary

The manifestation of a net positive lipoplex surface charge has been established as a means of promoting significant lipoplex–plasma membrane binding through electrostatic interactions (Felgner *et al.*, 1994; Gao and Huang, 1991; Zabner *et al.*, 1995). Lipoplex–target cell association is a time-dependent phenomenon that appears to plateau between 6 and 12 h postadministration (Felgner *et al.*, 1987; Legendre and Szoka, 1992; Zabner *et al.*, 1995). The amount of DNA associated with cells following lipoplex administration varies. Legendre and Szoka found a maximum of 4.5% of the administered dose to be taken up by cells after a vigorous washing procedure (Legendre and Szoka, 1992), whereas Zabner and colleagues and Labat-Moleur and colleagues found the same parameter to be 60% and 33%, respectively (Zabner *et al.*, 1995; Labat-Moleur *et al.*, 1996). The number of plasmid molecules internalized per cell was estimated to be 19,000 in the case of Legendre and Szoka's calculations and up to approximately 3×10^5 as per the Zabner group's assessment. The variation in intracellular lipoplex-mediated delivery of plasmid DNA may be a reflection of the different cell types or washing procedures used in these investigations. Based on these data, cell uptake may not be a significant rate-limiting step in *in vitro* gene delivery, but downstream barriers such as endosomal escape and nuclear uptake appear to limit expression. Capecchi and colleagues have shown that, upon microinjection of plasmid DNA into the cytoplasm, less than 0.01% of injected cells express the encoded protein, whereas nuclear microinjection resulted in 50–100% of injected cells expressing the injected DNA (Capecchi, 1980). These observations were subsequently confirmed by Zabner and colleagues (Zabner *et al.*, 1995). Pollard and co-workers, who also employed microinjection techniques, estimated that the efficiency of plasmid transit into the nucleus was 0.1% and that cationic lipoplexes failed to reach the nucleus in detectable amounts (Pollard *et al.*, 1998). It is clear that even if free plasmid DNA is released from the endosome (Xu and Szoka, 1996), major impediments to transgene expression exist. Dowty and colleagues have demonstrated that plasmid DNA, when microinjected near the nucleus, is expressed more efficiently than more distally administered DNA (Dowty *et al.*, 1995). This implies that the DNA is sequestered in the cytoplasm, perhaps by binding to cytoplasmic structures or by sieving through a cytoskeletal network. It is also possible that, prior to reaching the nuclear pore complex, DNase activity in the cytoplasm degrades free cytoplasmic DNA.

If gene delivery efficiency is expressed on a per genome basis, viral vectors are much more effective than nonviral systems. Adenoviruses, for instance, have been shown to be capable of quantitative transfection of cultured cells with as little as 10 MOI/cell (Lemarchand *et al.*, 1992; Zabner *et al.*, 1994). However, as mentioned previously, when several thousand plasmids per cell are delivered via lipoplex *in vitro*, cells yield much lower gene expression.

III. *IN VIVO* TRANSFECTION

Cationic lipid/DNA complexes have been employed in a number of contexts *in vivo*. Gene delivery has been realized following delivery to the airways, by systemic injection, and upon intraarterial administration via balloon catheterization (Table 1). Successful delivery of a lipoplex in an intact animal involves interactions with anatomical and physiological components located in the vasculature and extracellular space. These interactions with the biophase can alter lipoplex properties and/or prevent access to target cells. These phenomena will be discussed in more detail next.

A. PULMONARY DELIVERY

Gene delivery to the lung has applications in the treatment of genetic diseases such as cystic fibrosis and α-1-antitrypsin deficiency as well as acquired lung diseases where the expression of certain cytokines, surfactant proteins, antioxidants, or mucoproteins have potential therapeutic value. The principal objective of these therapeutic strategies is transfection of the pulmonary epithelial cell, which can be targeted most directly via intratracheal (itr) instillation of lipoplex in solution (Brigham *et al.*, 1989) or when administered as an aerosol (Stribling *et al.*, 1992; Canonico *et al.*, 1994). These approaches have the added benefit of limiting systemic exposure. The feasibility of direct administration to the airways has been demonstrated by the pioneering work of Brigham (Brigham *et al.*, 1989) and the initial experiments have been reviewed by Gao and Huang (Gao and Huang, 1995). More recently, Meyer and colleagues (Meyer *et al.*, 1995) showed that the complexation of plasmid DNA with DOTMA liposomes increased the amount of high-molecular-weight DNA present in the mouse lung following intratracheal injection, but DNA expression was not enhanced relative to DNA alone. The fact that naked DNA transfects as effectively as lipoplex-delivered DNA in the rodent lung is also observed for lipoplexes made from dioleoyltrimethylammoniumpropane (DOTAP) and dimethyldioctadecylammonium bromide (DDAB) (Tsan *et al.*, 1995). Lipoplexes have been applied to the nasal epithelium of cystic fibrosis patients in several clinical studies (Caplen *et al.*, 1995; Gill *et al.*, 1997; Porteous *et al.*, 1997; Zabner *et al.*, 1997). Only partial, transient correction of gene function was realized. Interestingly, GL-67/DOPE-based lipoplexes that proved 1000-fold more effective than naked DNA in transfecting murine lung tissue (Lee *et al.*, 1996) were no more effective than naked DNA when applied to human nasal epithelium (Zabner et al., 1997), emphasizing the complexity of extrapolating from animal experiments to clinical studies. Experiments by Meyer and co-workers have shown that instillation of DOTAP/cholesterol (chol)-based lipoplex mediated higher gene expression in the lung than when DOPE was included as a helper lipid or when DNA was administered alone. Cholesterol as a helper lipid might function to increase the stability of the complex in the lung, thus facilitating expression.

Freimark and colleagues reported an increase in cytokines and an influx of macrophages and neutrophils in the lung following the instillation of DOTMA/chol-based lipoplex into the airways (Freimark et al., 1998). Generally, the immune responses were more profound for the lipoplex than for the lipid or DNA components administered separately. The authors attributed this to three possibilities: first, the individual components may elicit the release of co-stimulatory cytokines; second, augmented induction of the immune response may be due to a lipid-facilitated increase in cell uptake of DNA; or finally, the less rapid degradation of pulmonary delivered DNA when complexed to cationic lipid (Meyer et al., 1995) may increase its persistence in the airways, thereby enhancing DNA's ability to stimulate the immune system.

Aerosol delivery of lipoplex to the lung provides a less invasive alternative to nasal or intratracheal instillation. A limitation of this approach has been the DNA shearing that transpires upon nebulization (Schwarz et al., 1996). In subsequent studies, Eastman and colleagues optimized a liposome preparation consisting of lipid GL-67 combined with dimyristoylphosphatidylethanolamine-polyethylene glycol (DMPE-PEG$_{5000}$) that prevented nebulization-based shearing of complexed DNA (Eastman et al., 1997b; Eastman et al., 1997c). Upon administration, this preparation yielded gene expression comparable to that of instilled lipoplex while invoking little or no inflammatory response (Eastman et al., 1997b). Similarly, McDonald and colleagues, utilizing aerosolized lipoplex, have demonstrated delivery and expression of the CFTR gene with EDMPC/chol-based lipids to the apical epithelium of rhesus monkeys without visible inflammation (McDonald et al., 1998). Chadwick and co-workers demonstrated the safety of aerosol delivery of lipid GL-67A to the lungs of healthy human volunteers (Chadwick et al., 1997). The lack of toxicity inherent in aerosolized preparations as compared to instilled lipoplex is presumably a reflection of more efficient distribution of the former, leading to less localized accumulation of lipoplex; consequently, aerosolization may provide a means of increasing gene expression in the lung through multiple dosing regimes.

B. INTRAVENOUS DELIVERY

The first example of successful intravenous (iv) lipoplex delivery was achieved by the Brigham group in 1989 (Brigham et al., 1989). Since that time, a variety of lipoplex formulations have been employed to effect systemic gene delivery (Table 1;

Gao and Huang, 1995). The fate of intravenously (iv) injected liposomes alone has been reviewed extensively (Patel, 1992; Semple and Chonn, 1996; Senior, 1987). Liposomes are cleared rapidly from circulation by the reticuloendothelial system (RES) unless they have a diameter of less than 100 nm and a neutral surface charge or have a surface polymer to sterically stabilize them (Woodle *et al.*, 1992). Virtually all lipoplex preparations used in systemic gene delivery are formulated with excess positive charge, which ensures that they will be rapidly eliminated from circulation. The utility of this composition when administered iv may be in its ability to facili-tate beneficial interactions with negatively charged cell surfaces and proteins. Alter-natively, the excess positive charge may aid in DNA protection or, as Yang and Huang have suggested, could neutralize negatively charged serum components that are deleterious to transfection (Yang and Huang, 1997). When formulated with excess positive charge, cationic lipid/DNA complexes are eliminated from circula-tion within 5 min of injection. Initially, both the lipid and the DNA are localized in the lung (Liu *et al.*, 1997a; Barron *et al.*, 1998a). As much as 85% of the lipid/ DNA dose administered may collect in the lung (Barron *et al.*, 1998a) as quantitated by radiolabeling, and a qualitative evaluation of lipoplex distribution by confocal fluorescent microscopy-demonstrated lipoplex delivery is principally limited to the pulmonary vasculature (McLean *et al.*, 1997).

Interestingly, based on radiolabeled systemic distribution experiments, the ratio of lipid to DNA in the lung decreases compared to that of the initially injected preparation (Barron *et al.*, 1998a). This suggests that a portion of the excess lipid is not entrapped in the lung as a part of the lipoplex and continues to circulate beyond the pulmonary microvasculature. The lipid to DNA ratio in the liver gradually in-creases, which might mean that the excess lipid accumulates in this organ.

Barron and co-workers demonstrated significant uptake of both lipid and DNA in the liver, but up to 24 h postinjection the percentage of administered dose in the liver remained less than that of the lung (Barron *et al.*, 1998a). Similarly, other groups have found the majority of the DNA localizing in the lung at early time points with the majority of the remaining label accumulating in the liver (Liu *et al.*, 1997a). McLean and colleagues, utilizing fluorescent imaging of the cationic lipid component of the lipoplex, found that relative to tissue mass, the lungs, lymph nodes, Peyer's patch, and ovaries accumulate the greatest amount of lipid (McLean *et al.*, 1997). The fluorescently labeled lipid generally was confined to endothelial cells, macrophages, and intravascular leukocytes. One exception to this circum-stance was the observation of diffuse fluorescence in the spleen, probably attribut-able to extravasation of lipoplex through the discontinuous endothelium in that organ (Poste *et al.*, 1982).

Transgene expression following iv lipoplex injection tends to reflect that of its distribution with the majority occurring in the lung (Barron *et al.*, 1998a; Liu *et al.*, 1997a; Liu *et al.*, 1997b). Also in accordance with distribution data, transgene expression was found primarily in endothelial cells and monocytes (Liu *et al.*,

1997a; Liu *et al.*, 1997b; Song *et al.*, 1997). Some expression was noted in interstitial macrophages (Liu *et al.*, 1997a).

Increasing pulmonary retention time for plasmid DNA correlates with higher reporter gene expression in the lung (Liu *et al.*, 1997a; Liu *et al.*, 1997b; Smyth-Templeton *et al.*, 1997). The inclusion of cholesterol as opposed to DOPE in the liposome formulation tends to increase lung expression following iv injection (Solodin *et al.*, 1995; Song *et al.*, 1997; Smyth-Templeton *et al.*, 1997), probably through increasing pulmonary retention (Smyth-Templeton *et al.*, 1997). This phenomenon is apparently a reflection of the increased stability of cholesterol-containing lipid bilayers in circulation (Senior and Gregoriadis, 1982).

The manner in which liposomes are prepared also affects transfection efficiency. Song and co-workers (1997) found that DOTAP-based lipoplex transfection was dependent on the size of the initial liposome and ensuing lipoplex formed. As liposome size increased, so did the size of the resulting lipoplex. This group demonstrated that as lipoplex diameter increased beyond approximately 450 nm, systemic transfection increased. We have noted a similar effect in that lipoplex of ca 550 nm diameter formed from extruded liposomes yielded 10-fold higher pulmonary transfection than lipoplex formed from sonicated liposomes that had a diameter of approximately 250 nm (Barron and Szoka unpublished). Interestingly, Smyth-Templeton and co-workers have demonstrated that serial extrusion of liposomes through membrane filters of decreasing pore size (down to 0.1 μm) yielded lipoplex with transfection activity twice that of unextruded liposomes of the same composition. This protocol appears to yield a liposome of a unique vase-like structure that may be implicated in its enhanced activity (Smyth-Templeton *et al.*, 1997).

Although a significant amount of the administered dose of lipoplex is delivered to the liver, transgene expression in this organ is generally low. This is likely a reflection of the fact that DNA is primarily internalized by nonparenchymal cells in the liver (Mahato *et al.*, 1995b; McLean *et al.*, 1997). These cells include professional phagocytes such as Kupffer cells that likely degrade most of the DNA before it can be expressed.

The higher ratio of gene expression to DNA uptake in the lung versus that in the liver has been suggested to be due to the presence of transactivating factors in the lung, which tend to promote transcription while such constituents are absent in the liver (Liu *et al.*, 1997a). Although most systemic studies employ CMV-based DNA constructs, the same expression pattern was observed with RSV-driven plasmid constructs (Thierry *et al.*, 1997), suggesting that if the variance in expression is explained by a molecular mechanism, it is not limited to a single promoter system.

The preferential distribution of systemically administered lipoplex to the lung and subsequent transgene expression therein has been explained as a first-pass phenomenon since, upon tail-vein injection, this is the first capillary bed encountered by the complex (Brigham *et al.*, 1989). This may happen by three mechanisms depicted in Figure 3. Cationic lipid/DNA complexes are inherently reactive with negatively charged blood proteins (Senior *et al.*, 1991). If the lipoplex is not completely coated by proteins, some positive residues will be exposed. On encountering the surface of the small-caliber blood vessels of the pulmonary microvasculature, these positive domains may interact with the negatively charged residues present on the pulmonary endothelium (Mislick and Baldeschwieler, 1996; Fig. 3A). The extent of uptake would then be a function of the large surface area of the pulmonary capillary bed. Alternatively, we have speculated that the localization and expression profile of lipoplexes may also represent plasma protein–lipoplex interactions in which associated proteins act to influence lipoplex pharmacokinetics and subsequent cell uptake (Barron *et al.*, 1998a). Although there is a paucity of studies addressing the interactions between cationic lipid/DNA complexes and blood proteins, Senior and colleagues have shown that upon incubating cationic lipids in serum, a dose-dependent increase in turbidity was observed, ultimately leading to the formation of clot-like masses (Senior *et al.*, 1991). Further evidence of the inherent reactivity of positively charged liposomes with negatively charged plasma proteins was provided by Semple and Chonn, who demonstrated that DOTMA-based liposomes accrue over 500 g of plasma protein per mole of lipid upon circulating *in vivo* for 5 min (Semple and Chonn, 1996). If the interaction between plasma proteins and the lipoplex results in cross linking and aggregate formation,

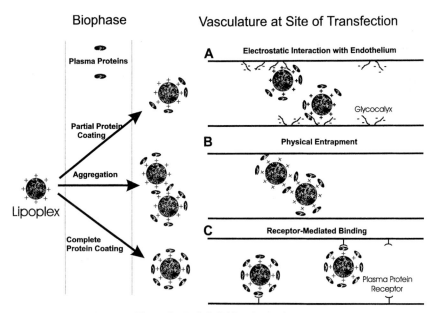

Figure 3   Endothelial lipoplex binding.

these aggregates could be trapped in the pulmonary capillary beds, thus passively targeting the lipoplex to the lung (Fig. 3B). Alternatively, lipoplex-associated proteins may indirectly target the lipoplex through binding their cognate receptor on a particular endothelium such as that of the lung (Fig. 3C). This latter phenomenon may explain the relatively avid binding and efficient transgene expression on a per gram tissue protein basis in the ovaries, lymph nodes, and anterior pituitary (McLean et al., 1997). McLean and colleagues speculated that these lipoplex distribution and expression patterns may be a reflection of as-yet-unelucidated regional differences in the vascular endothelium. These variations may include the expression of specific receptors such as that for low-density lipoprotein in different regions of the endothelium (Nistor and Simionescu, 1986). Transgene distribution and expression profiles achieved using lipoplexes could reflect a combination of the previously mentioned mechanisms: Pulmonary gene delivery may be a manifestation of the first pass effect followed by receptor-mediated uptake. In tissues "downstream" from the lung, receptor-mediated uptake may be the dominant mechanism. Candidate proteins that we thought might be implicated in *in vivo* transfection were the proteins of the complement system and/or albumin. The ability of cationic liposomes (alone or when complexed with plasmid DNA) to interact with plasma complement proteins has been demonstrated *in vitro* (Devine et al., 1994; Plank et al., 1996). In addition, Loughrey and colleagues have shown that systemically administered phosphatidylglycerol-based liposomes yield a transient thrombocytopenia that is mediated by complement activation (Loughrey et al., 1990). We hypothesized that the activation of complement *in vivo* could lead to the coating of lipoplex with complement proteins, thereby targeting it to complement receptors present on the pulmonary endothelium (Zhang et al., 1986). We found that cationic lipid/DNA complexes activate serum complement proteins upon iv injection in mice (Barron et al., 1998a). However, the elimination of serum complement proteins through ip injection of cobra venom factor and anticomplement antibodies did not influence the transfection efficiency or systemic distribution of the lipoplex (Fig. 4 & 5). These results make it highly unlikely that complement proteins are involved in transfection.

The high plasma concentration and low pI of albumin led us to speculate that albumin may interact with the lipoplex after its administration into the blood. In addition, there are albumin receptors on the pulmonary vasculature (Ghitescu et al., 1986). These receptors mediate transcytosis of fatty acid–laden albumin across the lung endothelium (Galis et al., 1988). We assumed that albumin associated with the lipoplex surface may target the lipoplex to the albumin receptor and lead to its uptake into the pulmonary endothelium. Our initial experiments involved the incubation of lipoplex in a purified albumin solution (30 mg/ml, 5 mM Hepes buffer, pH 7.4) for 15 min prior to administration. The net surface charge of the lipoplex changed from approximately $+35$ mV to -25 mV and it increased in diameter from ca 550 nm to several micron, indicating thorough interaction between protein

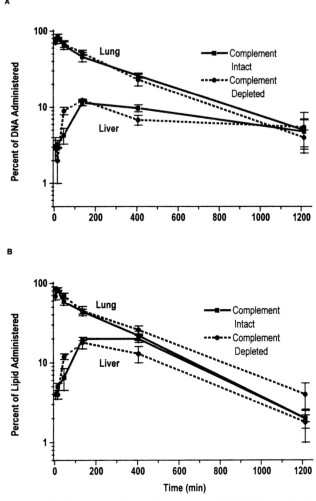

Figures 4a and 4b   The effect of complement depletion on radiolabeled plasmid DNA and lipid distribution following lipoplex injection. Distribution of iv-administered DOTAP/chol-plasmid DNA complex (5/1, +/−) labeled with ^{125}I -dCTP (plasmid DNA) and ^{131}I-BPE (liposomes) in mice as a function of time. Individual mice were either untreated or complement depleted as described (Barron et al., 1998).Complex was injected into the tail vein of female ICR mice and radioactivity measured in various tissues at time points given. Results are expressed as the mean percent of dose administered ± S.E.M. for 4 animals.

and lipoplex (unpublished data). Surprisingly, tail-vein injection of this complex did not result in altered gene expression or distribution versus that of native lipoplex. We tested the hypothesis that albumin influences lipoplex-mediated gene delivery

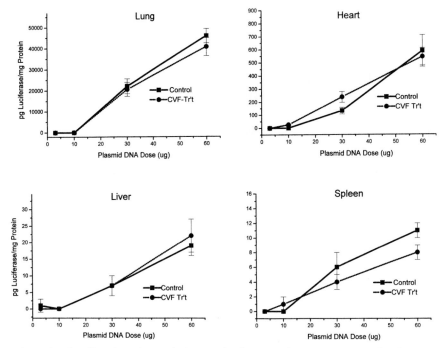

Figure 5   The effect of complement depletion on luciferase gene dose−response in transfected murine tissues. Luciferase reporter gene expression after intravenous complex injection in complement-intact and complement-depleted female ICR mice. Individual mice were either untreated or complement depleted as described in Barron *et al.* (1998). Animals then received the indicated dose of pCMVLuc plasmid DNA complexed with a corresponding DOTAP/chol (1/1 mol/mol) liposome preparation to yield a complex with a 5/1 charge ratio (+/−). Twenty-four hours later, animals were sacrificed and tissues analyzed for luciferase activity as described in Barron *et al.* (1998). Results are expressed as the mean of six samples ± S.E.M.

by administering lipoplexes into Nagase analbuminemic rats (Nagase *et al.,* 1979). These animals typically exhibit plasma albumin levels 7000-fold lower than normal. We found that pulmonary transfection in the Nagase animals was not reduced when compared to wild-type Sprague−Dawley rats. Moreover, when the complex was incubated *in vitro* with purified albumin and then injected into the Nagase rat, no change in transgene expression was observed (Fig. 6); therefore, albumin associated with the complex does not influence gene transfer.

Lipoprotein−liposome interactions have been the subject of research since the early days of liposome technology (Senior, 1987). Although the interactions between cationic lipid-based complexes and lipoproteins have not been directly studied, observations published for liposomes may also apply to lipoplex. Generally, the most profound interactions involve phospholipid exchange with high-density

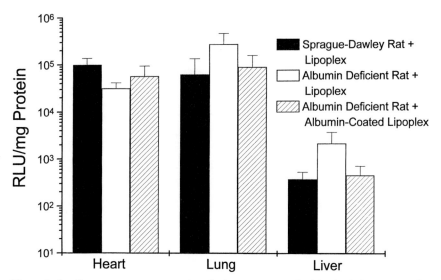

Figure 6 Luciferase reporter gene expression after intravenous complex injection in Sprague–Dawley and Nagase analbuminemic rats. Rats were injected via the tail vein with 150 μg of pCMVLuc plasmid DNA complexed with a DOTAP/chol (1/1 mol/mol) liposome preparation to yield a complex with a 5/1 charge ratio (+/−). Alternatively, lipoplex was incubated with a 30 mg/ml solution of purified albumin prior to injection. Twenty-four hours later, animals were sacrificed and tissues analyzed for luciferase activity as described in Barron *et al.* (1998). Results are expressed as the mean of six samples ± S.E.M.

lipoprotein (HDL) culminating in the reduction of liposome integrity. Strategies to reduce HDL-mediated liposome damage involve the formulation of liposomes from cholesterol and lipids with relatively high transition temperature. The fact that the inclusion of cholesterol in liposomes acts to increase pulmonary vasculature retention and subsequent transgene expression may be attributable to the ability of cholesterol to stabilize vesicle structure against HDL-mediated damage (Smyth-Templeton *et al.*, 1997). Given the potential for extensive lipoplex–lipoprotein interactions and the presence of endothelial lipoprotein receptors (Nistor and Simionescu, 1986), this topic warrants further study.

McLean and colleagues have documented a transient but significant thrombo-cytopenia and leukocytopenia upon iv administration of lipoplex to mice (McLean *et al.*, 1997). Transient lowering of plasma platelet levels are consistent with observations made by Reinish and colleagues following tail-vein injection of various liposome compositions into rats (Reinish *et al.*, 1988). Given the potential of platelet-mediated aggregate formation facilitating the delivery of lipoplex to the lung capillary beds, we hypothesized that platelets were involved in localizing the lipoplex in the lung. In addition, if the lipoplex–platelet interaction involved platelet activation, we thought it was possible that vascular permeability could be influenced, thus promoting endothelial uptake of the lipoplex. If platelets are required for lipoplex-mediated transfection in the lung, we reasoned that eliminating platelets would reduce transfection. We tested this hypothesis via the creation of a transiently thrombocytopenic mouse model by intraperitoneal injection of the alkylating agent busulfan (Kuter and Rosenberg, 1995). Although we successfully depleted platelets to less than 2% of control levels prior to lipoplex administration, transgene expression in the lung, heart, liver, and spleen was not affected (Fig. 7).

Upon airway delivery, reporter gene expression tends to peak at 2–3 days post-administration, with residual activity detected up to 21 days post-administration (Stribling et al., 1992; Meyer et al., 1995; Lee et al., 1996). The persistence of gene expression following iv administration tends to vary considerably. Zhu and colleagues (Zhu et al., 1993) were able to detect reporter gene activity for up to 21 weeks post-lipoplex injection, while others have observed that pulmonary expression peaked within 10 h of injection and then the activity declined to less than 1% of the peak value within 4 days (Liu et al., 1997b; Song et al., 1997). As Song and collaborators suggest, the discrepancy in activity upon systemic injection may

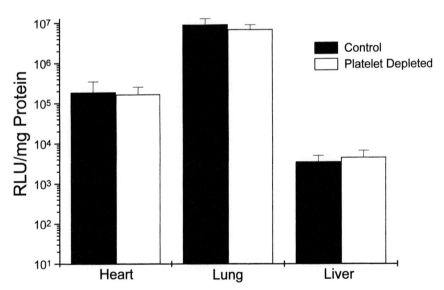

Figure 7 The effect of platelet depletion on luciferase gene expression murine tissues. Luciferase reporter gene expression after intravenous complex injection in control and platelet-depleted female ICR mice. Individual mice were either untreated or platelet depleted with busulfan (Kuter and Rosenberg, 1995). Animals then received 30 μg of pCMVLuc plasmid DNA complexed with a corresponding DOTAP/chol (1/1 mol/mol) liposome preparation to yield complex with a 5/1 charge ratio (+/−). Twenty-four hours later, animals were sacrificed and tissues analyzed for luciferase activity as described in Barron et al. (1998). Results are expressed as the mean of six samples ± S.E.M.

be based on the plasmid constructs and detection methods used (Song *et al.*, 1997). This short duration of gene expression is a significant limitation of the lipoplex as a gene carrier.

In comparing optimal formulations for itr versus iv administration, we have found that pulmonary gene expression following itr administration is 10% of that realized upon systemic injection (Meyer, Barron, and Szoka, unpublished results). The variation in expression levels following itr and iv administration may be a reflection of their different distribution characteristics and the cell types to which the lipoplex is delivered. Upon histological staining for transgene product following administration by these two modalities, we found expression to be the same at about 1% of the cells in a given field. However, gene expression from the iv dose was detected in all fields examined, whereas expression following itr delivery was more regional with many fields showing no expression. As indicated by several groups, the principal cell type transfected upon itr administration is the epithelial cell, whether expression is detected in the upper airways (Canonico *et al.*, 1994; McDonald *et al.*, 1998) or in the alveolar regions (Gorman *et al.*, 1997). Following iv injection, gene expression is found primarily in the endothelium but is also in circulating monocytes and interstitial macrophages (Liu *et al.*, 1997a; Liu *et al.*, 1997b). Canonico and co-workers have detected human α-1-antitrypsin (hAT) gene expression in the airway epithelium upon iv injection of lipoplexes (Canonico *et al.*, 1994). They explain that this observation may reflect the secretion of hAT from the endothelium and subsequent diffusion into the pulmonary epithelium. Lower gene expression following airway administration may also be explained by the mucin and surfactant-based milieu that covers the airways and alveoli, respectively. These media may limit intimate contact between lipoplexes and the epithelium. In support of this notion, it has been shown that *in vitro* transfection is decreased when lipoplex is exposed to natural and synthetic surfactants (Duncan *et al.*, 1997; Tsan *et al.*, 1997).

C. INTRAARTERIAL DELIVERY

Transgene expression in the arterial wall provides a strategy for the prevention and treatment of cardiovascular disease such as restenosis following balloon angioplasty. Regional delivery of corrective agents may facilitate the exposure of intimal lesions to concentrations of the agent that are not tolerated systemically; moreover, localized gene expression could generate therapeutic levels of a protein at the site (Nabel *et al.*, 1990). Initial gene delivery efforts involving the site-specific delivery of negatively charged lipoplex containing DOTMA/DOPE by Nabel and collaborators have demonstrated transgene activity in the intima, media, and adventitia of the iliofemoral arteries of pigs (Nabel *et al.*, 1990). Administration of lipoplex was carried out via double balloon catheter. Ectopic gene delivery was not detected upon PCR analysis of untreated tissues. Subsequent studies in dogs have established that lipoplex-mediated transgene expression can be achieved in the coronary arteries through balloon catheter administration (Chapman *et al.*, 1992). Interestingly, naked DNA yielded similar expression to that of the negatively charged DOTMA/DOPE-based lipoplex (Chapman *et al.*, 1992). The authors explain that this phenomenon may be a manifestation of the elevated pressure (4 atm) at which the intrarterial (ia) infusion is executed. The high pressure might damage the tissue and make transfection by naked DNA more probable. More recently, Stephan and colleagues found that by formulating lipoplex from GAP-DLRIE/DOPE and DNA at a positive molar charge ratio as opposed to the negatively charged preparations used in the preceding two studies, reporter gene expression in porcine arteries was approximately 15-fold higher than that of naked DNA (Stephan *et al.*, 1996). Transgene delivery mediated by direct arterial administration may be compared to iv injection as both modalities appear to transfect the endothelium, but important distinctions exist between the two methods. Intraarterial administration via balloon catheter typically involves the perfusion of the target arterial segment with buffer prior to infusion of DNA or lipoplex. This effectively renders the target site serum-free, thus reducing the potential for lipoplex–protein interactions. In addition, infusion is carried out under pressures of 1.5 to 4 atm, which may influence transfection by altering endothelial permeability.

D. Toxicity

The most common routes of administration for cationic lipids—iv and itr—are not associated with profound toxic effects when the lipoplex is administered in moderate doses. Serum chemistry and cardiac activity were not adversely affected by iv injection of lipoplexes; moreover, a histological examination of the organs showed no abnormalities (Stewart *et al.*, 1992). The doses tested in these studies were quite low. San and colleagues could detect no plasmid DNA in germ cells at the level of PCR sensitivity subsequent to iv injection of up to 50 μg of plasmid DNA complexed with 150 nmol of DMRIE/DOPE lipid (San *et al.*, 1993). One cautionary note is the observation of complement activation *in vitro* and *in vivo* in rodent models, which raises the potential for lipoplex toxicity due to complement activation in human subjects (Devine *et al.*, 1994; Plank *et al.*, 1996; Barron *et al.*, 1998a).

Canonico and colleagues have demonstrated that repeated iv or aerosol administrations of up to 500 μg of plasmid DNA complexed with 2.5 mg of DOTMA/DOPE lipids is possible in rabbits without eliciting inflammation in the lung. In addition, lung histology and function were unaffected (Canonico *et al.*, 1994). Conversely, Scheule and collaborators have shown significant short-term toxicity in the lung following nasal instillation of up to 300 nmol of lipid complexed with 400 nmol of plasmid DNA (Scheule *et al.*, 1997). This treatment resulted in multifocal lesions forming in the lung accompanied by inflammation and infiltrates of neutrophils, macrophages, and leukocytes. This study underscores differences in toxicity in the same organ when lipoplexes are administered by two different techniques. It also points out the promise of aerosolization of lipoplexes as it yields lower toxicity than instillation, possibly due to a more uniform pulmonary distribution (Eastman *et al.*, 1997b). Healthy human volunteers have had nebulized cationic liposomes delivered to their lungs in a dose proportional to that given to rodents (Chadwick *et al.*, 1997). No untoward clinical effects were observed in the human patients. It should be emphasized that the agent administered was the lipid component GL-67/DOPE/DMPE-PEG$_{5000}$ only and not the lipoplex.

Song and colleagues (Song *et al.*, 1997) have demonstrated that iv injection of lipoplexes can yield acute toxicity when doses in excess of 50 to 60 μg of DNA are complexed with lipid at high positive charge ratios. The effect of iv-administered lipoplexes may be more pronounced in larger animals since we have observed that the maximum tolerated dose of lipoplexes does not scale with body weight (unpublished data). Doses that are well tolerated when administered in mice become lethal in rats when the dose is increased proportionally to body weight. The mortality can be very rapid, implying that the lipoplex may be aggregating upon interaction with serum proteins and occluding the small-caliber pulmonary capillaries. This serves as a cautionary note for the intravenous administration of lipoplexes in humans and points out the need for additional toxicological evaluations in additional species prior to systemic use in humans.

The origin of toxicity is not understood, but cationic lipid-mediated toxicity may occur at the cell-signaling level. Bottega and Epand (1992) showed that cationic lipids consisting of quaternary amines and sterol hydrophobic groups (as opposed to tertiary amines conjugated to bilayer-stabilizing straight alkyl groups) are relatively potent inhibitors of protein kinase C activity (Bottega and Epand, 1992). The downstream effects of influencing cell signaling may be profound and warrant additional studies.

These observations may also point out that there is a correlation between transfection and toxicity; mechanisms that lead to cell transfection may lead to toxicity at marginally higher lipoplex doses. The role that cationic lipids play at the cellular level in transfection is not clear, but if it does involve the alteration of plasma membrane dynamics or intracellular trafficking, it is easy to imagine that cell function may be disrupted as the dose increases.

E. REMAINING QUESTIONS REGARDING CATIONIC LIPID-MEDIATED GENE THERAPY

If lipoplex-mediated gene therapy is to become a widely used therapeutic modality, gene expression must be augmented and the persistence of that expression must be increased. With few exceptions, transgene expression is an ephemeral phenomenon–typically reaching a peak within days of expression and then declining to less than 5% of peak levels within a week.

These challenges have been previously identified by Gao and Huang (Gao and Huang, 1995). Since that time, progress has been made in recognizing physiological, biochemical, and cellular processes governing the efficacy of gene delivery vehicles. However, several questions remain to be addressed. It is clear that serum proteins will have a role in manipulating distribution and, potentially, clearance of lipoplex upon systemic injection. In fact, preferential distribution of lipoplex and the subsequent transfection observed may be governed by interactions between lipoplexes and serum proteins. Most studies involving the administration of cationic lipid/DNA complexes formulated with a strong net positive charge document marked lipoplex uptake and transfection in the lung. The accumulation and subsequent transfection in the lung may be a reflection of the fact that the pulmonary microvasculature is the first capillary bed encountered by the lipoplex upon administration. It is important to note, however, that in the study by McLean and collaborators (McLean *et al.*, 1997) the ovaries and lymph nodes as well as the adrenal and pituitary glands, as a proportion of tissue mass, accrue a large amount of lipoplexes and undergo significant transfection. Since these tissues are not subject to a first-pass effect, there must be an alternative explanation for lipoplex-mediated transfection observed therein. McLean and co-workers have postulated that these differences in transgene expression are a manifestation of differences in endothelial properties such as the expression of tissue-specific receptors. We agree with this hypothesis and believe that regional uptake of the lipoplex may be dictated by as yet unidentified plasma components. We have eliminated two candidates for such a role in serum: complement and albumin. An *in vitro* study by the Huang group suggests that a fraction of serum proteins defined as neutral or positively charged promote transfection when included in culture medium, whereas a negatively charged fraction inhibited transfection (Yang and Huang, 1997). These data emphasize the complex nature of the lipoplex–protein interactions and the potential for deciphering the mechanism of systemic transfection if the crucial serum components can be identified.

There continues to be controversy concerning the reliance on cell division for efficient lipoplex-mediated transfection. The dissolution of the nuclear membrane during mitosis may permit plasmid DNA access to nuclear transcriptional machinery; indeed, Fasbender and colleagues have shown that actively dividing airway epithelial cells are more readily transfected by lipoplex than cells that were not mitotically active (Fasbender *et al.*, 1997). Conversely, successful transfection by naked DNA of post-mitotic myotubes (Dowty *et al.*, 1995) and neurons (Demeneix *et al.*, 1990) has been documented, indicating that cell division is not an absolute necessity for transfection. It is conceivable that the degree of dependence for transfection on the target cell's mitotic state may be governed by target cell type. It is also feasible that access to the nucleus occurs in the absence of cell division but uptake is enhanced during mitosis.

An explanation for the short-lived transgene expression associated with lipo-plex-mediated transfection has not been established. One possibility is that the transgene is degraded intracellularly. Although lipoplexes may be poor immune ef-fectors, it is conceivable that the transgene product is eliciting an immune response, possibly facilitating the destruction of the target cell via a cytotoxic t-lymphocyte (CTL) response. Decreasing transgene expression may also be a reflection of down regulation of viral promoter sequences frequently employed in plasmid vectors de-livered by cationic lipids.

The transitory expression of transgenes may be an indication of lipoplex-based cell toxicity discussed earlier. Song and collaborators have demonstrated that, upon iv administration, murine tissues remain refractory to subsequent systemic transfection for approximately 2 weeks posttreatment (Song et al., 1997). This re-fractory period may be a manifestation of cell damage or death mediated by the cationic lipid/DNA complex. It has been suggested that lipoplexes bind to the tar-get cell surface via proteoglycans (Mislick and Baldeschwieler, 1996), so it is con-ceivable that in the process of transfection, if the lipoplex does not damage the cell outright, it may alter the properties of the glycocalyx such that subsequent exposure to lipoplex does not yield uptake.

The fact that lipoplexes do not have to be preformed for effective *in vitro* transfection was originally noted by Felgner and Holm (1989). They demonstrated that the separate addition of DOTMA/DOPE liposomes and DNA to cells yielded gene expression comparable to that of cells to which preformed lipoplex had been applied. This observation has been ignored by investigators in the field until Liu and colleagues demonstrated virtually the same phenomenon *in vivo* (Liu, 1998). They delivered cationic liposomes iv 5 min prior to injection of a reporter gene; the level of transgene expression in the lung was the same as that obtained following injection of the preformed cationic lipid/DNA complex. This result has two inter-esting implications: First, preformation of the lipoplex *ex vivo* is not necessary for systemic transfection. Second, some degree of *in vivo* lipid–DNA interaction occurs upon addition of DNA even after the cationic lipid has resided in tissue or circula-tion for 5 min since iv injection of DNA alone does not yield measurable transfec-tion. These observations are important in that they suggest the mechanism of lipo-plex formation is not adversely affected by the biophase. Based on these data, Liu has suggested that DNA released from the lipoplex is the transfectionally active agent and that the cationic lipid component merely serves as a means to increase pulmonary retention of DNA (Liu, 1998).

We found this to be an intriguing possibility, particularly in light of the fact that naked plasmid DNA is capable of transfecting striated muscle (Wolff *et al.*, 1990; Buttrick *et al.*, 1992), the thyroid gland (Sikes *et al.*, 1994), the liver (Hickman *et al.*, 1994), the lung (Meyer *et al.*, 1995), solid tumors (Yang and Huang, 1996), and synovial tissue (Nita *et al.*, 1996). We repeated the cationic lipid preinjection protocol and confirmed Liu's observations.

In order to study this phenomenon further, we defined the time period within which intact lipoplex existed in the pulmonary vasculature following iv administration. Anionic liposomes have been demonstrated to cause lipoplex dissociation *in vitro* (Xu and Szoka, 1996), so they were injected at various time points postlipoplex injection. We found that lipoplex integrity is critical for lung transfection only in the first 60 min after its injection (Barron and Szoka, manuscript in preparation). Following this time period, clearance of lipoplex from the lung and other tissues by anionic liposome injection did not significantly influence transfection. This suggests that pulmonary lipoplex uptake occurs entirely within the first 60 min postinjection. We then iv infused an excess of naked luciferase plasmid over the course of 1 h such that steady-state levels of high-molecular-weight, transfectionally active plasmid were present in the plasma and measured gene expression in the lung at 24 h postinjection. We did not find significant luciferase activity in any tissues assayed (Table 3). This implies that free DNA released from the lipoplex is not responsible for gene transfer in the lung.

We reasoned further that if the intravascular release of naked plasmid DNA from the lipoplex was the active agent in mediating lung transfection, then a molecule capable of competing with released DNA, when administered in the first 60 min post-lipoplex injection, should decrease transfection. However, upon infusing an excess of noncoding plasmid DNA following lipoplex injection via the tail vein, there was no decrease in lipoplex-mediated transfection (Table 4). On the contrary, a significant increase in transfection in the lungs and livers of the treated animals was observed. This increase in expression may be evidence for saturation of the mechanism that is responsible for lipoplex elimination. These experiments led us to conclude that the cationic lipid component of the iv-administered lipoplex plays an essential role in lung transfection beyond that of a passive release matrix (Barron *et al.*, 1998b).

The cationic lipid preinjection protocol introduced by Liu also suggests that for up to 5 minutes subsequent to injection, cationic residues are exposed and able

Table III

Effect on Transfection of IV Infusion of Noncoding Plasmid DNA 5 Min after Lipoplex Injection

Tissue	Lipoplex	Lipoplex with noncoding infusion
	10^5 RLU/mg protein Mean ± S.D.	10^5 RLU/mg protein Mean ± S.D.
Heart	0.06 ± 0.03	0.17 ± 0.07
Lung	29.0 ± 5.0	$81.0 \pm 18*$
Liver	0.01 ± 0.003	$1.0 \pm 0.40*$

Table IV

Effect of IV Infusion of Free Luciferase Plasmid DNA
on Control Lipoplexes

Tissue	Lipoplex	Plasmid DNA infusion
	10^5 RLU/mg protein Mean $\pm$ S.D.	10^5 RLU/mg protein Mean $\pm$ S.D.
Heart	0.03 $\pm$ 0.01	0.01 $\pm$ 0.04*
Lung	42.0 $\pm$ 0.09	0.03 $\pm$ 0.03*
Liver	0.07 $\pm$ 0.05	0.004 $\pm$ 0.003*

*$p < 0.05$.

to bind DNA. We sought to corroborate this observation *in vitro* by incubating cationic liposomes in a buffered solution with serum for 5 min at 37°C. Then the liposomes were mixed with DNA in quantities comparable to those in the *in vivo* injection protocol and subjected to agarose gel electrophoresis. Subsequent to serum incubation, the cationic lipids were capable of retarding the migration of DNA into the gel (unpublished data). This indicates that positively charged domains still exist on the lipoplex following exposure to serum.

These *in vitro* and *in vivo* experiments are important because they imply that there are exposed positively charged residues on the lipoplex that may also facilitate electrostatic interactions with negative charges on endothelium. This mechanism for lipoplex–endothelium interaction is contrary to the model suggested by several lines of evidence produced in our lab and others (Table 5). We have hypothesized that the lipoplexes interact with serum proteins upon injection into the blood and that these interactions govern the distribution and subsequent transfection mediated by lipoplex.

First, as noted, the incubation of lipoplex with an albumin solution at physiologic concentrations yielded complexes with drastically increased size and net negative surface potential suggesting complete coating; however, tail-vein injection of this complex did not exhibit modified gene expression or distribution versus that of native lipoplex. The negative surface potential and size increase suggest that the lipoplex is thoroughly coated. The ability of albumin-coated lipoplexes to distribute and transfect in a manner similar to the native lipoplex indicates that a net positive charge on the lipoplex surface is not important for the gene delivery mechanism.

We attempted to examine the effect of neutralizing the negative charge inherent on the endothelium. This was done by injecting a cationic polymer, protamine sulphate, into the tail vein of mice prior to lipoplex administration. The amount of polymer injected was either a 10- or 100-fold charge excess over that of the ensuing injected lipoplex. Protamine sulphate was injected at 3, 6, or 18 min

Table V

Observations Supporting Mechanisms by Which Lipoplex Interacts with Endothelium

Direct pathway	Indirect pathway
Electrostatic interaction between lipoplex and endothelium	Protein-mediated interaction between lipoplex and endothelium
Cationic liposomes following incubation with serum can still retard DNA migration upon gel electrophoresis	Albumin coating of lipoplex reduces zeta potential to −25mV but does not alter lipoplex-mediated transfection or distribution
Plasmid DNA upon iv injection binds in lung to previously injected cationic liposomes and transfects pulmonary tissue	IV preinjection of protamine sulphate does not affect subsequent iv transfection via lipoplex
	Lipoplex uptake and transfection occur in a tissue specific manner
	Association of ligands to lipoplex surface influences distribution and transfection

prior to the injection of lipoplex. This protocol had no observed effect on systemic gene expression. Assuming that sufficient quantities of the polymer were injected to obviate the endothelial charge, these data suggest that the interaction between the lipoplex and the cells lining the vasculature is not electrostatic. In addition, the work of McLean and colleagues points out that lipoplexes have a tissue-specific localization and transfection. Transfection in discrete regions seems to us unlikely to occur if the principal factors governing binding of the lipoplex to the endothelium are nonspecific electrostatic interactions. Finally, there is support for the notion that surface modification of the lipoplex can lead to specific lipoplex targeting and transfection. Cheng associated transferrin with the lipoplex and showed increased transfection *in vitro* (Cheng, 1996). Smyth-Templeton and colleagues have shown that the complexation of succinylated asialofetuin to lipoplexes partially redirects transfection from the lung to the liver upon tail-vein injection (Smyth-Templeton *et al.*, 1997), although lung transfection remains significantly higher than liver transfection for the asialofetuin-modified complexes.

One of the limitations of nonviral gene therapy is the low efficiency of transgene expression realized with these systems. To gain insight on the amount of reporter gene expression in endothelial cells that could be expected under optimal conditions for cell–lipoplex interactions, we transfected human umbilical vein endothelial cells (HUVEC) *in vitro*. Upon transfecting the endothelial cells with a DOTAP/chol-based lipoplex identical to that employed *in vivo* studies (Barron *et al.*, 1998a), the mean luciferase activity was approximately 0.2 relative light units (RLU) per cell. We observe about 5×10^8 RLU/lung in mice after iv administration of this formulation. If we assume the number of endothelial cells in a 30 g mouse is approximately 5×10^9 (Weibel, 1985) and activity was confined primarily to these cells, one computes a value of 0.1 RLU/endothelial cell. We acknowledge that comparisons between *in vitro* and *in vivo* systems are not ideal; however, this comparison suggests that gaining access to and uptake into the lung endothelium may not be a rate-limiting process. Thus improvements in transfection may be best realized through addressing downstream events such as endosomal release or nuclear uptake of DNA.

The viability of nonviral systems for gene therapy has been demonstrated through the use of cationic lipid/DNA complexes. We perceive the problem of gene transfer as a linear process composed of multiple barriers arranged in series. The development of effective gene delivery vectors necessitates addressing the individual barriers (Fig. 1), conceiving strategies to overcome them, and combining the various components into a viable, multifaceted gene transfer vehicle. Overcoming these barriers necessitates generating vectors of more elegant design that do not simply depend on the delivery of astronomical numbers of plasmids to achieve expression of a therapeutic protein. Thus an important research strategy for improving nonviral gene delivery is to unravel the mechanisms by which current systems function *in vivo*. Unless this is accomplished, the field has little chance of improving lipoplex gene delivery efficiency.

ACKNOWLEDGMENTS

We gratefully acknowledge the members of the Szoka laboratory, as well as John McLean and Gavin Thurston for helpful discussion on the mechanisms of lipoplex function *in vivo*. This work was supported by TDRDP 6RT-0109 and NIH DK 46052.

REFERENCES

Akao, T., Osaki, T., Mitoma, J., Ito, A., and Kunitake, T. (1991). Correlation between physicochemical characteristics of synthetic cationic amphiphiles and their DNA transfection ability. *Bull. Chem. Soc. Japan* **64**, 3677–3681.

Balasubramaniam, R. P., Bennett, M. J., Aberle, A. M., Malone, J. G., Nantz, M. H., and Malone, R. W. (1996). Structural and functional analysis of cationic transfection lipids: The hydrophobic domain. *Gene Ther.* **3**, 163–172.

Barron, L. G., Meyer, K. B., and Szoka, F. C., Jr. (1998a). Effects of complement depletion on the pharmacokinetics and gene delivery mediated by cationic lipid–DNA complexes. *Human Gene Ther.* **9**, 315–323.

Barron, L. G., Uyechi, L. S., and Szoka, F. C., Jr. (1998b). Mechanism of cationic lipid–DNA complex distribution and transfection in the lung following IV administration in mice. Abstract #398. In G. Stomatoyannopoulos, (Ed.) Seattle, WA: American Society of Gene Therapy.

Behr, J. P., Demeneix, B., Loeffler, J. P., and Perez-Mutul, J. (1989). Efficient gene transfer into mammalian primary endocrine cells with lipopolyamine-coated DNA. *Proc. Na. Acad. Sci. U.S.A.* **86**, 6982–6986.

Bottega, R., and Epand, R. M. (1992). Inhibition of protein kinase C by cationic amphiphiles. *Biochemistry* **31**, 9025–9030.

Boussif, O., Zanta, M. A., and Behr, J. P. (1996). Optimized galenics improve *in vitro* gene transfer with cationic molecules up to 1000-fold. *Gene Ther.* **3**, 1074–1080.

Brigham, K. L., Meyrick, B., Christman, B., Magnuson, M., King, G., and Berry, L. (1989). *In vivo* transfection of expression in murine lungs with a functioning prokaryotic gene using a cationic liposome vehicle. *Am. J. Med. Sci.* **298**, 278–281.

Buttrick, P. M., Kass, A., Kitsis, R. N., Kaplan, M. L., and Leinwand, L. A. (1992). Behavior of genes directly injected into the rat heart *in vivo*. *Circ. Res.* **70**, 193–198.

Canonico, A. E., Plitman, J. D., Conary, J. T., Meyrick, B. O., and Brigham, K. L. (1994). No lung toxicity after repeated aerosol or intravenous delivery of plasmid-cationic liposome complexes. *J. App. Physiol.* **77**, 415–419.

Capecchi, M. R. (1980). High efficiency transformation by direct microinjection of DNA into cultured mammalian cells. *Cell* **22**, 479–488.

Caplen, N. J., Alton, E. W., Middleton, P. G., Dorin, J. R., Stevenson, B. J., Gao, X., Durham, S. R., Jeffery, P. K., Hodson, M. E., Coutelle, C., *et al.* (1995). Liposome-mediated CFTR gene transfer to the nasal epithelium of patients with cystic fibrosis [published erratum appears in *Nature Med.* (1995, Mar) 1(3);272]. *Nature Med.* **1**, 39–46.

Chadwick, S. L., Kingston, H. D., Stern, M., Cook, R. M., O'Connor, B. J., Lukasson, M., Balfour, R. P., Rosenberg, M., Cheng, S. H., Smith, A. E., Meeker, D. P., Geddes, D. M., and Alton, E. W. (1997). Safety of a single aerosol administration of escalating doses of the cationic lipid GL-67/DOPE/DMPE-PEG5000 formulation to the lungs of normal volunteers. *Gene Ther.* **4**, 937–942.

Chapman, G. D., Lim, C. S., Gammon, R. S., Culp, S. C., Desper, J. S., Bauman, R. P., Swain, J. L., and Stack, R. S. (1992). Gene transfer into coronary arteries of intact animals with a percutaneous balloon catheter. *Circulation Res.* **71**, 27–33.

Cheng, P. W. (1996). Receptor ligand-facilitated gene transfer: Enhancement of liposome-mediated gene transfer and expression by transferrin. *Human Gene Ther.* **7**, 275–282.

Demeneix, B. A., Kley, N., and Loeffler, J. P. (1990). Differentiation to a neuronal phenotype in bovine chromaffin cells is repressed by protein kinase C and is not dependent on *c-fos* oncoproteins. *DNA Cell Biol.* **9**, 335–345.

Devine, D. V., Wong, K., Serrano, K., Chonn, A., and Cullis, P. R. (1994). Liposome-complement interactions in rat serum: Implications for liposome survival studies. *Biochim. Biophys. Acta* **1191**, 43–51.

Dowty, M. E., Williams, P., Zhang, G., Hagstrom, J. E., and Wolff, J. A. (1995). Plasmid DNA entry into postmitotic nuclei of primary rat myotubes. *Proc. Natl. Acad. Sci. U.S.A.* **92**, 4572–4576.

Duncan, J. E., Whitsett, J. A., and Horowitz, A. D. (1997). Pulmonary surfactant inhibits cationic liposome-mediated gene delivery to respiratory epithelial cells *in vitro*. *Human Gene Ther.* **8**, 431–438.

Düzgünes, N., Goldstein, J. A., Friend, D. S., and Felgner, P. L. (1989). Fusion of liposomes containing a novel cationic lipid, *N*-[2,3-(dioleyloxy)propyl]-*N,N,N*-trimethylammonium: Induction by multivalent anions and asymmetric fusion with acidic phospholipid vesicles. *Biochemistry* **28**, 9179–9184.

Eastman, S. J., Lukason, M. J., Tousignant, J. D., Murray, H., Lane, M. D., St George, J. A., Akita, G. Y., Cherry, M., Cheng, S. H., and Scheule, R. K. (1997a). A concentrated and stable aerosol formulation of cationic lipid:DNA complexes giving high-level gene expression in mouse lung. *Human Gene Ther.* **8**, 765–773.

Eastman, S. J., Siegel, C., Tousignant, J., Smith, A. E., Cheng, S. H., and Scheule, R. K. (1997b). Biophysical characterization of cationic lipid:DNA complexes. *Biochim. Biophys. Acta* **1325**, 41–62.

Eastman, S. J., Tousignant, J. D., Lukason, M. J., Murray, H., Siegel, C. S., Constantino, P., Harris, D. J., Cheng, S. H., and Scheule, R. K. (1997c). Optimization of formulations and conditions for the aerosol delivery of functional cationic lipid:DNA complexes. *Human Gene Ther.* **8**, 313–322.

Farhood, H., Serbina, N., and Huang, L. (1995). The role of dioleoyl phosphatidylethanolamine in cationic liposome mediated gene transfer. *Biochim. Biophys. Acta* **1235**, 289–295.

Fasbender, A., Zabner, J., Zeiher, B. G., and Welsh, M. J. (1997). A low rate of cell proliferation and reduced DNA uptake limit cationic lipid-mediated gene transfer to primary cultures of ciliated human airway epithelia. *Gene Ther.* **4**, 1173–1180.

Felgner, J. H., Kumar, R., Sridhar, C. N., Wheeler, C. J., Tsai, Y. J., Border, R., Ramsey, P., Martin, M., and Felgner, P. L. (1994). Enhanced gene delivery and mechanism studies with a novel series of cationic lipid formulations. *J. Biol. Chem.* **269,** 2550–2561.

Felgner, P., and Holm, M. (1989). Cationic-liposome-mediated transfection. *Focus* **11,** 21–25.

Felgner, P. L., Barenholz, Y., Behr, J. P., Cheng, S. H., Cullis, P., Huang, L., Jessee, J. A., Seymour, L., Szoka, F., Thierry, A. R., Wagner, E., and Wu, G. (1997). Nomenclature for synthetic gene delivery systems [editorial]. *Human Gene Ther.* **8,** 511–512.

Felgner, P. L., Gadek, T. R., Holm, M., Roman, R., Chan, C., Wenz, M., Northrop, J. P., Ringold, G. M., and Danielson, M. (1987). Lipofection: A highly efficient, lipid-mediated DNA transfection procedure. *Proc. Natl. Acad. Sci. USA* **84,** 7413–7417.

Felgner, P. L., and Ringold, G. M. (1989). Cationic liposome-mediated transfection. *Nature* **337,** 387–388.

Floch, V., Audrezet, M. P., Guillaume, C., Gobin, E., Le Bolch, G., Clement, J. C., Yaouanc, J. J., Des Abbayes, H., Mercier, B., Leroy, J. P., Abgrall, J. F., and Ferec, C. (1998). Transgene expression kinetics after transfection with cationic phosphonolipids in hematopoietic nonadherent cells. *Biochim. Biophys. Acta* **1371,** 53–70.

Freimark, B. D., Blezinger, H. P., Florack, V. J., Nordstrom, J. L., Long, S. D., Deshpande, D. S., Nochumson, S., and Petrak, K. L. (1998). Cationic lipids enhance cytokine and cell influx levels in the lung following administration of plasmid: Cationic lipid complexes. *J. Immunol.* **160,** 4580–4586.

Friend, D. S., Papahadjopoulos, D., and Debs, R. J. (1996). Endocytosis and intracellular processing accompanying transfection mediated by cationic liposomes. *Biochim. Biophys. Acta* **1278,** 41–50.

Galis, Z., Ghitescu, L., and Simionescu, M. (1988). Fatty acids binding to albumin increases its uptake and transcytosis by the lung capillary endothelium. *Eur. J. Cell Biol.* **47,** 358–365.

Gao, X., and Huang, L. (1995). Cationic liposome-mediated gene transfer. *Gene Ther.* **2,** 710–722.

Gao, X., and Huang, L. (1993). Cytoplasmic expression of a reporter gene by co-delivery of T7 RNA polymerase and T7 promoter sequence with cationic liposomes. *Nucleic Acids Res.* **21,** 2867–2872.

Gao, X., and Huang, L. (1991). A novel cationic liposome reagent for efficient transfection of mammalian cells. *Biochem. Biophys. Res. Commun.* **179,** 280–285.

Gershon, H., Ghirlando, R., Guttman, S. B., and Minsky, A. (1993). Mode of formation and structural features of DNA–cationic liposome complexes used for transfection. *Biochemistry* **32,** 7143–7151.

Ghitescu, L., Fixman, A., Simionescu, M., and Simionescu, N. (1986). Specific binding sites for albumin restricted to plasmalemmal vesicles of continuous capillary endothelium: Receptor-mediated transcytosis. *J. Cell Biol.* **102,** 1304–1311.

Gill, D. R., Southern, K. W., Mofford, K. A., Seddon, T., Huang, L., Sorgi, F., Thomson, A., MacVinish, L. J., Ratcliff, R., Bilton, D., Lane, D. J., Littlewood, J. M., Webb, A. K., Middleton, P. G., Colledge, W. H., Cuthbert, A. W., Evans, M. J., Higgins, C. F., and Hyde, S. C. (1997). A placebo-controlled study of liposome-mediated gene transfer to the nasal epithelium of patients with cystic fibrosis. *Gene Ther.* **4,** 199–209.

Gorman, C. M., Aikawa, M., Fox, B., Fox, E., Lapuz, C., Michaud, B., Nguyen, H., Roche, E., Sawa, T., and Wiener-Kronish, J. P. (1997). Efficient *in vivo* delivery of DNA to pulmonary cells using the novel lipid EDMPC. *Gene Ther.* **4,** 983–992.

Gustafsson, J., Arvidson, G., Karlsson, G., and Almgren, M. (1995). Complexes between cationic liposomes and DNA visualized by cryo-TEM. *Biochim. Biophys. Acta* **1235,** 305–312.

Hickman, M. A., Malone, R. W., Lehmann-Bruinsma, K., Sih, T. R., Knoell, D., Szoka, F. C., Walzem, R., Carlson, D. M., and Powell, J. S. (1994). Gene expression following direct injection of DNA into liver. *Human Gene Ther.* **5,** 1477–1483.

Hofland, H. E., Shephard, L., and Sullivan, S. M. (1996). Formation of stable cationic lipid/DNA complexes for gene transfer. *Proc. Natl. Acad. Sci. USA* **93,** 7305–7309.

Hui, S. W., Stewart, T. P., Boni, L. T., and Yeagle, P. L. (1981). Membrane fusion through point defects in bilayers. *Science* **212,** 921–923.

Kuter, D. J., and Rosenberg, R. D. (1995). The reciprocal relationship of thrombopoietin (c-Mpl ligand) to changes in the platelet mass during busulfan-induced thrombocytopenia in the rabbit. *Blood* **85,** 2720–2730.

Labat-Moleur, F., Steffan, A. M., Brisson, C., Perron, H., Feugeas, O., Furstenberger, P., Oberling, F., Brambilla, E., and Behr, J. P. (1996). An electron microscopy study into the mechanism of gene transfer with lipopolyamines. *Gene Ther.* **3,** 1010–1017.

Lasic, D. D., Strey, H., Stuart, M. C. A., Podgornik, R., and Frederik, P. M. (1997). The structure of DNA-liposome complexes. *J. Am. Chem. Soc.* **119,** 832–833.

Lee, E. R., Marshall, J., Siegel, C. S., Jiang, C., Yew, N. S., Nichols, M. R., Nietupski, J. B., Ziegler, R. J., Lane, M. B., Wang, K. X., Wan, N. C., Scheule, R. K., Harris, D. J., Smith, A. E., and Cheng, S. H. (1996). Detailed analysis of structures and formulations of cationic lipids for efficient gene transfer to the lung. *Human Gene Ther.* **7,** 1701–1717.

Legendre, J. Y., and Szoka, F. C., Jr. (1992). Delivery of plasmid DNA into mammalian cell lines using pH-sensitive liposomes: Comparison with cationic liposomes. *Pharm. Res.* **9,** 1235–1242.

Lemarchand, P., Jaffe, H. A., Danel, C., Cid, M. C., Kleinman, H. K., Stratford-Perricaudet, L. D., Perricaudet, M., Pavirani, A., Lecocq, J. P., and Crystal, R. G. (1992). Adenovirus-mediated transfer of a recombinant human alpha 1-antitrypsin cDNA to human endothelial cells. *Proc. Natl. Acad. Sci. USA* **89,** 6482–6486.

Leventis, R., and Silvius, J. R. (1990). Interactions of mammalian cells with lipid dispersions containing novel metabolizable cationic amphiphiles. *Biochim. Biophys. Acta* **1023,** 124–132.

Liu, D. (1998). Mechanism of cationic liposome-mediated gene transfer to cells in mouse lung by intravenous administration. In P. Felgner, F. Szoka, L. Huang and E. Wagner (Eds.) *Keystone symposia on molecular and cellular biology,* Keystone, CO: .

Liu, Y., Mounkes, L. C., Liggitt, H. D., Brown, C. S., Solodin, I., Heath, T. D., and Debs, R. J. (1997a). Factors influencing the efficiency of cationic liposome-mediated intravenous gene delivery. *Nature Biotechnol.* **15,** 167–173.

Liu, F., Qi, H., Huang, L., and Liu, D. (1997b). Factors controlling the efficiency of cationic lipid-mediated transfection *in vivo* via intravenous administration. *Gene Ther.* **4,** 517–523.

Loughrey, H. C., Bally, M. B., Reinish, L. W., and Cullis, P. R. (1990). The binding of phosphatidylglycerol liposomes to rat platelets is mediated by complement. *Thromb. Haemostasis* **64,** 172–176.

Mahato, R. I., Kawabata, K., Nomura, T., Takakura, Y., and Hashida, M. (1995a). Physicochemical and pharmacokinetic characteristics of plasmid DNA/cationic liposome complexes. *J. Pharm. Sci.* **84,** 1267–1271.

Mahato, R. I., Kawabata, K., Takakura, Y., and Hashida, M. (1995b). *In vivo* disposition characteristics of plasmid DNA complexed with cationic liposomes. *J. Drug Targeting* **3,** 149–157.

McDonald, R. J., Liggitt, H. D., Roche, L., Nguyen, H. T., Pearlman, R., Raabe, O. G., Bussey, L. B., and Gorman, C. M. (1998). Aerosol delivery of lipid:DNA complexes to lungs of rhesus monkeys. *Pharm. Res.* **15,** 671–679.

McLachlan, G., Ho, L. P., Davidson-Smith, H., Samways, J., Davidson, H., Stevenson, B. J., Carothers, A. D., Alton, E. W., Middleton, P. G., Smith, S. N., Kallmeyer, G., Michaelis, U., Seeber, S., Naujoks, K., Greening, A. P., Innes, J. A., Dorin, J. R., and Porteous, D. J. (1996). Laboratory and clinical studies in support of cystic fibrosis gene therapy using pCMV-CFTR-DOTAP. *Gene Ther.* **3,** 1113–1123.

McLean, J. W., Fox, E. A., Baluk, P., Bolton, P. B., Haskell, A., Pearlman, R., Thurston, G., Umemoto, E. Y., and McDonald, D. M. (1997). Organ-specific endothelial cell uptake of cationic liposome–DNA complexes in mice. *Am. J. Physiol.* **273,** H387–H404.

Meyer, K. B., Thompson, M. M., Levy, M. Y., Barron, L. G., and Szoka, F. C., Jr. (1995). Intratracheal gene delivery to the mouse airway: characterization of plasmid DNA expression and pharmacokinetics. *Gene Ther.* **2,** 450–460.

Meyer, K. E. B., Uyechi, L. S., and Szoka Jr., F. C. (1997). Manipulating the intracellular trafficking of nucleic acids. In K. Brigham (Ed.), *Gene therapy for diseases of the lung.* New York: Dekker.

Mislick, K. A., and Baldeschwieler, J. D. (1996). Evidence for the role of proteoglycans in cation-mediated gene transfer. *Proc. Natl. Acad. Sci. USA* **93,** 12349–12354.

Nabel, E. G., Plautz, G., and Nabel, G. J. (1990). Site-specific gene expression *in vivo* by direct gene transfer into the arterial wall. *Science* **249,** 1285–1288.

Nabel, G. J., Chang, A. E., Nabel, E. G., Plautz, G. E., Ensminger, W., Fox, B. A., Felgner, P., Shu, S., and Cho, K. (1994). Immunotherapy for cancer by direct gene transfer into tumors. *Human Gene Ther.* **5,** 57–77.

Nabel, G. J., Nabel, E. G., Yang, Z. Y., Fox, B. A., Plautz, G. E., Gao, X., Huang, L., Shu, S., Gordon, D., and Chang, A. E. (1993). Direct gene transfer with DNA-liposome complexes in melanoma: Expression, biologic activity, and lack of toxicity in humans. *Proc. Natl. Acad. Sci. USA* **90,** 11307–11311.

Nagase, S., Shimamune, K., and Shumiya, S. (1979). Albumin-deficient rat mutant. *Science* **205,** 590–591.

Nistor, A., and Simionescu, M. (1986). Uptake of low density lipoproteins by the hamster lung. Interactions with capillary endothelium. *Am. Rev. Respir. Dis.* **134,** 1266–1272.

Nita, I., Ghivizzani, S. C., Galea-Lauri, J., Bandara, G., Georgescu, H. I., Robbins, P. D., and Evans, C. H. (1996). Direct gene delivery to synovium. An evaluation of potential vectors *in vitro* and *in vivo. Arthritis Rheum.* **39,** 820–828.

Patel, H. M. (1992). Serum opsonins and liposomes: Their interaction and opsonophagocytosis. In *CRC critical reviews in therapeutic drug carrier systems.* (pp. 39).

Plank, C., Mechtler, K., Szoka, F. C., Jr., and Wagner, E. (1996). Activation of the complement system by synthetic DNA complexes: A potential barrier for intravenous gene delivery. *Human Gene Ther.* **7,** 1437–1446.

Pollard, H., Remy, J. S., Loussouarn, G., Demolombe, S., Behr, J. P., and Escande, D. (1998). Polyethylenimine but not cationic lipids promotes transgene delivery to the nucleus in mammalian cells. *J. Biol. Chem.* **273,** 7507–7511.

Porteous, D. J., Dorin, J. R., McLachlan, G., Davidson-Smith, H., Davidson, H., Stevenson, B. J., Carothers, A. D., Wallace, W. A., Moralee, S., Hoenes, C., Kallmeyer, G., Michaelis, U., Naujoks, K., Ho, L. P., Samways, J. M., Imrie, M., Greening, A. P., and Innes, J. A. (1997). Evidence for safety and efficacy of DOTAP cationic liposome mediated CFTR gene transfer to the nasal epithelium of patients with cystic fibrosis. *Gene Ther.* **4,** 210–218.

Poste, G., Bucana, C., Raz, A., Bugelski, P., Kirsh, R., and Fidler, I. J. (1982). Analysis of the fate of systemically administered liposomes and implications for their use in drug delivery. *Cancer Res.* **42,** 1412–1422.

Rädler, J. O., Koltover, I., Salditt, T., and Safinya, C. R. (1997). Structure of DNA-cationic liposome complexes: DNA intercalation in multilamellar membranes in distinct interhelical packing regimes. *Science* **275,** 810–814.

Reinish, L. W., Bally, M. B., Loughrey, H. C., and Cullis, P. R. (1988). Interactions of liposomes and platelets. *Thromb. Haemostasis* **60,** 518–523.

Remy, J. S., Sirlin, C., Vierling, P., and Behr, J. P. (1994). Gene transfer with a series of lipophilic DNA-binding molecules. *Bioconjugate Chem.* **5,** 647–654.

San, H., Yang, Z. Y., Pompili, V. J., Jaffe, M. L., Plautz, G. E., Xu, L., Felgner, J. H., Wheeler, C. J., Felgner, P. L., Gao, X., *et al.* (1993). Safety and short-term toxicity of a novel cationic lipid formulation for human gene therapy. *Human Gene Ther.* **4,** 781–788.

Scheule, R. K., St. George, J. A., Bagley, R. G., Marshall, J., Kaplan, J. M., Akita, G. Y., Wang, K. X., Lee, E. R., Harris, D. J., Jiang, C., Yew, N. S., Smith, A. E., and Cheng, S. H. (1997). Basis of pulmonary toxicity associated with cationic lipid-mediated gene transfer to the mammalian lung. *Human Gene Ther.* **8,** 689–707.

Schwarz, L. A., Johnson, J. L., Black, M., Cheng, S. H., Hogan, M. E., and Waldrep, J. C. (1996). Delivery of DNA-cationic liposome complexes by small-particle aerosol. *Human Gene Ther.* **7,** 731–741.

Seddon, J. M. (1990). Structure of the inverted hexagonal (HII) phase, and nonlamellar phase transitions of lipids. *Biochim. Biophys. Acta* **1031,** 1–69.

Semple, S. C., and Chonn, A. (1996). Liposome-blood protein interactions in relation to liposome clearance. *J. Liposome Res.* **6,** 33–60.

Senior, J., and Gregoriadis, G. (1982). Stability of small unilamellar liposomes in serum and clearance from the circulation: The effect of the phospholipid and cholesterol components. *Life Sci.* **30,** 2123–2136.

Senior, J. H. (1987). Fate and behavior of liposomes *in vivo:* A review of controlling factors. In *CRC critical reviews of therapeutic drug carrier systems,* **3,** 123–193. Boca Raton: CRC Press.

Senior, J. H., Trimble, K. R., and Maskiewicz, R. (1991). Interaction of positively-charged liposomes with blood: Implications for their application *in vivo. Biochim. Biophys. Acta* **1070,** 173–179.

Sikes, M. L., O'Malley, B. W., Jr., Finegold, M. J., and Ledley, F. D. (1994). *In vivo* gene transfer into rabbit thyroid follicular cells by direct DNA injection. *Human Gene Ther.* **5,** 837–844.

Smyth-Templeton, N. S., Lasic, D. D., Frederik, P. M., Strey, H. H., Roberts, D. D., and Pavlakis, G. N. (1997). Improved DNA:liposome complexes for increased systemic delivery and gene expression. *Nature Biotechnol.* **15,** 647–652.

Solodin, I., Brown, C., Bruno, M., Chow, C.-Y., Jang, E.-J., Debs, R., and Heath, T. (1995). A novel series of amphiphilic imidazolinium compounds for *in vitro* and *in vivo* gene delivery. *Biochemistry* **34,** 13537–13544.

Song, Y. K., Liu, F., Chu, S., and Liu, D. (1997). Characterization of cationic liposome-mediated gene transfer *in vivo* by intravenous administration. *Human Gene Ther.* **8,** 1585–1594.

Sorscher, E. J., Logan, J. J., Frizzell, R. A., Lyrene, R. K., Bebok, Z., Dong, J. Y., Duvall, M. D., Felgner, P. L., Matalon, S., Walker, L., *et al.* (1994). Gene therapy for cystic fibrosis using cationic liposome mediated gene transfer: A phase I trial of safety and efficacy in the nasal airway. *Human Gene Ther.* **5,** 1259–1277.

Stephan, D. J., Yang, Z. Y., San, H., Simari, R. D., Wheeler, C. J., Felgner, P. L., Gordon, D., Nabel, G. J., and Nabel, E. G. (1996). A new cationic liposome DNA complex enhances the efficiency of arterial gene transfer *in vivo. Human Gene Ther.* **7,** 1803–1812.

Sternberg, B., Sorgi, F. L., and Huang, L. (1994). New structures in complex formation between DNA and cationic liposomes visualized by freeze–fracture electron microscopy. *FEBS Letters* **356,** 361–366.

Stewart, M. J., Plautz, G. E., Del Buono, L., Yang, Z. Y., Xu, L., Gao, X., Huang, L., Nabel, E. G., and Nabel, G. J. (1992). Gene transfer *in vivo* with DNA–liposome complexes: Safety and acute toxicity in mice. *Human Gene Ther.* **3,** 267–275.

Stopeck, A. T., Hersh, E. M., Akporiaye, E. T., Harris, D. T., Grogan, T., Unger, E., Warneke, J., Schluter, S. F., and Stahl, S. (1997). Phase I study of direct gene transfer of an allogeneic histocompatibility antigen, HLA-B7, in patients with metastatic melanoma. *J. Clin. Oncol.* **15,** 341–349.

Stribling, R., Brunette, E., Liggitt, D., Gaensler, K., and Debs, R. (1992). Aerosol gene delivery *in vivo. Proc. Natl. Acad. Sci. USA* **89,** 11277–11281.

Szoka, F. C., Xu, Y., and Zelphati, O. (1996). How are nucleic acids released in cells from cationic lipid–nucleic acid complexes. *J. Liposome Res.* **6,** 567–897.

Thierry, A. R., Rabinovich, P., Peng, B., Mahan, L. C., Bryant, J. L., and Gallo, R. C. (1997). Characterization of liposome-mediated gene delivery: Expression, stability and pharmacokinetics of plasmid DNA. *Gene Ther.* **4,** 226–237.

Tsan, M. F., Tsan, G. L., and White, J. E. (1997). Surfactant inhibits cationic liposome-mediated gene transfer. *Human Gene Ther.* **8,** 817–825.

Tsan, M. F., White, J. E., and Shepard, B. (1995). Lung-specific direct *in vivo* gene transfer with recombinant plasmid DNA. *Am. J. Physiol.* **268,** L1052–L1056.

Wang, J. K., Gou, X., Xu, Y., Barron, L., and Szoka, F. C. (1998). Synthesis and characterization of long chain alkyl acyl carnitine esters. Potentially biodegradable cationic lipids for use in gene delivery. *J. Med. Chem.* **41,** 2207–2215.

Weibel, E. R. (1985). Lung cell biology. In *Handbook of physiology* (pp. 47–91). Baltimore: Williams and Wilkins.

Wolff, J. A., Malone, R. W., Williams, P., Chong, W., Acsadi, G., Jani, A., and Felgner, P. L. (1990). Direct gene transfer into mouse muscle *in vivo*. *Science* **247,** 1465–1468.

Woodle, M. C., Collins, L. R., Sponsler, E., Kossovsky, N., Papahadjopoulos, D., and Martin, F. J. (1992). Sterically stabilized liposomes. Reduction in electrophoretic mobility but not electrostatic surface potential. *Biophys. J.* **61,** 902–910.

Wrobel, I., and Collins, D. (1995). Fusion of cationic liposomes with mammalian cells occurs after endocytosis. *Biochim. Biophys. Acta* **1235,** 296–304.

Xu, Y., Hui, S. W., Frederik, P., and Szoka, F., Jr. (1999). Physico-chemical characterization and purification of cationic lipoplexes. *Biophys. J.* In press.

Xu, Y., and Szoka, F. C., Jr. (1996). Mechanism of DNA release from cationic liposome/DNA complexes used in cell transfection. *Biochemistry* **35,** 5616–5623.

Yang, J. P., and Huang, L. (1996). Direct gene transfer to mouse melanoma by intratumor injection of free DNA. *Gene Ther.* **3,** 542–548.

Yang, J. P., and Huang, L. (1997). Overcoming the inhibitory effect of serum on lipofection by increasing the charge ratio of cationic liposome to DNA. *Gene Ther.* **4,** 950–960.

Yang, J. P., and Huang, L. (1998). Time-dependent maturation of cationic liposome–DNA complex for serum resistance. *Gene Ther.* **5,** 380–387.

Zabner, J., Cheng, S. H., Meeker, D., Launspach, J., Balfour, R., Perricone, M. A., Morris, J. E., Marshall, J., Fasbender, A., Smith, A. E., and Welsh, M. J. (1997). Comparison of DNA-lipid complexes and DNA alone for gene transfer to cystic fibrosis airway epithelia *in vivo*. *J. Clin. Invest.* **100,** 1529–1537.

Zabner, J., Couture, L. A., Smith, A. E., and Welsh, M. J. (1994). Correction of cAMP-stimulated fluid secretion in cystic fibrosis airway epithelia: Efficiency of adenovirus-mediated gene transfer *in vitro*. *Human Gene Ther.* **5,** 585–593.

Zabner, J., Fasbender, A. J., Moninger, T., Poellinger, K. A., and Welsh, M. J. (1995). Cellular and molecular barriers to gene transfer by a cationic lipid. *J. Biol. Chem.* **270,** 18997-19007.

Zelphati, O., and Szoka, F. C., Jr. (1997). Cationic liposomes as an oligonucleotide carrier: Mechanism of action. *J. Liposome Res.* **7,** 31–49.

Zelphati, O., and Szoka, F. C., Jr. (1996). Mechanism of oligonucleotide release from cationic liposomes. *Proc. Natl. Acad. Sci. USA* **93,** 11493–11498.

Zelphati, O., Uyechi, L. S., Barron, L. G., and Szoka, F. C., Jr. (1998). Effect of serum components on the physico-chemical properties of cationic lipid/oligonucleotide complexes and on their interactions with cells. *Biochim. Biophys. Acta* **1390,** 119–133.

Zhang, S. C., Schultz, D. R., and Ryan, U. S. (1986). Receptor-mediated binding of C1q on pulmonary endothelial cells. *Tissue Cell* **18,** 13–18.

Zhou, X., and Huang, L. (1994). DNA transfection mediated by cationic liposomes containing lipopolylysine: Characterization and mechanism of action. *Biochim. Biophys. Acta* **1189,** 195–203.

Zhou, X. H., Klibanov, A. L., and Huang, L. (1991). Lipophilic polylysines mediate efficient DNA transfection in mammalian cells. *Biochim. Biophys. Acta* **1065,** 8–14.

Zhu, N., Liggitt, D., Liu, Y., and Debs, R. (1993). Systemic gene expression after intravenous DNA delivery into adult mice. *Science* **261,** 209–211.

CHAPTER 12

Biopolymer–DNA Nanospheres

Kam W. Leong
Department of Biomedical Engineering, School of Medicine,
The Johns Hopkins University, Baltimore, Maryland

Nonviral vectors are increasingly being proposed as alternatives to viral vectors for *in vivo* gene transfer because of their potential advantages in addressing the pharmaceutical issues of applying gene as a drug. We have studied the complexation of DNA with gelatin or chitosan in forming nanospheres. In addition to the benefits common to other nonviral gene delivery systems such as protecting the DNA from nuclease degradation and allowing active targeting, characteristics unique to these biodegradable DNA nanospheres include coencapsulation of bioactive agents and sustained release of the DNA. The former raises the possibility of combining drug

Nonviral Vectors for Gene Therapy

and gene therapy in one single vehicle, and the latter may improve the tissue bio-availability of DNA. Positive gene transfer has been observed *in vivo* in the lung, muscle, and gastrointestinal tissues in animal models. While the transfection efficiency of these DNA nanospheres remains low, their application in DNA vaccination, where high antigen expression may not be required, is promising because of their ability to encapsulate and deliver cytokines in a local and sustained manner to stimulate the infiltrating immune cells.

I. INTRODUCTION

As the initial euphoria of gene therapy subsides, a general consensus of the need to better understand the basic aspects of gene transfer emerges. Of critical importance are fundamental issues involving interaction of the gene vectors with the host, the basic aspects of disease pathophysiology targeted for gene therapy, and mechanistic information that can lead to a rational design of safe and effective gene transfer vectors (Crystal, 1995a; Orkin and Molulsky, 1995; Rolland, 1998).

Viruses remain the vectors of choice in achieving high efficiency of gene transfer *in vivo*. The retroviral vector is the only reliable means of stably integrating the foreign gene into the host genome of mitotically active cells. However, issues of immune response to the expressed viral proteins, the limit of DNA size that can be packaged into the viral vectors, and scale-up difficulties remain to be addressed. Nonviral vectors, although achieving only transient and lower gene expression level compared to viral vectors, may be able to compete on potential advantages of ease of synthesis, low immune response, and unrestricted plasmid size. Cationic lipids are efficient vectors for gene transfer in cell culture. Their performance *in vivo*, however, remains to be optimized. Polymers cationic in nature are increasingly being proposed as potential vectors because of the versatility in fine-tuning the physicochemical properties of the carriers. Rigidity, hydrophobicity/hydrophilicity, charge density, biodegradability, and molecular weight of the polymer chain are all parameters that in principle can be adjusted to effect an optimal complexation with DNA. Admittedly, the desirable features for optimal complexation other than efficient condensation into small particle size are unclear. It is likely that for different cells or tissues or different routes of administration *in vivo* the desirable characteristics of the carrier/DNA complex would differ. Polymeric carriers with their multiple degrees of freedom for optimization are nevertheless well positioned to meet the challenge. Early work with polylysine has provided the proof of concept of the polymeric approach (Wu and Wu, 1988; Wagner *et al.*, 1990; Wagner *et al.*, 1993; Wagner *et al.*, 1992; Cotten and Wagner, 1993). However, the toxicity of polylysines, especially the high molecular ones, diminishes their utility *in vivo*. Recent generations of dendrimers with primary amino groups on the exterior appear to be efficient in gene transfer and less toxic than polylysine (Service, 1995). Other synthetic polycations, such as polyethylenimine, have also produced promising results (Zanta *et al.*, 1997; Ferrari *et al.*, 1997; Tang and Szoka, 1997).

In this brief review, we focus instead on our experience with the natural biopolymers of gelatin and chitosan functioning as gene carriers (Truong-Le *et al.*, in press; Leong *et al.*, 1998; Truong-Le *et al.*, 1998, 1997; Mao *et al.*, 1996, 1997; Roy *et al.*, 1997, 1998; Walsh *et al.*, 1996, 1997). Related approaches are being followed by others (Erbacher *et al.*, 1998). There is an increasing sentiment that gene should be viewed as a drug for gene therapy to be ultimately practical (Crystal, 1995b). Viewed from that angle, gene delivery, at least in the extracellular space, shares many of the hurdles faced by protein delivery. The bioavailability issue has limited the therapeutic efficacy of many potent but unstable proteins. Controlled-release technology has vastly improved the potential of many of these compounds. Hypothesizing that similar principles can be applied to gene delivery, we have formulated DNA nanoparticles by complexing DNA with either gelatin or chitosan. We will discuss the features of these DNA nanoparticles in relation to the hurdles barring the delivery of an intact gene to the nucleus. Attempts will also be made to highlight the differences and similarities of these DNA nanoparticles with other polymeric and liposomal gene delivery systems in addressing these rate-limiting steps.

II. PROPERTIES OF CARRIERS

A. GELATIN

Gelatin, the denatured form of collagen, is a polyampholyte that gels below 35–40°C. It is widely used in food, pharmaceutical, and photographic industries (Rose, 1990). The major anionic side groups are aspartic acid (0.50 mmol/g, pKa = 4–4.5) and glutamic acid (0.78 mmol/g, pKa = 4.5), and the major cationic side groups are lysine (0.30 mmol/g, pKa = 10–10.4) and arginine (0.53 mmol/g, pKa > 12). At pH below 5, it is positively charged and can complex with DNA to form a coacervate. Complex coacervation refers to the process of spontaneous phase separation that occurs when two oppositely charged polyelectrolytes are mixed in an aqueous solution. The electrostatic interaction between the two species of macromolecules results in the separation of a coacervate (polymer-rich phase) from the supernatant (polymer-poor phase). This phenomenon can be used to form microspheres or nanospheres for the encapsulation of a variety of compounds. The encapsulation process can be performed entirely in aqueous solution and at low temperatures, and has a good chance, therefore, of preserving the bioactivity of the encapsulant. Furthermore, encapsulation of protein by this method is attractive because charged proteins can actively participate in the coacervation process, leading to high drug-loading levels. Gelatin has been studied extensively for forming drug-loaded micropheres (Rao et al., 1994; Narayani and Rao, 1994; Nastruzzi et al., 1994; Azhari and Leong, 1992, 1991).

Commonly obtained from hydrolytic denaturation of collagen, gelatin is a heterogeneous material with respect to size (Saddler and Horsey, 1987). The molecular weight distribution and purity of gelatin are influenced by the method of denaturing collagen. Alkaline treatment may cleave the amide bonds of glutamine and asparagine, yielding free carboxyl groups that lower the pI. Mild acid treatment in contrast would yield a gelatin of higher pIs.

At 37°C, gelatin is most likely devoid of secondary structure and is therefore quite susceptible to proteolytic degradation. Gelatin also undergoes hydrolytic degradation in aqueous solution. The degradation rate is influenced by temperature and pH. At the isoelectric pH of gelatin prepared from calf skin (pI = 4.75), the Mn decreases in 50 h from 65,000 to 7,000 at 80°C, and to 17,500 at 40°C (Veis, 1964).

B. CHITOSAN

Chitosan is the fully or partially N-deacetylated derivative of chitin, which as one of the most abundant glycans is comprised of $1{\rightarrow}4$-β-linked N-acetyl-D-glucosamine units in the backbone (Brine et al., 1992). It is found in the exoskeletons of crustaceans such as shrimp and crab. The synthetic modification of chitin and chitosan has been extensively studied. The 2-amine-2-deoxy function of chitosan affords a convenient site for a variety of derivatization, ranging from amidation to Schiff base reaction or reductive alkylation. This allows conjugation of polar residues to render the polymer more water soluble at neutral pH or attachment of ligands for targeting or receptor-mediated endocytosis. For instance, a degree of substitution of 25% or above of quaternary alkyl ammonium would render chitosan water soluble.

Carrying a positive charge in the GI tract, chitosan is commercially available in fiber form and touted as an effective oral adsorbent for removing fat, lipids, and bile acids (Kas, 1997). Chitosan-coated dialdehyde cellulose has also been developed similarly to adsorb urea and ammonia (Yagi et al., 1998). Like gelatin, chitosan has been widely studied as a matrix for controlled-release microsphere formulation. It has also been used as a stabilizer for alginate capsules encapsulating cells (Yoshioka et al., 1990).

Chitosan undergoes no hydrolytic chain cleavage but is degraded by chitinase and chitosanase and—more relevant to biomedical applications—by lysozyme. The enzymatic degradation rate is quite sensitive to the degree of N-acetylation. A 150-μm film of chitosan with 31% of N-acetylation loses almost 50% of its mass in the subcutaneous space of rats within 2 weeks, whereas a fully deacetylated chitosan shows no mass loss in 12 weeks (Varum et al., 1997; Tomihata and Ikada, 1997).

The preceding primer probably has alerted readers to the diverse properties of gelatin and chitosan. The properties of the DNA nanoparticles are naturally determined by the characteristics of these biopolymers. The following discussion refers to a porcine Type A gelatin with a bloom strength of 60 and a chitosan with a weight average molecular weight of 390,000 and an average amino group density of 0.837 per disaccharide unit.

III. SYNTHESIS

The gelatin–DNA or chitosan–DNA nanospheres are synthesized by mixing the DNA solution with an aqueous solution of gelatin or chitosan (Truong-Le *et al.*, in press; Leong *et al.*, 1998; Truong-Le *et al.*, 1998). Important parameters affecting the quality of the nanospheres include (1) salt concentration, (2) temperature, (3) pH, and (4) reactant concentration.

1. Salt, such as sodium sulfate with a concentration ranging from 5 to 50 mM, is added to induce desolvation of the local water environment of the polyelectrolytes to favor phase separation. Other salts should have similar effects. In fact, if charged compounds are to be coencapsulated in the DNA nanospheres the salt concentration should be adjusted to yield optimal nanospheres. There are no a priori criteria of what features would constitute an ideal DNA nanosphere. The nanospheres are grossly judged by their size, size distribution, aggregation, shape, stability during isolation, and yield after workup. The assumption is that a compact size would facilitate cellular internalization. Narrow polydispersity, regular spherical shape, and nonaggregation would also be favored. The assumption of effects of size and shape on the efficiency of gene transfer has yet to be verified.

2. Temperature would affect the thermodynamics of the coacervation and ultimately the quality of the nanospheres. The relationship between the two variables, however, is unknown at this point. The temperature should not be so high as to denature the double-stranded DNA, but should be above 40°C to prevent gellation of the gelatin solution. There is no such temperature limit for the chitosan solution. For example, chitosan / DNA complexes have been prepared at room temperature for *in vitro* transfection (Erbacher *et al.*, 1998).

3. The pH of either the gelatin or the chitosan solution should be acidic. For gelatin, a pH (~5) below the pI imparts the polycationicity; for chitosan, the low pH ionizes the primary amino groups and solubilizes the hydrophobic polysaccharide. Since complex coacervation is driven by charge neutralization, pH would also affect the reactant concentration for nanosphere formation.

4. The range of reactant concentration is quite narrow for the nanosphere formation. At 55°C, pH 5.5, and a sodium sulfate concentration of 50 mM, distinct chitosan–DNA nanospheres would form at the concentration range of 0.005 to

0.02% (w/v) for chitosan and 0.004—0.008% for the DNA. The corresponding values for gelatin—DNA nanospheres are 1.0—3.0% (w/v) for gelatin and 0.004—0.006% for the DNA. Outside this range, there would be either no phase separation or just precipitation forming a lump. The three-phase diagram for DNA nanosphere formation at this temperature, pH, and salt concentration is applicable to plasmid size between 7 and 12 kb. Multiple plasmids of comparable sizes can be incorporated into the same nanoparticles readily.

The gelatin—DNA nanospheres have to be stabilized by crosslinking of the gelatin matrix, but the chitosan—DNA nanospheres are stable in neutral pHs and even in solutions of high ionic strength. The nanospheres are isolated by dialysis and sucrose gradient ultracentrifugation. Typically, over 95% of the DNA can be captured by the phase separation. Macromolecules or charged compounds could be coencapsulated by adding the compound to the chitosan or gelatin solution before coacervation. Interaction with DNA or gelatin/chitosan through ionic interaction or simple chain entanglement would lead to the entrapment. Shown in Table 1 are the encapsulation efficiency and loading level of different compounds encapsulated in gelatin—DNA nanospheres. The trend follows the expectation that the higher the molecular weight of the encapsulant the higher is the encapsulation efficiency, probably due to the chain entanglement effect. Low-molecular-weight compounds, if charged, can also be encapsulated, although generally at a lower encapsulation efficiency. Other compounds that have been encapsulated but with no data on encapsulation efficiency and loading level include sodium 4-phenylbutyric acid, GM-CSF, IL-2, and g-INF. This coencapsulation feature raises the possibility of combining drug and gene delivery in one single vehicle.

Table I

Encapsulation of Different Agents by Gelatin—DNA Nanospheres

Encapsulant	Size/M,	Encapsulation efficiency (%)	Loading level, (%, w/w)
p42-clacZ DNA	7 kb	>98	27
pcRELuc DNA	12 kb	>98	30
RNA (sperm whale)	0.7 kb	~75	18
BSA	68 kDa	>70	4
Interleukin-4	13.5 kDa	>80	<1
Chloroquine	319 Da	~60	2
Calcium	35 Da	<2	1.8

Encapsulation efficiency is defined as the percent of the agent added to the coacervation reaction that is encapsulated. Loading level is the weight percent of the agent in the nanosphere.

IV. LIGAND CONJUGATION

Ligands can be conjugated to the nanospheres after they are formed. The primary groups in the gelatin or chitosan allow convenient covalent coupling schemes for ligand attachment. Figure 1 shows the general scheme for conjugating PEG and transferrin to the chitosan–DNA nanospheres (Mao *et al.*, 1997). Attachment of transferrin to the nanospheres is achieved via the sulfhydryl groups introduced into transferrin. This has the advantage of minimizing any self-crosslinking of transferrin induced by the nondiscriminating carbodiimide conjugation chemistry, and it facilitates the placement of a spacer between the nanosphere and the ligand. This approach should be applicable to most ligands polypeptidic in nature. The stability of the disulfide bond, however, may not be as stable as the amide linkage. In our earlier work we used the avidin–biotin conjugation scheme, which has the advantage that biotinylation of a large number of proteins has been studied and is known to preserve their bioactivity. The avidin-coated nanospheres, however, are more prone to aggregation; the antigenicity of avidin has also been questioned. Other ligands conjugated to the nanospheres via the succinimidyl chemistry include KNOB (fiber protein of adenovirus), mannose-6-phosphate, and folic acid.

Compared to other nonviral vectors, derivatization after the complex is

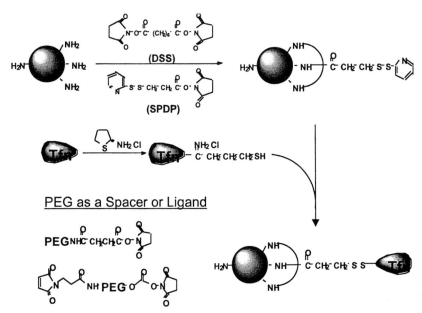

Figure 1 General chemical scheme for conjugation of ligands to DNA nanospheres.

formed may be convenient. For liposome/DNA complexes, active targeting is also possible. However, if the ligand is macromolecular, such as an antibody, the ligand might dominate the lipid and interfere with the complexation of the lipid with DNA. Conjugation post lipid/DNA complex formation is challenging because the DNA is adsorbing on the exterior of the liposome and the stability of the complex would be sensitive to the perturbation of the lipid composition. With a polymeric carrier high in molecular weight where the complex is held together by both electrostatic forces and chain entanglement, there is more room to accommodate different ligands and conjugation chemistries after the complex is formed. However, if the ligand is of low molecular weight, conjugation before the complex formation does present the advantages of a better characterized carrier, a more predictable nanosphere, and pragmatically a much simpler workup in the synthesis. While tissue targeting *in vivo* has yet to be proven with any of these nanospheres, the surface modification of the nanospheres can facilitate cellular internalization and improve storage stability, which will be discussed next.

V. PHYSICO-CHEMICAL PROPERTIES

A. SIZE AND ZETA POTENTIAL

The nanospheres range in size between 200 and 750 nm as measured both by dynamic light scattering and differential interference contrast microscopy. The size distribution of a typical batch of chitosan–DNA nanospheres is shown in Figure 2. With an average size of 288 nm, there is a small fraction hovering around 100 nm and another fraction, probably the aggregated particles, at 720 nm. The zeta potential for chitosan–DNA nanospheres without transferrin is −2 mV at pH 7.4 (Fig. 2); the value is between 2 and 5 mV for gelatin–DNA nanospheres. A typical batch of gelatin–DNA nanospheres would have an average size of 400 ± 120 nm. As observed under confocal microscopy by ethidium bromide staining, the DNA is uniformly distributed in either type of nanosphere. The nanospheres would remain stable in water for at least 3 months. Uncrosslinked gelatin–DNA nanospheres, however, would spontaneously dissociate in PBS or serum culture media, but uncrosslinked chitosan–DNA nanospheres would remain intact under the same conditions. Hydrophobic in nature, the chitosan we used is insoluble above neutral pH. The chitosan–DNA nanospheres do slowly aggregate when incubated in cell culture medium for a few hours, probably induced by this hydrophobic character. This aggregation in solution can be greatly reduced by surface attachment of PEG to the chitosan–DNA nanospheres. Such nanospheres also can be lyophilized without any anticaking agent and resuspended readily.

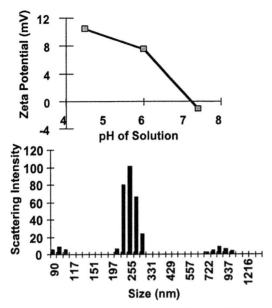

Figure 2 Size distribution and zeta potential of chitosan–DNA nanospheres.

B. INTEGRITY OF ENCAPSULATED DNA

Embedded in the polymeric matrix, the DNA is partially protected from nuclease degradation. Gel electrophoretic mobility analysis indicates that the DNA in the nanosphere remains intact in the first 4 h of incubation in 10% fetal bovine serum (FBS). Degradation of the DNA is evident after that period, and complete by 12 h. In contrast, naked DNA is degraded within the first 30 min of serum exposure, and totally decomposed by 1 h. Electrophoretic mobility analysis of the released DNA reveals no change in the molecular weight of the DNA, suggesting that there is minimal, if any, DNA–matrix and DNA–ligand crosslinking. This is understandable since the reactivity of the primary amino groups of the lysine or arginine side chains is higher than the reactivity of the secondary amines in the nitrogenous bases of DNA. The analysis does show that the released DNA has shifted from the supercoiled state to the relaxed state as a result of the nanosphere formation. This may be unavoidable because of the forced interaction of the plasmid with another polymer that is also constricted in its equilibrium configuration. A synthetic protocol that minimizes this shift in conformation of the DNA may nevertheless improve the stability of the plasmid and presumably improve the transfection efficiency.

C. Toxicology of Carrier

While the *in vivo* biocompatibility and toxicity of gelatin and chitosan as gene carriers remain to be definitively determined, these two natural biopolymers have been presented as bioabsorbable, nontoxic, and low-immunogenic drug carriers (Rose, 1990; Brine *et al.*, 1992). We previously determined the cytotoxicity of different cationic vectors on COS7 cells using the MTT dye reduction assay (Brown *et al.*, 1995). Among the four carriers tested—gelatin, lipofectamine, polylysine, and DEAE dextran—gelatin is the least toxic; no appreciable cytotoxicity is detected for at least up to 250 μg/ml. Chitosan is equally innocuous in a separate evaluation (Roy *et al.*, 1997). Possible repeated administration is one of the advantages of nonviral gene delivery systems. A strong antibody response against the human transferrin is seen in mice, but there is minimal immune response against the gelatin (Truong-Le *et al.*, 1997).

D. Release of DNA

Chitosan is stable in the absence of enzymes such as chitinase, chitosanase, and lysozyme. The chitosan–DNA nanospheres are therefore stable in PBS and cell culture medium. Release of DNA from the nanospheres in these media is negligible. Presumably it is in the endolysosomes that the DNA can be freed.

Release of DNA from the gelatin–DNA nanospheres is predominantly effected by proteolytic degradation of gelatin, leading to diffusion of the DNA through the loosened gelatin matrix. The stability of gelatin nanospheres has been examined by following the release of [^{35}S]-CTP-labeled DNA (Truong-Le *et al.*, 1998). Gelatin nanospheres without crosslinking are stable in a 20% sucrose solution for more than a week at 4°C without significant release of DNA (<10%), but, as expected of a coacervate, the nanospheres are unstable in high-ionic-strength (>0.25 M) solutions. Nanospheres with the gelatin crosslinked by 1-ethyl-3-(3-dimethylaminopropyl) carbodiimide hydrochloride (EDC) (0.1 mg/ml) slowly release the DNA or the DNA fragments at 37°C. After an initial burst of approximately 5–10% of the DNA in the nanospheres in the first few hours, the average release rate per day of DNA in the first week is 0.9% in water, 2.2% in PBS, and 10% in serum. The release of DNA from nanospheres can be adjusted by varying the EDC crosslinking concentration, indicating the possibility of controlled release of DNA. *In vivo,* similar crosslinked gelatin/polyanion complexes have been shown to be completely degradable in the extracellular space. Near complete degradation of gelatin–chondroitin sulfate microspheres is observed when they are incubated for 2 days in joint fluids exhibiting gelatinase activities (Brown *et al.*, in press). The same microspheres containing GM-CSF in the subcutaneous space of mice also disappear within 2 weeks (Golumbek *et al.*, 1993). The microspheres are degraded by proteolytic enzymes in the interstitial fluid and engulfed by phagocytic cells attracted to the site. Histology shows that the microspheres decrease in size and their surface becomes more and more irregular and ruffled with time, with only fragments and small eosinophilic traces remaining at the site by Day 12.

Although the DNA nanospheres are designed to be internalized by cells, the portion that is not phagocytosed rapidly may be due to aggregation *in situ* and may release the DNA extracellularly in a sustained manner. Controlled delivery can also be achieved by encapsulating these nanospheres in biodegradable matrices such as microspheres. Microencapsulation of free DNA faces the technical challenging of attaining a high loading level, a problem that may not be so acute for the nanospheres because there are various processing techniques for incorporating particulates into a matrix. Akin to the therapeutic benefits realized by controlled delivery of delicate drugs, there are potential advantages offered by controlled gene delivery:

1. Improved bioavailability: While the mechanism of gene transfer by naked DNA *in vivo* remains mysterious, as long as the mechanism is not injury induced, the sustained release should be advantageous over free DNA. As a bolus injection of free DNA is subject to extracellular degradation and tissue clearance, a local and sustained delivery of the DNA should increase the fraction of intact DNA taken up by the target cell/tissue. Passive and active targeting would also improve the bioavailability. The former may be achieved by size variation of the nanospheres or microspheres and the latter by the attached ligands to alter the biodistribution.

2. Prolonged gene expression: The continuous delivery may alleviate the disadvantage of transient expression inherent in all nonviral gene transfer vectors. Related to this is the advantage of reduced frequency of administration. Implicit in this claim is the assumption that there is no threshold concentration the DNA must exceed in order to be internalized by the cells. Also uncertain is how the sustained low dose would affect the gene expression level.

The controlled-release feature of these nanospheres also raises the intriguing question of whether there can be sustained release of DNA intracellularly. Thus far we have not been able to prove this point in cultures of 293 cells. Exocytosis of the complexes or rapid degradation of the released DNA in the cytosol may have neutralized any intracellular sustained-release effect.

VI. *IN VITRO* TRANSFECTION

Most of the *in vitro* transfection comparison studies have been done in the luciferase—293 cells system. Plain gelatin—DNA nanospheres effect a gene expression level barely above the background. Attachment of transferrin on the surface enhances the level by over two orders of magnitude in terms of relative light units, presumably by engaging the transferrin receptor to facilitate entry of the nanospheres into the cell. Further encapsulation of chloroquine in the gelatin–DNA nanospheres boosts the level another 10- to 50-fold. The improvement in transfection efficiency for the nanospheres is equal to or higher than that achieved by adding chloroquine exogenously at a concentration of 100 μM to the culture medium. The concentration of chloroquine (from the nanospheres) in the cell culture media would be approximately 5 μM if all the encapsulated chloroquine were released extracellularly. Chloroquine added exogenously at this concentration fails to elicit enhancement in transfection. This is to be compared to concentrations of 50–200 μM and an increase of 5- to 20-fold in transfection level when chloroquine is included in the incubation medium for the liposome, poly-L-lysine, or CaPO$_4$ precipitate transfection techniques (Coll *et al.,* 1997). These results suggest that the potentiation effect of chloroquine stems from the encapsulated form and not chloroquine that might have leached from the nanospheres into the culture medium. This is consistent with the fact that the nanospheres were dialyzed against water before use and that the chloroquine remaining in the nanospheres probably represents the stable portion complexing with DNA. While exogenous chloroquine is effective in improving the transfection efficiency of different nonviral gene delivery systems *in vitro,* it is impractical *in vivo* because of the toxicity of the compound. Encapsulated in these DNA nanospheres, however, chloroquine is applicable since only a low dose would be required and would affect only the transfected cells, assuming chloroquine is released only in the endolysosomal compartment. There are other endosomolytic or membrane-modifying agents that can disrupt the endosomes or influence the intracellular trafficking (Wagner, 1998). The possibility of incorporating these compounds into the nanospheres offers interesting opportunities of studying the rate-determining step intracellularly and of optimizing the delivery system design.

The enhancement effects of transferrin and encapsulated chloroquine are not observed in the transfection by chitosan–DNA nanospheres. The plain chitosan–DNA nanospheres produce a gene expression level comparable to that achieved by gelatin–DNA nanospheres with conjugated transferrin and encapsulated chloroquine. The reason for this lack of potentiating effect of transferrin is not clear. We have also observed that conjugation of mannose-6-phosphate did not significantly enhance the transfection potency of the DNA–chitosan nanospheres (Mao *et al.,* 1997). These data suggest that the DNA–chitosan nanospheres might be entering the 293 cells through a mechanism different from that used by the DNA–gelatin nanospheres. The recent report of a lactosylated chitosan/DNA complex also indicates the lack of potentiating effects of the ligand and exogenous chloroquine (Erbacher *et al.,* 1998). The lack of dependence on chloroquine probably reflects the ionization of chitosan in the low-pH endolysosomal compartment. The effectiveness of gene carriers such as cationic dendrimers and polyethylenimine has been attributed to their buffering capacity, acting as a "proton sponge" in the lysosomes. A similar phenomenon might be at work in this case.

Interestingly, conjugation of PEG_{5000} to the surface of the chitosan–DNA nanospheres has no detrimental effect on the transfection efficiency. These PEGylated nanospheres can be lyophilized without any anticaking agent and resuspended; after 4 weeks of storage in the lyophilized state, the transfection efficiency remains unaffected (Mao *et al.*, 1997). PEG coating has been used to create "stealth" liposomes and drug-loaded polymeric nanospheres to minimize opsonization and prolong their systemic circulation. It will be interesting to see if these PEGylated chitosan–DNA nanospheres can similarly evade the reticuloendothelial system and still maintain their ability to transfect cells *in vivo*.

For the marker genes, the gene expression level achieved by the nanospheres is consistently lower than that obtained by lipofectamine and $CaPO_4$ precipitate techniques. The same trend is true regarding the percentage of cells transfected. For gelatin–DNA nanospheres, generally greater than 10% of 293, HeLa, and human tracheal epithelial cells (9-HTE cells) can be transfected with the luciferase or GFP construct, but less than 5% of COS7, CHO, and endothelial cells (HUVec). A systematic study of the transfection efficiency of the chitosan–DNA nanospheres has yet to be done.

Other than marker genes, several other genes with potential therapeutic value have also been evaluated. Table 2 shows the representative transfection levels of the nanospheres compared with lipofectamine and $CaPO_4$ methods. Other than the gelatin–CFTR–DNA nanospheres, which appear to be particularly effective against the HTE cells, the nanospheres are consistently less efficient than the liposome and $CaPO_4$–DNA complexes.

The rate-limiting steps responsible for the low *in vitro* transfection efficiencies of these nanospheres are unclear at this point. Preliminary experiments on cellular uptake based on dual-label confocal analysis suggest that at 1 h posttransfection

Table II

Transfection Efficiency of DNA Nanospheres in Various Gene–Cell Combinations

Plasmid	Cell	Transfection efficiency, (% cells transfected)		
		Nsp	Lipofectin or lipofectamine	CaPO4
LAMP-1[g]	293, U937	6	20	15
CFTR[g]	HTE	>80	50–60	10–20
Nitric oxide synthase[c]	SMC	Positive	Positive	Positive
B7-1[g]	293	10	37	50
γ-INF[c]	293	Nsp: 8 ng/10^6 cells/day		
GM-CSF[g]	293, B16-F1	Nsp: 4–40 ng/10^6 cells/day		

SMC = human intestinal smooth muscle cells, g = gelatin–DNA, c = chitosan–DNA, Nsp = Nanospheres.

many nanospheres are localized on the cell surface and in the lysosomes. The intracellular FITC-labeled DNA-nanospheres colocalize with the Texas Red-labeled lysosomes, indicating that the nanospheres were sequestered into the endolysosomal compartment once they were internalized. In contrast, there is little association of the two labels intracellularly when cells are transfected by the CaPO$_4$/DNA complexes, indicating most of the complexes are localized in the cytosol rather than the endolysosomal compartment. The images of the lipofectamine/DNA complexes resemble those of CaPO4/DNA complexes in the transfected cells, although localization in liposomes can also be faintly seen. While detailed mechanistic studies remain to be done, these preliminary data suggest that the liposome/DNA complexes are more capable of escaping from the endolysosomes probably because of the membrane fusion property of the lipid. One of the intricate parameters that must be optimized for the nanospheres is their stability in the endolysosomal compartment. Too stable a complex would not allow release into the cytosol; the other extreme would have the DNA degraded before they can be internalized and escape from the endolysosomal compartment. The stability of the nanospheres can be adjusted by varying the crosslinking density of the gelatin or modifying the degree of deacetylation of chitosan.

VII. *IN VIVO* TRANSFECTION

A. LUNG

The potential of delivering the CFTR gene to the lung airways was demonstrated in instilling gelatin–DNA nanospheres into the right lower lobes of adult rabbits using a pediatric bronchoscope with fiber-optic camera, at a dose of 100–350 μg of DNA per animal (Walsh *et al.,* 1996, 1997). Delivery of the CFTR gene to the airway epithelia was determined by specifically amplifying the pSA306 DNA without amplification of endogenous rabbit CFTR DNA. This is made possible by choosing one of the PCR primers in the fusion peptide region of pSA306, which is not present in any native CFTR sequence. Rabbits treated with nanospheres show a strong positive signal for the presence of pSA306 CFTR DNA compared to rabbits treated with a saline control. The DNA, persisting in airway nuclei for at least 28 days, is observed in a high percentage of airway epithelia. Expression of the CFTR protein as observed by anti-CFTR immunostaining is highly localized to the apical membrane surface of airway cells, demonstrating that the expressed CFTR is correctly trafficked to its site of function. The protein expression is transient, detectable for up to 2 weeks. Positive cells with CFTR expression appear to be limited only to the site of delivery. The expression level of the whole lung is therefore likely to be very low. Most likely the transfection efficiency of this delivery system needs to be significantly improved because it can be therapeutically beneficial for this particular application.

B. MUSCLE

Injection of gelatin–DNA nanospheres containing 1 μg of the LacZ gene into the tibialis muscle bundle of mice produces β-gal expression for at least 21 days (Truong-Le *et al.*, 1998). Transfected cells are present throughout the dermis region. Positive cells are also observed in the local ipsilateral popliteal draining lymph nodes. The level for an equivalent dose of naked DNA is 10- to 30-fold lower at Day 7 in terms of relative light units and declines to background level by Day 21. In contrast to their relative performance *in vitro,* the lipofectamine/DNA complexes are not as efficient as the nanospheres. The expression level at Day 7 is even lower than that of naked DNA. By gross observation, there is acute inflammatory response in the muscle tissue treated by the lipofectamine complexes, which might account for the poor result. The AAV vector is the most efficient, eliciting a β-gal expression of 50–100 times higher than that of the nanospheres at Day 7, and three orders of magnitude higher at Day 21.

The delivery of naked DNA into muscle has proven to be remarkably effective in several model systems for transgene expression (Donnelly *et al.*, 1997). This has stimulated a strong interest of DNA immunization. Hypothesizing that these DNA nanospheres would be efficiently taken up by antigen-presenting cells, we evaluated their efficacy in delivering DNA vaccines, using β-gal as the model immunogen (Truong-Le *et al.*, 1997). Intramuscular injection of 1 μg of Lac-Z DNA–gelatin nanospheres into mice, followed by two booster injections at Weeks 3 and 5, elicits a modest antibody response. Vaccination by naked DNA following the same protocol generates a strong antibody response, comparable to that attained by the protein administered with complete Freund's adjuvant. The DNA nanospheres does elicit a strong CTL response equal to, if not better than, the naked DNA administration.

Genetic immunization offers the advantage of expressing the antigen in its native form, which may lead to optimal processing and presentation to antigen-presenting cells for induction of both humoral and cellular immune responses. Direct injection of the plasmid DNA into the muscular or dermal tissues or by bombardment of DNA-coated gold particles through a high pressure gene gun has produced promising results. Transfection efficiency is typically low, however, and the immune response is often suboptimal. One proposal has been to coadminister cytokines and adjuvants. This approach has attracted intense attention, particularly in the field of cancer immunotherapy. The cytokines either enhance the presentation of antigens to T cells or provide additional costimulatory signals for T cell activation. In many cases, the locally secreted cytokines elicit an inflammatory reaction that leads to the rejection of the injected tumor cells. In some cases, these genetically altered tumor cells can generate systemic immunity against subsequent challenge of parental tumor cells, and occasionally even against established micrometastases. Applying to HIV vaccines, GM-CSF and TNF-α have been found to synergize with IL-12 to enhance induction of cytotoxic T lymphocytes against

HIV-1 MN vaccine constructs (Ahlers *et al.,* 1997). In a separate study, codelivery of IL-12 with HIV-1 vaccines in mice results in splenomegaly and reduced humoral response, while GM–CSF has the opposite effect (Kim *et al.,* 1997). Both cytokines stimulated antigen-specific T cell responses, with a dramatic increase in CTL response observed for the codelivery of IL-12. Noteworthy is the potential of this strategy to optimize the desired immune response against HIV infection by emphasizing the humoral or cellular arm. In our previous studies, we have shown that the cytokines need to be present at the vaccine site for days in order to be effective (Golumbek *et al.,* 1993). The DNA nanosphere approach provides the option of encapsulating the cytokine gene or cytokine protein in the nanospheres for this sustained delivery. Preliminary studies did show that coencapsulation of IL-4 or γ-INF in the nanospheres would enhance the T_H-2 or T_H-1 type immune response, respectively. While the exact mechanism—whether the cytokines are exerting their effect extra- or intracellularly—remains to be determined, these cytokine–DNA-nanospheres suggest intriguing possibilities of manipulating the immune response of a DNA vaccine.

C. Gastrointestinal Tract

The possibility of using these DNA nanospheres to elicit mucosal immunity was explored by evaluating oral immunization of a DNA encoding a model antigen, Arah2, a major peanut allergen (Roy *et al.,* 1998). Chitosan–DNA nanospheres were administered intragastrically to AKR/J mice at a dose of 50 μg per animal by feeding needles. After one booster dose 2 weeks later, followed by sensitization of the animals with weekly intraperitoneal injections of crude extracts of Arah2, the mice were challenged with a large dose of 1 mg of pure Arah2 per animal. The animals were scored for their anaphylactic response, ranging from 0 to 5, according to the following system: 0, no sign of reaction; 1, scratching and rubbing around the nose and head; 2, decreased activity with an increasing respiratory rate, pilar erecti and/or puffing around the eyes; 3, labored respirations, cyanosis around the mouth and tail; 4, slight or no activity after prodding or tremors, convulsion; 5, death.

The anaphylactic response is clearly delayed and reduced in the nanosphere-immunized animals. Naked DNA does not provide any protection. At 30 min postchallenge, the average anaphylactic score is 1 for the immunized groups versus 3 for the control group. This preliminary study suggests the efficacy of oral allergen-gene immunization with chitosan–DNA nanospheres.

Oral delivery is always attractive because of the ease of administration. Gene transfer by naked DNA in the gastrointestinal tract has been largely unsuccessful, presumably because of degradation in transit through the stomach. Chitosan is a particularly attractive gene carrier for this purpose because it is a potent absorption enhancer for large hydrophilic compounds across mucosal surfaces (Kotze *et al.,* 1998, 1997). It exhibits mucoadhesive characteristics and has been studied for peroral peptide delivery (Bernkop-Schnwich *et al.,* 1997a,b; Bernkop-Schnwich and Pasta, 1998). Its ability to enhance transepithelial permeability is reportedly through opening of the tight junctions to allow for paracellular transport (Kotze *et al.,* 1998). In microsphere form, chitosan can also improve the uptake of hydrophilic substances across the epithelial layer (Mooren *et al.,* 1998). For instance, prednisolone adsorbed to chitosan microspheres is transported across the polarized HT-29B6 cells with a fourfold increase of permeability compared to the free drug solution. An effective oral gene delivery system for DNA vaccination would be particularly attractive for mass immunization.

VIII. CONCLUSIONS AND SUMMARY

By way of summary, the features of the proposed gelatin– and chitosan–DNA-nanospheres are compared with the viral and liposomal vectors and presented in Table 3. Naturally the qualitative comparison does not cover exceptions. Cationic liposome research is progressing at such a feverish pace that possibly some recent lipid compositions have overcome many of the hurdles highlighted in the table. Intuitively, a polymeric approach is more versatile for optimization. The charge density as well as the molecular chain length can be varied to interact with DNA. The biopolymer–DNA nanospheres discussed in the review may be advantageous particularly in the context of gene medicine. Their ease of synthesis, improved bioavailability through matrix protection and ligand targeting, storage stability, and biocompatibility suggest that they are well positioned to address the pharmaceutical issues. For therapeutic applications where the gene expression needs to be maintained at a high level, the low transfection efficiency of these nanospheres needs to be improved. For DNA vaccination, where high antigen level may not be required, these nanospheres are promising, particularly considering their ability to deliver cytokines in a local and sustained manner at the site of vaccination to stimulate the infiltrating immune cells.

There is ample room to improve on the nanospheric gene delivery system. Many elegant approaches of synthetic polymeric designs have appeared in the literature recently. They include block copolymers (Toncheva *et al.,* 1998), RGD-oligolysine peptide (Harbottle *et al.,* 1998), galactosylated polyethylenimine (Zanta *et al.,* 1997), chimeric multidomain protein (Fominaya and Wels, 1996), and polyamidoamine dendrimers (Bielinska *et al.,* 1997; Haensler and Haensler, 1993). Nevertheless, a definitive favorite nonviral vector has yet to emerge. Most likely, the breakthrough will have to come from input derived from mechanistic studies understanding the extracellular transport and intracellular fate of the nonviral vector/DNA complex. In time the initial optimism of gene therapy will be justified.

Table III

**Features of Viral Vectors, Liposome / DNA Complex,
and Proposed Biopolymer–DNA–Nanosphere**

Properties/vectors	Virus (retro-, adeno-, AAV, etc.)	Liposome/DNA complex	Biopolymer–DNA nanosphere
Physical form	Protein shell/DNA core	Soluble and particulate; submicron	Particulate; 150–700 nm
Interaction between DNA and carrier	Physical entrapment	Ionic	Ionic and physical entrapment
DNA size limit	<10 kb; one copy per virion	Unlimited; multiple copies per complex	Unlimited; multiple copies per complex
Targetability	Limited by natural tropism to certain tissues	Ligands may be conjugated to lipid, but may affect the interaction of lipid with DNA	Ligands may be conjugated to nanospheres after they are formed
Cellular uptake	Receptor-mediated endocytosis	Endocytosis; may involve membrane fusion	Receptor-mediated endocytosis and non-specific phagocytosis
Endosome disruption	Fusogenic sequences on viral coat proteins	pH sensitive lipid may destabilize endosomal membrane	Encapsulation of endosomolytic agents in nanospheres
Transfection efficiency	High *in vitro* and *in vivo*	High in cell culture	Low
Persistence of gene expression	Possible stable integration with retroviral vectors involving mitotically active cells, or with AAV Rep proteins	Transient	Transient
Codelivery of bioactive agents	Polypeptides possible with chimeric constructs	Difficult	Macromolecules or charged compounds
Controlled release capability	No	Has not been demonstrated	Yes
Stability in storage	Good in suitable medium and low temperature	Generally low	Slowly aggregate in solution, but can be lyophilized with PEG coating
Immune response/ biocompatibility	Generally immunogenic; repeated administration problematic	Difficult to couple high transfection efficiency and noninflammatory characteristics in a lipid composition	Relatively nontoxic and tissue biocompatible

ACKNOWLEDGMENTS

The research results discussed in this review are the contributions of V. Truong-Le, H. Q. Mao, K. Roy, and S. Walsh. The author would like to thank collaborators J. T. August, W. Guggino, P. Zeitlin, and S. Huang. Support to this research is provided by NIH-CA68011 and the Cystic Fibrosis Foundation.

REFERENCES

Ahlers, J. D., et al. (1997). Cytokine-in-adjuvant steering of the immune response phenotype to HIV-1 vaccine constructs. J. Immunol., **158,** 3947.

Azhari, R., and Leong, K. W. Protein release from enzymatically-degradable chondroitin-sulfate/gelatin microspheres. In Intern. Symp. Control. Rel. Bioact. Mater. Amsterdam, Netherlands: Controlled Release Society.

Azhari, R., and Leong, K. W. Enzymatically-triggered release of recombinant soluble complement receptor 1 (sCR1) from chondroitin sulfate/gelatin microspheres. In Intern. Symp. Control. Rel. Bioact. Mater. Orlando, FL: Controlled Release Society.

Bernkop-Schnurch, A., and Pasta, M. Intestinal peptide and protein delivery: novel bioadhesive drug-carrier matrix shielding from enzymatic attack. J. Pharm. Sci., **87,** 430–434.

Bernkop-Schnurch, A., Paikl, C., and Valenta, C. (1997a). Novel bioadhesive chitosan–EDTA conjugate protects leucine enkephalin from degradation by aminopeptidase N. Pharm. Res., **14,** 917–922.

Bernkop-Schnurch, A., Paikl, C., and Valenta, C. (1997b). Mucoadhesive polymers as platforms for peroral peptide delivery and absorption: Synthesis and evaluation of different chitosan–EDTA conjugates. Pharm. Res., **14,** 917–922.

Bielinska, A. U., Kukowska-Latallo, J. F., and J. R. B. Jr. (1997). The interaction of plasmid DNA with polyamidoamine dendrimers: Mechanism of complex formation and analysis of alterations induced in nuclease sensitivity and transcriptional activity of the complexed DNA. Biochim. Biophys. Acta, **1353,** 180.

Brine, C. J., Sandford, P. A., and Zikakis, J. P. (Eds.) (1992). Advances in chitin and Chitosan. London: Elsevier Applied Science.

Brown, K. E., et al. (1995). Cationic gelatin as a gene carrier. Mater. Res. Soc. Symp. Ser., **394,** 331.

Brown, K., et al., Gelatin/chondroitin sulfate microspheres for the delivery of therapeutic proteins to the joint. J. Rheumatol., in press.

Coll, J. L., et al., (1997). In vitro targeting and specific transfection of human neuroblastoma cells by chCE7 antibody-mediated gene transfer. Gene Ther., **4,** 156.

Cotten, M., and Wagner, E. (1993). Non-viral approaches to gene therapy. Curr. Opin. Biotechnol., **4,** 705–710.

Crystal, R. G., (1995a). Transfer of genes to humans: Early lessons and obstacles to success. Science, **270,** 404.

Crystal, R. G. (1995b). The gene as the drug. Nature Med., **1,** 15.

Donnelly, J. J., Ulmer, J. B., and Liu, M. A. (1997). DNA vaccines. Life Sci., **60,** 163.

Erbacher, P., et al. (1998). Chitosan-based vector/DNA complexes for gene delivery: Biophysical characteristics and transfection ability. Pharm. Res., **15,** 1332.

Ferrari, S., et al. (1997). ExGen 500 is an efficient vector for gene delivery to lung epithelial cells. Gene Ther., **4,** 1100 1106.

Fominaya, J., and Wels, W. (1996). Target cell-specific DNA transfer mediated by a chimeric multi-domain protein. J. Biol. Chem., **271,** 10560.

Golumbek, P., et al. (1993). Controlled release, biodegradable cytokine depots: A new approach in cancer vaccine design. Cancer Res., **53,** 5841.

Haensler, J., and Szoka, F. K. (1993). Polyamidoamine cascade polymers mediate efficient transfection of cells in culture. *Bioconjugate Chem.*, **4**, 372.

Harbottle, R. P., *et al.* (1998). An RGD-oligolysine peptide: A prototype construct for integrin-mediated gene delivery. *Human Gene Ther.*, **9**, 1037–1047.

Kas, H. S. (1997). Chitosan: Properties, preparations and application to microparticulate systems. *J. Microencapsulation*, **14**, 689–711.

Kim, J. J., *et al.* (1997). *In vivo* engineering of a cellular immune response by coadministration of IL-12 expression vector with a DNA immunogen. *J. Immunol.*, **158**, 816.

Kotze, A. F., *et al.* (1997). *N*-trimethyl chitosan chloride as a potential absorption enhancer across mucosal surfaces: In vitro evaluation in intestinal epithelial cells (Caco-2). *Pharm. Res.*, **14**, 1197–1202.

Kotze, A. F., *et al.* (1998). Comparison of the effect of different chitosan salts and *N*-trimethyl chitosan chloride on the permeability of intestinal epithelial cells (Caco-2). *J. Controlled Release*, **51**, 35–46.

Leong, K. W., *et al.* (1998). DNA nanospheres as non-viral gene delivery vehicles. *J. Controlled Release*, **53**, 183.

Mao, H. Q., *et al.* (1996). DNA–Chitosan nanospheres for gene delivery. In *Intern. Symp. Control. Rel. Bioact. Mater.* Kyoto, Japan: Controlled Release Society.

Mao, H. Q., *et al.* (1997). DNA–Chitosan nanospheres: Derivatization and storage stability. In *Intern. Symp. Control. Rel. Bioact. Mater.* Stockholm, Sweden: Controlled Release Society.

Mooren, F. C., *et al.* (1998). Influence of chitosan microspheres on the transport of prednisolone sodium. *Pharm. Res.*, **15**, 58–65.

Narayani, R., and Rao, K. (1994). Controlled release of anticancer drug methotrexate from biodegradable gelatin microspheres. *J. Microencapsul.*, **11**, 69–77.

Nastruzzi, C., *et al.* (1994). Production and in vitro evaluation of gelatin microspheres containing an antitumour tetra-amidine. *J. Microencapsul.*, **11**, 249–260.

Orkin, S. H., and Motulsky, A. G. (1995). Report and recommendations of the panel to assess the NIH investment in research on gene therapy.

Rao, J., Ramesh, D., and Rao, K. (1994). Controlled release systems for proteins based on gelatin microspheres. *J. Biomater. Sci. Polym. Ed.*, **6**, 391–398.

Rolland, A. P. (1998). From genes to gene medicines: Recent advances in non-viral gene delivery. *Crit. Rev. Ther. Drug Carrier Sys.*, **15**, 143–198.

Rose, P. I. (1990). Gelatin. In J. I. Kroschwitz (Ed.), *Concise Encyclopedia of Polymer Science and Engineering* (p. 430). New York: Wiley.

Roy, K., Mao, H. Q., and Leong, K. W. (1997). DNA–Chitosan nanospheres: Transfection efficiency and cellular uptake. In *Intern. Symp. Control. Rel. Bioact. Mater.* Stockholm, Sweden: Controlled Release Society.

Roy, K., *et al.* (1998). Oral gene delivery with chitosan–DNA nanoparticles. In *Intern. Symp. Control. Rel. Bioact. Mater.* Las Vegas, NV: Controlled Release Society.

Saddler, J. M., and Horsey, P. J. (1987). The new generation gelatins. A review of their history. *Anaesthesia*, **42**, 998-1004.

Service, R. F., (1995). Dendrimers: Dream molecules approach real applications. *Science*, 458.

Tang, M. X., and Szoka, F. C. (1997). The influence of polymer structure on the interactions of cationic polymers with DNA and morphology of the resulting complexes. *Gene Ther.*, **4**, 823–832.

Tomihata, K., and Ikada, Y. (1997). *In vitro* and *in vivo* degradation of films of chitin and its deacetylated derivatives. *Biomaterials*, **18**, 567.

Toncheva, V., *et al.* (1998). Novel vectors for gene delivery formed by self-assembly of DNA with poly(L-lysine) grafted with hydrophilic polymers. *Biochim. Biophysic. Acta*, **1380**, 354.

Truong-Le, V. L., *et al.* (1997). Delivery of DNA vaccine using gelatin–DNA nanospheres. In *Intern. Symp. Control. Rel. Bioact. Mater.* Stockholm, Sweden: Controlled Release Society.

Truong-Le, V., August, J. T., and Leong, K. W. (1998). Gene delivery by DNA-gelatin nanospheres. *Human Gene Ther.*, **9**, 1709.

Truong-Le, V., *et al.,* Gene transfer by gelatin-DNA nanospheres. *Arch. Biochem. Biophys.,* in press.

Varum, K. M., *et al.* (1997). *In vitro* degradation rates of partially *N*-acetylated chitosans in human serum. *Carbohydrate Res., 299,* 99.

Veis, A., (1964). The macromolecular chemistry of gelatin.

Wagner, E., (1998). Effects of membrane-active agents in gene delivery. *J. Controlled Release, 53,* 155.

Wagner, E., *et al.* (1990). Transferrin-polycation conjugates as carriers for DNA uptake into cells. *Proc. Natl. Acad. Sci. USA, 87,* 3410–3414.

Wagner, E., *et al.* (1992). Influenza virus hemagglutinin HA-2 terminal fusogenic peptides augment gene transfer by transferrin-polylysine–DNA complexes: Toward a synthetic virus-like gene-transfer vehicle. *Proc. Natl. Acad. Sci. USA, 89,* 7934–7938.

Wagner, E., Curiel, D., and Cotten, M. (1993). Delivery of drugs, proteins and genes into cells using transferrin as a ligand for receptor-mediated endocytosis. *Adv. Drug Del. Rev., 14,* 113–135.

Walsh, S. M., *et al.* (1996). Delivery of CFTR Gene to rabbit airways by gelatin–DNA microspheres. In *Intern. Symp. Control. Rel. Bioact. Mater.* Kyoto, Japan: Controlled Release Society.

Walsh, S. M., *et al.* (1997). Combination of drug and gene delivery by gelatin nanospheres for the treatment of cystic fibrosis. In *Proc. Intern. Conf. Controlled Rel. Bioactive Agents.* Stockholm, Sweden: Controlled Release Society.

Wu, G. Y., and Wu, C. H. (1988). Receptor-mediated gene delivery and expression *in vivo. J. Biol. Chem., 263,* 14621.

Yagi, M., *et al.* (1998). Effects of chitosan-coated dialdehyde cellulose, a newly developed oral adsorbent, on glomerulonephritis induced by anti-Thy-1 antibody in rats. *Nephron, 78,* 433–439.

Yoshioka, T., *et al.* (1990). Encapsulation of mammalian cell with chitosan-CMC capsule. *Biotechnol. Bioeng., 35,* 66.

Zanta, M. A., *et al.* (1997). *In vitro* gene delivery to hepatocytes with galactosylated polyethylenimine. *Bioconjugate Chem., 8,* 839–844.

CHAPTER 13

Novel Lipidic Vectors for Gene Transfer

Song Li and Leaf Huang
Laboratory of Drug Targeting, Department of Pharmacology, University of Pittsburgh School of Medicine, Pittsburgh, Pennsylvania

Intensive work in the last 10 years has convincingly demonstrated the potential of cationic lipidic vectors in gene therapy. However, the efficiency of first-generation lipidic vectors needs to be improved before patients can benefit from gene therapy. In addition, these vectors are not suitable for targeted gene delivery via intravenous administration. In this chapter, we discuss several major problems that are associated with these vectors and show that at least some of the problems can be resolved by the use of virus-like particles. These novel lipidic vectors, each containing a condensed genome as the core and a lipidic shell as the envelope, are much more efficient than the first-generation cationic lipidic vectors in transfecting

Nonviral Vectors for Gene Therapy

cells *in vitro* and *in vivo*. The novel vectors can also serve as a prototype to develop a vector for tissue-specific gene delivery.

I. INTRODUCTION

Gene therapy could represent an important advance in the treatment of both genetic and acquired diseases (Miller, 1992; Anderson, 1992). With the discovery of many genes capable of correcting diseased phenotypes, the success of gene therapy is largely dependent on the development of a vehicle, or vector, that can efficiently deliver a gene to target cells with minimal toxicity. Viral vectors are efficient in transfecting cells, yet they suffer from a number of problems, such as immunogenicity (Herz and Gerard, 1993), toxicity (Simon *et al.*, 1993) and potential recombination or complementation (Ali *et al.*, 1994). As a result of these limitations, there have been substantial efforts focused on nonviral vectors, particularly on the use of lipidic systems (Ledley, 1995; Gao and Huang, 1995). Lipidic systems (liposomes, micelles, and other organized structures of lipids) are attractive due to their favorable characteristics such as low immunogenicity, ease in large-scale production, and simplicity of use. There are several cationic lipids commercially available and more lipids are being reported. Currently, cationic liposomes are widely used for the transfection of eukaryotic cells in research laboratories. Several liposomal formulations have also undergone clinical evaluation as vectors for gene therapy in cancer (Nabel *et al.*, 1993; Hui *et al.*, 1997) and cystic fibrosis (Caplen *et al.*, 1995; McLachlan *et al.*, 1996; Gill *et al.*, 1997).

While early laboratory studies and clinical trials have convincingly shown the potential of cationic liposomes in gene therapy, they have also revealed several problems that are associated with these vectors. Apparently, improvement in efficiency must be met before patients can benefit from cationic liposome-mediated gene therapy. Recent studies have also shown that cationic liposome/DNA complexes are not without toxicity when administered systemically (Li and Huang, unpublished data). In this chapter, we will address several major problems that are associated with lipidic vectors. Recent progress in this laboratory toward solving these problems will also be discussed. Emphasis will be placed on several novel lipidic vectors that were recently developed in this laboratory.

II. CELLULAR AND MOLECULAR BARRIERS FOR CATIONIC LIPID-MEDIATED GENE TRANSFER

1. Entry of DNA into Cells

Cationic lipid/DNA complexes are generally believed to interact with cells through a nonspecific charge interaction. A study by Zabner *et al.* (1995) demonstrated that the process of DNA entry into cells was relatively slow. There was a correlation between the percentage of cells taking up DNA and the percentage of cells expressing transgene. Cell lines taking up more DNA gave a higher level of gene expression than the cell lines that take up less DNA. This result suggests that in some cells the uptake of liposome/DNA complexes may be an important barrier to transfection. The factors governing the binding of liposome/DNA complexes to cells and their subsequent cellular uptake are not clearly understood. A recent study suggests that the amount of negative charge on cell surface plays an important role in determining the interaction between cationic liposome/DNA complexes and cells and therefore influences the efficiency of gene expression (Matsui *et al.*, 1997). Cells with more negative charge on the surface are easier to transfect. However, it should be noted that other factors besides cellular uptake of DNA also play important roles in transfection.

2. Escape of DNA into the Cytoplasm

Fusion of cationic liposomes with the cell membrane was initially proposed as the major pathway of internalization of liposome/DNA complexes (Felgner and Ringold, 1989). Later studies with electron microscopy (EM) and other biological assays indicate that cationic liposome/DNA complexes are taken up by cells primarily via an endocytosis mechanism (Zhou and Huang, 1994; Farhood *et al.*, 1995; Wrobel and Collins, 1995). Zhou and Huang have studied the intracellular trafficking of DNA complexed with cationic liposomes composed of dioleoylphosphatidylethanolamine (DOPE, Fig. 1) and lipopolylysine (LPLL). LPLL liposomes condense DNA to form electron dense particles that can be positively identified by negative stain or thin-section EM. A majority of the dense particles were found in the vesicular compartments. Disruption of the endosomal compartment was also visualized, the frequency of which was much higher than that of direct penetration thorough the plasma membrane (15% vs 0.7%). When cells were treated with chloroquine, about 57% of the observed endosomes displayed destabilized morphology, a 42% increase compared with cells without treatment. This was in accordance with the result of a bioassay that showed a six-fold increase in transfection efficiency after chloroquine treatment. Interestingly, if dioleoylphosphatidylcholine (DOPC) instead of DOPE was used as a helper lipid, less than 1% of complex-containing endosomes were destabilized. Membrane enriched with DOPE has a strong tendency to form an inverted hexagonal phase, a structure frequently seen in regions where membrane fusion takes place. Thus, it is believed that DOPE facilitates the disruption of endosomes and the release of free DNA or liposome/DNA complexes into the cytoplasm. Currently, several other approaches have also been attempted to facilitate cytoplasmic release of DNA or cationic lipid/DNA complexes in an effort to further improve transfection efficiency. One of the approaches is the use of a fusogenic peptide. Fusogenic peptides are usually anionic, water soluble, and random coil in conformation at physiological pH, but undergo a transition to an

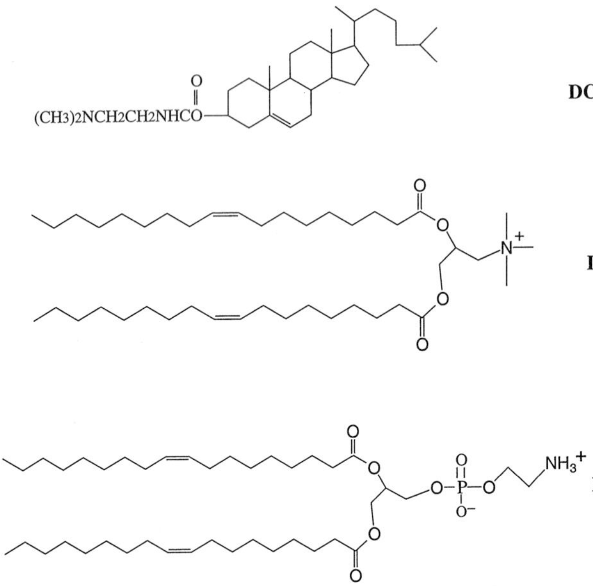

Figure 1 Structures of DC-chol, DOTAP, and DOPE.

amphipathic α-helix when the pH is reduced. It was shown that at low pH values the peptides became associated with lipid bilayer and caused pH-dependent fusion of small liposomes (Torchilin *et al.*, 1993). It is expected that such peptides facilitate the process of endosome disruption when codelivered with the liposome/DNA complexes, resulting in an improved cytoplasmic delivery of DNA. We found that *in vitro* lipofection by several cationic liposomes can be enhanced by 6- to 14-fold when codelivered with the fusogenic peptide, GLFEALLELLESLWELLLEA (Li and Huang, unpublished data). Similar results were found by other groups (Kichler *et al.*, 1997; Simoes *et al.*, 1998).

3. Entry of DNA into Nucleus

The mechanisms involved in the nuclear transport of macromolecules, particularly proteins and RNA/nucleoproteins, have been extensively investigated in the last few years (Hicks and Raikhel, 1995; Davis, 1995). The transport of macromolecules across the nuclear envelope occurs through the nuclear pore. The pore size is approximately 55 Å in diameter, and molecules with a molecular weight of 40 KD or smaller can diffuse freely across the pore. For larger molecules, nuclear transport must depend on a facilitated process involving the nuclear pore complex (NPC). The NPC is a supramolecular aggregate with a molecular weight of 124 megadaltons. Generally, the NPC binds proteins that carry consensus basic amino acid sequences called nuclear localization signal (NLS), such as GKKRSKA. The second step in the nuclear import of NLS-bearing macromolecules is the translocation across the NPC. The study by J. A. Wolff and his colleagues suggest that the import of pDNA into the nucleus in gene delivery also occurs via the same pathway in classical NLS-mediated nuclear transport (Dowty et al., 1995). Most studies indicate that nuclear transport of pDNA is a key limiting step for cationic lipid-mediated gene transfer especially for transfection of nonmitotic cells. A recent study by Coonrod et al. (1997) suggest that the rate of pDNA exclusion from the nucleus also plays an important role in determining the residency of pDNA in the nucleus. In contrast to most other studies, they found, using a PCR-amplified fragment of about 1 Kb, that DNA can efficiently be translocated into the nucleus in both primary human fibroblasts and HeLa cells. DNA rapidly disappeared from the nucleus in fibroblasts, while it stayed for a much longer time in HeLa cells. This observation agrees with the fact that HeLa cells are much more transfectable than fibroblasts. Whether the rapid nuclear import of DNA is also true for plasmid DNA is not clear from the study.

Attempts to improve nuclear transport of DNA by incorporation of karyophilic peptide signal sequences or nucleotide signal sequences into the vector system have only met with limited success thus far. For example, conjugation of a NLS peptide to pDNA at a molar ratio greater than 40 can significantly improve nuclear transport of pDNA but also inactivates gene expression (Sebestyen et al., 1998). Recently, several groups have tried to develop plasmid DNA with improved expression activity to compensate for its inefficient nuclear transport. For example, the Vical group reported a 1,000-fold increase in gene expression by modifying the uncoding sequences of a plasmid DNA (Hartikka et al., 1996). Inclusion of a replication/persistence system may further increase the persistence of vector DNA in transfected cells. Another alternative is the use of a cytoplasmic expression system, that is, a T7-based expression system (Gao and Huang, 1993; Gao et al., 1994) as detailed in the introduction. Eventually, a gene transfer protocol employing both expression systems could be utilized to further improve transfection efficiency. This possibility is being examined in our laboratory.

4. Uncoating of DNA

It is hypothesized that DNA must dissociate from the polycation before transcription can take place. Using microinjection, Zabner et al. (1995) has clearly demonstrated that DNA is transcriptionally inactive when complexed with cationic liposomes. Cationic liposomes inhibit gene expression in a dose-dependent manner. The exact mechanism by which DNA is released from a liposome/DNA complex is not clear. It was thought previously that release of DNA takes place in the nucleus due to displacement of the plasmid DNA from cationic lipids by genomic DNA (Remy et al., 1995). However, a recent in vitro study showed that neither polycations such as spermidine and histone nor nucleic acids can bring about dissociation of preformed liposome/DNA complex (Xu and Szoka, 1996). Instead, plasmid DNA can be readily released from a liposome/DNA complex by anionic liposomes containing compositions that mimic the cytoplasmic-facing monolayer of the plasma membrane. It is speculated that DNA might be similarly released by anionic lipids inside cells (Xu and Szoka, 1996). Our study with the T7 cytoplasmic expression system also indicated the importance of DNA uncoating but only when DNA is condensed by a polycation (Li *et al.*, 1997a). Simply coating plasmid DNA with a monovalent cationic lipid did not inhibit its accessibility to T7 RNA polymerase (Li *et al.*, 1997a). This result suggests a difference between the two different expression systems in their requirement for a delivery vehicle.

The barriers for *in vivo* gene delivery are poorly understood and may vary with administration routes. For example, a formulation suitable for intratracheal injection may not work when administered intravenously, and vice versa. When a direct intratissue injection method is employed, some tissues are more efficiently transfected with naked plasmid DNA, for example, muscle (Wolff *et al.*, 1990) and certain types of solid tumor (Yang and Huang, 1996). This result suggests that optimization of a vector has to be individualized according to each clinical setting. It also stresses the importance of including naked plasmid DNA as a control when the efficiency of any new vector is evaluated.

III. STRUCTURES OF CATIONIC LIPID/DNA COMPLEXES AND THE ROLE OF POLYCATIONS IN DNA CONDENSATION

Characterization of the structure of cationic lipid/DNA complexes is important in understanding the mechanism of lipofection and in designing more efficient formulations. Detailed description of structures of cationic lipid/DNA complexes is not the focus of this chapter; however, we would like to point out here that unlike polycations such as polylysine, which form a highly condensed torroid structure (around 50 nm) with DNA (Wilson and Bloomfield, 1979), the structures formed between DNA and a monovalent cationic lipid such as 3β-[N-N'_iN'-dimethylaminoethane)carbamoyl] cholesterol (DC-chol, Fig. 1) were heterogeneous (Sternberg *et al.*, 1994). Aggregated and fused complexes, together with tubular "spaghetti-like" structures, are abundant (Sternberg *et al.*, 1994). Whether any of these structures is responsible for *in vitro* lipofection is not clear. With the large size of the aggregated complexes, however, uptake of these particles via the mechanism of endocytosis would be inefficient. Interaction of DNA with a multivalent cationic lipid such as lipofectamine results in the formation of small condensed particles (Sorgi *et al.*, unpublished data). This might explain why *in vitro* lipofection by lipofectamine is more efficient than that by DC-chol/DOPE lipo-somes in most cell lines.

IV. CATIONIC LIPOSOME-ENTRAPPED, POLYCATION-CONDENSED DNA (LPD-I)

A. PREPARATION AND CHARACTERIZATION OF LPD-I

Considering the mechanism of lipofection and the difference in the ability to condense DNA between monovalent cationic lipids and polyamines or multivalent cationic lipids, Gao and Huang (1996) hypothesized that introduction of cationic polymers at appropriate ratios to the mixture of the DC-chol/DOPE cationic lipo-somes and DNA might alter the overall structure of the liposome/DNA complex, and thus change the biological activity of the complex. This idea was tested by examining the transfection efficiency of a reporter gene complexed with cationic liposomes alone or cationic liposomes plus various amounts of polymer. Vari-ous liposomes including lipofectin, lipofectamine, DC-chol/DOPE liposomes, and several different polymers were evaluated. The use of a cationic polymer consis-tently improved *in vitro* lipofection. Among several polymers tested, polylysine with a molecular weight of 25 KD was found to be the most efficient in enhancing transfection. The potentiation effect (2- to 28-fold) was observed in a number of cell lines *in vitro*. In particular, cells difficult to transfect with liposome/DNA complex could be transfected when a cationic polymer was included (Gao and Huang, 1996).

Analysis of the mixture on a sucrose gradient ultracentrifugation showed the existence of several populations that differ in the amount of lipids associated with the complex. Those fractions enriched with lipids were more efficient in transfec-tion than those with lower amounts of lipids. However, the activity of the fractions lacking sufficient amounts of lipids could be significantly improved when free DC-chol/DOPE liposomes were added. The purified complexes were severalfold more efficient than the unpurified mixture in transfecting cells *in vitro*. In addition, the purified complexes were less toxic (Gao and Huang, 1996).

Negative-stain EM study of purified complexes showed various shapes of electron-dense structures ranging from elongated rod-shaped to ball-shaped parti-cles. This was related to the amount of lipids associated with the particles. Typically, the complexes giving the highest level of activity appeared as compact spherical particles with a mean diameter of <100 nm (Fig. 2). The core of the particles was heavily stained and might represent polylysine-condensed DNA. Some of the ring structures had typical characteristics of a membrane staining pattern (Gao and Huang, 1996). The purified complexes have been named LPD-I particles. A

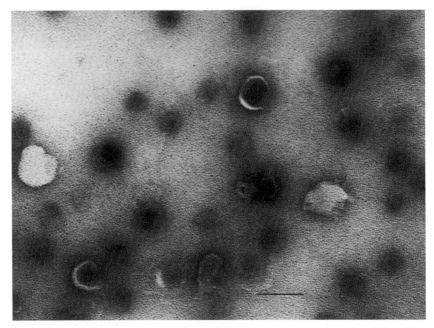

Figure 2 Electron micrograph of LPD-I particles. The complex was prepared from 20 μg DNA, 10 μg PLL (MW 26,500) and 800 nmol DC-chol/DOPE liposomes (2/3, m/m), purified with sucrose gradient ultracentrifugation, and examined with negative-staining EM. Bar = 100 nm. [From "Potentiation of Cationic Liposome-Mediated Gene Delivery by Polycations," by X. Gao and L. Huang, 1996, *Biochemistry, 35,* p. 1035. Copyright 1996 American Chemical Society. Reprinted with permission.]

possible explanation for the formation of such particles is that the highly efficient interaction between DNA and PLL results in the formation of a partially condensed particle. Association of liposomes with the condensed particles then leads to lipid structural rearrangement and the formation of a lipidic shell on the surface of the particle.

The improved activity in transfection by LPD-I particles might be primarily due to their condensed structures. Unlike the DC-chol liposome/DNA complexes, which are large and heterogeneous in size, LPD-I particles are highly compact with a size of less than or close to 100 nm. These particles would be more favorable for entering cells via an endocytosis pathway, which is the major mechanism responsible for the cellular uptake of liposome/DNA complexes. Thus, this formulation is particularly suitable for transfecting those cells that lack other DNA-uptake mechanisms. This might explain why some cells difficult to transfect with liposome/DNA complexes could be efficiently transfected by LPD-I particles. LPD-I particles also offer better protection of DNA from enzymatic digestion. DC-chol liposomes at suboptimal ratio did not protect the DNA from DNase in the serum.

Even at the optimal ratio, DNA was not completely protected. In contrast, LPD-I offered complete protection of the supercoiled conformation of plasmid DNA (Gao and Huang, 1996). Therefore, greater amounts of active DNA could be delivered to cells by LPD-I. Finally, PLL may mimic the nuclear localization signal and facilitate the nuclear transport of DNA. All these favorable characteristics make LPD-I a novel, highly efficient nonviral gene delivery vehicle.

As part of a continuous effort to improve LPD-I formulations, protamine sulfate was later found to be more efficient than PLL in enhancing gene transfer (Sorgi *et al.*, 1997). It is interesting to note that protamine phosphate or protamine-free base worked poorly in enhancing lipofection (Sorgi *et al.*, 1997). The difference among various forms of protamine appears to be attributable to structural differences among the protamines and not to differences in the net charge of the molecules. Nevertheless, the data clearly suggest that protamine sulfate is a better candidate for the preparation of LPD-I. There are additional advantages of utilizing protamine sulfate and these are related to its biocompatibility. Protamine sulfate is nontoxic and only weakly immunogenic in humans and has a record of wide clinical applications. As detailed next, this novel LPD-I shows great promise as a vector for *in vivo* gene delivery.

B. Gene Delivery by LPD-I
via Intracranial Administration

The possibility of using LPD-I to deliver a gene to the brain was tested in collaboration with Drs. Leone and During. LPD-I containing a LacZ reporter gene was injected into one side of a rat brain. The other side received lipofectamine/LacZ complex as a control. A high level of gene expression was found at the side receiving LPD-I injection. Although gene expression was mainly localized in the cells along the track of needle injection, cells several mm away from the needle injection are also efficiently transfected. Gene expression lasted for several months following the injection. In contrast, only weak gene expression was found at the side receiving lipofectamine/LacZ complexes (During *et al.*, unpublished data). Following extensive toxicity studies, LPD-I was used in a clinical trial for the treatment of Canavan disease, a leukodystrophy caused by mutation in the aspartoacylase (ASPA) gene. The loss of ASPA activity leads to an elevation in the brain concentration of N-acetylaspartate (NAA) and spongiform degeneration of oligodendrocytes leading to neurodevelopment retardation and childhood death. A total of seven children with Canavan disease were enrolled in the trial; two of them received an injection of ASPA gene-containing LPD-I. LPD-I was delivered directly into the ventricular fluid. No significant adverse events were found. At 1 month post-surgery, both children had normal levels of NAA in frontal and parietal regions and, in one child, NAA levels remained in the normal range for 12 months. This child, on repeated MRI examinations, had a less abnormal myelin signal for 12 months that was associated with improved neurodevelopmental scores and normalization of visual evoked potentials. In contrast, the five control Canavan children had stable, high NAA levels and no change in the MRI myelin signal (During *et al.*, unpublished data). These data suggest that nonviral, direct *in vivo* gene therapy of Canavan disease is safe and may be associated with biochemical, radiological, and clinical benefit.

C. Intravenous Gene Delivery by LPD-I

1. Reoptimization of LPD-I for Intravenous Gene Delivery

Recent efforts in this laboratory toward the development of a lipidic vector suitable for intravenous gene delivery have led to development of another LPD-I formulation that is composed of protamine, DNA, and 1,2-dioleoyl-3-trimethyl-ammonium-propane (DOTAP, Fig. 1)/cholesterol liposomes (Li and Huang, 1997b; Huang and Li, 1997; Li *et al.*, 1998a). Upon iv administration, this formulation gives a high level of systemic gene expression. All major organs are transfected including lung, heart, liver, spleen, and kidney while the lung is the organ with the highest level of gene expression. The cells that are transfected are primarily vascular endothelial cells, especially lung endothelial cells, while some monocytes are also transfected. Furthermore, our studies as well as studies from other groups have revealed a dramatic difference in the optimal formulation between intravenous gene delivery and gene delivery via other routes of administration. The following are two identified factors that are important for achieving a high level of gene expression by intravenous administration.

(a). Charge Ratio

A recent study in this laboratory has demonstrated that serum sensitivity of *in vitro* lipofection can be overcome by increasing the charge ratio $(+/-)$ between cationic lipids and DNA (Yang and Huang, 1997). Later studies have shown that a high charge ratio $(+/-)$ is also important for achieving a high level of gene expression via iv administration of LPD-I (Li and Huang, 1997b). Similar results are found in an independent study by Liu *et al.* (1997) using cationic lipid/DNA complexes without a cationic polymer. The optimal charge ratio varies with the cationic lipid and also depends on the choice of a helper lipid. Inclusion of cholesterol in DOTAP liposomes significantly decreased the dose of cationic lipids required for achieving a maximal level of gene expression (Li *et al.*, 1998a). It can be envisioned that some of the liposomes will stay as free liposomes at high $+/-$ charge ratios. Excess cationic liposomes might help to protect the integrity of LPD-I by preventing the release of DNA by anionic molecules in the serum. Excess cationic liposomes also prolong the residence of LPD-I in microvasculature, especially the lung capillary. Taken together, these factors provide a more efficient interaction of intact DNA with target cells, resulting in a high level of gene expression.

(b). Lipid Compositions

The *in vivo* transfection efficiency of LPD-I varies with different cationic lipids. Thus far, it is not known what kind of structure is ideal for intravenous gene delivery. Results from our studies have shown that double-chain hydrocarbon-anchored lipids are more favorable than cholesterol-anchored cationic lipids for intravenous gene delivery. The study by Gao demonstrates that monovalent cationic lipids are more efficient than multivalent cationic lipids in transfecting cells *in vivo* (Gao, personal communication). The choice of a helper lipid also affects greatly the *in vivo* activity of LPD-I. Particularly, the use of DOPE as a helper lipid significantly decreases the *in vivo* activity of LPD-I while the use of cholesterol as a helper lipid significantly improves its *in vivo* transfection (Li et al., 1998a). Similar results are reported by other groups using cationic lipid/DNA complexes without a cationic polymer (Hong et al., 1997; Liu et al., 1997). It appears that a lipid vector with a more rigid bilayer is more favorable for iv administration. As detailed later, the rigidity of a lipid vector affects its interaction with serum and eventually its transfection efficiency.

2. Interactions between LPD-I and Mouse Serum

As an approach to understand why LPDs of different lipid compositions have a dramatic difference in the level of transfection, we have investigated the interactions of LPD-I with mouse serum with an emphasis on how serum affects biophysical and biological properties of LPD-I (Li et al., 1998a; Li and Huang, unpublished data). As shown in Table 1, exposure of LPD-I to mouse serum resulted in an immediate increase in size, suggesting that LPD-I forms aggregates in the presence of serum. Interactions of LPD-I with serum also led to changes in surface charge, and all LPDs became negatively charged following exposure to serum. Interestingly, DOPE-containing LPD-I recruits much more serum proteins than the other two LPD-I formulations. The associated serum proteins are mainly albumin and some other proteins of higher molecular weight (Li et al., 1998a). Further studies have shown that prolonged incubation of LPD-I with serum leads to disintegration of the vector (Li and Huang, unpublished data). As shown in Figure 3, the turbidity of LPDs undergoes dynamic changes following exposure to mouse serum. Addition of LPDs into serum results in an immediate increase in their turbidity. The DOPE-containing LPD-I is the most turbid one. Prolonged incubation of LPDs with serum is then associated with a decrease in turbidity. The rate of decrease in

Table 1

Biophysical Characteristics of LPDs Before and After Exposure to Serum[a]

	Particle diameter (nm)		Zeta potential (mV)		
Lipid	Before	After	Before	After	μg protein/μg DNA
DOTAP	120 ± 30	389 ± 89	22.5 ± 1.8	-21.7 ± 2.3	1.32 ± 0.47[b]
DOTAP/DOPE	146 ± 36	647 ± 109	29.5 ± 1.2	-22.8 ± 2.5	2.13 ± 0.96
DOTAP/cholesterol	135 ± 42	490 ± 102	24.7 ± 2.0	-25.4 ± 2.0	1.09 ± 0.68[b]

[a]LPDs composed of different lipid compositions were mixed with serum at 1/2 ratio (v/v) and the mixtures were subjected to a sucrose-gradient ultracentrifugation. The purified complexes were then analyzed for size, zeta potential, and the concentration of associated serum proteins, respectively.

[b]$P < 0.05$ (vs DOTAP/DOPE) ($n = 5$). [From "Characterization of Cationic Lipid-Protamine-DNA (LPD) Complexes for Intravenous Gene Delivery," by S. Li, M. A. Rizzo, S. Bhattacharya, and L. Huang, 1998, *Gene Therapy*, *5*, p. 934. Copyright 1998 by Stockton Press. Reprinted with permission.]

turbidity is much faster for the DOPE formulation than for the other two formulations. The decrease with time in turbidity after exposure to serum suggests a process of vector disintegration. Serum-induced disintegration of lipidic vector is further confirmed by sucrose gradient analysis of LPDs following exposure to serum. Vector disintegration is associated with DNA release and degradation, which is correlated with a decrease in the level of transfection (Li and Huang, unpublished data).

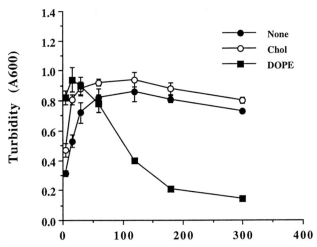

Figure 3 Dynamic changes in the turbidity of LPD-I after exposure to mouse serum. LPD-I that contain different helper lipids were mixed with mouse serum at a ratio of 1/2 (v/v) and the mixtures were incubated at 37°C with gentle shaking. The absorbance of the mixtures at 600 nm was recorded at different times with serum alone as a blank control.

Table 2

Effect of Serum on Cationic Lipidic Vectors of Different Lipid Compositions

Lipid composition	Aggrega-tion rate	Disassembly/ DNA degra-dation rate	Lung		
			Accumu-lation	Reten-tion	Trans-fection
DOTAP	+	+	++	+++	+++
DOTAP/chol	+++	++	+++	+++	++++
DOTAP/DOPE	++++	++++	++++	+	+

Table 2 summarizes how the serum affects the biophysical properties of LPD-I and their transfection efficiency. The immediate effect of serum is aggregation. Aggregation plays an important role in determining the initial accumulation of LPD-I in the microvasculatures of injected mice, particularly in the lung capillary. Subsequent interactions of LPD-I with serum lead to vectors disintegration, the consequences of which are poor retention of the vector in the lung, DNA release, and degradation. The balance between the initial aggregation and the subsequent disintegration plays an important role in determining the amount of functional DNA that interacts with target cells (endothelial cells). DOPE-containing LPD, in spite of its initial, efficient accumulation, is poorly retained in the lung due to its rapid rate of disintegration. This, together with the rapid degradation of plasmid DNA, explains why DOPE-containing formulations poorly transfect cells *in vivo*. The formulations without any helper lipid, although highly resistant to the inactivation by serum, have a relatively slow rate of initial aggregation. These formulations have an intermediate level of gene expression. Cholesterol-containing formulations have the best balance between the initial rapid aggregation and the subsequent slow disintegration, which leads to the highest level of gene expression. This study well explains why cationic lipidic vectors of different lipid compositions have a dramatic difference in their *in vivo* transfection efficiency. These results also suggest that the study of the interactions of lipidic vectors with serum may serve as a model for predicting the *in vivo* efficiency of a lipidic vector.

3. Application of LPD-I in Treating Pulmonary Tumor Metastasis

The fact that the lung is the major organ transfected by LPD-I raises the possibility of using LPD-I to deliver a therapeutic gene to metastatic tumors in the lung. This question was answered in collaboration with Dr. Wen-Hua Lee at the University of Texas at San Antonio using Rb $(+/-)$ mice. These mice develop spontaneous, primary tumor in the pituitary and later develop lung metastases with high frequency. When the Rb gene was formulated in LPD-I and iv injected to Rb $(+/-)$ mice, significant expression of Rb protein was found in tumors as shown by immunohistochemical staining (Nikitin *et al.*, unpublished data). A single injection of Rb gene formulated in LPD-I led to a significant reduction in the proliferative activity in metastatic tumors as shown in a BrdU incorporation study (Nikitin *et al.*, unpublished data). Repeated injections of Rb-containing LPD-I twice a week for 3 weeks decreased the number of lung tumor metastases from 84 to 12% (Nikitin *et al.*, unpublished data), clearly demonstrating that iv delivery of a therapeutic gene by LPD-I can be used in the treatment of pulmonary tumor metastasis.

4. Toxicity of LPD-I

It is generally thought that cationic lipidic vectors are nontoxic and weakly immunogenic. This conclusion is largely drawn from studies in which only limited amounts of cationic lipid/complexes were used. A recent study by the Genzyme group demonstrated that intranasal instillation of lipid 67/DNA complexes into BALB/c mouse lungs induced a dose-dependent pulmonary inflammation (Scheule *et al.*, 1997). Systemic administration of cationic lipid/DNA complexes also shares a similar problem of toxicity (Li *et al.*, 1998a). Injection of lipid/DNA complexes in large doses can cause the death of animals (Li *et al.*, 1998a). Interestingly, the acute toxicity is primarily related to cationic lipid/DNA complexes and not the lipid alone, although the long-term toxicity of lipid accumulation in animals remains to be determined. In an effort to understand the mechanism by which cationic lipid/DNA complexes cause toxicity, we have found that the LPD-I—not each of the three components alone—triggers a high level of cytokine production including IFN-γ and TNF-α (Li and Huang, unpublished data). Cytokines can cause not only toxicity to the animals but also gene inactivation by either directly inhibiting gene transcription or triggering apoptosis of vascular endothelial cells. Further studies have shown that cytokines are primarily induced by the unmethylated CpG sequences in the plasmid DNA. Methylation of plasmid DNA by CpG SssI methylase greatly decreased the level of cytokine induction (Li and Huang, unpublished data). These studies suggest that chemical modification of plasmid DNA may represent a new and important approach to improve cationic lipid-mediated iv gene delivery.

V. ANIONIC LIPOSOME-ENTRAPPED, POLYCATION-CONDENSED DNA (LPD-II)

While the preceding studies indicated the potential of LPD-I in the treatment of pulmonary diseases including lung tumor metastases, they also support the long-believed notion that positively charged lipidic vectors are not suitable for active targeting of a gene to distant tissues such as a subcutaneous tumor. In an effort to seek an iv-injectable, targetable vector for gene transfer, another new lipidic vector was then developed in this laboratory (Lee and Huang, 1996). It is similar to

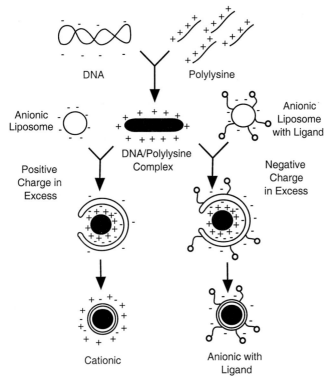

Figure 4 Possible mechanism for the formation of LPD-II particles. The targeting ligand is folate. [From "Folate-Targeted, Anionic Liposome-Entrapped Polylysine condensed DNA (LPD-II) for Tumor-Specific Gene Transfer," by R. J. Lee and L. Huang, 1996, *J. Biol. Chem.*, *271*, p. 8485. Copyright 1996 by The American Society for Biochemistry & Molecular Biology. Reprinted with permission.]

LPD-I in structure. Both contain a highly condensed core composed of PLL and DNA and a lipid shell; therefore, this new vector is named LPD-II. The protocol and the proposed mechanism for the formation of LPD-II particles are shown in Figure 4. DNA is first condensed with PLL with positive charge in moderate excess. The resulting complex is then mixed with anionic liposomes carrying a targeting ligand. In the initial study, pH-sensitive liposomes composed of DOPE/CHEMS/folate-PEG-DOPE were used. Folate was chosen as the targeting ligand because folate receptors are overexpressed in many human tumors, especially ovarian cancer (Coney *et al.*, 1991; Campbell *et al.*, 1991). Folate is a low-molecular-weight ligand with little or no immunogenicity and has a high affinity and specificity for its receptor. Folate-mediated targeting of anticancer agents have been well demonstrated *in vitro* and in animal models (Lee and Low, 1995; Wang *et al.*, 1995).

Depending on the lipid to DNA ratio, either cationic (at low lipid to DNA ratios) or anionic (at high lipid to DNA ratios) LPD-II particles could be generated.

No purification was needed. LPD-II transfection of KB cells, a cell line that over-expresses the folate receptor, was affected by the overall net charge and the lipid composition. Transfection and uptake of cationic LPD-II were independent of the folate receptor and did not require a pH-sensitive lipid composition. Meanwhile, transfection and uptake of anionic LPD-II were folate receptor dependent, partially blockable by 1 mM free folate, and required a pH-sensitive lipid composition. Both cationic and anionic LPD-II were 20−30 times higher in transfecting cells *in vitro* than DC-chol liposome/DNA complexes and were much less toxic (Lee and Huang, 1996).

Like LPD-I, the formation of LPD-II particles is likely due to fusion of liposomes and lipid rearrangement after mixing the cationic PLL/DNA complex with anionic liposomes. The presence of a targeting ligand did not appear to hinder the formation of LPD-II particles. Recently, another LPD-II was developed by mixing a cationic PLL/DNA complex with anionic liposomes containing a high-molecular-weight ligand, transferrin. Transferrin-targeted LPD-II selectively deliv-ered a reporter gene to myoblasts and myotubes *in vitro* (Feero *et al.*, 1997). LPD-II can also be employed to deliver antisense oligodeoxynucleotides (ODN) to target cells (Li and Huang, 1998b). However, the formation of ODN-containing LPD-II requires the use of greater amounts of PLL as compared with DNA-containing LPD-II. This might be related to the linear structure and low molecular weight of ODN. Delivery of an ODN against an epidermal growth factor receptor (EGFR) to KB cells by folate-targeted LPD-II resulted in down regulation of EGFR and growth inhibition in treated cells in an *in vitro* study (Li and Huang, 1998b).

The advantages of LPD-II are that, besides being highly compact, LPD-II preparation does not require purification and is a single-vial formulation. Com-pared with traditional anionic and neutral liposomal vectors, DNA is highly con-densed and is quantitatively encapsulated without the use of excess amounts of lipids. Finally, since anionic liposomes are relatively compatible with biological fluids, LPD-II potentially can be used for targeted gene delivery via systemic ad-ministration. The existing problem with LPD-II is its sensitivity to serum, as incor-poration of serum components into LPD-II may stabilize the bilayer resulting in the loss of pH-dependent fusogenic activity. This problem might be resolved by the use of a pH-sensitive but serum−insensitive fusogenic peptide. The possibility is cur-rently being examined in this laboratory.

VI. RECONSTITUTED CHYLOMICRON REMNANTS FOR *IN VIVO* GENE DELIVERY

Recently, it was shown that a hydrophobic lipid/DNA complex can be pre-pared in the absence of preformed liposomes (Reimer *et al.*, 1995; Wong *et al.*, 1996). This hydrophobic complex can be isolated in an organic phase and used as an intermediate in the preparation of well-defined particles. A typical process for preparation of a hydrophobic lipid/DNA complex is as follows: Cationic lipids and plasmid DNA are solubilized in a Bligh and Dyer monophase consisting of chloroform/methanol/water (1/2.1/1). Subsequently, the sample is partitioned into an aqueous phase and an organic phase by further addition of chloroform and water. It is believed that binding of DNA to cationic lipids results in charge neutralization and extraction of the complex into the organic phase. The lipid/DNA complex might serve as a unique intermediate in the preparation of lipidic gene delivery systems. For example, the complex can be used in the preparation of oil/water emulsions. The complex can also be dissolved in alternative solvents for the preparation of membrane structures via a reverse-phase evaporation technique. There are several advantages of this system over liposome/DNA complexes for gene delivery. First, interaction of cationic lipids with DNA is a better-defined process and the complex might be more homogeneous in structure. Second, formation of particles between the lipid/DNA complex and other lipids or among the complexes themselves is driven primarily by hydrophobic interactions and, therefore, aggregation caused by electrostatic interactions is minimized. Finally, the surface properties of particles can be easily manipulated. Neutral or anionic particles can be prepared by incorporating appropriate lipids into the complex. A targeting ligand can also be used for targeted *in vivo* gene delivery. These favorable characteristics of this novel system are not shared by cationic liposome/DNA complexes.

The reconstituted chylomicron remnants (RCR) recently developed in our laboratory (Hara *et al.*, 1997) are an example of the use of a hydrophobic lipid/DNA complex for the preparation of well-defined particles. 3β-[N',N',N'-trimethylaminoethane)-cholesterol iodide (TC-chol), a quaternary ammonium derivative of DC-chol, was employed to form a hydrophobic complex with DNA. DC-Chol was not chosen because it contains a tertiary amino head group that is only partially ionized at neutral pH. The hydrophobic TC-chol/DNA complex extracted from the organic phase was then incorporated in RCR by emulsifying with appropriate amounts of triglyceride, 1-α-phosphatidylcholine (PC), lysophosphatidylcholine (lyso PC), cholesterol (chol), and cholesteryl oleate in a 70/22.7/2.3/3.0/2.0 weight ratio. After extrusion, the size of RCR was approximately 100 nm with a DNA incorporation efficiency of about 60% or greater. The RCR are stable and give a high level of gene expression upon injection from the portal vein. When a luciferase reporter gene was used, gene expression was found in all major organs including lung, heart, liver, spleen, and kidney with the highest level of gene expression found in the liver. At a dose of 100 μg DNA per mouse, approximately 10 ng luciferase protein per mg extracted tissue protein could be detected in the liver. The level of gene expression by RCR was about 100-fold higher than the level when naked plasmid DNA was used (Hara *et al.*, 1997). Gene expression is transient and lasts for about 1 week. However, long-term expression can be achieved by repeated injections and/or by using a plasmid with extended lifetime in transfected cells (Hara and Huang, unpublished data). The existing problem with this formulation is the low efficiency when injected intravenously, only limited expression is found in the liver. This might be due to the lack of a targeting ligand. Addition of apo E or other ligands may improve the efficiency of iv delivery of gene to the liver by an efficient mechanism of receptor-mediated cellular uptake. This possibility is currently being examined in this laboratory.

VII. CONCLUSIONS

Recent efforts in this laboratory have led to the development of several novel lipidic formulations that are more efficient than the first-generation cationic lipidic vectors in transfecting cells *in vitro* and *in vivo*. Structurally, these formulations are virus-like particles, each containing a condensed genome as the core and a lipidic shell as the envelope. Further improvements in these formulations rely on better understanding of cellular and *in vivo* barriers for lipofection and deeper appreciation of how viruses escape these barriers. The problems of immunogenicity and endotoxin contamination that are associated with plasmid DNA will eventually be resolved by the use of synthetic genes, that is, genes that are produced by totally chemical synthesis. Finally, advances in the understanding of regulation of gene expression will help to achieve the ultimate goal of gene therapy.

ACKNOWLEDGMENTS

The original work in this laboratory was supported by NIH grants CA 59327, DK 44935, CA 64654, CA 71731, and a contract from Targeted Genetics Corporation.

REFERENCES

Ali, M., Lemoine, N. R., and Ring, C. J. (1994). The use of DNA viruses as vectors for gene therapy. *Gene Ther.* **1,** 367–384.

Anderson, W. F. (1992). Human gene therapy. *Science* **256,** 808–813.

Campbell, I. G., Jones, T. A., Foulkes, W. D., and Trowsdale, J. (1991). Folate-binding protein is a marker for ovarian cancer. *Cancer Res.* **51,** 5329–5338.

Caplen, N. J., Alton, E. W., Middleton, P. G., Dorin, J. R., Stevenson, B. J., Gao, X., Durham, S. R., Jeffery, P. K., Hodson, M. E., Coutelle, C., Huang, L., Porteous, D. J., Williamson, R., and Geddes, D. M. (1995). Liposome-mediated CFTR gene transfer to the nasal epithelium of patients with cystic fibrosis. *Nature Med.* **1,** 39–46.

Coney, L. R., Tomassetti, A., Carayannopoulos, L., Frasca, V., Kamen, B. A., Colnaghi, M. I., and Zurawski, V. R., Jr. (1991). Cloning of a tumor-associated antigen: MOv18 and MOv19 antibodies recognize a folate-binding protein. *Cancer Res.* **51,** 6125–6132.

Coonrod, A., Li, F. Q., and Horwitz, M. (1997). On the mechanism of DNA transfection: Efficient gene transfer without viruses. *Gene Ther.* **4,** 1313–1321.

Davis, L. I. (1995). The nuclear pore complex. *Annu. Rev. Biochem.* **102,** 1183–1190.

Farhood, H., Serbina, N., and Huang, L. (1995). The role of dioleoyl phosphatidylethanolamine in cationic liposome mediated gene transfer. *Biochim. Biophys. Acta* **1235,** 289–295.

Feero, W. G., Li, S., Rosenblatt, J. D., Sirianni, N., Morgan, J. E., Partridge, T. A., Huang, L., and Hoffman, E. P. (1997). Selection and use of ligands for receptor-mediated gene delivery to myogenic cells. *Gene Ther.* **4,** 664–674.

Felgner, P. L., and Ringold, G. M. (1989). Cationic liposome-mediated transfection. *Nature* **337,** 387–388.

Gao, X., and Huang, L. (1991). A novel cationic liposome reagent for efficient transfection of mammalian cells. *Biochem. Biophys. Res. Commun.* **179,** 280–285.

Gao, X., and Huang, L. (1993). Cytoplasmic expression of a reporter gene by co-delivery of T7 RNA polymerase and T7 promoter sequence with cationic liposomes. *Nucleic Acids Res.* **21,** 2867–2872.

Gao, X., Jaffurs, D., Robbins, P. D., and Huang, L. (1994). A sustained, cytoplasmic transgene expression system delivered by cationic liposomes. *Biochem. Biophys. Res. Commun.* **200,** 1201–1206.

Gao X., and Huang, L. (1995). Cationic liposome-mediated gene transfer. *Gene Ther.* **2,** 710–722.

Gao, X., and Huang, L. (1996). Potentiation of cationic liposome mediated gene delivery by polycations. *Biochemistry* **35,** 1027–1036.

Gill, D. R., Southern, K. W., Mofford, K. A., Seddon, T., Huang, L., Sorgi, F., Thomson, A., MacVinish, L. J., Ratcliff, R., Bilton, D., Lane, D.J., Littlewood, J. M., Webb, A. K., Middleton, P. G., Colledge, W. H., Cuthbert, A. W., Evans, M. J., Higgins, C. F., and Hyde, S. C. (1997). A placebo-controlled study of liposome-mediated gene transfer to the nasal epithelium of patients with cystic fibrosis. *Gene Ther.* **4,** 199–209.

Hara, T., Tan, Y., and Huang, L. (1997). *In vivo* gene delivery to the liver using reconstituted chylomicron remnants as a novel non-viral vector. *Proc. Natl. Acad. Sci. USA* **94,** 14547–14552.

Hartikka, J., Sawdey, M., Cornefert-Jensen, F., Margalith, M., Barnhart, K., Nolasco, M., Vahlsing, H. L., Meek, J., Marquet, M., Hobart, P., Norman, J., and Manthorpe, M. (1996). *Human Gene Ther.* **7,** 1205–1217.

Herz, J., and Gerard, R. D. (1993). Adenovirus-mediated transfer of low density lipoprotein receptor gene acutely accelerates cholesterol clearance in normal mice. *Proc. Natl. Acad. Sci. USA* **90,** 2812–2816.

Hicks, G. R., and Raikhel, N. V. (1995). Protein import into the nucleus: An integrated view. *Annu. Rev. Cell Dev. Biol.* **11,** 155–188.

Hong, K., Zheng, W., Baker, A., and Papahadjopoulos, D. (1997). Stabilization of cationic liposome-plasmid DNA complexes by polyamines and poly(ethyleneglycol)-phospholipid conjugates for efficient *in vivo* gene delivery. *FEBS Lett.* **400,** 233–237.

Huang, L., and Li, S. (1997). Liposomal gene delivery: A complex package. *Nature Biotechnol.* **15,** 620–621.

Hui, K. M., Ang, P. T., Huang, L., and Tay, S. K. (1997). Phase I study of immunotherapy of cutaneous metastases of human carcinoma using allogeneic and xenogeneic MHC DNA-liposome complexes. *Gene Ther.* **4,** 783–790.

Kichler, A., Mechtler, K., Behr, J.P., and Wagner, E. (1997). Influence of membrane-active peptides on lipospermine/DNA complex mediated gene transfer. *Bioconjugate Chem.* **8,** 213–221.

Ledley, F. D. (1995). Nonviral gene therapy: The promise of genes as pharmaceutical products. *Human Gene Ther.* **6,** 1129–1144.

Lee, R. J., and Low, P. S. (1995). Folate-mediated tumor cell targeting of liposome-entrapped doxorubicin *in vitro*. *Biochim. Biophys. Acta* **1233,** 134–144.

Lee, R. J., and Huang, L. (1996). Folate-targeted, anionic liposome-entrapped polylysine-condensed DNA (LPDII) for tumor cell-specific gene transfer. *J. Biol. Chem.* **271,** 8481–8487.

Li, S., Brisson, M., He, Y., and Huang, L. (1997a). Delivery of a PCR amplified DNA fragment into cells: A model for using synthetic genes for gene therapy. *Gene Ther.* **4**, 449–454.

Li, S., and Huang, L. (1997b). *In vivo* gene transfer via intravenous administration of cationic lipid–protamine–DNA (LPD) complexes. *Gene Ther.* **4**, 891–900.

Li, S., Rizzo, M. A., Bhattacharya, S., and Huang, L. (1998a). Characterization of cationic lipid–protamine–DNA (LPD) complexes for intravenous gene delivery. *Gene Ther.* **5**, 930–937.

Li, S., and Huang, L. (1998b). Targeted delivery of antisense oligodeoxynucleotides formulated in a novel lipidic vector. *J. Liposome Res.* **8**, 239–250.

Liu, F., Qi, H., Huang, L., and Liu, D. (1997). Factors controlling efficiency of cationic lipid-mediated transfection *in vivo* via intravenous administration. *Gene Ther.* **4**, 517–523.

Liu, Y., Mounkes, L. C., Liggitt, H. D., Brown, C. S., Solodin, I., Heath, T. D., and Debs, R. J. (1997). Factors influencing the efficiency of cationic liposome-mediated intravenous gene delivery. *Nature Biotechnol.* **15**, 167–173.

Matsui, H., Johnson, L. G., Randell, S. H., and Boucher, R. C. (1997). Loss of binding and entry of liposome–DNA complexes decreases transfection efficiency in differentiated airway epithelial cells. *J. Biol. Chem.* **272**, 1117–1126.

McLachlan, G., Ho, L. P., Davidson-Smith, H., Samways, J., Davidson, H., Stevenson, B. J., Carothers, A. D., Alton, E. W., Middleton, P. G., Smith, S. N., Kallmeyer, G., Michaelis, U., Seeber, S., Naujoks, K., Greening, A. P., Innes, J. A., Dorin, J. R., and Porteous, D. J. (1996). *Gene Ther.* **3**, 1113–1123.

Miller, A. D. (1992). Human gene therapy comes of age. *Science* **357**, 455–460.

Nabel, G. J., Nabel, E., Yang, Z. Y., Fox, B., Plautz, G., Gao, X., Huang, L., Shu, S., Gordon, D., and Chang, A. E. (1993). Direct gene transfer with DNA–liposome complexes in melanoma: Expression, biologic activity, and lack of toxicity in humans. *Proc. Natl. Acad. Sci. USA* **90**, 11307–11311.

Reimer, D. L., Zhang, Y., Kong, S., Wheeler, J. J., Graham, R. W., Bally, M. B. (1995). Formation of novel hydrophobic complexes between cationic lipids and plasmid DNA. *Biochemistry* **34**, 12877–12883.

Remy, J. S., Kichler, A., Mordvinov, V., Schuber, F., and Behr, J. P. (1995). Targeted gene transfer into hepatoma cells with lipopolyamine-condensed DNA particles presenting galactose ligands: A stage toward artificial viruses. *Proc. Natl. Acad. Sci. USA* **92**, 1744–1748.

Scheule, R. K., St George, J. A., Bagley, R. G., Marshall, J., Kaplan, J. M., Akita, G. Y., Wang, K. X., Lee, E. R., Harris, D. J., Jiang, C., Yew, N. S., Smith, A. E., and Cheng, S. H. (1997). Basis of pulmonary toxicity associated with cationic lipid-mediated gene transfer to the mammalian lung. *Human Gene Ther.* **8**, 689–707.

Sebestyen, M. G., Ludtke, J. J., Bassik, M. C., Zhang, G., Budker, V., Lukhtanov, E. A., Hagstrom, J. E., and Wolff, J. A. (1998). DNA vector chemistry: The covalent attachment of signal peptides to plasmid DNA. *Nature Biotechnol.* **16**, 80–85.

Simoes, S., Slepushkin, V., Gaspar, R., Delima, M.C.P., and Duzgunes, N. (1998). Gene delivery by negatively charged ternary complexes of DNA, cationic liposomes and transferrin or fusigenic peptides. *Gene Ther.* **5**, 955–964.

Simon, R. H., Engelhardt, J. F., Yang, Y., Zepeda, M., Weber-Pendleton, S., Grossman, M., and Wilson, J. M. (1993). Adenovirus-mediated transfer of the CFTR gene to lung of nonhuman primates: Toxicity study. *Human Gene Ther.* **4**, 771–780.

Sorgi, F. L., Bhattacharya, S., and Huang, L. (1997). Protamine sulfate enhances lipid mediated gene transfer. *Gene Ther.* **4**, 961–968.

Sternberg, B., Sorgi, F. L., and Huang, L. (1994). New structures in complex formation between DNA and cationic liposomes visualized by freeze–fracture electron microscopy. *FEBS Lett.* **356**, 361–366.

Torchilin, V. P., Zhou, F., and Huang, L. (1993). pH-sensitive liposomes. *J. Liposome Res.* **3**, 201–255.

Xu, Y., and Szoka, F. C., Jr. (1996). Mechanism of DNA release from cationic liposome/DNA complexes used in cell transfection. *Biochemistry* **35**, 5616–5623.

Wang, S., Lee, R. J., Cauchon, G., Gorenstein, D. G., and Low, P. S. (1995). Delivery of antisense oligodeoxyribonucleotides against the human epidermal growth factor receptor into cultured KB cells with liposomes conjugated to folate via polyethylene glycol. *Proc. Natl. Acad. Sci. USA* **92**, 3318–3322.

Wilson, R. W., and Bloomfield, V. A. (1979). Counterion-induced condesation of deoxyribonucleic acid: A light-scattering study. *Biochemistry* **18**, 2192–2196.

Wolff, J. A., Malone, R. W., Williams, P., Chong, W., Acsadi, G., Jani, A., and Felgner, P. L. (1990). Direct gene transfer into mouse muscle *in vivo. Science* **247**, 1465–1468.

Wong, F. M., Reimer, D. L., and Bally, M. B. (1996). Cationic lipid binding to DNA: Characterization of complex formation. *Biochemistry* **35**, 5756–5763.

Wrobel, I., and Collins, D. (1995). Fusion of cationic liposomes with mammalian cells occurs after endocytosis. *Biochim. Biophys. Acta* **1235**, 296–304.

Yang, J. P., and Huang, L. (1996). Direct gene transfer to mouse melanoma by intratumor injection of free DNA. *Gene Ther.* **3**, 542–548.

Yang, J. P., and Huang L. (1997). Overcoming the inhibitory effect of serum on lipofection by increasing the charge ratio of cationic liposome to DNA. *Gene Ther.* **4**, 950–960.

Zabner, J., Fasbender, A. J., Moninger, T., Poellinger, K. A., and Welsh, M. J. (1995). Cellular and molecular barriers to gene transfer by a cationic lipid. *J. Biol. Chem.* **270**, 18997–19007.

Zhou, X., and Huang, L. (1994). DNA transfection mediated by cationic liposomes containing lipo-polylysine: Characterization and mechanism of action. *Biochim. Biophys. Acta* **1189**, 195–203.

PART IV

Animal Models
and Clinical Trials

CHAPTER 14

Mechanisms of Cationic Liposome-Mediated Transfection of the Lung Endothelium

Dexi Liu, Joseph E. Knapp, and Young K. Song
Department of Pharmaceutical Sciences, School of Pharmacy,
University of Pittsburgh, Pittsburgh, Pennsylvania

Nonviral Vectors for Gene Therapy

Cationic liposomes have been successfully used to transfect lung endothelium by systemic administration. A variety of physicochemical parameters have been found to affect the level of transgene expression in the lung, including cationic lipid structure, liposome composition, liposome size, cationic lipid to DNA ratio in DNA/liposome complexes, and administered dose. Different strategies have been used to elucidate the mechanisms underlying cationic liposome-mediated transfection of the lung endothelium. Studies using these new strategies reveal that the transfection efficiency of cationic liposomes to the lung endothelium is mainly determined by three factors: their ability to bring the DNA molecules to the maximal number of endothelial cells in the lung, their ability to retain the DNA in the lung endothelium for an extended period of time, and their ability to release the DNA from the complexes. Different properties of DNA/liposome complexes exert their effects on the transfection efficiency of the lung endothelium by affecting one or more of these three factors *in vivo*. The interaction between the injected cationic liposomes (free or complexed with DNA) and blood components may also contribute significantly to how these physicochemical parameters of liposomes determine their ultimate efficiency in transfecting the lung endothelium.

I. INTRODUCTION

Cationic liposome technology has become well established for introducing DNA into cells and these liposomes have been considered to be among the most promising carriers for gene therapy. Many *in vitro* studies have shown that a variety of cell types can be transfected by cationic lipid-based transfection reagents (Behr *et al.*, 1989; Felgner *et al.*, 1987; Felgner *et al.*, 1994; Gao and Huang, 1991; Hawley-Nelson *et al.*, 1993; Lee *et al.*, 1996; Leventis and Silvius, 1990; Rose *et al.*, 1991; Solodin *et al.*, 1994, Wang *et al.*, 1998). Systemically administered complexes of cationic liposomes and plasmid DNA have also been shown to transfect cells *in vivo* (Hong *et al.*, 1997; Li *et al.*, 1997; Liu *et al.*, 1997; Liu *et al.*, 1997; Song *et al.*, 1997; Templeton *et al.*, 1997; Thierry *et al.*, 1995; Zhu *et al.*, 1993). However, in spite of the effectiveness of cationic liposomes in transfecting cells *in vitro*, their transfection efficiency *in vivo* is still fairly low compared to that of viral vectors. Thus, there has been increasing attention focused on improving the current liposome systems. Significant improvements in this area will be based on a firm understanding of the mechanisms by which the *in vivo* transfection efficiency of cationic liposomes is regulated.

A common phenomenon observed in systemic transfection by cationic liposomes is that the lung appears to be the most transfectable organ (Hong *et al.*, 1997; Li *et al.*, 1997; Liu *et al.*, 1997; Liu *et al.*, 1997; Templeton *et al.*, 1997; Thierry *et al.*, 1995; Zhu *et al.*, 1993). In general, the level of transgene expression obtained in the lung is two to four orders of magnitude higher than that obtained in other organs such as the heart, spleen, kidney, and liver. The cells transfected in the lung are mainly the endothelial cells (Li *et al.*, 1997; McLean *et al.*, 1997; Song *et al.*, 1997). The transfection efficiency of DNA/liposome complexes has been found to depend on the structure of the specific cationic lipid used, the ratio of cationic lipid to DNA, liposome composition, the particle diameter, and the injected dose (Song *et al.*, 1997). It is believed that these physicochemical parameters exert their effects on transfection efficiency by affecting the interactions between liposome/DNA complexes and the lung endothelium. In this chapter, we will attempt to explain how these physicochemical parameters affect the ultimate transfection activity of DNA/liposome complexes. The focus of our discussion is on transfection of the lung endothelium by systemically administered DNA and DNA/liposome complexes (systemic transfection). Efforts have also been made to provide some thoughts on means to further enhance the level of gene expression in lung endothelium through systemic gene delivery.

II. ANATOMIC AND PHYSIOLOGICAL CHARACTERISTICS OF THE LUNG

A. LUNG ANATOMY

With some minor exceptions, there is no difference between different mammalian species, other than size, in the functional anatomy of the lungs. Their primary function is to bring air and blood into sufficiently close proximity to facilitate gas exchange. The alveoli are the functional units in which this exchange takes place. Alveoli are sac-like structures arising from the respiratory bronchioles and in large numbers from the alveolar ducts. Surrounding each alveolus is an extensive network of capillaries, which provide a rich blood supply. The air–blood interface occurs via the respiratory membrane, a delicate structure comprising three layers. The air side layer consists of the alveolar epithelium while the center layer is formed from the fused basement membranes of the alveoli and the capillaries. The final (blood side) layer is the capillary endothelium. The minimum thickness (0.2 μm) of the respiratory membrane is nearly identical in mammalian species (Biven *et al.*, 1979) and provides an extensive surface area for gas exchange between the alveoli and the associated pulmonary capillary bed.

Since the individual pulmonary capillaries are closely adjacent to each other to form an almost continuous layer, the lung endothelium can be considered functionally as an unicellular layer made up of the walls of the pulmonary capillaries. The internal capillary diameter is small (4–9 μm in the human) and is just large enough to allow for the single-file passage of erythrocytes and other cells (Guyton, 1991). Adjacent endothelial cells are in close contact but are separated from each other by a thin space called the intercellular cleft. The cleft is bridged at intervals by a ridge of protein that connects adjacent cells together while providing gaps for the passage of water and small solutes. The width of the cleft (6–7 nm in the human) limits the passage of proteins and other molecules the size of albumin or larger (Guyton, 1991). Except in certain pathologic conditions, pulmonary capillaries have no pores or fenestrations.

B. HEMODYNAMICS AND PHYSIOLOGY OF THE LUNG

The first capillary beds encountered by substances injected into the peripheral circulation are found in the lungs. The inner diameter of the pulmonary capillaries is such that large complexes can be readily trapped during the first pass through the pulmonary circuit. Blood supplied to the lungs by the pulmonary arteries is directed via the lobar arteries into pulmonary capillary beds and is carried away via the pulmonary venules and veins to the left atrium of the heart. Due to the distensibility of the pulmonary blood vessels, the blood flow to the lungs is essentially equal to the cardiac output (Guyton, 1991). It has been estimated that, at normal cardiac output, blood passes through the pulmonary capillaries in about 0.8 seconds (Guyton, 1991). Under conditions of increased cardiac output this can decrease to as low as 0.3 seconds. Rather than being continuous, blood flow through the pulmonary capillaries is discontinuous. On inspiration, as the alveoli fill and swell, flow slows and stops. On expiration, flow resumes. Capillary blood flow is also dependent on alveolar oxygen levels. In contrast to other tissues, as the level of oxygen in a particular alveolar area falls, blood flow to the area is decreased and is shunted to areas more highly oxygenated. Both of these mechanisms function to ensure that maximal oxygenation of the blood occurs during its residence in the lung. These same mechanisms may also play a role in the transport of substances from the blood into the endothelial tissues.

III. PHYSICOCHEMICAL FACTORS AFFECTING THE SYSTEMIC TRANSFECTION EFFICIENCY OF DNA / LIPOSOME COMPLEXES

The standard systemic transfection procedure using cationic liposomes as a vehicle involves the preparation of liposomes and the DNA/liposome complexes, administration into animals, and analysis of transgene expression. The following is a summary of the effects of some physicochemical parameters of liposomes and DNA/liposome complexes on the efficiency of cationic liposome-mediated lung endothelium transfection.

A. CATIONIC LIPID STRUCTURE

The most commonly used cationic lipids can be divided into two major groups based on the structure of their hydrophobic moiety. In one group this moiety is long hydrocarbon chain based and in the other it is cholesterol based. Among the many cationic lipids developed in the past, the most commonly used of the long hydrocarbon chain-based type includes 2,3-dioleyloxyl-1-propyltri-methylammonium chloride (DOTMA) (Felgner et al., 1987) and 1,2-dioleoyloxy-3-trimethylammonium propane (DOTAP) (Leventis and Silvius, 1990). The most commonly used cationic lipid with the cholesterol moiety as the hydrophobic anchor is $3\beta[N-(N',N'-$dimethylaminoethane)carbamoyl] cholesterol (DC-chol) (Gao and Huang, 1991). Although they may vary in their transfection efficiency, both types of cationic lipids are active in transfecting many type of cells *in vitro* (Gao and Huang, 1995). However, *in vivo*, liposomes prepared with these two groups of cationic lipids have shown very different activities in systemic transfection of the lung endothelium. Compared to many published studies that reported the systemic transfection efficiency of long hydrocarbon chain-based cationic lipids (Hong et al., 1997; Li et al., 1997; Liu et al., 1997; Liu et al., 1997; Templeton et al., 1997; Thierry et al., 1995; Zhu et al., 1993), reports of the success of cholesterol-based cationic lipids in this area are few. In our laboratory, using cholesterol-based cationic liposomes and the luciferase gene as a reporter, we found that, under optimal conditions, the level of gene expression in the lungs of animals transfected by DNA complexed with either DC-chol or trimethylaminoethane carbamoyl cholesterol (TC-chol) liposomes is much lower than that of complexes made with either DOTMA or DOTAP liposomes (Figure 1). These data further suggest that cholesterol-based cationic liposomes are less active in the systemic transfection of the lung endothelium. As illustrated in Figure 1, it is also evident that the level of transgene expression in the lungs of animals transfected with DNA/DOTMA

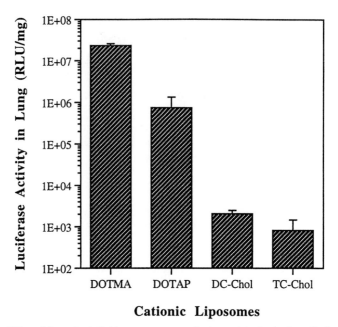

Cationic Liposomes

Figure 1   Effect of the cationic lipid structure on transfection activity in the lung. Each mouse was injected via tail vein with DNA/liposome complexes containing 25 μg of pCMV-luc plasmid DNA and 900 nmol of cationic liposomes in 200 μl PBS. Luciferase activity was assayed 8 h post-injection. Error bar represents SEM from three mice.

liposome complexes is approximately 10-fold higher than that of animals transfected with DNA/liposome complexes prepared with DOTAP liposomes, suggesting that the presence of the ether linkage between the cationic head group and the alkyl chain in DOTMA is functionally better than the ester linkage in DOTAP.

B. Cationic Lipid to DNA Ratio

The ratio of cationic lipid to DNA in DNA/liposome complexes has been shown to be one of the most important factors affecting transfection efficiency. For *in vitro* transfection the optimal ratio found is in the range of 3.6 to 9 (cationic lipid/DNA, nmol/μg) depending on the type of cationic liposomes and cell types used (unpublished data). However, for systemic transfection in mice, a higher cationic lipid to DNA ratio appears to be required for an efficient transfection of the lung endothelium. As shown in Figure 2, the level of transgene expression in the lung increases with increasing cationic lipid to DNA ratio in the complexes. The optimal cationic lipid to DNA ratio for transfection of the lung is approximately 36 to 1 or greater under the experimental conditions employed.

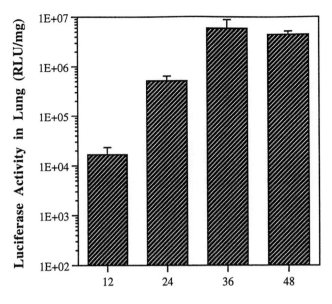

Figure 2 Effect of the cationic lipid to DNA ratio on transfection activity in the lung. Each mouse was injected via tail vein with DNA/liposome complexes containing 25 μg of pCMV-luc plasmid DNA and different amounts of DOTMA liposomes in 200 μl PBS. Luciferase activity was assayed 8 h post-injection. Error bar represents SEM from three mice.

C. HELPER LIPIDS AND SOLUTIONS FOR COMPLEX PREPARATION

Inclusion of a neutral lipid such as dioleoylphosphatidylethanolamine (DOPE) into cationic liposomes has been a common practice in cationic liposome-mediated transfection (Behr *et al.*, 1989; Felgner *et al.*, 1987; Felgner *et al.*, 1994; Gao and Huang, 1991; Hawley-Nelson *et al.*, 1993; Lee *et al.*, 1996; Leventis and Silvius, 1990; Rose *et al.*, 1991; Solodin *et al.*, 1995; Wang *et al.*, 1998). In fact, most commercially available cationic liposomes contain DOPE. It is generally believed that, once inside a cell, DOPE in DNA/liposome complexes facilitates the transfer of DNA across the endosomal membrane and thereby, enhances transfection activity (Farhood *et al.*, 1995; Legendre and Szoka, 1992; Wrobel and Collins, 1995). In addition, several studies have also shown that inclusion of cholesterol into cationic liposomes can also enhance the transfection activity at lower cationic lipid to DNA ratios (Bennett *et al.*, 1995; Hong *et al.*, 1997; Liu *et al.*, 1997; Templeton *et al.*, 1997). In addition, different aqueous solutions have been used in the preparation of DNA/liposome complexes. While some of the reported studies use 5% dextrose (Hong *et al.*, 1997; Li *et al.*, 1997; Liu *et al.*, 1997; Templeton *et al.*, 1997; Zhu *et al.*, 1993), we have used phosphate buffered saline (PBS) (pH 7.4) for the preparation of DNA/liposome complexes (Liu *et al.*, 1997; Song *et al.*, 1997). The data presented in Figure 3 show the level of transgene expression in the lung of animals injected with DNA/liposome complexes, with or without the helper lipid and prepared in different solutions. For DOTMA-based liposomes, the highest level of gene expression was observed with complexes of DNA and DOTMA liposomes prepared in PBS. Inclusion of 50 mole% of DOPE or cholesterol into the liposomes decreased the level of gene expression by a factor of 10 in terms of the luciferase activity per mg of protein extracted from the lung. A similar decrease in efficiency was also observed for liposomes containing DOTMA alone when 5% glucose was used for preparation of liposomes and the DNA/liposome complexes. In terms of DOTAP liposomes, inclusion of DOPE appears to decrease the overall level of transgene expression in the lung compared to that of animals transfected with liposomes prepared with DOTAP alone or with 50% cholesterol. These results suggest that, despite the fact that DOPE and cholesterol have previously been shown to be effective in enhancing the transfection activity of liposomes *in vitro*, inclusion of DOPE or cholesterol into the cationic liposomes may not enhance the systemic

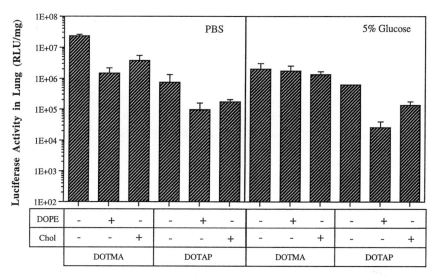

Liposome Composition

Figure 3   Effect of the neutral lipid and solution for complex preparation on the transfection activity of cationic liposomes in the lung. Each mouse was injected via tail vein with DNA/liposome complexes containing 25 μg of pCMV-luc plasmid DNA. For the solution effect, the same solution was used in preparation of liposome, DNA, and DNA/liposome complexes. The cationic lipid to neutral lipid ratio, if present, in liposomes was 1/1 (molar ratio). The cationic lipid to DNA ratio was 36/1 (nmol/μg). Luciferase activity was determined 8 h post-injection. Error bar represents SEM from three mice.

transfection efficiency under conditions where liposomes prepared with cationic lipid alone exhibit an optimal activity.

D. Diameter of Lipid Particles

The particle diameter of liposomes and DNA/liposome complexes is another parameter that was examined for its effect on the transfection activity of cationic liposomes. It is apparent from the data presented in Figure 4 that the diameter of DNA/liposome complexes is directly related to the liposome diameter. For example, the average diameter of DNA/liposome complexes was averaged approximately 400 to 500 nm when small-sized liposomes (87 nm) were used, compared to a complex diameter of more than 1 μm when the average liposome diameter was

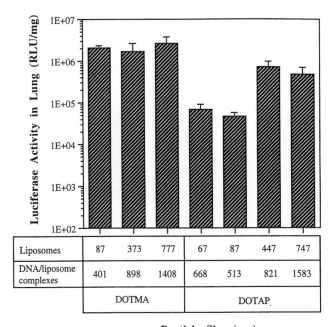

Liposomes	87	373	777	67	87	447	747
DNA/liposome complexes	401	898	1408	668	513	821	1583
	DOTMA			DOTAP			

Particle Size (nm)

Figure 4 Size effect on liposome-mediated transfection in the lung. Each mouse was injected via tail vein with DNA/liposome complexes containing 25 μg of pCMV-luc plasmid DNA and 900 nmol of cationic liposomes in 200 μl PBS. DOTMA or DOTAP liposomes with different diameter were prepared by extrusion through the polycarbonate membranes with defined pore size. The size of liposomes and DNA/liposome complexes represents an average size of particles measured by laser light scattering. Luciferase activity was assayed 8 h post-injection. Error bar represents SEM from three mice. [Data taken from Song *et al.* (1997).]

around 750 nm. This pattern did not seem to depend on the cationic lipid structure since similar increases in the particle size of DNA/liposome complexes were obtained for both DOTMA and DOTAP liposomes. Interestingly, the transfection activity of DOTAP liposomes appears to depend on liposome size. An approximately 10-fold increase of luciferase activity in the lung was seen when the size of DOTAP liposomes increased from below 100 to 450 nm or greater. As such size dependence was unique to DOTAP liposomes and was not observed with DOTMA liposomes, these results may indicate that the ester linkage bonds in DOTAP may be hydrolyzed faster in small liposomes.

E. Administered Dose

The dose–response curve for DOTMA liposomes complexed with the pCMV-Luc plasmid DNA is shown in Figure 5. These data clearly show that the level of luciferase activity in the lung increases with increasing the injected dose. An

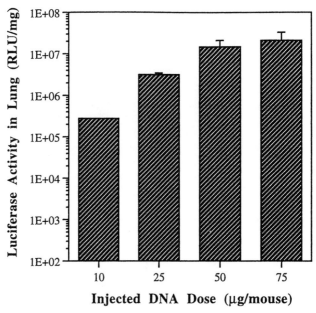

Figure 5 Dose-dependent transfection activity in the lung. Each mouse was injected via tail vein with DNA/liposome complexes containing various amount of pCMV-luc plasmid DNA with DOTMA liposomes in 200 μl PBS. The cationic lipid to DNA ratio was 36/1 (nmol/μg). Luciferase activity was assayed 8 h post-injection. Error bar represents SEM from three mice. [Data taken from Song *et al.* (1997).]

approximately 10- to 100-fold increase in luciferase activity was seen in the lung when the injected DNA dose was increased from 10 to 75 μg/mouse. Under these conditions, the level of luciferase activity appeared to be saturated at 50 μg DNA/ mouse.

IV. MECHANISMS OF CATIONIC LIPOSOME-MEDIATED GENE TRANSFER INTO THE LUNG ENDOTHELIUM

Progress in elucidating the mechanisms involving in cationic liposome-mediated transfection into lung endothelium *in vivo* has been slow in the past. This could largely be due to the lack of appropriate methods allowing one to dissect the complicated transfection process involving multiple steps and many factors. Despite such difficulties, many intriguing observations have been made recently toward the mechanistic aspects of cationic liposome-mediated systemic transfection *in vivo*. These observations, as summarized next, have led us to put forward a hypothesis that may prove to be useful for future development.

A. SUCCESSFUL TRANSFECTION OF THE LUNG ENDOTHELIUM DOES NOT REQUIRE A SPECIFIC DNA / LIPOSOME COMPLEX STRUCTURE

It has been long believed that the transfection efficiency of DNA/liposome complexes is determined by their physical structure. In fact, some unique structures in mixtures of cationic liposomes and DNA have been reported, including the so-called "spaghetti and meatballs" (Sternberg *et al.*, 1994), the invaginated liposomes (Templeton *et al.,* 1997), and the DNA/lipid sandwich structures (Lasic *et al.*, 1997; Radler *et al.*, 1997). Data obtained from these structural studies have been interpreted as evidence that these structures may play an important role in determining the overall level of transgene expression. While it may be true that the transfection efficiency of DNA/liposome complexes is dictated by the DNA/liposome complex structure *in vitro,* the data presented in Figure 6 suggest that, for lung endothelium transfection, the physical structure of DNA/liposome complexes may not be as important as one may have imagined. It is evident that the level of transgene expression in the lung is the same in animals injected with plasmid DNA in either the free form or complexed with cationic liposomes as long as each animal receives the same amount of cationic liposomes. This suggests that it is the total amount of cationic liposomes each animal receives that determines the ultimate level of gene expression in the lung.

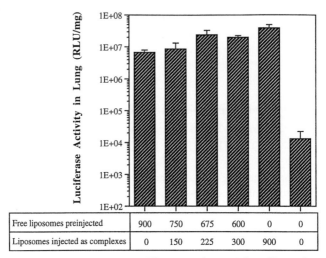

Free liposomes preinjected	900	750	675	600	0	0
Liposomes injected as complexes	0	150	225	300	900	0

Liposome Amount (nmol/mouse)

Figure 6 Dependence of gene expression on the total cationic lipid to DNA ratio. DNA/liposome complexes in 200 μl with different cationic lipid to DNA ratios were administered to animals pre-injected with different amounts of DOTMA liposomes in 100 μl. The total amount of cationic lipid and DNA each mouse received was 900 nmol and 25 μg, respectively. Two injections were 1 minute apart. Luciferase activity was assayed 8 h post-injection. Error bar represents SEM from three to six mice. [Data taken from Song and Liu (1998).]

B. RELATIONSHIP BETWEEN THE DNA UPTAKE AND THE LEVEL OF TRANSGENE EXPRESSION

Assuming that all the lung endothelial cells are capable of binding and internalizing DNA molecules, one would predict that the more DNA taken up by these cells, the more gene product would be generated. The study summarized in Figure 7 provides direct evidence in support of this relationship. In these experiments, a sequential injection strategy was used in which free cationic liposomes were first injected into mice followed by naked plasmid DNA 5 min later. The level of lung uptake of DNA was determined 5 min after the injection of plasmid DNA. Figure 7(A) illustrates the dose effect of cationic liposomes on DNA uptake by the lung. It is clear that the long hydrocarbon chain-based lipids (DOTMA and DOTAP) exhibit much higher activity than cholesterol derivatives (DC-chol and TC-chol) in retaining the plasmid DNA in the lung. Except for animals preinjected with TC-chol liposomes, the level of DNA accumulation in the lung was proportional to the dose of cationic liposomes administered. At the highest dose of cationic liposomes used (1050 nmol/mouse), approximately 50–60% of injected DNA was taken up by the lung in animals receiving either DOTMA or DOTAP liposomes,

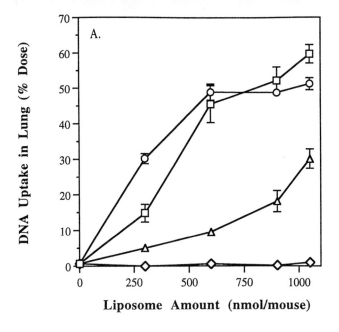

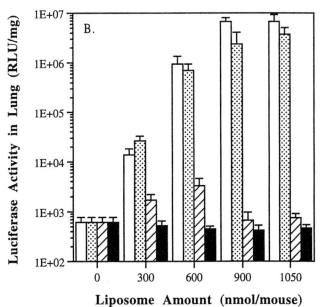

Figure 7 Dose-dependent effect of cationic liposomes on DNA uptake and transgene expression in the lung. Each mouse was injected via tail vein with various amounts of cationic liposomes in 100 μl PBS. Five minutes later, the same mouse was injected via tail vein with 25 μg of pCMV-luc plasmid DNA in 200 μl PBS with or without inclusion of a trace amount of ^{125}I-labeled same plasmid. For DNA uptake (A), animals were sacrificed 5 minutes after DNA administration and the ^{125}I radioactivity in the lung was analyzed. Animals were preinjected with DOTMA (□), DOTAP (○), DC-chol (△), or TC-chol (◇). For measurement of gene expression (B), animals were sacrificed 8 h post-injection of DNA. Luciferase activity in the lung of animals preinjected with liposomes consisted of DOTMA (□), DOTAP (▦), DC-chol (▨), and TC-chol (■). Error bar represents SEM from three to six mice. [Taken from Song *et al.* (1998).]

compared to about 30% in those animals receiving the same amount of DC-chol liposomes. Preinjection with TC-chol liposomes prior to DNA administration within the tested dose range did not result in any significant enhancement in DNA uptake by the lung. This suggests that the cholesterol-based liposomes are not efficient in retaining the DNA molecules in the lung.

The level of gene expression, as determined by luciferase activity, was examined 8 h after DNA injection. As shown in Figure 7(B), with the exception of animals preinjected with DC-chol liposomes, the level of gene expression in the lung correlates to DNA uptake (Fig. 7A). Animals receiving either DOTMA or DOTAP liposomes exhibited a similar level of gene expression. In animals preinjected with either DOTMA or DOTAP liposomes prior to DNA administration, the level of gene expression appeared to be saturated at 900 nmol cationic lipid per mouse. The highest level of gene expression observed was approximately 10^7 RLU per mg of protein from the lung extract, equivalent to 1 ng luciferase protein per mg of extracted protein. The luciferase activity in animals preinjected with either DC-chol or TC-chol liposomes was the same as that of control animals preinjected with PBS. The level of gene expression by the sequential injection of DOTMA liposomes and plasmid DNA is almost identical to those shown in Figure 2 in which the DNA/liposome complexes were used for transfection.

C. ARE DNA/LIPOSOME COMPLEXES FORMED IN THE LUNG AFTER A SEQUENTIAL INJECTION OF CATIONIC LIPOSOMES AND NAKED DNA?

While it is evident from the data presented in Figures 6 and 7 that formation of DNA/liposome complexes prior to administration is not necessary for a successful systemic transfection of the lung endothelium, the immediate question raised from these results is whether DNA/liposome complexes similar in structure to those formed in vitro can be formed in vivo after the sequential injection of cationic liposomes and naked DNA. A direct answer to this question is difficult considering that large amounts of negatively charged blood components may bind to injected liposomes and that it is technically challenging to differentiate between liposomes bound with blood components and those bound with sequentially injected DNA. To arrive at an answer, we have taken a fluorescence microscopic approach using both labeled DNA and labeled cationic liposomes. The rationale of this approach is simple and straightforward. If DNA/liposome complexes are formed in the lung after the sequential injection of fluorescence-labeled cationic liposomes and DNA, both probes should be colocalized in the lung with a pattern identical to that of animals injected with preformed DNA/liposome complexes. Utilizing this experimental approach, 1,1'-dioctadecyl-3,3,3',3'-tetramethylindocarbocyanine per-

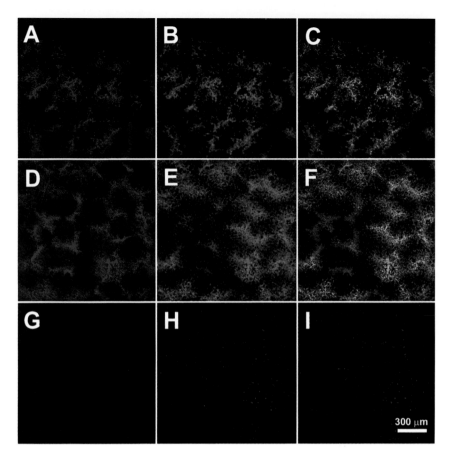

Figure 8 Fluorescence images of lungs in animals injected with DiI-labeled DOTMA liposomes and psoralen fluorescein-labeled plasmid DNA. DiI fluorescence was imaged with excitation at 550 nm and emission at 620–680 nm. Psoralen fluorescein was imaged with excitation at 488 nm and emission at 505–530 nm. The images were taken from the lungs of animals 5 min after the injection of psoralen fluorescein-labeled DNA. To prevent the collapse, airspace of the lung was filled with saline through the trachea before observation under a confocal microscope (Zeiss LS410, Germany). A, B, and C are representative images of DiI, psoralen fluorescein, and overlay of both fluorescences, respectively, of a lung from a mouse injected with DNA/liposome complexes containing 25μg of psoralen fluorescein-labeled DNA and 900 nmol DiI-labeled DOTMA liposomes. D, E, and F are the images of DiI, psoralen fluorescein, and overlay of both fluorescences, respectively, of a lung from a mouse sequentially injected with 900 nmol DiI-labeled DOTMA liposomes (100 μl) followed 2 min later by 25 μg of psoralen fluorescein-labeled plasmid DNA (100μl). G, H, and I are the images of DiI, psoralen fluorescein, and overlay of both fluorescences, respectively, of a lung from a normal mouse without injection.

chlorate (DiI)-labeled DOTMA liposomes were first injected into a mouse followed 2 min later by psoralen fluorescein-labeled plasmid DNA. As a control, DNA/liposome complexes prepared with the same fluorescence-labeled liposomes and DNA were also injected into animals. The animals were sacrificed 5 min after DNA injection, the lungs removed, and the intensity of both florescence markers in the lung immediately recorded by a fluorescence confocal microscopy (Song *et al.,* 1997). The major advantage of this type of confocal microscopic analysis is that it allows a direct observation of the fluorescence distribution within the intact lung without the need for tissue staining or sectioning, and the original distribution patterns for DiI liposomes and psoralen fluorescein-labeled DNA in the lung is, therefore, fully preserved. In Figure 8 are presented the fluorescence images of the lungs of animals injected with plasmid DNA either in liposome complexed form or as naked DNA following the injection of DiI-labeled cationic liposomes. The fluorescence images were recorded for DiI (red color, Figs. 8A and D), psoralen fluorescein-labeled DNA (green color, Figs. 8B and E) and with double exposure for both probes (Figs. 8C and F). Figures 8G to I are the images from the control animal into which no fluorescence probe was injected. As expected, the gold-colored image seen in Figure 8(C) indicates that liposomes are well colocalized with DNA in the lungs of animals injected with DNA/liposome complexes. However, the fluorescence image due to DiI-labeled liposomes (red color) was only partially colocalized with that of psoralen fluorescein-labeled DNA (green color) in the lungs of animals into which cationic liposomes and DNA were sequentially injected (Fig. 8F). The red and greenish images seen in Figure 8F suggest that the degree of cationic liposome and DNA colocalization in the lungs is much less in animals receiving sequential injection. These results suggest that, although some of the DNA molecules may become associated with cationic liposomes in the lung after sequential injection, it is not likely that DNA/liposome complex structures similar to those formed *in vitro* were formed in the lung as a consequence of this sequential injection of free liposomes and naked DNA.

D. Naked DNA Is Effective in Lung Endothelium Transfection

The data presented in Figures 6 to 8 have prompted us to test the hypothesis that free DNA may be the active form responsible for the successful transfection that appears to be mediated by DNA/liposome complexes. To test this hypothesis, transfection was performed in mice using an approach involving *in situ* perfusion of an intact lung followed by an *in vitro* organ culture system. To mimic *in vivo* conditions, a DNA solution was perfused into the lung *in situ* via the pulmonary artery. An incision previously made in the left ventricle of the heart allowed for drainage of the perfusate. The lung was then removed from the animal and cultured in RPMI 1640 medium with 10% fetal bovine serum (FBS) under standard cell culture conditions. The unique feature of this technique is that plasmid DNA, rather than being washed out of the lung capillary bed by the normal circulation as it would be *in vivo*, remains in the lung since circulation through the lung was stopped at the beginning of the experiment. Figure 9(A) shows the time-dependent transgene expression obtained with this technique. In these experiments, 100 μg of pCMV-luc

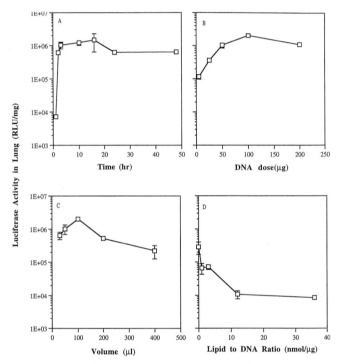

Figure 9   Free DNA-mediated gene transfer to lung *in vitro*. pCMV-luc plasmid was directly perfused into the lung through the pulmonary artery in anesthetized mice. The perfused lung was dissected from the animal and placed in a 12-well plate containing 2.5 ml of medium (RPMI1640, 10% FBS) for culture. Luciferase gene expression in the lung was determined by a standard luciferase assay using protein extract of the lung. (A) Time-dependent gene expression. Each lung was perfused with 100 μg plasmid DNA in 50 μl medium. Luciferase activity was determined at various times after culture at 37°C. (B) DNA dose–dependent gene expression. Each lung was perfused with various amounts of plasmid DNA in 100 μl medium. Luciferase activity was determined 3 h after culture at 37°C. (C) Perfusion volume effect on gene expression. Each lung was perfused with 100 μg plasmid DNA in different volume of medium. Luciferase activity was determined 3 h after culture at 37°C. (D) Effect of cationic liposomes on free DNA-mediated transfection. Each lung was perfused with 15 μg plasmid DNA with various amounts of DOTAP liposomes in 100 μl medium. Luciferase activity was determined 3 h after culture at 37°C. Error bar represents SEM from three mice.

plasmid in 50 μl of medium were perfused into the lungs and the perfused lungs were immediately placed into 2.5 ml culture medium with 10% FBS in a 12-well plate for culture. Luciferase activity was analyzed at different times during the incubation period. A DNA dose–response curve (Fig. 9B) was also established under identical experimental conditions except that the level of gene expression was determined 3 h after the perfusion. The perfusion volume effect on the level of gene expression was also tested and it is evident from the results presented in Figure 9(C) that the level of gene expression was not significantly affected by the perfusion volume ranging from 20 to 400 μl. For additional evidence in support of the conclusion that naked DNA is the active form transfecting the lung endothelium, we also tested the transfection activity of DNA/liposome complexes employing DOTAP liposomes as the carrier. As can be seen in Figure 9(D), luciferase activity in the lung decreases with increasing cationic liposome to DNA ratio used in preparation of DNA/liposome complexes. The level of gene expression decreased to a minimal activity of 10^4 RLU/mg, approximately two orders of magnitude lower than the level obtained with naked DNA, at cationic lipid to DNA ratios of 12/1 or greater. This low level of gene expression may result from the transfection by a small amount of plasmid DNA released from the DNA/liposome complexes due to the interaction between DNA/liposomes complexes and the surface of the lung endothelium.

E. The Function of Cationic Liposomes in Liposome-Mediated Transfection in the Lung Is to Prolong the Retention Time of DNA in the Lung

One possible explanation for the data given in Figures 6, 7, and 9, and the fact that injection of naked DNA alone does not result in a significant level of gene expression is that cationic liposomes assist in retaining DNA molecules in the lung for a time sufficient for gene transfer into the endothelium to be complete before they can be washed out of the capillary bed by normal blood flow. Thus, it appears that the retention time of the DNA molecules in the lung is likely to play a critical role in determining the level of gene expression. Figure 10 shows the effect of different cationic liposomes in retaining plasmid DNA in the lung. Depending on the type of cationic liposomes, the retention time of DNA in the lungs of animals pre-injected with either DOTMA or DOTAP liposomes prior to injection of DNA was significantly longer than that of animals receiving the same amount of either DC-chol or TC-chol liposomes. These results suggest that the high level of trans-gene expression obtained in animals preinjected with either DOTMA or DOTAP liposomes is due not only to their capability to enhance the uptake of DNA (Fig. 7) but also to their activity in retaining the DNA molecules in the lung for a prolonged period of time.

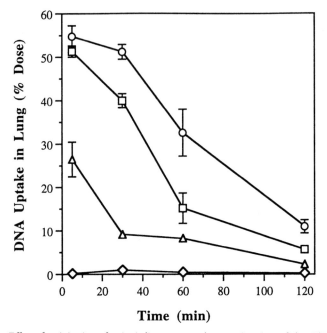

Figure 10   Effect of preinjection of cationic liposomes on the retention time of plasmid DNA in the lung. Each mouse was injected via tail vein with 900 nmol of cationic liposomes in 100 μl PBS. Five minutes later, each mouse was injected via tail vein with 25 μg of pCMV-luc plasmid DNA containing trace amounts of ^{125}I-labeled DNA in 200 μl PBS. DNA uptake in the lung was measured at various times after administration of plasmid DNA. Preinjected liposomes were DOTMA ($\square$), DOTAP ($\bigcirc$), DC-chol ($\triangle$), or TC-chol ($\Diamond$). Error bar represents SEM from three mice. [Data taken from Song *et al.* (1998).]

F. CATIONIC LIPOSOMES DO NOT POTENTIATE THE TRANSFECTION ACTIVITY OF NAKED DNA

While it was clear that the retention time in the lung of DNA molecules provided by cationic liposomes is critical in determining the level of transgene expression (Figs. 7 and 10), the possibility that these cationic liposomes may make the lung endothelial cells more sensitive to naked DNA transfection also exists. To investigate this possibility, animals were injected with 900 nmol DOTMA liposomes, the amount that results in a highest level of gene expression in the lung (Figs. 2 and 7). Following this injection, animals were sacrificed at different times and the combined *in situ/in vitro* lung perfusion experiments with plasmid DNA were performed. The level of gene expression in the lung was determined after 5 h in culture. Figure 11 is a plot of the level of gene expression observed as a function of the time after injection of cationic liposomes. It is clear that the level of gene expression

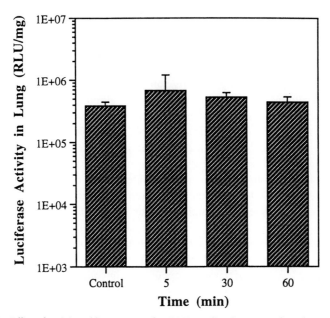

Figure 11   Effect of preinjected liposomes on free DNA-mediated gene transfer to lung *in vitro*. Each mouse was injected via tail vein with 100 μl PBS alone (control) or 900 nmol of DOTAP liposomes in 100 μl PBS. At different times, 50 μg of pCMV-luc plasmid DNA in 100 μl RPMI 1640 medium containing 10% FBS were perfused via pulmonary artery to the lung of an anesthetized mouse. Luciferase activity was determined 5 h after culturing the lungs in RPMI 1640 medium containing 10% FBS. Error bar represents SEM from three mice. [Data taken from Song *et al.* (1998).]

in the lung is almost identical regardless of whether animals were preinjected with free liposomes, suggesting that cationic liposomes do not make the lung endothelium more transfectable.

G. A Hypothetical Model for Cationic Liposome-Mediated Transfection of the Lung Endothelium

Based on data previously summarized, we hypothesize that the transfection efficiency of liposomes for the lung endothelium is mainly determined by three major factors: their ability to bring the DNA molecules to the maximal number of endothelial cells in the lung, their ability to retain the DNA in the lung for an extended period of time, and finally, their ability to release DNA from the complexes. The physicochemical parameters described in Section II, including the cationic lipid structure, the cationic lipid to DNA ratio, the diameters of liposomes and DNA/liposome complexes, the helper lipids in the liposomes, the solutions used for sample preparation and the injection dose, all exert their effect on the transfection efficiency by affecting one or more of these three factors.

In addition, considering that large amounts of negatively charged components (both cellular and noncellular) are present in blood, the interaction between the injected cationic liposomes, either free or complexed with DNA, and the blood components may contribute significantly to how these physicochemical parameters of liposomes might affect their ultimate efficiency in transfecting the lung endothelium. Theoretically, three different outcomes may result from the interaction of cationic liposomes with blood components. Depending on the physicochemical characteristics of the DNA/liposome preparation, it is possible that such interaction will generate large aggregates composed of the injected lipid particles and blood components. These types of aggregates, if larger than the diameter of the capillaries, are likely to be trapped in larger blood vessels before reaching the capillary bed, resulting in a limited exposure of DNA molecules to the lung endothelial cells and, consequently, a low level of transgene expression. In contrast to this outcome, it is also possible that the blood components may provide a solubilization effect on cationic liposomes through many well-illustrated mechanisms concerning the liposome interaction with serum proteins (Senior, 1987). Alternatively, the negatively charged blood components may compete with DNA for binding to cationic liposomes such that DNA molecules will be released from the DNA/liposome complexes prematurely, resulting in a transient exposure of DNA molecules to the lung endothelium with a consequent minimal success of transfection. The final possible outcome of the interaction between the cationic liposomes and blood components is that the interaction could generate appropriate lipid particles in the blood that are able to reach a large area of the lung capillary bed and these particles are able to retain the transfectionally active DNA molecules on the surface of the endothelial cells for sufficient time that a successful gene transfer into cells can be accomplished.

While data presented in this chapter and those in previous publications (Liu et al., 1997; Song et al., 1998; Song and Liu, 1998) clearly indicate the importance of the factors described in controlling the efficiency of liposome-mediated transfection to the lung endothelium, it is still too early and difficult to provide a quantitative analysis on how much each physicochemical parameter as described in Section II contributes to each or all of these controlling factors. Nevertheless, time-dependent tissue distribution studies with ^{125}I-labeled plasmid DNA in mice have shown that a higher cationic lipid to DNA ratio in the DNA/liposome mixture injected favors a prolonged residence time of DNA in the lung (Liu et al., 1997, Song and Liu, 1998). It was also noted in these studies that a long residence time of DNA in the lung does not necessarily produce a high level of gene expression. A possible explanation for this phenomenon is that these DNA molecules may be trapped in the lung in such a way that there is a limited exposure to the endothelial cells and/or the complexes are so formed that DNA molecules cannot be released from the complexes.

The importance of retention time of DNA molecules in the lung with respect to the transfection efficiency cannot be overemphasized considering that the normal blood flow is capable of carrying efficiently the DNA molecules out of the lung endothelium. For a successful transfection, DNA molecules must remain in contact with endothelial cells for a sufficient time that successful internalization into a cell is complete before these molecules can be washed out via the blood flow. Theoretically, a prolonged retention time of DNA in the lung may have its greatest contribution if a maximal number of endothelial cells are reached.

The hypothesis that the DNA molecules must be released from the complex with cationic liposomes in the lung is derived from data shown in Figures 6, 7, 8, and 9. While it is evident that preformed DNA/liposome complexes are not necessary for transfecting the lung endothelium systemically and naked DNA is more active than DNA/liposome complexes in transfecting the lung by intrapulmonary perfusion, the detailed mechanism of DNA release from the DNA/liposome complexes in the lung is still unknown. It is not clear how DNA release from the complexes is coordinated with DNA entrance to the cell. More importantly, it will be interesting to know what causes the DNA release and how the blood components affect the release process. From a basic research point of view, it will be interesting to elucidate the mechanisms by which the entrance of DNA into the endothelial cells is regulated.

V. CONCLUSIONS

The anatomy and physiology of the lung favors its use as a target for systemic gene delivery. An obvious advantage of this system is that, once intravenously injected, DNA molecules may have full access to the lung endothelium as the entire blood volume circulates through the lung. An additional advantage is that the large capillary bed of the lung provides a extensive cell surface area to allow DNA to bind and enter the endothelial cells. For optimal transfection activity, the cationic liposome-based carrier system needs to deliver the DNA molecules to a large number of endothelial cells, keep these molecules for a prolonged period of time on the surface of these cells, and finally, release the DNA at an appropriate rate that the removal of the DNA molecules from the lung by either blood flow or/and nuclease-mediated DNA degradation can be minimized. Toward this end, additional studies are needed for the development of ways to predict the potential outcome of the interaction between the cationic liposomes, DNA/liposome complexes, and blood components.

ACKNOWLEDGMENTS

The original work described in this chapter was supported by a grant from NIH CA72925 and by Targeted Genetics Corporation. We are grateful to Dr. Leaf Huang for his help in providing us with TC-chol and pCMV-luc plasmid. DOTMA was provided by Roche Bioscience. Psoralen fluorescein-labeled DNA was kindly provided by Gene Medicine, Inc. We thank Dr. Shaoyu Chu for performing the DNA and liposome colocalization study using confocal microscopy.

REFERENCES

Behr, J. P., Demeneix, B., Loeffler, J. P., and Mutul, J. P. (1989). Efficient gene transfer into mammalian primary endocrine cells with lipopolyamine-coated DNA. *Proc. Natl. Acad. Sci. USA* **86,** 6982–6986.

Bennett, M. J., Nantz, M. H., Balasubramaniam, R. P., Gruenert, D. C., and Malone, R. W. (1995). Cholesterol enhances cationic liposome-mediated DNA transfection of human respiratory epithelial cells. *Biosci. Rep.* **15,** 47–53.

Biven, W. S., Crawford, M. P., and Brewer, N. R. (1979). Morphophysiology. In H. J. Baker, J. R. Lindsey, and S. H. Weisbroth (Eds.), *The Laboratory Rat* (p. 85). New York: Academic Press.

Farhood, H., Serbina, N., and Huang, L. (1995). The role of dioleoylphosphatidylethanolamine in cationic liposome mediated gene transfer. *Biochim. Biophys. Acta* **1235,** 289–295.

Felgner, P. L., Gadek, T. R., Holm, M., Roman, R., Chan, H. W., Wenz, M., Northrop, J. P., Ringold, G. M., and Danielsen, M. (1989). Lipofectin: A highly efficient, lipid-mediated DNA-transfection procedure. *Proc. Natl. Acad. Sci. USA* **84,** 7413–7417.

Felgner, J. H., Kumar, R., Sridhar, C. N., Wheeler, C. J., Tsai, Y. J., Border, R., Ramsey, P., Martin, M., and Felgner, P. L. (1994). Enhanced gene delivery and mechanism studies with a novel series of cationic lipid formulations. *J. Biol. Chem.* **269,** 2550–2561.

Gao, X., and Huang, L. (1991). A novel cationic liposome reagent for efficient transfection of mammalian cells. *Biochem. Biophyhs. Res. Commun.* **179,** 280–285.

Gao, X., and Huang, L. (1995). Cationic liposome-mediated gene transfer. *Gene Ther.* **2,** 710–722.

Guyton, A. C. (1991). *Textbook of medical physiology* (8th ed. pp. 171–415). Philadelphia: Saunders.

Hawley-Nelson, P., Ciccarone, V., Gebeyehu, G., and Jessee, J. (1993). Lipofectamine reagent: A new, higher efficiency polycationic liposome transfection reagent. *Focus* **15,** 73–79.

Hong, K., Zheng, W., Baker, A., and Papahadjorpoulos, D. (1997). Stabilization of cationic liposome-plasmid DNA complexes by polyamines and poly(ethylene glycol)-phospholipid conjugates for efficient *in vivo* gene delivery. *FEBS Lett.* **400,** 233–237.

Lasic, D. D., Strey, H., Stuart, M. C. A., Podgornik, R., and Frederic, P. M. (1997). The structure of DNA-liposome complexes. *J. Am. Chem. Soc.* **119,** 832–833.

Lee, E. R., Marshall, J., Siegel, C. S., Jiang, C., Yew, N. S., Nichols, M. R., Nietupski, J. B., Ziegler, R. J., Lane, M. B., Wang, K. X., Wan, N. C., Scheule, R. K., Harris, D., Smith, A. E., and Cheng, S. H. (1996). Detailed analysis of structures and formulations of cationic lipids for efficient gene transfer to the lung. *Human Gene Ther.* **7,** 1701–1717.

Legendre, J. Y., and Szoka, F. C. (1992). Delivery of plasmid DNA into mammalian cell lines using pH-sensitive liposomes: Comparison with cationic liposomes. *Pharm. Res.* **9,** 1235–1242.

Leventis, R., and Silvius, J. R. (1990). Interactions of mammalian cells with lipid dispersions containing novel metabolizable cationic amphiphiles. *Biochim. Biophys. Acta* **1023,** 124–132.

Li, S., and Huang, L. (1997). *In vivo* gene transfer via intravenous administration of cationic lipid-protamine–DNA (LPD) complexes. *Gene Ther.* **4,** 891–900.

Liu, F., Qi, H., Huang, L., and Liu, D. (1997). Factors controlling efficiency of cationic lipid-mediated transfection *in vivo* via intravenous administration. *Gene Ther.* **4,** 517–523.

Liu, Y., Mounkes, L. C., Liggitt, H. D., Brown, C. S., Solodin, I., Heath, T. D., and Debs, R. J. (1997). Factors influencing the efficiency of cationic liposome-mediated intravenous gene delivery. *Nature Biotechnol.* **15,** 167–173.

McLean, J. W., Fox, E. A., Baluk, P., Bolton, P. B., Haskell, A., Pearlman, R., Thurston, G., Umemoto, E. Y., and McDonald, D. M. (1997). Organ-specific endothelial cell uptake of cationic liposome–DNA complexes in mice. *Am. J. Phys.* **273,** H387-H404.

Radler, J. O., Koltover, I., Salditt, T., and Safinya, C. R. (1997). Structure of DNA–cationic liposome complexes: DNA intercalation in multilamellar membranes in distinct interhelical packing regimes. *Science* **275,** 810–814.

Rose, J. K., Buonocore, L., and Whitt, M. A. (1991). A new cationic liposome reagent mediating nearly quantitative transfection of animal cells. *Biotechniques* **1065,** 8–14.

Senior, J. H. (1987). Fate and behavior of liposomes *in vivo:* A review of controlling factors. *Crit. Rev. Ther. Drug Carrier Sys.* **3,** 123–193.

Sternberg, B., Sorgi, F. L., and Huang, L.(1994). New structures in complex formation between DNA and cationic liposomes visualized by freeze–fracture electron microscopy. *FEBS Lett.* **356,** 361–366.

Solodin, I., Brown, C. S., Bruno, M. S., Chow, C. Y., Jang, E. H., Debs, R. J., and Heath, T. D. (1995). A novel series of amphiphilic imidazolinium compounds for *in vitro* and *in vivo* gene delivery. *Biochem.* **34,** 13537–13544.

Song, Y. K., and Liu, D. (1998). Free liposomes enhance the transfection activity of DNA/lipid complexes *in vivo* by intravenous administration. *Biochim. Biophys. Acta* **1372,** 141–150.

Song, Y. K., Liu, F., Chu, S. Y., and Liu, D. (1997). Characterization of cationic liposome-mediated gene transfer *in vivo* by intravenous administration. *Human Gene Ther.* **8,** 1585–1594.

Song, Y. K., Liu, F., and Liu, D. (1998). Enhanced gene expression in mouse lung by prolonging the retention time of intravenously injected plasmid DNA. *Gene Ther.* **5,** 1531–1537.

Templeton, N. S., Lasic, D. D., Frederik, P. M., Strey, H. H., Roberts, D. D., and Pavlakis G. N. (1997). Improved DNA:liposome complexes for increased systemic delivery and gene expression. *Nature Biotechnol.* **15,** 647–652.

Thierry, A. R., Lunardi-Iskandar, Y., Bryant, J. L., Rabinovich, P., Gallo, R. C., and Mahan, L. C. (1995). Systemic gene therapy: Biodistribution and long-term expression of a transgene in mice. *Proc. Natl. Acad. Sci. USA* **92,** 9742–9746.

Wang, J. K., Guo, X., Xu, Y. H, Barron, L., and Szoka, F. C. (1998). Synthesis and characterization of long chain alkyl acyl carnitine esters. Potentially biodegradable cationic lipids for use in gene delivery. *J. Med. Chem.* **41,** 2207–2215.

Wrobel, I., and Collins, D. (1995). Fusion of cationic liposomes with mammalian cells occurs after endocytosis. *Biochim. Biophys. Acta* **1235,** 296–304.

Zhu, N., Liggitt, D., Liu, Y., and Debs, R. (1993). Systemic gene expression after intravenous DNA delivery into adult mice. *Science* **261,** 209–211.

CHAPTER 15

Cystic Fibrosis Gene Therapy

Uta Griesenbach, Duncan M. Geddes, and Eric W. F. W. Alton
Imperial College School of Medicine, at the National Heart & Lung Institute, London, England

I. Introduction
II. Research Leading to Preclinical Studies for CF
    Gene Therapy
    A. Analysis of CFTR Expression in the Lung
    B. CFTR Promoter Studies
    C. Animal Models
III. Preclinical Studies for CF Gene Therapy
    A. *In Vitro* CFTR Gene Transfer
    B. Gene Transfer into Animal Models
    C. Gene Transfer Agents
IV. Considerations Proceeding Phase I Clinical Trials
    A. Predicted Expression Levels Necessary to Correct the
       Ion Transport Defect
    B. Safety Considerations
V. Clinical Trials
    A. Virus-Mediated Gene Therapy Trials
    B. Liposome-Mediated Gene Therapy Trials
VI. Considerations before Proceeding into Phase II
    and III Clinical Trials
    A. Development of More Relevant Clinical End
       Point Assays
    B. Improving Gene Transfer Efficiency
    C. CF Patients Suitable for Gene Transfer Studies
VII. Summary and Outlook
    References

Nonviral Vectors for Gene Therapy

Cystic fibrosis (CF) is caused by mutations in the cystic fibrosis transmembrane conductance regulator gene (CFTR). The disease presents with a complex phenotype involving multiple organs. However, progressive lung failure is the major cause of death in CF patients. Gene therapy for CF has been particularly attractive for two reasons. First, the disease arises from a single gene defect and second, the lung is an organ that is accessible through noninvasive techniques. More clinical trials have been approved and carried out for CF than for any other inherited disease. Gene transfer to the lung has been demonstrated with viral and nonviral vectors in phase I clinical trials and partial correction of the ion transport defect has been achieved.

I. INTRODUCTION

Cystic fibrosis (CF) is the most common lethal, autosomal, recessive, single-gene disorder among Caucasians. The cystic fibrosis gene was cloned in 1989 and the gene product was called cystic fibrosis transmembrane conductance regulator (CFTR) (Rommens et al., 1989; Riordan et al., 1989). The CFTR gene spans 250 kb of genomic DNA and contains 27 exons that encode a polypeptide of 1480 amino acids (Zielenski et al., 1991). The amino acid sequence suggests a tandem repeat structure with two identical halves, each consisting of six putative α-helices and an intracellular nucleotide binding domain. The two halves of the protein are linked through a highly charged intracellular domain (Riordan et al., 1989). The proposed structure assigns the protein to the ATP-binding cassette transporter family (ABC-transporters) (Ames and Lecar, 1992). Other members of this family include the yeast α-mating factor exporter (STE6) and the mammalian P-glycoprotein (Higgins and Gottesman, 1992). Despite having the characteristic structure of a transporter, numerous studies have indicated that CFTR functions as a cAMP-regulated chloride channel. Thus, expression of CFTR in cells that do not normally express this protein such as Chinese hamster ovary cells (CHO) or Xenopus oocytes leads to the generation of a cAMP-dependent chloride channel (Bear et al., 1991; Tabcharani et al., 1991). Further, reconstitution of purified CFTR protein in planar lipid bilayers also demonstrated the cAMP-dependent chloride channel activity of the protein (Bear et al., 1992).

Apart from functioning as a chloride channel, CFTR has been implicated in the regulation of a number of other epithelial ion channels—among them the amiloride-sensitive sodium channel (ENaC) (Stutts et al., 1995); the outwardly rectifying Cl^- channel (ORCC) (Gabriel et al., 1993); ROM K2, a potassium channel (McNicholas et al., 1996); and certain Ca^{2+}-dependent chloride channels (Grubb et al., 1994b). The mechanism of interaction between CFTR and the other channels is not clear. There is some evidence that CFTR transports ATP out of cells, which is consistent with the observation that most members of the ABC family actively transport substrates across the cell membrane (Cantiello *et al.*, 1998; Reisin *et al.*, 1994). These findings, however, have been disputed by other studies (Li *et al.*, 1996; Reddy *et al.*, 1996).

To date, more than 800 mutations have been identified in the CFTR gene (Zielenski and Tsui, 1995). A 3 base pair deletion that causes the loss of a phenylalanine residue at position 508 (ΔF508) is found in approximately 70% of CF chromosomes. The relative frequency of the other mutations varies among different populations, but most of them are very rare. The ΔF508 mutation leads to a defect in posttranslational processing and prevents the CFTR protein from reaching the apical cell membrane. Other mutations can cause absent or truncated proteins or alter the function of CFTR in the apical membrane.

The incidence of CF varies from 1 in 1500 to 1 in 6500 depending on the population. Several factors might account for this high incidence. Increased resistance to tuberculosis (Meindl, 1987), cholera (Morral *et al.*, 1994), and typhoid fever (Pier *et al.*, 1998) of heterozygote carriers has been suggested. Founder effects and inbreeding have been shown to account for the high incidence of CF in some Caucasian populations (Super, 1979).

The disease presents with a complex phenotype involving the pancreas, intestine and reproductive system. However, progressive lung failure is the major cause of death. The lungs of CF patients are often colonized with *Staphylococcus aureus* and *Hemophilus influenzae,* but most characteristically with *Pseudomonas aeruginosa* (Gilligan, 1991). In addition to chronic infection, inflammation is a major contributing factor in the pathogenesis of the lung disease. At present it is unknown how the clinical phenotype in the CF lung links with abnormalities in CFTR. Impaired mucociliary clearance because of altered ion (decreased chloride secretion and increased sodium absorption) and water movement across the airway epithelium has been suggested (Welsh, 1987). However, other defects in host defense may also play a role including increased bacterial adherence (Saiman *et al.*, 1992), reduced ingestion of bacteria by epithelial cells (Pier *et al.*, 1996), and impaired antimicrobial activity of surface defensins (Smith *et al.*, 1996).

Conventional therapy for CF includes the use of physiotherapy, antibiotics, and pancreatic supplements. This combination of treatments has helped to increase life expectancy from approximately 1 year in the 1930s to the current 30 years of age. Although further improvements in both quality of life and life expectancy are likely to occur from conventional therapy, gene therapy may significantly contribute to the treatment of the CF lung disease.

II. RESEARCH LEADING TO PRECLINICAL STUDIES FOR CF GENE THERAPY

A. ANALYSIS OF CFTR EXPRESSION IN THE LUNG

The expression of CFTR has been analyzed in both rodents and humans by mRNA *in situ* hybridization and protein detection during different stages of lung development and was shown to be regulated in both a cell-specific and developmental manner (Tizzano *et al.*, 1994; Turnberg *et al.*, 1970; Engelhardt *et al.*, 1992; Trezise *et al.*, 1993; Trezise and Buchwald, 1991). In the lung overall expression of CFTR is not only much higher during fetal development than in the adult but is also different in spatial distribution. In the early fetus CFTR is expressed in the primordial lung epithelium. During later stages of development expression decreases in cells of the future alveolar spaces. The highest level of CFTR expression after birth occurs in the submucosal glands, which predominate in the proximal airways. Expression is also seen more peripherally in a subpopulation of both ciliated and nonciliated cells, although the precise cell types have not yet been identified. Earliest pathological changes occur in the peripheral airways which indicates that gene therapy should target these cells. However, a significant contribution of the submucosal glands to the disease phenotype cannot be ruled out.

B. CFTR PROMOTER STUDIES

Gene therapy ideally requires gene delivery systems that ensure expression of the therapeutic cDNA specifically in the affected organs and cell types. The endogenous CFTR promoter should therefore be used to ensure tissue-specific regulation of the CFTR transgene. However, despite extensive studies *in vitro* (Yoshimura *et al.*, 1991; Koh *et al.*, 1993; Chou *et al.*, 1991) and *in vivo* (Griesenbach *et al.*, 1994), it has proven difficult to identify a fully functional CFTR promoter. Recently, a 300 kb yeast artificial chromosome (YAC) containing the CFTR genomic region has been shown to ensure tissue-specific regulation of CFTR expression in many of the cells where CFTR is normally expressed (Manson *et al.*, 1997). In the absence of a functional homologous CFTR promoter, most studies have used strong viral promoters such as SV40, CMV, and RSV. Lung-specific heterologous promoters such as the SP-C (surfactant protein C) (Whitsett *et al.*, 1992) and the CC10 (clara cell protein) regulatory elements have also been used in some studies (Ray *et al.*, 1997).

C. ANIMAL MODELS

Over the past 5 years several mouse models have been created through gene targeting strategies, including mice carrying the ΔF508 and the G551D mutations

(Zeiher et al., 1995; Colledge *et al.*, 1995; Dorin *et al.*, 1992; Snouwaert *et al.*, 1992; Rozmahel *et al.*, 1996; Ratcliff *et al.*, 1993; Hasty *et al.*, 1995). The phenotype of most mouse models is similar. Intestinal blockage due to mucus accumulation and subsequent rupture of the intestines leads to premature death in 50–95% of mice. In contrast to the intestinal pathology, all mouse models fail to develop the severe lung pathology that is characteristic of CF patients. The presence of an alternative non-CFTR chloride channel in the lung but not in the intestine could explain these findings (Clarke *et al.*, 1994). In addition, murine lungs largely lack submucosal glands—a major site of CFTR expression in man—and further, they do not demonstrate the characteristic increase in sodium absorption. Thus, a number of factors may be involved in this difference in phenotype. Despite their limitations, the CF mouse models have been valuable for electrophysiological, pharmacological, and gene therapy research strategies.

III. PRECLINICAL STUDIES FOR CF GENE THERAPY

A. IN VITRO CFTR GENE TRANSFER

Following the cloning of the CFTR gene it was soon demonstrated that expression of CFTR could correct the chloride transport defect in CF cell lines. Retrovirus-mediated transfer of CFTR cDNA into a CF pancreatic carcinoma cell line reversed the chloride transport defect of the cell line as measured through patch clamp and ion efflux analysis (Drumm *et al.*, 1990). Similar results were obtained when a CF airway epithelial cell line was transfected with CFTR cDNA using a vaccinia virus vector (Rich *et al.*, 1990). Rosenfeld *et al.* demonstrated the transfection of freshly isolated human respiratory epithelial cells with an adenoviral vector (Rosenfeld *et al.*, 1994). Interestingly, it was also shown that restoration of the CFTR chloride defect following CFTR cDNA transfer is linked to correction of other CF ion transport defects, providing further evidence that CFTR has regulatory functions (Egan *et al.*, 1992).

B. GENE TRANSFER INTO ANIMAL MODELS

Successful CFTR gene transfer experiments were first reported in cotton rats and wild-type mice using adenoviral (Rosenfeld *et al.*, 1992) and liposomal vectors (Yoshimura *et al.*, 1992), respectively. The presence of CFTR mRNA was detectable for up to 4 weeks in the mouse lung. Subsequently, a number of studies were carried out in nonhuman primates using various adenoviral vector systems to transfer CFTR cDNA or reporter genes to the airway epithelium (Engelhardt *et al.*, 1993; Goldman *et al.*, 1995; Bout *et al.*, 1994). Expression was seen throughout the airways including the alveoli, but was generally patchy in distribution. Local signs of inflammation were seen in some but not all studies and correlated with increasing number of delivered virus particles (Brody *et al.*, 1994; Simon *et al.*, 1993).

Following the development of the CF knockout mice it was quickly demonstrated that CFTR gene transfer can partially correct the CF chloride transport defect. Hyde *et al.* (1993) instilled CFTR cDNA complexed to the cationic liposome DOTMA into the trachea and Alton *et al.* (1993) nebulized DNA/liposome complexes (CFTR cDNA/DC-chol/DOPE) into the nose of Cftr^{m1HGU} knockout mice (Alton *et al.*, 1993). In both studies correction of the chloride transport defect was variable, suggesting inefficient gene transfer. Grubb *et al.* instilled adenoviral vectors expressing CFTR into the nasal cavity of knockout mice and demonstrated a 50% correction in chloride transport after multiple- but not single-dose administration of the virus. There was no change in sodium transport measurable (Grubb *et al.*, 1994a).

In addition to the described animal studies, a number of xenograft models have been developed and have been used for the assessment of various gene transfer agents. Pilewski *et al.* transplanted human bronchial segments into severe combined-immunodeficient mice and demonstrated the transfection of human airway submucosal glands with adenoviral vectors (Pilewski *et al.*, 1995). Goldman *et al.* reported almost complete correction of the cAMP-stimulated chloride transport, but only partial and variable correction of sodium absorption in a human CF bronchial xenograft model (Goldman *et al.*, 1995b).

C. Gene Transfer Agents

A direct comparison of the efficiency of liposome- and adenovirus-mediated gene transfer systems has not been carried out yet *in vivo*. Adenoviral vectors possess some natural tropism for the human airway epithelium. However, recently it has been shown that columnar airway epithelium cells are relatively resistant to adenoviral transfer because they do not express $\alpha_v\beta_5$ integrins, which are necessary for the efficient uptake of adenovirus into cells (Goldman and Wilson, 1995; Grubb *et al.*, 1994). It has been suggested that undifferentiated, regenerating airway epithelial cells (basal cells) are preferentially transfected with adenoviral vectors (Dupuit *et al.*, 1995). These cells, however, do not express CFTR protein and might therefore not be suitable targets for gene therapy. Adenoviral vectors are limited with respect to the length of cDNA they can take up. This might become a problem if potentially larger endogenous CFTR regulatory elements need to be included into the constructs. Another concern regarding adenoviral vectors is related to its inflammatory and immunogenic properties. The development of a humoral immune response against virus particles as well as a cellular immune response directed against viral gene products is currently limiting the success of repeated virus administration (Kaplan *et al.*, 1996). Local inflammatory responses after virus administration are dose dependent. Providing that a therapeutic window can be defined, inflammation will not be a limiting factor (Simon *et al.*, 1993).

Cationic liposomes are less efficient at gene transfer *in vitro* in comparison with viral vectors, but are less likely to provoke inflammation or immune reactions. In addition, any size of cDNA can be transfected via liposomal vectors. Recently, Zhang *et al.* demonstrated that transfection with adenoviral or liposomal vectors leads to differences in the correction of two primary defects of CF (Cl⁻ secretion and mucus sulfation) in a xenograft model. Adenovirus transduction led to a correction of Cl⁻ secretion but not mucus sulfation, whereas liposome transfection resulted in a correction of mucus sulfation but only minimal Cl⁻ secretion (Zhang *et al.*, 1998). The choice of the vector system might therefore depend on the specific defect that needs most urgent correction.

IV. CONSIDERATIONS PROCEEDING PHASE I CLINICAL TRIALS

A. PREDICTED EXPRESSION LEVELS NECESSARY TO CORRECT THE ION TRANSPORT DEFECT

Overall expression of CFTR in the lung is very low (Crawford *et al.*, 1991), which is encouraging in suggesting that possibly only low levels of expression will be required for clinical benefit. However, it is uncertain whether all cells or only a subset of cells will need to be corrected to achieve an improvement of the clinical phenotype. Several studies have begun to suggest that only small increments in CFTR expression may be required in the airways to produce much larger changes in function. An *in vitro* study has shown that if "corrected" and "uncorrected" CF cells are mixed within a monolayer of purely "corrected" cells, approximately 6 to 10% of the former produce the same effect as a monolayer of 100% "corrected" cells (Johnson *et al.*, 1992). Goldman *et al.* demonstrated in a xenograft model that correction of 5% of human CF airway epithelial cells is sufficient to correct 100% of the chloride transport defect and partially correct the sodium transport defect (Goldman *et al.*, 1995b). Finally, interbreeding of compound herterozygote, wild type, and complete null CF knockout mice has shown that the presence of 1 to 5% of the normal CFTR mRNA within each cell is able to completely prevent the intestinal problems and to produce a marked correction of the chloride defect (Alton *et al.*, 1993).

B. SAFETY CONSIDERATIONS

It has already been mentioned that airway inflammation developed with increasing adenoviral titers in nonhuman primates (Simon *et al.*, 1993; Zabner *et al.*,

1994). The production of adenovirus-neutralizing antibodies causes reduced reporter gene expression after multiple applications (Yei *et al.*, 1994).

Cationic liposomes demonstrate cytotoxicity at high doses *in vitro* but have not caused problems *in vivo*. DC–chol/DOPE (see Chapter 1 in this book) was first used in a clinical gene therapy trial for malignant melanoma and has not demonstrated any safety problems (Nabel *et al.*, 1993). In preparation for our most recent CF lung trial we completed a study nebulizing Genzyme lipid 67 into the lungs of normal volunteers and have not detected any clinical adverse side effects in the 15 subjects receiving ascending doses of liposomes (Chadwick *et al.*, 1997).

Recently it has been suggested that plasmid DNA as such has proinflammatory properties. Plasmid DNA is propagated in bacteria and carries the bacterial methylation pattern. Bacterial DNA, specifically the CpG islands, are frequently unmethylated, in contrast to eukaryotic DNA. Schwartz *et al.* have shown that DNA with unmethylated CpG islands cause inflammation in mice and may do the same in humans (Schwartz *et al.*, 1997). Changing the methylation pattern as well as removing as much bacterial backbone DNA from plasmids as possible might overcome this problem.

V. CLINICAL TRIALS

Ten gene therapy trials for CF (five using viral and five using liposomal vectors) have been completed and published. A further four trials are in progress and expected to be published soon. The majority of the published studies used the nasal epithelium as surrogate for the lung epithelium. Adenovirus administration led to dose-dependent mild local inflammation in some studies. These symptoms were not noted after liposomal gene transfer. Apart from one patient who developed systemic and local inflammation after receiving the highest vector dose, administration of virus to the lung had no adverse clinical side effects. The first liposomal lung trial noted the development of mild flu-like symptoms that did not require any special treatment. Correction of the electrical abnormalities were partial, transient, and not detected in all studies. However, the proof of concept for gene transfer into airway epithelium was reproducible in all studies. The result of the published trials will now be discussed in detail in chronological order.

A. VIRUS-MEDIATED GENE THERAPY TRIALS

The first CF clinical trial was published by Zabner (Zabner *et al.*, 1993), who administered adenovirus carrying CFTR cDNA to the nose of three CF subjects in escalating doses. A correction of the chloride transport defect as measured by basal voltage as well as the response to a cAMP agonist was measured for at least 10 days.

CFTR protein or mRNA could not be detected in any of the treated subjects. Mild inflammation in the nasal epithelium was detected. It was unclear whether the inflammation was virus- or procedure-related and it should be noted that inflammation itself has the potential to reduce baseline voltage. Further interpretation of the data is difficult because the protocol used to measure electrical changes did not readily discriminate between CF and normal subjects.

Crystal (Crystal *et al.*, 1995) administered an adenoviral vector to the nose and to a restricted area of the lung (via a bronchoscope) of four CF subjects in escalating doses. Vector-specific CFTR mRNA was detected in nasal samples in one out of four patients, but could not be detected in the three patients from whom bronchial samples were obtainable. CFTR protein was also detected in one out of four nasal samples as well as in one out of three bronchial samples. Vector replication or a rise in neutralizing antibodies was not detected. Transient systemic and local inflammation was seen in one patient receiving the highest virus dose (2×10^9) to the lung. The trial was subsequently extended to include electrical measurements in a total of nine CF subjects (Hay *et al.*, 1995). The authors reported a decrease in baseline PD and amiloride response toward normal and an increase of low Cl^-/cAMP agonist response toward normal in the majority of subjects over a period of 2 weeks.

A third study administered four logarithmically increasing doses of a CFTR containing adenovirus or vehicle alone, to the nasal epithelium of 12 patients in a randomized double-blind study (Knowles *et al.*, 1998). The vector was detected in nasal fluid by culture, PCR, or both for up to 8 days after administration. There was molecular evidence of gene transfer by RT-PCR or *in situ* hybridization in five of six patients treated with the two highest doses. However, less than 1% of cells were estimated to have been transfected and no significant changes in electrophysiological measurements were detected. At the highest dose there was mucosal inflammation in two of three patients. One patient had an earache and an inflamed tympanic membrane and the second patient complained about jaw pain and mandibular-angle tenderness. This study did not reproduce the results of the two previous nasal trials with respect to electrical changes in the CF nose.

Bellon *et al.* delivered escalating doses of a replication-deficient adenovirus expressing the human CFTR protein to the nose (instillation) and to the lung (aerosolization) of six CF subjects (Bellon *et al.*, 1997). No acute toxic side effects were observed. CFTR mRNA was detected in all nasal samples at Day 15 after virus administration, but only in one of six bronchial brushing samples. Recombinant CFTR protein was also detectable in all nasal samples at Day 15 after administration, but only in two out of six bronchial brushing samples. Measurements of the nasal potential difference were attempted but due to spontaneous variations were difficult to interpret.

The first repeated administration of adenovirus vectors encoding CFTR cDNA to the nasal epithelium was carried out in six CF subjects receiving five logarithmically increasing doses (Zabner et al., 1996). The authors did not detect any adverse side effects, but an increase in neutralizing antibodies was measured following the first administration. A partial correction of the chloride transport defect was detected in a subset of patients, which decreased with subsequent administration of the vector. The authors attributed this result to the development of an immune response that decreases gene transfer efficiency.

A clinical maxillary sinus trial using an adeno-associated virus vector (AAV) encoding CFTR cDNA was carried out by Wagner (Wagner et al., 1998). Vector-specific mRNA was not detectable, but persistent DNA transfer was noted for up to 41 days in patients receiving the highest doses of AAV. There were no significant changes in the inflammation already present in the sinuses nor an increase in neutralizing antibodies. All patients treated with the two highest AAV doses showed transient electrical responses to the cAMP-agonist isoproterenol. However, baseline potential differences were not altered after virus treatment. It must be noted that the electrical response of CF sinuses has never been compared to that of non-CF controls.

B. Liposome-Mediated Gene Therapy Trials

The first liposome-mediated CF gene therapy trial was carried out by our group in 1995 (Caplen et al., 1995) as a double-blind placebo controlled study. A eukaryotic expression plasmid with a SV40 promoter regulating CFTR cDNA expression was complexed to the cationic lipid DC-chol/DOPE and was instilled onto the nasal epithelium of 15 ΔF508 homozygous CF subjects (9 CFTR cDNA, 6 placebo). No safety problems were encountered, either in the routine clinical assessment or by a blind, semiquantitative analysis of nasal biopsies. Both plasmid DNA and CFTR mRNA were detected from the nasal biopsies in five of the eight treated patients. Chloride secretion showed a significant 20% increase toward normal values, a change well outside the variation in these measurements. In two subjects, these chloride responses reached values within the non-CF range with the changes lasting for approximately 7 days.

In a similar double-blind randomized study Porteous (Porteous et al., 1997) administered a single dose of a CFTR plasmid or buffer alone to 16 CF volunteers. The plasmid contained a CMV promoter and was complexed to the cationic liposome DOTAP. There was no evidence of nasal inflammation on biopsy, circulating inflammatory markers, or other adverse events related to active treatment. Transgene DNA was detected in seven of the eight treated patients up to 28 days after treatment and vector derived CFTR mRNA in two of the seven patients 3 and 7 days after administration. Partial correction of CFTR-related chloride transport was detected in two treated patients, sustained for up to 4 weeks.

A third double-blind placebo controlled study was performed by Gill (Gill *et al.,* 1997) using a CFTR cDNA plasmid containing a RSV promoter. Eight patients received the plasmid complexed with DC-chol/DOPE and four received buffer alone. Biopsies of the nasal epithelium taken 7 days after dosing were normal. No significant changes in any clinical parameters were observed. Functional expression of CFTR assessed by nasal PD measurements showed transient correction of the chloride transport abnormality in two patients. Fluorescence microscopy showed evidence for CFTR function *ex vivo* in cells from nasal brushings in a further four patients. In total, evidence of functional CFTR gene transfer was obtained in six of the eight treated patients.

Zabner *et al.* used a plasmid that carried the CFTR cDNA under the control of the CMV promoter (Zabner *et al.,* 1997). The plasmid was either complexed to Genzyme lipid 67 (see Chapter 3 in this book) and administered to one nostril or administered as DNA alone to the other nostril of nine CF subjects in a randomized double-blind study. There were little if any systemic or local adverse side effects. However, it must be noted that serum IL-6 levels were elevated in three out of nine subjects at day 4 after treatment. Plasmid DNA was detected by PCR in all nostrils treated with DNA alone and in eight out of nine nostrils treated with DNA/liposome complexes. The detection of mRNA was hindered by the severe degradation of RNA in most samples and RT-PCR could only be performed on three samples, of which one was positive for the nostrils treated with complexed DNA and two were positive for the nostrils treated with DNA alone. Nasal PD measurements were variable between subjects but indicated a statistically significant partial correction of the cAMP-stimulated chloride transport. This study suggests that transfection of the nasal epithelium with DNA alone is at least as effective as transfection with DNA/liposome complexes. However, cross contamination from one nostril to the other cannot be excluded and might have affected the results.

In our most recent study we nebulized liposome/DNA complexes to the lung and 7 days later to the nose of eight CF patients. A further eight patients received only liposomes (placebo group) (Alton *et al.,* 1999). The liposome/DNA complexes were identical to those used by Zabner (Zabner *et al.,* 1997). Because this was the first time that liposome DNA complexes were administered to the lungs of CF patients, safety parameters were carefully studied. The majority of patients (seven of eight) receiving liposome/DNA complexes developed mild flu-like symptoms including a temperature and headache about 6 h after nebulization that resolved within 36 h. Six of eight patients in both groups showed signs of mild bronchial hyperactivity, which lasted for about 48 h. No specific treatment was required for either event. Interestingly, the same dose of lipid when applied to the lungs of normal volunteers failed to demonstrate such responses, suggesting that the bacterial-derived plasmid DNA has inflammatory properties in the lung (Chadwick *et al.,* 1997). IL-6 and C-reactive protein, both plasma markers of inflammation, were slightly increased 48 h after administration of liposome/DNA, but not liposome alone, to the lungs. We did not detect the development of antibodies against the CFTR protein or against double-stranded DNA. There was no difference in the degree of histological bronchial inflammation before and after treatment. However, a reduction in inflammatory cells and IL-8 in the sputum of subjects treated with liposome/DNA complexes was noted. Molecular analysis of lung biopsies and nasal brushings at Day 2 after treatment enabled the detection of plasmid DNA in all samples. Vector-specific mRNA could not be detected in any of the samples. This is most likely due to the presence of contaminating plasmid DNA in the cDNA preparations, which leads to competing reactions during the PCR and preferential amplification of the DNA, but not of the much less abundant vector-specific cDNA.

Potential difference measurements in the lung at Day 2 showed a correction of the cAMP-stimulated chloride secretion toward normal values in six of eight patients receiving DNA/liposome complexes but not in placebo-treated subjects. In the nose the PD values were also significantly altered toward normal for up to 21 days. An *ex vivo* chloride efflux assay (SPQ-assay) (Stern *et al.*, 1995) performed on lung biopsies showed an increase in cAMP-mediated chloride efflux in five of six subjects in the liposome/DNA group but not in the placebo group. A bacterial adherence assay (see following) measured a decrease in bacterial binding in five of six liposome/DNA-treated lung samples, but only in one of six placebo samples. This study provides proof of principle that cationic liposome-mediated transfection can alter the electrophysiological defect in the CF lung.

VI. CONSIDERATIONS BEFORE PROCEEDING INTO PHASE II AND III CLINICAL TRIALS

A. DEVELOPMENT OF MORE RELEVANT CLINICAL END POINT ASSAYS

Measurements of ion transport provide a useful end point of CFTR function. However, it is at present unclear how these relate to disease pathology. The development of end point assays that are more closely related to disease pathology would be of value.

A number of studies have indicated that both *Pseudomonas aeruginosa* and *Staphylococcus aureus* show increased adherence to CF as compared to non-CF airway epithelial cells (Schwab *et al.*, 1993; Saiman and Prince, 1993). Our group has developed an *ex vivo* assay that measures *Pseudomonas aeruginosa* adherence prior to and following gene transfer within the same patients. Changes in the degree of lung inflammation after gene therapy treatment might also provide a suitable end point measure. The detection of inflammatory markers such as nitrite in exhaled breath would be a useful noninvasive approach (Ho *et al.*, 1998). It has also been shown that certain proinflammatory cytokines, for example, interleukin-8 (IL-8), are increased in the sputum samples of CF subjects (Dean *et al.*, 1993).

The development of reliable techniques to measure changes in mucus rheology and mucociliary clearance might also be beneficial.

B. Improving Gene Transfer Efficiency

New lipids and plasmids are constantly being synthesized, but to date *in vivo* gene transfer is still suboptimal. The respiratory epithelium presents a particular challenge for these gene transfer vectors, since one function of the upper respiratory tract is to keep foreign particles, including liposome/DNA complexes, out of the lung. In addition to mucocilliary clearance, mucus coverage of the lung represents a significant extracellular barrier for transfection. Transfection efficiency can be increased some 25-fold on the *ex vivo* model after mucus depletion. It might be feasible to achieve a similar effect through pretreatment of the airway epithelium with mucolytics or drugs, which suppress mucus secretion. Apart from mucus the CF lung is covered in sputum. *In vitro* studies have shown that sputum reduces the transfection efficiency of Cos7 cells. Recombinant DNase was able to reverse the inhibitory effect of sputum significantly (Stern *et al.*, 1998). This enzyme degrades the genomic DNA derived from necrosing inflammatory cells and bacteria, which is a major component of CF sputum, but not the liposome-complexed plasmid DNA.

Alternatively, it might be more efficient to transfect lung epithelial cells through their basolateral membrane. Intravenous injection of liposome/DNA complexes has also been suggested as a route to transfect the airway epithelium via the basolateral membrane (Goula *et al.*, 1998; Griesenbach *et al.*, 1998). Substances that increase paracellular permeability by loosening tight junctions, such as certain fatty acids (Lindmark *et al.*, 1998) or calcium depletion (Bhat *et al.*, 1993), could also be utilized to help liposome/DNA complexes to migrate from the apical to the basolateral membrane when delivered through nebulization.

An alternative to the conventional gene therapy that involves replacing the defective gene is the repair of the gene through homologous recombination or mismatch repair. Such techniques involve transfection of the target tissue (e.g., the lung) with short oligonucleotides that anneal specifically to the target sequence in the gene and lead to a replacement of nucleotides. Kren *et al.* recently used a mismatch repair strategy to introduce mutations in the rat factor IX gene in the liver (Kren *et al.*, 1998).

Various improvements in adenoviral vector design have promise for decreasing the limitations of the vector. Modifications of the vector backbone include the removal of all viral genes, which will potentially prevent the generation of a humoral immune response against viral proteins (Chen *et al.*, 1997). Recently, it was demonstrated that the E4 region, which is deleted in most first- and second-generation vectors, will increase persistent expression (Armentano *et al.,* 1997). This region has now been reintroduced in some vector backbones. Another route for preventing the generation of an immune response is to transiently disable certain pathways in the immune system. Antibodies against the CD40 ligand, for example, have been shown to inhibit both humoral and cellular responses to adenoviral vectors in the mouse airway epithelium (Scaria *et al.,* 1997). In addition to adeno-associated vectors, lentiviral vectors are currently being evaluated for the transduction of differentiated airway epithelium (Goldman *et al.,* 1997).

C. CF PATIENTS SUITABLE FOR GENE TRANSFER STUDIES

Currently, adult patients with established lung disease are enrolled in CF clinical trials. However, these patients are least likely to gain any benefit from the therapy. Clinical benefit is more likely to occur in patients with relatively normal airways. The lungs of these patients are not yet damaged and most likely not coated with mucus and sputum, which might improve transfection efficiencies in the airway epithelium. In addition, it is likely that these patients will tolerate the treatment better. Therefore, the most obvious group of patients to treat are children. However, ethical considerations concerning the treatment of presymptomatic individuals and the considerable problems of the length of follow-up required to demonstrate clinical benefit remain to be solved.

VII. SUMMARY AND OUTLOOK

Over the past 10 years a tremendous amount of gene therapy-related research has been carried out successfully and has led to a large number of phase I clinical trials which unanimously provide proof of principle that gene transfer into the airway epithelium is possible and safe. However, many problems still exist. Low transfection efficiency of the airway epithelium appears to be one of the most prominent problems in all trials. Major research efforts are now underway to increase the transfection efficiency of the lung epithelium. To date, the development for CF gene therapy falls well into the predicted time frame for pharmaceutical drug development.

REFERENCES

Alton, E. W. F. W., Middleton, P. G., Caplen, N. J., Smith, S. N., Steel, D. M., Munkonge, F. M., Jeffery, P. K., Geddes, D. M., Hart, S. L., Williamson, R., Fasold, K. I., Miller, A. D., Dickinson, P.,

Stevenson, B. J., McLachlan, G., Dorin, J. R., and Porteous, D. J. (1993). Non-invasive liposome-mediated gene delivery can correct the ion transport defect in cystic fibrosis mutant mice. *Nature Genet.* **5**, 135–142.

Alton, E. W. F. W., Stern, M., Farley, R., Jaffe, A., Chadwick, S. L., Phillips, J., Davies, J., Smith, S. N., Browning, J., Davies, M. G., Hodson, M. E., Durham, S. R., Li, D., Jeffrey, P. K., Scallan, M., Balfour, R., Eastman, S. J., Cheng, S. H., Smith, A. E., Meeker, D., and Geddes, D. M. (1999). A double-blinded placebo-controlled trial of cationic lipid-mediated CFTR gene transfer to the lungs and nose of CF subjects. *Lancet* **353**, 947–954.

Ames, G. F., and Lecar, H. (1992). ATP-dependent bacterial transporters and cystic fibrosis: Analogy between channels and transporters [Review]. *FASEB J.* **6**, 2660–2666.

Armentano, D., Zabner, J., Sacks, C., Sookdeo, C. C., Smith, M. P., St. George, J. A., Wadsworth, S. C., Smith, A. E., and Gregory, R. J. (1997). Effect of the E4 region on the persistence of transgene expression from adenovirus vectors. *J. Virol.* **71**, 2408–2416.

Bear, C. E., Duguay, F., Naismith, A. L., Kartner, N., Hanrahan, J. W., and Riordan, J. R. (1991). Cl⁻ channel activity in Xenopus oocytes expressing the cystic fibrosis gene. *J. Biol. Chem.* **266**, 19142–19145.

Bear, C. E., Li, C., Kartner, N., Bridges, R. J., Jensen, T. J., Ramjeesingh, M., and Riordan, J. R. (1992). Purification and functional reconstitution of the cystic fibrosis transmembrane conductance regulator (CFTR). *Cell* **68**, 809–818.

Bellon, G., Michel-Calemard, L., Thouvenot, D., Jagneaux, V., Poitevin, F., Malcus, C., Accart, N., Layani, M. P., Aymard, M., Bernon, H., Bienvenu, J., Courtney, M., Doring, G., Gilly, B., Gilly, R., Lamy, D., Levrey, H., Morel, Y., Paulin, C., Perraud, F., Rodillon, L., Sene, C., So, S., Touraine-Moulin, F., Pavirani, A., et al. (1997). Aerosol administration of a recombinant adenovirus expressing CFTR to cystic fibrosis patients: A phase I clinical trial. *Human Gene Ther.* **8**, 15–25.

Bhat, M., Toledo-Velasquez, D., Wang, L., Malanga, C. J., Ma, J. K., and Rojanasakul, Y. (1993). Regulation of tight junction permeability by calcium mediators and cell cytoskeleton in rabbit tracheal epithlium. *Pharm. Res.* **10**, 991–997.

Bout, A., Perricaudet, M., Baskin, G., Imler, J.-L., Scholte, B. J., Pavirani, A., and Valerio, D. (1994). Lung gene therapy: *In vivo* adenovirus-mediated gene transfer to rhesus monkey airway epithelium. *Human Gene Ther.* **5**, 3–10.

Brody, S. L., Metzger, M., Danel, C., Rosenfeld, M. A., and Crystal, R. G. (1994). Acute responses of non-human primates to airway delivery of an adenovirus vector containing the human cystic fibrosis transmembrane conductance regulator cDNA. *Human Gene Ther.* **5**, 821–836.

Cantiello, H. F., Jackson, G. R. Jr., Grosman, C. F., Prat, A. G., Borkan, S. C., Wang, Y., Reisin, I. L., O'Riordan, C. R., and Ausiello, D. A. (1998). Electrodiffusional ATP movement through the cystic fibrosis transmembrane conductance regulator. *Am. J. Physiol.* **274**, C799–809.

Caplen, N. J., Alton, E. W., Middleton, P. G., Dorin, J. R., Stevenson, B. J., Gao, X., Durham, S. R., Jeffery, P. K., Hodson, M. E., Coutelle, C., et al. (1995). Liposome-mediated CFTR gene transfer to the nasal epithelium of patients with cystic fibrosis [see comments] [published erratum appears in *Nature Med.* (1995) **1**, 272]. *Nature Med.* **1**, 39–46.

Chadwick, S. L., Kingston, H. D., Stern, M., Cook, R. M., O'Connor, B. J., Lukasson, M., Balfour, R. P., Rosenberg, M., Cheng, S. H., Smith, A. E., Meeker, D. P., Geddes, D. M., and Alton, E. W. (1997). Safety of a single aerosol administration of escalating doses of the cationic lipid GL-67/DOPE/DMPE-PEG5000 formulation to the lungs of normal volunteers. *Gene Ther.* **4**, 937–942.

Chen, H. H., Mack, L. M., Kelly, R., Ontell, M., Kochanek, S., and Clemens, P. R. (1997). Persistence in muscle of an adenoviral vector that lacks all viral genes. *PNAS USA* **94**, 1645–1650.

Chou, J. L., Rozmahel, R., and Tsui, L. C. (1991). Characterization of the promoter region of the cystic fibrosis transmembrane conductance regulator gene. *J. Biol. Chem.* **266**, 24471–24476.

Clarke, L. L., Grubb, B. R., Yankaskas, J. R., Cotton, C. U., McKenzie, A., and Boucher, R. C. (1994). Relationship of a non-cystic fibrosis transmembrane conductance regulator-mediated chloride conductance to organ-level disease in Cftr(−/−) mice. *PNAS USA* **91**, 479–483.

Colledge, W. H., Abella, B. S., Southern, K. W., Ratcliff, R., Jiang, C., Cheng, S. H., MacVinish, L. J., Anderson, J. R., Cuthbert, A. W., and Evans, M. J. (1995). Generation and characterization of a delta F508 cystic fibrosis mouse model. *Nature Genet.* **10,** 445–452.

Crawford, I., Maloney, P. C., Zeitlin, P. L., Guggino, W. B., Hyde, S. C., Turley, H., Gatter, K. C., Harris, A., and Higgins, C. F. (1991). Immunocytochemical localisation of the cystic fibrosis gene product CFTR. *PNAS USA* **88,** 9262–9266.

Crystal, R. G., Jaffe, A., Brody, S., Mastrangeli, A., McElvaney, N. G., Rosenfeld, M, Chu, C. S., Danel, C., Hay, J., and Eissa, T. (1995). A phase 1 study, in cystic fibrosis patients, of the safety, toxicity, and biological efficacy of a single administration of a replication deficient, recombinant adenovirus carrying the cDNA of the normal cystic fibrosis transmembrane conductance regulator gene in the lung. *Human Gene Ther.* **6,** 643–666.

Dean, T. P., Dai, Y., Shute, J. K., Church, M. K., and Warner, J. O. (1993). Interleukin-8 concentrations are elevated in bronchoalveolar lavage, sputum, and sera of children with cystic fibrosis. *Pediatr. Res.* **34,** 159–161.

Dorin, J. R., Dickinson, P., Alton, E. W. F. W., Smith, S. N., Geddes, D. M., Stevenson, B. J., Kimber, W. L., Fleming, S., Clarke, A. R., Hooper, M. L., Anderson, L., Beddington, R. S. P., and Porteous, D. J. (1992). Cystic fibrosis in the mouse by targeted insertional mutagenesis. *Nature* **359,** 211–215.

Drumm, M. L., Pope, H. A., Cliff, W. H., Rommens, J. M., Marvin, S. A., Tsui, L.-C., Collins, F. S., Frizzell, R. A., and Wilson, J. M. (1990). Correction of the cystic fibrosis defect *in vitro* by retrovirus-mediated gene transfer. *Cell* **62,** 1227–1233.

Dupuit, F., Zahm, J. M., Pierrot, D., Brezillon, S., Bonnet, N., Imler, J. L., Pavirani, A., and Puchelle, E. (1995). Regenerating cells in human airway surface epithelium represent preferential targets for recombinant adenovirus. *Human Gene Ther.* **6,** 1185–1193.

Egan, M., Flotte, T., Afione, S., Solow, R., Zeitlin, P. L., Carter, B. J., and Guggino, W. B. (1992). Defective regulation of outwardly rectifying Cl^- channels by protein kinase A corrected by insertion of CFTR. *Nature* **358,** 581–584.

Engelhardt, J. F., Yankaskas, J. R., Ernst, S. A., Yang, Y., Marino, C. R., Boucher, R. C., Cohn, J. A., and Wilson, J. M. (1992). Submucosal glands are the predominant site of CFTR expression in the human bronchus. *Nature Genet.* **2,** 240–248.

Engelhardt, J. F., Simon, R. H., Yang, Y., Zepeda, M., Weber-Pendleton, S., Doranz, B., Grossman, M., and Wilson, J. M. (1993). Adenovirus-mediated transfer of the CFTR gene to lung of nonhuman primates: Biological efficacy study. *Human Gene Ther.* **4,** 759–769.

Gabriel, S. E., Clarke, L. L., Boucher, R. C., and Stutts, M. J. (1993). CFTR and outward rectifying chloride channels are distinct proteins with a regulatory relationship. *Nature* **363,** 263–266.

Gill, D. R., Southern, K. W., Mofford, K. A., Seddon, T., Huang, L., Sorgi, F., Thomson, A., MacVinish, L. J., Ratcliff, R., Bilton, D., Lane, D. J., Littlewood, J. M., Webb, A. K., Middleton, P. G., Colledge, W. H., Cuthbert, A. W., Evans, M. J., Higgins, C. S., and Hyde, S. C. (1997). A placebo controlled study of liposome-mediated gene transfer to the nasal epithelium of patients with cystic fibrosis. *Gene Ther.* **4,** 199–209.

Gilligan, P. H. (1991). Microbiology of airway disease in patients with cystic fibrosis [Review] [213 refs]. *Clin. Microbiol. Rev.* **4,** 35–51.

Goldman, M. J., Litzky, L. A., Engelhardt, J. F., and Wilson, J. M. (1995a). Transfer of the CFTR gene to the lung of nonhuman primates with E1-deleted, E2a-defective recombinant adenoviruses: A preclinical toxicology study. *Human Gene Ther.* **6,** 839–851.

Goldman, M. J., Yang, Y., and Wilson, J. M. (1995b). Gene therapy in a xenograft model of cystic fibrosis lung corrects chloride transport more effectively than the sodium defect. *Nature Genet.* **9,** 126–131.

Goldman, M. J., Lee, P. S., Yang, J. S., and Wilson, J. M. (1997). Lentiviral vectors for gene therapy of cystic fibrosis. *Human Gene Ther.* **8,** 2261–2268.

Goldman, M. J., and Wilson, J. M. (1995). Expression of $alpha_v beta_5$ integrin is necessary for efficient adenovirus-mediated gene transfer in the human airway. *J. Virol.* **69,** 5951–5958.

Goula, D., Benoist, C., Mantero, S., Merlo, G., Levi, G., and Demeneix, B. A. (1998). Polyethyl-enimine-based intravenous delivery of transgenes to mouse lung. *Gene Ther.* **5,** 1291–1295.

Griesenbach, U., Suen, T.-C., Chamberlain, J., Olek, K., and Tsui, L.-C. (1994). Study of the human CFTR promoter in transgenic mice. *Pediatr. Pulmonol.* [Supplement] **10,** 199 [Abstract].

Griesenbach, U., Chonn, A., Cassady, R., Hannam, V., Ackerley, C., Post, M., Tanswell, A. K., Olek, K., O'Brodovich, H., and Tsui, L. (1998). Comparison between intratracheal and intra-venous administration of liposome-DNA complexes for cystic fibrosis lung gene therapy. *Gene Ther.* **5,** 181–188.

Grubb, B. R., Pickles, R. J., Ye, H., Yankaskas, J. R., Vick, R. N., Engelhardt, J. F., Wilson, J. M., Johnson, L. G., and Boucher, R. C. (1994a). Inefficient gene transfer by adenovirus vector to cystic fibrosis airway epithelia of mice and humans. *Nature* **371,** 802–806.

Grubb, B. R., Vick, R. N., and Boucher, R. C. (1994b). Hyperabsorption of Na$^+$ and raised Ca^{2+}-mediated Cl$^-$ secretion in nasal epithelia of CF mice. *Am. J. Physiol.* **266,** C1478–C1483.

Hasty, P., O'Neal, W. K., Liu, K. Q., Morris, A. P., Bebok, Z., Shumyatsky, G. B., Jilling, T., Sorscher, E. J., Bradley, A., and Beaudet, A. L. (1995). Severe phenotype in mice with termination mutation in exon 2 of cystic fibrosis gene. *Somatic Cell Mol. Genet.* **21,** 177–187.

Hay, J. G., McElvaney, N. G., Herena, J., and Crystal, R. G. (1995). Modification of nasal epithelial potential differences of individuals with cystic fibrosis consequent to local administration of a normal CFTR cDNA adenovirus gene transfer vector. *Human Gene Ther.* **6,** 1487–1496.

Higgins, C. F., and Gottesman, M. M. (1992). Is the multidrug transporter a flippase? [Review] [35 refs]. *Trends Biochem. Sci.* **17,** 18–21.

Ho, L. P., Innes, J. A., and Greening, A. P. (1998). Nitrite levels in breath condensate of patients with cystic fibrosis is elevated in contrast to exhaled nitric oxide. *Thorax* **53,** 680–684.

Hyde, S. C., Gill, D. R., Higgins, C. F., Trezise, A. E. O., MacVinish, L. J., Cuthbert, A. W., Ratcliff, R., Evans, M. J., and Colledge, W. H. (1993). Correction of the ion transport defect in cystic fibrosis transgenic mice by gene therapy. *Nature* **362,** 250–255.

Johnson, L. G., Olsen, J. C., Sarkadi, B., Moore, K. L., Swanstrom, R., and Boucher, R. C. (1992). Efficiency of gene transfer for restoration of normal airway epithelial function in cystic fibrosis. *Nature Genet.* **2,** 21–25.

Kaplan, J. M., St. George, J. A., Pennington, S. E., Keyes, L. D., Johnson, R. P., Wadsworth, S. C., and Smith, A. E. (1996). Humoral and cellular immune responses of nonhuman primates to long-term repeated lung exposure to Ad2/CFTR-2. *Gene Ther.* **3,** 117–127.

Knowles, M. R., Noone, P. G., Hohneker, K., Johnson, L. G., Boucher, R. C., Efthimiou, J., Crawford, C., Brown, R., Schwartzbach, C., and Pearlman, R. (1998). A double-blind, placebo controlled, dose ranging study to evaluate the safety and biological efficacy of the lipid-DNA complex GR213487B in the nasal epithelium of adult patients with cystic fibrosis. *Human Gene Ther.* **9,** 249–269.

Koh, J., Sferra, T. J., and Collins, F. S. (1993). Characterization of the cystic fibrosis transmembrane conductance regulator promoter region. Chromatin context and tissue-specificity. *J. Biol. Chem.* **268,** 15912–15921.

Kren, B. T., Bandyopadhyay, P., and Steer, C. J. (1998). *In vivo* site-directed mutagenesis of the factor IX gene by chimeric RNA/DNA oligonucleotides [see comments]. *Nature Med.* **4,** 285–290.

Li, C., Ramjeesingh, M., and Bear, C. E. (1996). Purified cystic fibrosis transmembrane conductance regulator (CFTR) does not function as an ATP channel. *J. Biol. Chem.* **271,** 11623–11626.

Lindmark, T., Kimura, Y., and Artursson, P. (1998). Absorption enhancement through intracellular regulation of tight junction permeability by medium chain fatty acids in Caco-2. *J. Pharmacol. Exper. Ther.* **284,** 362–369.

Manson, A. L., Trezise, A. E., MacVinish, L. J., Kasschau, K. D., Birchall, N., Episkopou, V., Vassaux, G., Evans, M. J., Colledge, W. H., Cuthbert, A. W., and Huxley, C. (1997). Complementation of null CF mice with a human CFTR YAC transgene. *EMBO J.* **16,** 4238–4249.

McNicholas, C. M., Guggino, W. B., Schwiebert, E. M., Hebert, S. C., Giebisch, G., and Egan, M. E.

(1996). Sensitivity of a renal K+ channel (ROMK2) to the inhibitory sulfonylurea compound glibenclamide is enhanced by coexpression with the ATP-binding cassette transporter cystic fibrosis transmembrane regulator. *PNAS USA* **93,** 8083–8088.

Meindl, R. S. (1987). Hypothesis: A selective advantage for cystic fibrosis heterozygotes. *Am. J. Phys. Anthropol.* **74,** 39–45.

Morral, N., Bertranpetit, J., Estivill, X., Nunes, V., Casals, T., Gimenez, J., Reis, A., Varon-Mateeva, R., Macek, M., Jr., Kalaydjieva, L., *et al.* (1994). The origin of the major cystic fibrosis mutation (delta F508) in European populations [see comments]. *Nature Genet.* **7,** 169–175.

Nabel, G. J., Nabel, E. G., Yang, Z.-Y., Fox, B. A., Plautz, G. E., Gao, X., Huang, L., Shu, S., Gordon, D., and Chang, A. E. (1993). Direct gene transfer with DNA-liposome complexes in melanoma: Expression, biologic activity, and lack of toxicity in humans. *Proc. Natl. Acad. Sci. USA* **90,** 11307–11311.

Pier, G. B., Grout, M., Zaidi, T. S., Olsen, J. C., Johnson, L. G., Yankaskas, J. R., and Goldberg, J. B. (1996). Role of mutant CFTR in hypersusceptibility of cystic fibrosis patients to lung infections. *Science* **271,** 64–67.

Pier, G. B., Grout, M., Zaidi, T., Meluleni, G., Mueschenborn, S. S., Banting, G., Ratcliff, R., Evans, M. J., and Colledge, W. H. (1998). *Salmonella typhi* uses CFTR to enter intestinal epithelial cells. *Nature* **393,** 79–82.

Pilewski, J. M., Engelhardt, J. F., Bavaria, J. E., Kaiser, L. R., Wilson, J. M., and Albelda, S. M. (1995). Adenovirus-mediated gene transfer to human bronchial submucosal glands using xenografts. *Am. J. Physiol.* **268,** L657-L665.

Porteous, D. J., Dorin, J. R., McLachlan, G., Davidson-Smith, H., Davidson, H., Stevenson, B. J., Carothers, A. D., Wallace, A. J., Moralee, S., Hoenes, C., Kallmeyer, G., Michaelis, U., Naujoks, K., Ho, L., Samways, J. M., Imrie, M., Greening, A. P., and Innes, J. A. (1997). Evidence for safety and efficacy of DOTAP cationic liposome mediated CFTR gene transfer to nasal epithelium of patients with cystic fibrosis. *Gene Ther.* **4,** 210–218.

Ratcliff, R., Evans, M. J., Cuthbert, A. W., MacVinish, L. J., Foster, D., Anderson, J. R., and Colledge, W. H. (1993). Production of a severe cystic fibrosis mutation in mice by gene targeting. *Nature Genet.* **4,** 35–41.

Ray, P., Tang, W., Wang, P., Homer, R., Kuhn, C., III, Flavell, R. A., and Elias, J. A. (1997). Regulated overexpression of interleukin 11 in the lung. Use to dissociate development-dependent and -independent phenotypes. *J. Clin. Invest.* **100,** 2501–2511.

Reddy, M. M., Quinton, P. M., Haws, C., Wine, J. J., Grygorczyk, R., Tabcharani, J. A., Hanrahan, J. W., Gunderson, K. L., and Kopito, R. R. (1996). Failure of the cystic fibrosis transmembrane conductance regulator to conduct ATP. *Science* **271,** 1876–1879.

Reisin, I. L., Prat, A. G., Abraham, E. H., Amara, J. F., Gregory, R. J., Ausiello, D. A., and Cantiello, H. F. (1994). The cystic fibrosis transmembrane conductance regulator is a dual ATP and chloride channel. *J. Biol. Chem.* **269,** 20584–20591.

Rich, D. P., Anderson, M. P., Gregory, R. J., Cheng, S. H., Paul, S., Jefferson, D. M., McCann, J. D., Klinger, K. W., Smith, A. E., and Welsh, M. J. (1990). Expression of cystic fibrosis transmembrane conductance regulator corrects defective chloride channel regulation in cystic fibrosis airway epithelial cells. *Nature* **347,** 358–363.

Riordan, J. R., Rommens, J. M., Kerem, B.-S., Alon, N., Rozmahel, R., Grzelczak, Z., Zielenski, J., Lok, S., Plavsic, N., Chou, J.-L., Drumm, M. L., Iannuzzi, M. C., Collins, F. S., and Tsui, L.-C. (1989). Identification of the cystic fibrosis gene: Cloning and characterization of complementary DNA. *Science* **245,** 1066–1073.

Rommens, J. M., Iannuzzi, M. C., Kerem, B.-S., Drumm, M. L., Melmer, G., Dean, M., Rozmahel, R., Cole, J. L., Kennedy, D., Hidaka, N., Zsiga, M., Buchwald, M., Riordan, J. R., Tsui, L.-C., and Collins, F. S. (1989). Identification of the cystic fibrosis gene: Chromosome walking and jumping. *Science* **245,** 1059–1065.

Rosenfeld, M. A., Yoshimura, K., Trapnell, B. C., Yoneyama, K., Rosenthal, E. R., Dalemans, W., Fukayama, M., Bargon, J., Stier, L. E., Stratford-Perricaudet, L., Perricaudet, M., Guggino, W. B., Pavirani, A., Lecocq, J.-P., and Crystal, R. G. (1992). In vivo transfer of the human cystic fibrosis transmembrane conductance regulator gene to the airway epithelium. *Cell* **68,** 143–155.

Rosenfeld, M. A., Chu, C.-S., Seth, P., Danel, C., Banks, T., Yoneyama, K., Yoshimura, K., and Crystal, R. G. (1994). Gene transfer to freshly isolated human respiratory epithelial cells *in vitro* using a replication-deficient adenovirus containing the human cystic fibrosis transmembrane conductance regulator cDNA. *Human Gene Ther.* **5,** 331–342.

Rozmahel, R., Wilschanski, M., Matin, A., Plyte, S., Oliver, M., Auerbach, W., Moore, A., Forstner, J., Durie, P., Nadeau, J., Bear, C., and Tsui, L. C. (1996). Modulation of disease severity in cystic fibrosis transmembrane conductance regulator deficient mice by a secondary genetic factor. *Nature Genet.* **12,** 280–287.

Saiman, L., Cacalano, G., Gruenert, D., and Prince, A. (1992). Comparison of adherence of *Pseudomonas aeruginosa* to respiratory epithelial cells from cystic fibrosis patients and healthy subjects. *Infect. Immun.* **60,** 2808–2814.

Saiman, L. and Prince, A. (1993). *Pseudomonas aeruginosa* pili bind to asialoGM1 which is increased on the surface of cystic fibrosis epithelial cells. *J. Clin. Invest.* **92,** 1875–1880.

Scaria, A., St. George, J. A., Gregory, R. J., Noelle, R. J., Wadsworth, S. C., Smith, A. E., and Kaplan, J. M. (1997). Antibody to CD40 ligand inhibits both humoral and cellular immune responses to adenoviral vectors and facilitates repeated administration to mouse airway. *Gene Ther.* **4,** 611–617.

Schwab, U. E., Wold, A. E., Carson, J. L., Leigh, M. W., Cheng, P.-W., Gilligan, P. H., and Boat, T. F. (1993). Increased adherence of *Staphylococcus aureus* from cystic fibrosis lungs to airway epithelial cells. *Am. Rev. Respir. Dis.* **148,** 365–369.

Schwartz, D. A., Quinn, T. J., Thorne, P. S., Sayeed, S., Yi, A. K., and Krieg, A. M. (1997). CpG motifs in bacterial DNA cause inflammation in the lower respiratory tract. *J. Clin. Invest.* **100,** 68–73.

Simon, R. H., Engelhardt, J. F., Yang, Y., Zepeda, M., Weber-Pendleton, S., Grossman, M., and Wilson, J. M. (1993). Adenovirus-mediated transfer of the CFTR gene to lung of nonhuman primates: Toxicity study. *Human Gene Ther.* **4,** 771–780.

Smith, J. J., Travis, S. M., Greenberg, E. P., and Welsh, M. J. (1996). Cystic fibrosis airway epithelia fail to kill bacteria because of abnormal airway surface fluid [published erratum appears in *Cell* (1996), **87,** following 355]. *Cell* **85,** 229–236.

Snouwaert, J. N., Brigman, K. K., Latour, A. M., Malouf, N. N., Boucher, R. C., Smithies, O., and Koller, B. H. (1992). An animal model for cystic fibrosis made by gene targeting. *Science* **257,** 1083–1088.

Stern, M., Munkonge, F. M., Caplen, N. J., Sorgi, F., Huang, L., Geddes, D. M., and Alton, E. W. F. W. (1995). Quantitative flourescence measurements of chloride secretion in native airway epithelium from CF and non-CF subjects. *Gene Ther.* **2,** 766–774.

Stern, M., Caplen, N. J., Browning, J. E., Griesenbach, U., Sorgi, F., Huang, L., Gruenert, D. C., Marriot, C., Crystal, R. G., Geddes, D. M., and Alton, E. W. (1998). The effect of mucolytic agents on gene transfer across a CF sputum barrier *in vitro*. *Gene Ther.* **5,** 91–98.

Stutts, M., Canessa, C. M., Olsen, J. C., Hamrick, M., Cohn, J. A., Rossier, B. C., and Boucher, R. C. (1995). CFTR as a cAMP-dependent regulator of sodium channels. *Science* **269,** 847–850.

Super, M. (1979). Factors influencing the frequency of cystic fibrosis in South West Africa. *Monogr. Paediatr.* **10,** 106–113.

Tabcharani, J. A., Chang, X.-B., Riordan, J. R., and Hanrahan, J. W. (1991). Phosphorylation-regulated Cl⁻ channel in CHO cells stably expressing the cystic fibrosis gene. *Nature* **352,** 628–631.

Tizzano, E. F., O'Brodovich, H., Chitayat, D., Bènichou, J.-C., and Buchwald, M. (1994). Regional expression of CFTR in developing human respiratory tissues. *Am. J. Respir. Cell Mol. Biol.* **10,** 355–362.

Trezise, A. E. O., Chambers, J. A., Wardle, C. J., Gould, S., and Harris, A. (1993). Expression of the cystic fibrosis gene in human foetal tissues. *Human Mol. Genet.* **2,** 213–218.

Trezise, A. E. O. and Buchwald, M. (1991). *In vivo* cell-specific expression of the cystic fibrosis transmembrane conductance regulator. *Nature* **353,** 434–437.

Turnberg, L. A., Bieberdorf, F. A., Morawski, S. G., and Fordtran, J. S. (1970). Interrelationships of chloride, bicarbonate, sodium, and hydrogen transport in the human ileum. *J. Clin. Invest.* **49,** 557–567.

Wagner, J. A., Reynolds, T., Moran, M. L., Moss, R. B., Wine, J. J., Flotte, T. R., and Gardner, P. (1998). Efficient and persistent gene transfer of AAV-CFTR in maxillary sinus. *Lancet* **351,** 1702–1703.

Welsh, M. J. (1987). Electrolyte transport by airway epithelia. *Physiol. Rev.* **67,** 1143–1184.

Whitsett, J. A., Dey, C. R., Stripp, B. R., Wikenheiser, K. A., Clark, J. C., Wert, S. E., Gregory, R. J., Smith, A. E., Cohn, J. A., Wilson, J. M., and Engelhardt, J. (1992). Human cystic fibrosis transmembrane conductance regulator directed to respiratory epithelial cells of transgenic mice. *Nature Genet.* **2,** 13–20.

Yei, S., Mittereder, N., Tang, K., O'Sullivan, C., and Trapnell, B. C. (1994). Adenovirus-mediated gene transfer for cystic fibrosis: Quantitative evaluation of repeated *in vivo* vector administration to the lung. *Gene Ther.* **1,** 192–200.

Yoshimura, K., Nakamura, H., Trapnell, B. C., Dalemans, W., Pavirani, A., Lecocq, J.-P., and Crystal, R. G. (1991). The cystic fibrosis gene has a "housekeeping"-type promoter and is expressed at low levels in cells of epithelial origin. *J. Biol. Chem.* **266,** 9140–9144.

Yoshimura, K., Rosenfeld, M. A., Nakamura, H., Scherer, E. M., Pavirani, A., Lecocq, J.-P., and Crystal, R. G. (1992). Expression of the human cystic fibrosis transmembrane conductance regulator gene in the mouse lung after *in vivo* intratracheal plasmid-mediated gene transfer. *Nucleic Acids Res.* **20,** 3233–3240.

Zabner, J., Couture, L. A., Gregory, R. J., Graham, S. M., Smith, A. E., and Welsh, M. J. (1993). Adenovirus-mediated gene transfer transiently corrects the chloride transport defect in nasal epithelia of patients with cystic fibrosis. *Cell* **75,** 207–216.

Zabner, J., Petersen, D. M., Puga, A. P., Graham, S. M., Couture, L. A., Keyes, L. D., Lukason, M. J., St. George, J. A., Gregory, R. J., Smith, A. E., and Welsh, M. J. (1994). Safety and efficacy of repetitive adenovirus-mediated transfer of CFTR cDNA to airway epithelia of primates and cotton rats. *Nature Genet.* **6,** 75–83.

Zabner, J., Ramsey, B. W., Meeker, D. P., Aitken, M. L., Balfour, R. P., Gibson, R. L., Launspach, J., Moscicki, R. A., Richards, S. M., Standaert, T. A., *et al.* (1996). Repeat administration of an adenovirus vector encoding cystic fibrosis transmembrane conductance regulator to the nasal epithelium of patients with cystic fibrosis. *J. Clin. Invest.* **97,** 1504–1511.

Zabner, J., Cheng, S. H., Meeker, D., Launspach, J., Balfour, R., Perricone, M. A., Morris, J. E., Marshall, J., Fasbender, A., Smith, A. E., and Welsh, M. J. (1997). Comparison of DNA-lipid complexes and DNA alone for gene transfer to cystic fibrosis airway epithelium *in vivo*. *J. Clin. Invest.* **100,** 1529–1537.

Zeiher, B. G., Eichwald, E., Zabner, J., Smith, J. J., Puga, A. P., McCray, P. B., Jr., Capecchi, M. R., Welsh, M. J., and Thomas, K. R. (1995). A mouse model for the delta F508 allele of cystic fibrosis. *J. Clin. Invest.* **96,** 2051–2064.

Zhang, Y., Jiang, Q., Dudus, L., Yankaskas, J. R., and Engelhardt, J. F. (1998). Vector-specific complementation profiles of two independent primary defects in cystic fibrosis airways. *Human Gene Ther.* **9,** 635–648.

Zielenski, J., Rozmahel, R., Bozon, D., Kerem, B., Grzelczak, Z., Riordan, J. R., Rommens, J., and Tsui, L. C. (1991). Genomic DNA sequence of the cystic fibrosis transmembrane conductance regulator (CFTR) gene. *Genomics* **10,** 214–228.

Zielenski, J. and Tsui, L. C. (1995). Cystic fibrosis: Genotypic and phenotypic variations [Review]. *Annu. Rev. Genet.* **29,** 777–807.

CHAPTER 16

Targeting HER-2/neu-Overexpressing Cancer Cells with Transcriptional Repressor Genes Delivered by Cationic Liposome

Mien-Chie Hung,* Shao-Chun Wang,* and Gabriel Hortobagyi[†]

*Department of Cancer Biology, Section of Molecular Cell Biology, University of Texas M. D. Anderson Cancer Center, Houston, Texas

[†]Department of Breast Medical Oncology, University of Texas M. D. Anderson Cancer Center, Houston, Texas

HER-2/neu gene overexpression is a frequent molecular event in human cancers. Overexpression of HER-2/neu indicates an unfavorable prognosis and is correlated with the low survival rate of patients with multiple cancer types including breast and ovarian cancer. Down-regulation of the HER-2/neu gene overexpression in cancer cells has been shown to be a useful strategy in significantly reversing the malignancy induced by HER-2/neu overexpression. One way to repress

Nonviral Vectors for Gene Therapy

HER-2/neu overexpression is to attenuate the promoter activity of the HER-2/neu gene. We have identified a number of potent transcriptional regulators, including the ets family member PEA3, the SV40 large T antigen, and the adenovirus type 5 E1A, to test their ability to repress HER-2/neu gene expression. Expression of these transcriptional regulators resulted in down-regulation of the HER-2/neu promoter activity and reversed the transformation phenotype of the cancer cells *in vitro*. These observations were followed by a series of studies to investigate whether these HER-2/neu repressors can act therapeutically as tumor suppressor genes for cancers that overexpress HER-2/neu. The results of these preclinical studies clearly indicate that transcriptional repressors that down-regulate HER-2/neu can be an effective regimen for cancer treatment in a gene therapy format combined with an appropriate gene delivery system such as the cationic liposome DC-chol. Our results have demonstrated that the nonviral cationic liposome can effectively transfer the therapeutic genes into the cancer cells *in vivo* and can lead to significant suppressive effects on tumor growth. Furthermore, the tumor-free survival rate of treated animals is dramatically increased under nontoxic doses compared with the rate for nontreated animals. Data yielded from these animal studies have paved the way for the clinical trial of the tumor suppressor gene E1A delivered by cationic liposome. A phase I clinical trial using the E1A liposome on breast and ovarian patients has recently been completed. Down-regulation of HER-2/neu overexpression and a decreased number of cancer cells are readily detectable from the fluid of the treated cancer patients, suggesting the feasibility of using the E1A liposome in cancer gene therapy. These studies, together with others discussed in this book, have demonstrated the great potential for the clinical application of the cationic liposome as a nonviral gene delivery system.

I. INTRODUCTION

A. HER-2/NEU AS A CRITICAL TARGET FOR CANCER GENE THERAPY

The HER-2/neu (also known as c-erbB2) gene encodes a receptor tyrosine kinase (p185) with significant structural and functional homology to the epidermal growth-factor receptor (EGFR) (Bargmann *et al.*, 1986a; Hung *et al.*, 1986; Yamamoto *et al.*, 1986; Coussens *et al.*, 1985). Each protein member of the erbB receptor family contains an extracellular domain, a transmembrane domain, and an intracellular domain with intrinsic tyrosine kinase activity. Although the ligand for the HER-2/neu receptor has not been identified, the HER-2/neu receptor is known to mediate lateral signal transduction through all erbB receptor family members (Plowman *et al.*, 1993; Graus-Porta *et al.*, 1997; Sliwkowski *et al.*, 1994; Wallasch *et al.*, 1995; Carraway *et al.*, 1994), due to the preference for the HER-2/neu receptor as a heterodimerazation partner for all erbB receptors. After ligand binding, EGFR, HER-3 (also known as erbB3), and HER-4 (also known as erbB4) can heterodimerize with HER-2/neu, and can lead to the tyrosine phosphorylation of all these receptors (Graus-Porta *et al.*, 1997; Sliwkowski *et al.*, 1994; Wallasch *et al.*, 1995).

The oncogenic properties of the HER-2/neu oncogene, including cellular transformation, tumorigenicity, and metastasis-promoting activity, were first demonstrated in the rat neu oncogene (Hung *et al.*, 1986; Yu and Hung, 1991; Bargmann *et al.*, 1986b; Hung *et al.*, 1989). As a matter of fact, the mutation-activated rat neu oncogene, which contains a point mutation in the transmembrane domain of the protein resulting in a constitutive tyrosine kinase activity, was originally isolated from rat neuroglioblastoma because of its ability to transform mouse cells (Hung *et al.*, 1986; Bargmann *et al.*, 1986b). In humans, the HER-2/neu proto-oncogene is frequently amplified or overexpressed in many types of cancers including breast (Slamon *et al.*, 1987, 1989; Gusterson *et al.*, 1992; Toikkanen *et al.*, 1992), ovarian (Slamon *et al.*, 1989; Berchuck *et al.*, 1990, 1991), lung (Schneider *et al.*, 1989; Weiner *et al.*, 1990; Shi *et al.*, 1992), stomach (Yokota *et al.*, 1998; Park *et al.*, 1989), and oral (Xia *et al.*, 1997) cancers, suggesting that HER-2/neu overexpression plays a critical role in the development of human malignancy. The overall survival rate of cancer patients whose tumors have HER-2/neu overexpression is significantly shorter than for those patients whose tumors do not have HER-2/neu overexpression (Slamon *et al.*, 1987, 1989; Berchuck *et al.*, 1990; Weiner *et al.*, 1990; Xia *et al.*, 1997). Furthermore, increased expression of the HER-2/neu gene has been shown to correlate with the number of lymph node metastases in breast cancer patients (Slamon *et al.*, 1987); this observation is consistent with many laboratory studies and indicates that increased expression of the HER-2/neu gene enhanced metastatic potential in mouse fibroblast, human breast, ovarian, and non-small-cell lung carcinoma (NSCLC) (Yu and Hung, 1991; Yu *et al.*, 1994; Yu, Hamada, *et al.*, 1992; Tan *et al.*, 1997; Tsai *et al.*, 1995, 1993; Chazin *et al.*, 1992; Benz *et al.*, 1993). Transfection and expression of HER-2/neu in the human breast and ovarian cancer cells resulted in more malignant growth characteristics including higher DNA synthesis rate, faster growth rate, gain in the ability to grow in soft agar, as well as enhanced tumorigenicity and metastatic potential when transplanted in nude mice (Tsai *et al.*, 1995, 1993). In addition to the ability to enhance metastasis potential in cancer cells, HER-2/neu overexpression was also reported to induce chemoresistance of cancer cells in certain experimental conditions. A high level of HER-2/neu expression in human NSCLC appeared to result in enhanced resistance to a panel of chemotherapeutic agents (Tsai *et al.*, 1995, 1993). Similarly, overexpression of HER-2/neu in breast cancer cells induced chemoresistance to Taxol (Paclitaxel) (Yu *et al.*, 1996; Yu, Liu, Jing, Sun, *et al.*, 1998). However, the expression level of HER-2/neu seems to be critical for the development of chemo-resistance since in certain cell lines moderate p185 expression levels do not result in significant drug resistance (Pegram *et al.,* 1997). It is likely that HER-2/neu expression must be higher than a threshold level to induce significant drug resistance. Furthermore, in contrast with the results in lung cancer cells, the chemo-resistance developed in those HER-2/neu–overexpressing breast cancer cells is seen with Paclitaxel and Taxotere but not with other drugs (Yu *et al.,* 1996; Yu, Liu, Jing, Sun, *et al.,* 1998; Yu, D., and Hung, M.-C., unpublished results), suggesting a selective mechanism of resistance. It is not yet clear why HER-2/neu overexpression-mediated drug resistance behaves differently for lung and breast cancer cells. However, in the case of resistance to Paclitaxel by HER-2/neu overexpression in breast cancer cells, a molecular mechanism has recently been suggested (Yu, Liu, Jing, McDommell, *et al.,* 1998): up-regulation of p21 by HER-2/neu overexpression inhibits cyclin B/cdc2 kinase activity in the G2/M phase, which is required for Paclitaxel-induced apoptosis. This mechanism clearly indicates that HER-2/neu overexpression in breast cancer cells antagonizes Paclitaxel-induced apoptosis.

Since the HER-2/neu proto-oncogene overexpression significantly contributes to the malignant development of many types of human cancers in different aspects, molecular strategies that aim to down-regulate HER-2/neu gene expression have become highly attractive approaches to fight human cancer. One of the successful examples is the recently FDA-approved Herceptin, which is a humanized monoclonal antibody against the extracellular domain of the HER-2/neu-encoded p185 protein and is capable of down-regulation of HER-2/neu (Pegram *et al.,* 1998; Baselga *et al.,* 1998).

B. Transcriptional Repression as an Effective Means to Down-Regulate HER-2/neu Expression in Cancer Cells

HER-2/neu gene amplification can be detected in the majority of breast tumor tissues with overexpression of the HER-2/neu-encoded p185 protein (Slamon *et al.,* 1989). In established breast cancer cell lines, both gene amplification and transcriptional up-regulation are common scenarios accounting for the increased HER-2/neu gene expression in different breast cancer cells (Kraus *et al.,* 1987; Miller *et al.,* 1994; Bosher *et al.,* 1996; Hollywood and Hurst, 1993). It is therefore likely that both gene amplification and transcriptional up-regulation are involved in HER-2/neu overexpression in cancer cells.

The promoter of the HER-2/neu gene has been well characterized. In the past few years, knowledge about the *cis-* and *trans*-acting elements regulating the transcription of the HER-2/neu proto-oncogene have been rapidly accumulated. A number of *cis*-acting motifs are distributed along the HER-2/neu promoter, including the binding sites of transcription factors Sp1, OTF1, AP2, E4TF1, and PEA3 (White and Hung, 1992; Miller and Hung, 1995; Yan and Hung, 1991). In addition, a 13-bp element that is required for HER-2/neu transactivation has been identified while the cognate DNA-binding protein remains to be determined (Millar *et al.*, 1994). AP2 has been shown to be a strong activator of the HER-2/neu gene and is functionally activated in HER-2/neu–overexpressing breast cancer cell lines such as MDA-MB-361, MDA-MB-175, ZR-75-1, BT-474, and SK-BR-3 (Hollywood and Hurst, 1993). The high activity of AP2 in these cell lines has been correlated with the elevated HER-2/neu gene expression level in these cells. On the other hand, the HER-2/neu gene is subject to the negative regulation of a number of cellular or viral factors through different mechanisms (Fig. 1). For example, PEA3, a member of the ets family (X. Xing, S.-C. Wang, and M.-C. Hung, unpublished results; Xing, Miller, *et al.*, 1997), and the retinoblastoma tumor suppressor (RB) (Yu, Matin, and Hung, 1992) can repress the HER-2/neu gene expression. Interestingly, in addition to the cellular factors, the HER-2/neu gene transcription can also be repressed by a number of viral transcription factors such as the simian virus 40 (SV40) large T antigen and the adenovirus type 5 E1A (Yan *et al.*, 1991; Yu *et al.*, 1991; Xing *et al.*, 1996; Yu *et al.*, 1990). These findings have prompted us to examine whether transcriptional repression of the HER-2/neu gene may serve as an effective way to reverse the malignant transformation mediated by HER-2/neu overexpression as described in later sections. These studies have demonstrated the potential application of transcriptional repressors as therapeutic agents targeting HER-2/neu–overexpressing cancer cells.

II. PRECLINICAL STUDY USING DC-CHOL LIPOSOME COMPLEXING WITH THE E1A, SV40 LARGE T, OR PEA3 GENE

A. Gene Delivery System and the Animal Model

To establish a gene delivery system for the gene therapy experiments discussed next, we tested the DNA transfer efficiency of the cationic liposome, DC-chol/DOPE $\{3\beta[N-(N',N'-\text{dimethylaminoethane})\text{-carbamoyl}]$ cholesterol/ dioleoylphosphatidyl-ethanolamine $(3/2)\}$, by transfecting the lacZ gene into SK-OV-3 cells, an ovarian cancer cell line with overexpressed HER 2/neu that was used to establish animal models in our experiments. This liposome formula

A.

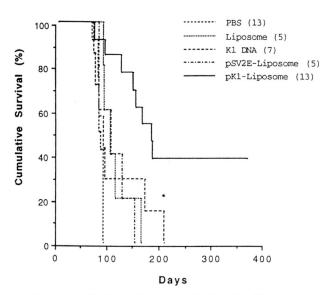

B.

Figure 1 Liposome-mediated gene therapy on animal models. (A) Liposome-mediated *in vivo* E1A gene transfer suppressed dissemination of ovarian cancer cells that overexpress HER-2/neu. Mice treated with wild-type E1A-liposome complex survived longer than untreated mice or mice injected with wild-type E1A DNA alone, with liposome alone, or with liposome plus a E1A frame-shift DNA (Efs). Mice were given intraperitoneal (ip) injection of 2×10^6 viable SK-OV-3 human ovarian cancer cells 5 days before treatment. The mice then received weekly ip injection of 200 μl of a reagent containing

has been shown to transfect mammalian cells efficiently with very low toxicity (Xing *et al.,* 1998; Xing, Liu, *et al.,* 1997; Gao and Huang, 1996; Nabel *et al.,* 1993). For *in vivo* gene transfer, we initially chose a DNA/liposome ratio of 1/13 (15 μg DNA/200 nmol liposome) (Yu *et al.,* 1995), based on a maximum ratio of transfection efficiency to cytotoxicity in tissue culture. However, other combinations with different DNA/liposome ratios such as 1/1 (15 μg DNA/15 nmol liposome) may be as effective as the 1/13 ratio in tumor tissues after intraperitoneal (ip) injection (Xing *et al.,* 1998).

The animal model used for the following studies including E1A, SV40 large T, and PEA3 gene therapy has nude mice bearing disseminated human ovarian tumors derived from the p185-overexpressing SK-OV-3 ovarian cancer cells (Yu *et al.,* 1995). Injection of these cells resulted in their dissemination into the peritoneal cavity, which is a natural dissemination site of ovarian cancer cells. Our observation showed that within 5 days tumors developed in the peritoneal cavity. These tumors were confirmed to overexpress the p185 oncoprotein by immunohistochemical staining. This animal model mimics stage III human ovarian cancers (Beahrs *et al.,* 1992), for which there is so far no effective therapeutic regimen. In the case of the E1A test, a breast cancer orthotopic model was also explored (Chang *et al.,* 1997). MDA-MB-361 cells that overexpress p185 were injected into the mammary fat pad of nude mice, and the mice were given a local injection of DNA/liposome complex when palpable solid tumors were detected. Tumor volume was monitored monthly, and cancer metastasis and the mouse survival rate were observed for up to 2 years.

A series of studies have been conducted to evaluate the safety of ip injection of E1A in normal mice (Xing, Liu, *et al.,* 1997). The cumulative doses used were from 5 to 40 times the DNA/liposome therapeutically effective dose (15 μg E1A DNA plus 200 nm of the DC-chol/DOPE liposome) in the animal model. The starting dose (1.8 mg/m²) in the phase I clinical trial was also converted from this dose. In this range of doses, the administration of the E1A/liposome complex did not have adverse effects on renal, hepatic, and hematological parameters studied.

---

15 μg of E1A DNA (or the same amount of the negative control DNA) complex with 200 nmol of liposome. [Yu *et al.* (1995), reproduced with permission.] (B) Survival of the mice was prolonged after treatment with K1-liposome complexes. Female nu/nu mice were injected intraperitoneally (ip) with 2×10^6 SK-OV-3 cells 5 days before treatment. The mice received weekly ip injection of 200 μl of a reagent containing 15 μg of K1 DNA complex with 200 nmol of liposome, the same amount of control DNA (pSV2E) complex with liposome, K1 DNA alone, liposome alone, or phosphate-buffered saline alone. The responses of the mice to treatment were observed for 1 year. The number of mice in each group is indicated. The asterisk marks the last injection. [Reprinted from *Cancer Gene Therapy,* 1996, Vol. 3 pages 168–174, by permission of Stockton Press ©.]

In both figures, the survival curves were obtained by recording the total survival days for each mouse in different groups from the day of injection with SK-OV-3 cells (Day 1, 100% survival) to the days they died.

No major organ pathologic changes were observed. No E1A DNA can be detected in the liver, heart, spleen, brain, uterus, and ovaries of the treated mice after 1½ years following the last E1A/liposome treatment (Xing *et al.*, 1998). The study concludes that ip administration of the E1A/liposome complex at the proposed dose did not produce significant toxicity, and the treatment may be relatively safe for human trials.

B. TUMOR SUPPRESSION BY E1A AND LARGE T ANTIGEN

Both E1A and T antigen are thought to be transforming viral proteins, and their ability to suppress HER-2/neu-mediated cell transformation is surely a surprising biological phenomenon. The adenovirus genome is about 36 kb in size. Among the proteins encoded by the adenovirus genome, E1A gene products are nuclear-localized phosphoproteins and have a special regulatory role in the adenoviral life cycle (Berk, 1986). E1A is the first region to be expressed after infection (Tooze, 1981). Other late adenoviral genes can then be turned on by E1A proteins through interacting and modifying the host transcriptional apparatus. There are two types of adenovirus E1A. One is the transforming E1A carried by the adenovirus type 12. This type of E1A gene alone can transform normal cell lines (Schrier *et al.*, 1983). Another type of E1A, such as the adenovirus type 2 or type 5 E1A, cannot transform cells by itself. However, the type 5 E1A can cooperate with the transforming ras or E1B genes to transform primary embryo cells (Ruley, 1983; Byrd *et al.*, 1988; Land *et al.*, 1983; Montell *et al.*, 1984). Therefore, the type 5 E1A was classified as an *immortalization oncogene*. However, it should be emphasized that expression of the E1A gene itself does not induce transforming phenotypes (Yu, Hamada, *et al.*, 1992). In the latter sections we use E1A to refer to the nontransforming E1A such as the type 5 or type 2 E1A.

We have first discovered that the adenovirus 5 E1A gene can repress HER-2/ neu overexpression through both transient transfection and adenovirus delivery systems (Yan *et al.*, 1991; Yu *et al.*, 1990). Transfection of the E1A gene into the genomic rat HER-2/neu oncogene-transformed mouse embryo fibroblast cell lines virtually abolishes the tumorigenicity and metastatic potential induced by the HER-2/neu oncogene through repression of HER-2/neu gene expression (Yu, Hamada, *et al.*, 1992; Yu *et al.*, 1991). Re-expression of the HER-2/neu-encoded p185 protein in these E1A transfectants by transfection of a HER-2/neu cDNA construct driven by a promoter that cannot be inhibited by E1A recovered virtually all of the transforming phenotypes including tumorigenicity, the ability to grow in soft agar, and a higher *in vitro* growth rate (Yu, Shi, *et al.*, 1993). The ability to induce experimental metastasis (measured by lung colonization through iv injection of the tumor cells) was only partially recovered. The incomplete regeneration of metastatic potential could be accounted for by the fact that E1A inhibits other metastasis-promoting functions such as urokinase plasminogen and gelatinolytic activities that were critical for invasive activity of metastatic cells (Young *et al.,* 1989; Frisch *et al.,* 1990). This result indicates that the suppression of metastasis by E1A is through multiple molecular mechanisms in addition to repressing the HER-2/neu gene expression. We have also demonstrated that E1A can indeed function as a tumor suppressor in the HER-2/neu-overexpressing human ovarian cancer cell line by down-regulating the expression of the HER-2/neu mRNA and the p185 protein product (Yu *et al.,* 1991, 1995, 1990; Yu, Wolf, *et al.,* 1993). The E1A-expressing ovarian cancer cell line had reduced malignancy, including a decreased ability to develop tumors in nude mice. Therefore, for the HER-2/neu-overexpressing transforming cells including fibroblasts and human cancer cells, E1A can function as a tumor suppressor through repression of the HER-2/neu oncogene. However, since E1A is not a DNA-binding protein, the transcriptional repression of HER-2/neu by E1A must be mediated through the targeting of other transcription factors. This is supported by our recent study demonstrating that E1A can abolish HER-2/neu overexpression by targeting the coactivator p300, which is required for efficient expression of HER-2/neu (Fig. 2; Pozzatti *et al.,* 1988).

To further investigate whether the E1A gene can be used as a therapeutic agent for HER-2/neu-overexpressing human ovarian cancers in a living host, a tumor-bearing mouse model was established and the E1A gene was delivered by the cationic liposome DC-chol to the tumor by ip injection. The liposome-mediated E1A gene therapy was able to effectively reduce the mortality of tumor-bearing mice and in some cases resulted in tumor-free survival, suggesting that liposome-mediated E1A gene therapy is a promising therapeutic regimen for later-stage ovarian cancers that overexpress HER-2/neu (Fig. 1A). In the orthotopic breast cancer animal model bearing HER-2/neu-overexpressing MDA-MB-361 breast cancer cells, E1A also significantly reduced cancer growth and prolonged survival for up to 2 years in 60–80% of the mice. Western blot analysis confirmed E1A protein expression, and immunohistochemical staining indicated that HER-2/ neu was suppressed in 35% of tumor sections. In addition, the number of mice with distant metastases was significantly reduced even though a local treatment protocol by mammary fat pad injection was used in the orthotopic breast cancer model (Chang *et al.,* 1997). Thus in spite of its immortalization function, the type 5 E1A apparently suppresses the malignancy of the HER-2/neu-overexpressing cancer cells. Therefore, it is a misconception to "tag" the type 5 E1A gene as an oncogene. As a matter of fact, there are a number of studies indicating that this type of E1A is associated with metastasis- or tumor-suppression activities. The E1A gene was reported to reduce the metastatic potential of *ras*-transformed rat embryo fibroblast cells (Pozzatti *et al.,* 1988). E1A has been shown to be able to repress a number of tumor progression marker genes in the metastatic human melanoma cell line BLM cells, including calcyclin, thymosin β-10, plasminogen activator inhibitors type 1 and 2, urokinase type and tissue type plasminogen activators, vimentin, tissue type transglutaminase, and interleukin-6 (van Groningen *et al.*, 1996). In addition, stable expression of the E1A gene reduced anchorage-independent growth and tumorigenic potential, induced flat morphology, and restored contact inhibition in some human tumor cell lines (Frisch, 1991; Frisch and Dolter, 1995; Deng *et al.*, 1998). It is also worth mentioning that E1A is able to induce apoptosis under some conditions (Rao *et al.*, 1992; Lowe and Ruley, 1993; Shao *et al.*, 1997). This property is similar in the well-known tumor suppressor gene p53, which also has the ability to induce apoptosis (Subramanian *et al.*, 1995; Symonds *et al.*, 1994). It was originally reported that E1A-induced apoptosis is mediated by wild-type p53 and can be inhibited by E1B that binds to p53 (Debbas and White, 1993). However, E1A proteins were recently shown to induce cell death by both p53-dependent and p53-independent mechanisms involving separate E1A functions (Teodoro *et al.*, 1995). Thus, although E1A possesses an immortalization function, it is also associated with multiple functions that may induce antitumor activities.

The simian virus 40 (SV40) large T antigen is a multifunctional protein required for the replication of the viral genome and for cell transformation (Lane and Crawford, 1979; Linzer and Levine, 1979). This viral protein contains transformation domains that can mediate binding to the retinoblastoma protein (pRb) and p53, respectively (Manfredi and Prive, 1994). Our previous studies showed that a mutant SV40 large T antigen can repress rat neu transcription in mouse fibroblast NIH 3T3 cells (Matin and Hung, 1993). The mutant large T antigen, named K1, contains a single amino acid change within the pRb-binding/transformation domain, which renders the viral protein unable to bind to pRb, and consequently it fails to induce cell transformation (Kalderon and Smith, 1984; Cherington *et al.*, 1988; DeCaprio *et al.*, 1988). Since the K1 mutant represses HER-2/neu expression as effectively as the wild-type counterpart (Matin and Hung, 1993), we further tested whether K1 can function as a tumor suppressor for HER-2/neu-overexpressing ovarian cancer cells. K1 did suppress cancer cell growth, resulting in a significant therapeutic effect on mice with ovarian cancer, with about 40% of treated mice alive after 1 year (Figure 1B; Xing *et al.*, 1996). The autopsies showed that the mice from the control groups had a larger volume of ascites and tumors within the peritoneal cavity or diaphragm or metastasis to the lungs. However, the mice that received the K1/liposome complex had more locally distributed tumor nodules in their peritoneal cavities. This difference indicates that K1 suppressed the growth of HER-2/neu-overexpressing tumor cells so that the tumors developed with longer latency. The K1-treated mice that survived for 1 year were sacrificed and examined for residual tumors, but no tumors were observed in the peritoneal cavity.

To further evaluate the treatment efficiency by SV40 large T, tissue sections from the major organs of the surviving mice were analyzed by hematoxylin-eosin staining and found to be pathologically normal and free of microscopic tumor cells (Xing *et al.*, 1996; Yu *et al.*, 1995; Chang *et al.*, 1997). The results suggest that similar to E1A, the large T K1 mutant may suppress HER-2/neu-overexpressing malignancy through its ability to repress HER-2/neu overexpression.

C. TUMOR SUPPRESSION BY PEA3

The PEA3 (Polyomavirus Enhancer Activator 3) gene was first cloned from a murine cDNA expression library due to its binding ability to the sequence 5'-AGGAAG-3' within the polyomavirus enhancer (Xin et al., 1992). The PEA3 protein contains a stretch of about 80 amino acids with extensive sequence homology with the ETS domain—a conserved region corresponding to the DNA-binding region shared by proteins of the ets family, which recognize similar binding sites with a central 5'-GGAA/ T-3' consensus motif—and regulates the expression of many genes involved in cell growth and differentiation (Ma et al., 1998a; Taylor et al., 1997). The PEA3 subfamily is composed of three members: PEA3, ERM, and ER81. In addition to the ETS domain, members of this subfamily share significant sequence similarity at an N-terminal acidic transcriptional activation domain (Ma et al., 1998a; Taylor et al., 1997; Nakae et al., 1995; Brown and McKnight, 1992; Karim et al., 1990; Ma et al., 1998b; Laudet et al., 1993).

It is interesting to investigate the role of PEA3 in HER-2/neu gene expression and HER-2/neu-mediated transformation since a consensus PEA3-binding site, 5'-AGGAAG-3', is present 26 nucleotide upstream from the major mRNA start site in the promoter of the human, rat, and mouse HER-2/neu gene (White and Hung, 1992). PEA3 binding motifs also occur in the promoters of a number of genes required for cancer cell metastasis (Mitsunori et al., 1996). It has been reported that PEA3 can directly activate transcription from a group of matrix metalloproteinase (MMP) genes (Mitsunori et al., 1996), which potentially can enhance the metastatic ability of cancer cells. However, the biological significance of these observations remains unclear since PEA3 or a PEA3-related protein, together with another transcription factor AP1, can transactivate the promoter of the maspin gene, a tumor suppressor that can suppress cancer cell invasion and metastasis (Zhang et al., 1997). Furthermore, decreased or nondetectable PEA3 RNA expression was found in breast cancer cell lines with HER-2/neu overexpression (such as BT 474, SK-BR-3, MDA-MB-361, and MDA-MB-453) (Baert et al., 1997). These results suggest that PEA3 may have a negative role in regulating HER-2/neu expression. This hypothesis was directly tested in our laboratory and the following results demonstrate that PEA3 is indeed a negative transregulator of the proto-oncogene HER-2/neu (Xing, Miller, et al., 1997).

1. The purified GST-PEA3 fusion protein can specifically recognize and bind to the consensus PEA3 binding motif on the HER-2/neu promoter.

2. Based on the cotransfection experiments performed on HER-2/neu-overexpressing human cancer cell lines, the HER-2/neu promoter activity can be down-regulated by PEA3 in a dose-dependent manner. However, destruction of the PEA3-binding site on the HER-2/neu promoter by site-directed mutagenesis abolished PEA3-mediated HER-2/neu down-regulation, indicating that PEA3-induced transrepression of the HER-2/neu promoter is indeed through the PEA3-binding site. It is also interesting to note that the HER-2/neu promoter activity is significantly reduced when the PEA3-binding site is mutated. The results indicate that the PEA3-binding site is required for efficient transcription of the HER-2/neu promoter. Thus, it is most likely that there is another ets-related transcription factor(s) activating the HER-2/neu promoter through the PEA3-binding site in the HER-2/neu-overexpressing cancer cells, and the PEA3 protein can compete with this transcriptional activator for the same PEA3-binding site. A model for this hypothesis is shown in Figure 2.
3. PEA3 can suppress the focus-forming ability of mouse embryonic fibroblast transformed by the mutation-activated genomic rat neu.
4. Expression of PEA3 can suppress the growth of HER-2/neu-overexpressing human cancer cell lines *in vitro* but not cell lines with a basal level of HER-2/neu expression.

These promising *in vitro* data prompted us to further test whether PEA3 can be used as a therapeutic agent *in vivo*. Tumors were induced in nude mice (nu/nu) with SK-OV-3-ip1, an ovarian cancer cell line derived from SK-OV-3 and with higher HER-2/neu expression. For mice treated with PEA3/DC-chol complex, 50% of the mice were alive and healthy without palpable tumors after 12 months. The mice of the control group, however, developed tumors and ascites and died within 6 months. The tumor suppression activity of PEA3 is correlated with HER-2/neu expression since another cell line 2774 c-10, an ovarian cancer cell line with the basal level of HER-2/neu expressed, did not respond to PEA3 treatment and the mice died of tumor within 5 months. Tumor samples were examined for the expression of HER-2/neu with immunoblot analysis. The results confirmed that PEA3 delivered by the cationic liposome down-regulated the expression of p185. The correlation between PEA3 expression and HER-2/neu down-regulation was further demonstrated by immunohistochemical staining of the tumor samples obtained from the PEA3-treated, moribund mice. Approximately 30% of the cancer cells in the tumor were positive for PEA3 protein expression, while the p185 staining was negative for about 50% of the cells. A similar level of PEA3 expression was observed for PEA3-treated 2774 c-10-derived tumors while no repression of p185 was detected in these tumors. These *in vitro* and *in vivo* data clearly demonstrate the

A.

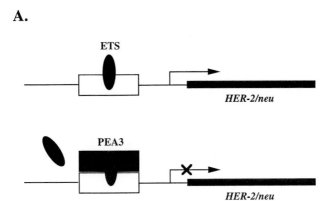

B.

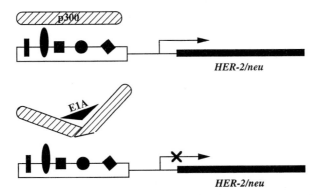

Figure 2 The overexpression of a proto-oncogene can be blocked by transcriptional repressors. (A) Transactivators like some ets family members can sustain the induction of HER-2/neu expression by binding to the particular binding motif on the promoter, while the PEA3 protein can repress the gene expression by competing for the same motif and consequently block transactivation. (B) Alternatively, the HER-2/neu promoter can be repressed by E1A, which can bind to the co-activator p300. p300 serves as an adaptor bridging the enhancer-like factors (represented by the rectangular, round, oval, and square objects on the promoter) to the basal transcriptional apparatus and mediating the up-regulation of the gene. Even though E1A cannot directly bind to DNA, binding of E1A to p300 can abolish HER-2/neu overexpression, presumably by hindering the interaction between p300 and other transcriptional factors. [From "Liposome-Mediated in vivo E1A Gene Transfer Suppressed Dissemination of Ovarian Cancer Cells That Overexpress HER-2/neu," by Yu, D., Matin, A., Xia, W., Sorgi, F., Huang, L., and Hung, M.-C. 1995, *Oncogene, 11,* pp. 1383–1388. Copyright 1995 by Stockton Press. Used with permission.]

tumor suppression activity of PEA3 and indicate the potential clinical application of PEA3/cationic liposome targeting the HER-2/neu overexpressing cancer cells.

Even though PEA3 as well as the viral proteins E1A and SV40 large T can all suppress HER-2/neu transcription, they are very likely functioning through different mechanisms. Both E1A and SV40 large T may suppress HER-2/neu in an indirect manner. Association of E1A with the transcriptional coactivator CBP/p300 inhibits the p300 transactivation activity, which is required for efficient expression of the HER-2/neu gene (Chen and Hung, 1997). On the other hand, PEA3 downregulates the HER-2/neu gene by directly binding to its cognate binding sequence on the promoter. This feature makes PEA3 a more attractive target for further molecular manipulation to develop therapeutic molecules with higher binding affinity and enhanced specificity. In addition, the data discussed here clearly indicate that the nonviral DC-chol/DOPE liposome can be used to deliver appropriate genes for cancer gene therapy and achieve effective therapeutic efficacy.

III. A PHASE I CLINICAL TRIAL USING THE E1A–LIPOSOME

Based on the data presented in earlier sections, it is clear that genes encoding transcriptional repressors such as E1A, large T, and PEA3 for HER-2/neu can be used to efficiently treat mice bearing HER-2/neu-overexpressing tumors. Since much more extensive studies have been performed using the E1A gene (Yu, Hamada, et al., 1992; Yu et al., 1995; Chang et al., 1997; Yu, Wolf, et al., 1993; Zhang et al., 1995) and safety studies using the E1A/liposome (Xing et al., 1998; Xing, Liu, et al., 1997) have shown that using E1A/liposome produces no adverted effects in immunocompetent mice, a phase I clinical trial was approved by FDA and the NIH Recombinant DNA Advising Committee and initiated at the M. D. Anderson Cancer Center in 1996. This trial, entitled "Phase I Study of E1A Gene Therapy for Patients with Metastatic Breast or Epithelial Ovarian Cancer That Overexpresses HER-2/neu," aims at breast and ovarian cancer patients whose tumors have HER-2/neu overexpression; this subset of patients is considered to have a poor prognosis as HER-2/neu overexpression is an unfavorable prognostic factor for cancer patients. The eligibility of patients enrolling in the trial involves these factors:

1. Patients with advanced metastatic breast and/or ovarian carcinoma refractory to standard chemotherapy and hormone therapy.
2. Patients must have easily accessible pleural effusion and/or ascites.
3. Patients must carry tumors that overexpress HER-2/neu oncoprotein.
4. Patients must be willing to undergo serial aspiration of pleural/peritoneal effusion.

5. Performance status ≤ 3.
6. Signed informed consent.

The objectives of the trial are

1. To demonstrate E1A gene transduction into malignant cells after the administration of E1A-lipid complex by intrapleural/intraperitoneal administration.
2. To determine whether E1A gene therapy can down-regulate HER-2/neu expression after intrapleural/intraperitoneal administration.
3. To determine the maximum biological active dose (MBAD) or the maximum tolerated dose (MTD) of E1A/lipid complex.
4. To determine the toxicity and tolerance of E1A/lipid complex administered into the pleural/peritoneal space and to assess the reversibility of such toxicity.
5. To evaluate tumor response.

The treatment plan is as follows:

1. Placement of a Tenckhoff catheter.
2. Intracavitary injection of E1A DNA by a cationic liposome (DC-chol) delivery system (DNA dose: $1.8 \rightarrow 3.6 \rightarrow 7.2$ mg/m^2).
3. Injection of the patients every week for 3 weeks, then take 1 week off, total of six cycles.

This trial has recently completed. There are a total 12 patients treated under this protocol. Among the six patients whose body fluids have been followed up for HER-2/neu gene expression, we observed down-regulation of HER-2/neu by immunohistochemical staining along the treatment route. We also detected E1A expression using RT-PCR. We have also reached the intolerable dose level at 7.2 mg/m^2. However, at doses of 1.8 and 3.6 mg/m^2 only minor toxicity such as fever and pain around the injection site were observed. The six patients who were analyzed for HER-2/neu expression and were found to have HER-2/neu down-regulation are those who were treated at a dose of either 1.8 or 3.6 mg/m^2. In addition, the number of tumor cells in the pleural effusion or ascites was found to be dramatically reduced posttreatment. Although further clinical trials are required to evaluate therapeutic efficiency for patients with less severe disease, the results from this phase I clinical trial are certainly encouraging and, together with the preclinical animal data, should provide the basis for the phase II clinical trial.

IV. CONCLUSIONS

Overexpression of the proto-oncogene HER-2/neu can lead to cell transformation and is tightly correlated with the development of malignant tumor growth in many tissue types. There are molecular approaches to target the promoter of HER-2/neu, which can down-regulate the gene expression, reverse the malignant phenotype, and retard tumor growth in animals. In addition to these molecular agents, vehicles with safe, efficient, and convenient properties are critical for transferring the therapeutic DNA into tumor cells. The results of our animal experiments demonstrate that cationic liposome combined with genes of transcriptional repressors can result in significant therapeutic effects on cancer cells. The phase I clinical trial also indicates that appropriate gene expression can be observed at tolerable doses. Although further clinical trials are required to evaluate therapeutic efficacy, the results are certainly encouraging. Since gene overexpression is a common mechanism of cancer as well as other types of diseases such as AIDS, the therapeutic strategy discussed here can have tremendous potential in clinical application.

ACKNOWLEDGMENTS

The authors are supported by NCI RO1 CA 58880 and CA 77858 (to M.C.H.).

REFERENCES

Baert, J. L., Monte, D., Musgrove, E. A., Albagli, O., Sutherland, R. L., and Launoit, Y. (1997). Expression of the *PEA3* group of EST-related transcription factors in human breast cancer cells. *Int. J. Cancer* **70**, 590–597.

Bargmann, C. I., Hung, M.-C., and Weinberg, R. A. (1986a). The *neu* oncogene encodes an epidermal growth factor receptor-related protein. *Nature* **319**, 226–230.

Bargmann, C. I., Hung, M.-C., and Weinberg, R. A. (1986b). Multiple independent activations of the *neu* oncogene by a point mutation altering the transmembrane domain of p185. *Cell* **45**, 649–657.

Baselga, J., Norton, L., Albanell, J., Kim, Y. M., and Mendelsohn, J. (1998). Recombinant humanized anti-HER2 antibody (Herceptin™) enhances the antitumor activity of paclitaxel and doxorubicin against *HER-2/neu* overexpressing human breast cancer xenografts. *Cancer Res.* **58**, 2825–2831.

Beahrs, O. H., Henson, D. E., Hutter, R. V. P., and Kennedy, B. J. (1992). *Manual for staging of cancer.* Philadelphia: Lippincott.

Benz, C. C., Scott, G. K., Sarup, J. C., Johnson, R. M., Tripathy, D., Coronado, E., Shepard, H. M., and Osborne, C. K. (1993). Estrogen-dependent, tamoxifen-resistant tumorigenic growth of MCF-7 cells transfected with HER2/neu. *Breast Cancer Res. Treat.* **24**, 85–95.

Berk, A. J. (1986). Adenovirus promoters and *E1A* transactivation. *Ann. Rev. Genet.* **20**, 45–79.

Bosher, J. M., Totty, N. F., Hsuan, J. J., Williams, T., and Hurst, H. C. (1996). A family of AP-2 proteins regulates *c-erbB-2* expression in mammary carcinoma. *Oncogene* **13**, 1701–1707.

Brown, T. A., and McKnight, S. L. (1992). Specificities of protein–protein and protein–DNA interaction of GABP alpha and two newly defined ets-related proteins. *Genes Dev.* **6**, 2502–2512.

Berchuck, A., Kamel, A., Whitaker, R., Kerns, B., Olt, G., Kinney, R., Soper, J., Dodge, R., Clark-Pearson, D., Marks, P., McKenzie, S., Yin, S., and Bast, R., Jr. (1990). Overexpression of *HER-2/neu* is associated with poor survival in advanced epithelial ovarian cancer. *Cancer Res.* **50**, 4087–4091.

Berchuck, A., Rodriguez, G., Kinney, R., Soper, J., Dodge, R., Clark-Pearson, D., and Bast, R. (1991).

Overexpression of *HER-2/neu* in endometrial cancer is associated with advanced stage disease. *Am. J. Obstet. Gynecol.* **164,** 15–21.

Byrd, P. J., Grand, R. J. A., and Gallimore, P. H. (1988). Differential transformation of primary human embryo retinal cells by adenovirus *E1A* regions and combination of *E1A* (+) *ras. Oncogene* **2,** 477–484.

Carraway, K. L., Sliwkowski, M. X., Akita, R. M., *et al.* (1994). The erbB-3 gene product is a receptor for heregulin. *J. Biol. Chem.* **269,** 14303–14306.

Chang, J. Y., Xia, W., Shao, R., Sorgi, F., Hortobagyi, G. N., Huang, L., and Hung, M.-C. (1997). The tumor suppression activity of *E1A* in *HER-2/neu*-overexpressing breast cancer. *Oncogene* **14,** 561–568.

Chazin, V. R., Kaleko, M., Miller, A. D., and Slamon, D. J. (1992). Transformation mediated by the human *HER-2* gene independent of the epidermal growth factor receptor. *Oncogene* **7,** 1859–1866.

Chen, H., and Hung M.-C. (1997). Involvement of co-activator p300 in the transcriptional regulation of the *HER-2/neu* gene. *J. Biol. Chem.* **272,** 6101–6104.

Cherington V., Brown, M., Paucha, E., Louis, J., Spiegelman, B. M., and Roberts, T. M. (1988). Separation of simian virus 40 large T antigen-transforming and origin-binding functions from the ability to block differentiation. *Mol. Cell Biol.* **8,** 1380–1384.

Coussens, L., Yang-Feng, T. L., Liao Y. C., Chen, E., Gray, A., McGrath, J., Seeburg, P. H., Libermann, T. A., Schlessinger, J., Francke, U., *et al.* (1985). Tyrosine kinase receptor with extensive homology to EGF receptor shares chromosomal location with *neu* oncogene. *Science* **230,** 1132–1139.

Debbas, M., and White, E. (1993). Wild-type p53 mediates apoptosis by *E1A,* which is inhibited by *E1B. Genes Dev.* **7,** 546–554.

DeCaprio, J. A., Ludlow, J. W., Figge, J., Shew, J.-Y., Huang, C.-M., Lee, W.-H., Marsilis, E., Paucha, E., and Livingston, D. M. (1988). SV40 large tumor antigen forms a specific complex with the product of the retinoblastoma susceptibility gene. *Cell* **54,** 275–283.

Degnan, B. M., Degnan, S. M., Naganuma, T., and Morese, D. E. (1993). The *ets* multigene family is conserved throughout the Metazoa. *Nucleic Acid Res.* **21,** 3479–3484.

Deng, J., Xia, W., and Hung, M.-C. (1998). Adenovirus 5 *E1A*-mediated tumor suppression associated with *E1A*-mediated apoptosis *in vivo. Oncogene* **17,** 2167–2175.

Frisch, S. M. (1991). Antioncogenic effect of adenovirus *E1A* in human tumor cells. *Proc. Natl. Acad. Sci. USA* **88,** 9077–9081.

Frisch, S. M., and Dolter, K. E. (1995). Adenovirus *E1A*-mediated tumor suppression by a *c-erbB2/neu*-independent mechanism. *Cancer Res.* **55,** 5551–5555.

Frisch, S. M., Reich, R., Collier, I. E., *et al.* (1990). Adenovirus *E1A* represses protease gene expression and inhibits metastasis of human tumor cells. *Oncogene* **5,** 75–83.

Gao, X., and Huang, L. (1996). Potentiation of cationic liposome-mediated gene delivery by polycations. *Biochemistry* **35,** 1027–1036.

Graus-Porta, D., Beerli, R. R., Daly, J. M., *et al.* (1997). ErbB-2, the preferred heterodimerazation partner of all ErbB receptors, is a mediator of lateral signalling. *EMBO J.* **16,** 1647–1655.

Gusterson, B., Gelber, R., Goldhirsch, A., Price, K., Save-Soderborgh, J., Anbazhagan, R., Styles, J., Rudenstam, C.-M., R., Reed, R., Martinez-Tello, F., Tiltman, A., Torhorst, J., Grigolato, P., Bettelheim, R., Neville, A., Burki, K., Castiglione, M., Collins, J., Lindtner, J., and Senn, H.-J. (1992). Prognostic importance of *c-erbB-2* expression in breast cancer, *J. Clin. Oncol.* **10,** 1049–1056.

Hollywood, D. P., and Hurst, H. C. (1993). A novel transcription factor, *OB2–1,* is required for overexpression of the proto-oncogene *c-erbB-2* in mammary tumor cell lines. *EMBO J.* **12,** 2369–2375.

Hung, M.-C., Schechter, A. L., Chevray, P. L., Stern, D. F., and Weinberg, R. A. (1986). Molecular cloning of the *neu* gene: Absence of gross structural alteration in oncogenic alleles. *Proc. Natl. Acad. Sci. USA* **83,** 261–264.

Hung, M.-C., Yan, D., and Xhao, X. (1989). Amplification of the proto-*neu* gene facilitates oncogenic activation by a single point mutation. *Proc. Natl. Acad. Sci. USA* **86,** 2545–2548.

Kalderon, D., and Smith, A. E. (1984). *In vitro* mutagenesis of a putative DNA binding domain of SV40 large T. *Virology* **39**, 109–137.

Karim, F. D., Urness, L. D., Thummel, C. S., *et al.* (1990). The ETS-domain: A new DNA-binding motif that recognizes a purine-rich core DNA sequence. *Genes Dev.* **4**, 1451–1453.

Kraus, M. H., Popescu, N. C., Ambaugh, S. C., and King, C. R. (1987). Overexpression of the EGF receptor-related proto-oncogene *erbB-2* in human mammary tumor cell lines by different molecular mechanisms. *EMBO J.* **6**, 605–610.

Land, H., Parada, L. F., and Weinberg, R. A. (1983). Tumorigenic conversion of primary embryo fibroblasts requires at least two cooperating oncogenes. *Nature* **304**, 596–602.

Lane, D. P., and Crawford, L. V. (1979). T antigen is bound to a host protein in SV40-transformed cells. *Nature* **278**, 261–262.

Laudet, V., Niel, C., Deuterque-Coquillard, M., Le Prince, D., and Stehelin, D. (1993). Evolution of the *ets* gene family. *Biochem. Biophys. Res. Comm.* **190**, 8–14.

Linzer, D. I. H., and Levine, A. J. (1979). Characterization of a 54K Dalton cellular SV40 tumor antigen present in SV40-transformed cells and uninfected embryonal carcinoma cells. *Cell* **17**, 43–52.

Lowe, S. W., and Ruley, H. E. (1993). Stabilization of the p53 tumor suppressor is induced by adenovirus 5 *E1A* and accompanies apoptosis. *Genes Dev.* **7**, 535–545.

Ma, Y., Su, Q., and Tempst, P. (1998a). Differentiation-stimulated activity binds an ETS-like, essential regulatory element in the human promyelocytic defensin-1 promoter. *J. Biol. Chem.* **273**, 8727–8740.

Ma, Y., Su, Q., and Tempst, P. (1998b). Differentiation-stimulated activity binds an ETS-like, essential regulatory element in the human promyelocytic defensin-1 promoter. *J. Biol. Chem.* **273**, 8727–8740.

Manfredi, J. J., and Prive, C. (1994). The transformation activity of simian virus 40 large tumor antigen. *Biochim. Biophys. Acta.* **1198**, 65–83.

Matin, A., and Hung M.-C. (1993). Negative regulation of the *neu* promoter by the SV40 large T antigen. *Cell Growth Differ.* **4**, 1051–1056.

Miller, S. J. and Hung, M.-C. (1995). Regulation of *HER-2/neu* gene expression. *Oncol. Rep.* **2**, 497–503.

Miller, S. J., Suen, T. C., Sexton, T. B., Hung, M.-C. (1994). Mechanisms of deregulated *HER2/nue* expression in breast cancer cell lines. *Int. J. Oncol.* **4**, 599–608.

Mitsunori, K., Yoshida, K., Higashino, F., Mitaka, T., Ishii, S., and Fujinaga, K. (1996). A single ets-related transcription factor, E1AF, confers invasive phenotype on human cancer cells. *Oncogene* **12**, 221–227.

Montell, C., Courtois, G., Eng, C., *et al.* (1984). Complete transformation by adenovirus 2 requires both *E1A* proteins. *Cell* **36**, 951–961.

Nabel, G. J., Nabel, E. G., Yang, Z. Y., Fox, B. A., Plautz, G. E., Gao, X., Huang, L., Shu, S., Gordon, D., and Chang, A. E. (1993). Direct gene transfer with DNA-liposome complexes in melanoma: Expression, biologic activity, and lack of toxicity in humans. *Proc. Natl. Acad. Sci. USA* **90**, 11307–11311.

Nakae, K., Nakajima, K., Inazawa, J., Kitaoka, T., and Hirano, T. (1995). *ERM,* a *PEA3* subfamily of Ets transcription factors, can cooperate with c-Jun. *J. Biol. Chem.* **270**, 23795–23800.

Park, J. B., Rhim, J. S., Park, S. C., Kimm, S. W., and Kraus, M. H. (1989). Amplification, overexpression, and rearrangement of the *c-erbB-2* proto-oncogene in primary human stomach carcinomas. *Cancer Res.* **49**, 6005–6009.

Pegram, M. D., Finn, R. S., Arzoo, K., Beryt, M., Pietras, R. J., and Slamon, D. J. (1997). The effect of *HER-2/neu* overexpression on chemotherapeutic drug sensitivity in human breast and ovarian cancer cells. *Oncogene* **15**, 537–547.

Pegram, M. D., Lipton, A., Hayes, D. F., Weber, B. L., Baselga, J. M., Tripathy, D., Baly, D., Baughman, S. A., Twaddell, T., Glaspy, J. A., and Slamon, D. J. (1998). Phase II study of receptor-enhanced chemosensitivity using recombinant humanized anti-p185HER2/neu monoclonal antibody plus cisplatin in patients with HER2/neu-overexpressing metastatic breast cancer refractory to chemotherapy treatment. *J. Clin. Oncol.* **16**, 2659–2671.

Plowman, G. D., Grenn, J. M., Culouscou, J. M., *et al.* (1993). Heregulin induces tyrosine phosphorylation of HER4/p180 [erbB-4]. *Nature* **366**, 473–475.

Pozzatti, R., McCormick, M., Thompson, M. A., and Khoury, G. (1988). The *E1A* gene of adenovirus type 2 reduces the metastatic potential of *ras*-transformed rat embryo cells. *Mol. Cell Biol.* **8**, 2984–2988.

Rao, L., Debbas, M., Sabbatini, P., Hockenbery, D., Korsmeyer, S., and White, E. (1992). The adenovirus E1A proteins induce apoptosis, which is inhibited by the E1B 19-kDa and Bcl-2 proteins. *Proc. Natl. Acad. Sci. USA* **89**, 7742–7746.

Ruley, H. E. (1983). Adenovirus early region 1A enables viral and cellular transforming genes to transform primary cells in culture. *Nature* **304**, 602–606.

Schneider, P. M., Hung, M.-C., Chiocca, S. M., Manning, J., Zhao, X. Y., Fang, K., and Roth, J. A. (1989). Differential expression of the *c-erbB-2* gene in human small cell and non-small cell lung cancer. *Cancer Res.* **49**, 4968–4971.

Schrier, P. I., Bernards, R., Vaessen, R. T. M. J., *et al.* (1983). Expression of class I major histocompatibility antigens switched off by highly oncogenic adenovirus 12 in transformed rat cells. *Nature* **305**, 771–775.

Shao, R., Karunagaran, D., Zhou, B. P., Li, K., Lo, S. S., Deng, J., Chiao, P., and Hung, M.-C. (1997). Inhibition of nuclear factor-kappa B activity is involved in *E1A*-mediated sensitization of radiation-induced apoptosis. *J. Biol. Chem.* **272**, 32739–32742.

Shi, D., He, G., Cao, S., Pan, W., Zhang, H. Z., Yu, D., and Hung, M.-C. (1992). Overexpression of the *c-erbB-2/neu*-encoded p185 protein in primary lung cancer. *Mol. Carcinog.* **5**, 213–218.

Slamon, D. J., Clark, G. M., Wong, S. G., Levin, W. J., Ullrich A., and McGuire, W. L. (1987). Human breast cancer: Correlation of relapse and survival with amplification of the *HER-2/neu* oncogene. *Science* **235**, 177–182.

Slamon, D. J., Godolphin, W., Jones, L. A., Holt, J. A., Wong, S. G., Keith, D. E., Levin, W. J., Stuart, S. G., Udove, J., Ullrich, A., and McGuire, W. L. (1989). Studies of the *HER-2/neu* proto-oncogene in human breast and ovarian cancer. *Science* **244**, 707–712.

Sliwkowski, M. X., Schaefer, G., Akita, R. W., *et al.* (1994). Coexpression of erbB2 and erbB3 proteins reconstitutes a high affinity receptor for heregulin. *J. Biol. Chem.* **269**, 14661–14665.

Subramanian, T., Tarodi, B., and Chinnadurai, G. (1995). p53-independent apoptotic and necrotic cell deaths induced by adenovirus infection: Suppression by E1B 19K and Bcl-2 proteins. *Cell Growth Diff.* **6**, 131–137.

Symonds, H., Krall, L., Remington, L., Saenz-Robles, M., Lowe, S., Jacks, T., and Van Dyke, T. (1994). p53-dependent apoptosis suppresses tumor growth and progression *in vivo*. *Cell* **78**, 703–711.

Tan, M., Yao, J., and Yu, D. (1997). Overexpression of the *c-erbB-2* gene enhanced intrinsic metastasis potential in human breast cancer cells without increasing their transformation abilities. *Cancer Res.* **57**, 1199–1205.

Taylor, J. M., Dupont-Versteegden, E. E., Davies, J. D., Hassell, J. A., Houle, J. D., Gurley, C. M., and Peterson, C. A. (1997). A role for the ETS domain transcription factor *PEA3* in myogenic differentiation. *Mol. Cell. Biol.* **17**, 5550–5558.

Teodoro, J. G., Shore, G. C., and Branton, P. E. (1995). Adenovirus E1A proteins induce apoptosis by both p53-dependent and p53-independent mechanisms. *Oncogene* **11**, 467–474.

Toikkanen, S., Helin, H., Isola, J., and Joensuu, H. (1992). Prognostic significance of *HER-2* oncoprotein expression in breast cancer: A 30 year follow-up. *J. Clin. Oncol.* **10**, 1044–1048.

Tooze, J. (1981). DNA tumor viruses. In *Molecular biology of tumor viruses* (2nd ed.). Cold Spring Harbor, New York: Cold Spring Harbor Laboratory.

Tsai, C. M., Chang, K.-T., Perng, R.-P., Mitsudomi, T., Chen, M.-H., and Gazdar, A. F. (1993). Correlation of intrinsic chemoresistance of non-small-cell lung cancer cell lines with *HER-2/neu* gene expression but not with *ras* gene mutations. *J. Natl. Cancer Inst.* **85,** 897–901.

Tsai, C. M., Yu, D., Chang, K. T., Wu, L. H., Perng, R. P., Ibrahim, N. K., and Hung, M.-C. (1995). Enhanced chemoresistance by elevation of the level of p185^{neu} in *HER-2/neu* transfected human lung cancer cells. *J. Natl. Cancer Inst.* **87,** 682–684.

van Groningen, J. J., Cornelissen, I. M., van Muijen, G. N., *et al.* (1996). Simultaneous suppression of progression marker genes in the highly malignant human melanoma cell line BLM after transfection with the adenovirus-5 *E1A* gene. *Biochem. Biophys. Res. Commun.* **225,** 808–816.

Wallasch, C., Weiss, F. U., Niederfellner, G., *et al.* (1995). Heregulin-dependent regulation of HER2/neu oncogenic signaling by heterodimerization with HER3. *EMBO J.* **14,** 4267–4275.

Weiner, D. B., Nordberg, J., Robinson, R., Nowell, P. C., Gazdar, A., Green, M. I., Williams, W. V., Cohen, J. A., and Kern, J. A. (1990). Expression of the *neu* gene-encoded protein (p185^{neu}) in human non-small cell carcinomas of the lung. *Cancer Res.* **50,** 421–425.

White, M. R., and Hung, M.-C. (1992) Cloning and characterization of the mouse *neu* promoter. *Oncogene* **7,** 677–683.

Xia, W., Lau, Y.-K., Zhang, H.-Z., Liu, A.-R., Kiyokawa, N., Clayman, G. L., Katz, R., L., and Hung, M.-C. (1997). Strong correlation between *c-erbB-2* overexpression and overall survival of patients with oral squamous cell carcinoma. *Clin. Cancer Res.* **3,** 3–9.

Xin, J. H., Cowie, A., Lachance, P., and Hassell, J. A. (1992). Molecular cloning and characterization of PEA3, a new member of the *Ets* oncogene family that is differentially expressed in mouse embryonic cells. *Genes Dev.* **6,** 481–496.

Xing, X., Liu, V., Xia, W., Stephens, L. C., Huang, L., Lopez-Berestein, G., and Hung, M.-C. (1997). Safety studies of the intraperitoneal injection of *E1A*-liposome complex in mice. *Gene Ther.* **4,** 238–243.

Xing, X., Matin, A., Yu, D., Xia, W., Sorgi, F., Huang, L., and Hung, M.-C. (1996). Mutant SV40 large T antigen as a therapeutic agent for *HER-2/neu*-overexpressing ovarian cancer. *Cancer Gene Ther.* **3,** 168–174.

Xing, X., Miller, S. J., Xia, W., and Hung, M.-C. (1997). PEA3 as a therapeutic agent for HER-2/neu-overexpressing human cancers. Abstract of the Department of Defense Breast Cancer Research Program Meeting, Washington, DC, 1997 (Volume II, pp. 549–550).

Xing, X., Zhang, S., Chang, J., Tucker, S. D., Chen, H., Huang, L., and Hung, M.-C. (1998). Safety study and characterization of *E1A*-liposome complex gene-delivery protocol in an ovarian cancer model. *Gene Ther.* **5,** 1538–1544.

Yamamoto, T. M., Ikawa, S., Akjiyana, T., Semba, K., Normura, N., Miyajima, N., Saito, T., and Toyoshiman, K. (1986). Similarity of protein encoded by the human *c-erbB-2* gene to the epidermal growth factor receptor. *Nature* **319,** 230–234.

Yan, D. H., Chang, L. S., and Hung, M.-C. (1991). Repressed expression of the *HER-2/neu/c-erbB-2* proto-oncogene by the adenovirus *E1A* gene products. *Oncogene* **6,** 343–345.

Yan, D. H., and Hung, M.-C. (1991). Identification and characterization of a novel enhancer for the rat *neu* promoter. *Mol. Cell Biol.* **11,** 1875–1882.

Yokota, J., Yamamoto, T., Miyajima, N., Toyoshima, K., Nomura, N., Sakamoto, H., Yoshida, T., Terada, M., and Sugimura, T. (1988). Genetic alterations of the *c-erbB-2* oncogene occur frequently in tubular adenocarcinoma of the stomach and are often accompanied by amplification of the *v-erbA* homologue. *Oncogene* **2,** 283–287.

Young, K. S., Weigel, R., Hiebert, S., *et al.* (1989). Adenovirus *E1A*-mediated negative control of genes activated during F9 differentiation. *Mol. Cell Biol.* **9,** 3109–3113.

Yu, D., Hamada, J., Zhang, H., Nicolson, G. L., and Hung, M.-C. (1992). Mechanisms of *c-erbB2/neu* oncogene-induced metastasis and repression of metastatic properties by adenovirus 5 *E1A* gene products. *Oncogene* **6,** 2263–2270.

Yu, D., and Hung, M.-C. (1991). Expression of activated rat *neu* oncogene is sufficient to induce experimental metastasis in NIH3T3 cells. *Oncogene* **6,** 1991–1996.

Yu, D., Liu, B., Jing, T., McDommell, T. J., Sun, D., El-Deiry, W. S., and Hung, M.-C. (1998). Over-expression of c-erbB2 blocks Taxol-induced apoptosis by upregulation of p21[cip1] which inhibits p34[cdc2] kinase. *Molecular Cell* (November).

Yu, D., Liu, B., Jing, T., Sun, D., Price, J. E., Singletary, S. E., Ibrahim, N., Hortobagyi, G. N., and Hung, M.-C. (1998). Overexpression of both p185[c-erbB-2] and p170[mdr-1] renders breast cancer cells highly resistant to taxol. *Oncogene* **16**, 2087–2094.

Yu, D., Liu, B., Tan, M., Li, J., Wang, S.-S., and Hung, M.-C. (1996). Overexpression of *c-erbB-2/neu* in breast cancer cells confers increased resistance to Taxol via *mdr-1*-independent mechanisms. *Oncogene* **13**, 1359–1365.

Yu, D., Matin, A., and Hung, M.-C. (1992). The retinoblastoma gene product suppresses neu oncogene-induced transformation via transcriptional repression of *neu**. *J. Biol. Chem.* **267**, 10203–10206.

Yu, D., Matin, A., Xia, W., Sorgi, F., Huang, L., and Hung, M.-C. (1995). Liposome-mediated *E1A* gene transfer as therapy for ovarian cancers that overexpress *HER-2/neu*. *Oncogene* **11**, 1383–1388.

Yu, D., Scorsone, K., and Hung, M.-C. (1991). Adenovirus Type 5 *E1A* products acts as transformation suppressors of the *neu* oncogene. *Mol. Cell Biol.* **11**, 1745–1750.

Yu, D., Shi, D., Scanlon, M., and Hung, M.-C. (1993). Re-expression of *neu*-encoded oncoprotein counteracts the tumor-suppressing activity of *E1A*. *Cancer Res.* **53**, 5784–5790.

Yu, D., Suen, T. C., Yan, D. H., Chang, L. S., and Hung, M.-C. (1990). Transcriptional repression of the *neu* proto-oncogene by the adenovirus 5 *E1A* gene products. *Proc. Natl. Acad. Sci. USA* **87**, 4499–4503.

Yu, D., Wang, S. S., Dulski, K. M., Tsai, C.-M., Nicolson, G. L., and Hung, M.-C. (1994). *c-erbB2/neu* overexpression enhances metastatic potential of human lung cancer cells by induction of metastasis-associated properties. *Cancer Res.* **54**, 3260–3266.

Yu, D., Wolf, J. K., Scanlon, M., Price, J. E., and Hung, M.-C. (1993). Enhanced *c-erbB-2/neu* expression in human ovarian cancer cells correlates with more severe malignancy that can be suppressed by *E1A*. *Cancer Res.* **53**, 891–898.

Zhang, Y., Yu, D., Xia, W., and Hung, M.-C. (1995). *HER-2/neu*-targeting via adenovirus-mediated *E1A* delivery in an animal model. *Oncogene* **10**, 1947–1954.

Zhang, M., Maass, N., Magit, D., and Sager, R. (1997). Transactivation through Ets and Ap1 transcription sites determines the expression of the tumor-suppressing gene maspin. *Cell Growth Diff.* **8**, 179–186.

CHAPTER 17

Immune Pathways Used in Nucleic Acid Vaccination

Gary H. Rhodes

Department of Medical Pathology, University of California, Davis, California

I. INTRODUCTION

Nucleic acid vaccination (NAV) is a collection of techniques whose common theme is the introduction of an antigen gene into an animal for the purpose of inducing an immune response to the protein antigen. Expression of the antigen gene *in vivo* leads to a long-lived humoral and cellular immune responses to the antigen. In the 7 years since its discovery, nucleic acid vaccination has been shown to produce protective immunity in several animal species to challenge with viral, bacterial, and protozoal pathogens. Advances have also been made in illuminating some of the mechanisms involved in the generation of immunity. However, despite this progress, many questions remain. This review focuses on the immune mechanisms that are involved in the generation of an immune response after nucleic acid vaccination. The goal is to develop a tentative model that outlines the immune pathways involved in generating an immune response for each form of NAV and that can be used to predict ways to enhance and modify the immune response.

A. NOMENCLATURE

The process called nucleic acid vaccination here has been given many names including genetic vaccination (Tang *et al.*, 1992), DNA vaccination, and polynucleotide vaccination. None of these names are perfect but NAV will be used here in describing the process. This name was also selected in an informal vote at a meeting at the World Health Organization in 1994.

B. A SHORT HISTORY

(See Felgner, 1997.) Nucleic acid vaccination was discovered independently in four different laboratories. The first published paper appeared in early 1992 and it utilized ballistic technology to induce antibody responses in mice (Tang *et al.*, 1992). Three other independent labs reported results at the Cold Spring Harbor Vaccine meeting in August 1992 (Rhodes *et al.*, 1993; Robinson *et al.*, 1993; Wang *et al.*, 1993b; Liu *et al.*, 1993). Data presented at this meeting showed induction of humoral and cellular immune responses following DNA injection at a number of anatomical sites. Protection of vaccinated animals from influenza virus challenge was also demonstrated. These data appeared in follow-up publications in the following months (Ulmer *et al.*, 1993; Fynan *et al.*, 1993; Wang *et al.*, 1993a) along with data from another group (Davis *et al.*, 1993). The years that have followed have produced an explosion of publications in this field to the point where it is no longer practical to do a comprehensive review of the entire field. Several other excellent recent reviews of NAV are available (Robinson and Torres, 1997; Bagarazzi *et al.*, 1998; Donnelly *et al.*, 1997; Tighe *et al.*, 1998; Johnston and Barry, 1997; Haynes *et al.*, 1996; Vogel and Sarver, 1995; Cohen *et al.*, 1998) and more specific ones will be cited after paragraph headings.

C. VECTORS

Many of the vectors used for vaccination derive from one that was designed for high-level expression of HIV gp120 in tissue culture cells (Chapman *et al.*,

1991). This vector has the IE1 promoter and intron from human cytomegalovirus (CMV), a cloning site, a second intron from SV40, and a polyadenylation and transcription terminator region also from SV40. It also contains a signal sequence from the human tissue plasminogen activator gene that permits cloning of an antigen gene as a fusion to this heterologous signal sequence. Deletion of the second intron and replacement of the SV40 polyadenylation region by that from bovine growth hormone (BGH) increases expression and has been used in several immunization studies (Yankauckas *et al.*, 1993; Raz, Carson, *et al.*, 1994; Ulmer *et al.*, 1993; Montgomery *et al.*, 1994). Other labs have successfully used commercially available vectors that contain the CMV promoter and the BGH terminator without an intron (Barry and Johnston, 1997; Martins *et al.*, 1995). Expression of cDNA sequences and immune responses after NAV are generally higher when introns are present (Barry and Johnston, 1997; Chinsangaram *et al.*, 1998).

More than a dozen different promoters have been tested for muscle expression. The highest expression is provided by the CMV promoter. Slightly lower levels are found in vectors containing the Rouse sarcoma virus (RSV) promoter. Other tested promoters, including some from muscle specific genes and some from housekeeping genes, functioned at levels of 5% or less than that provided by CMV after injection into muscle (Manthorpe *et al.*, 1993). All of the tested vectors gave high-level expression after transfection into tissue culture cells. More recent studies have utilized a lentivirus and phospho-glycerate kinase (Wagener *et al.*, 1996), creatine kinase (Bartlett *et al.*, 1996), and the MHC class I and class II promoters (Xiang *et al.*, 1997) for nucleic acid vaccination.

Vector studies on *in vivo* expression in nonmuscle tissues have been more limited. The RSV promoter, although efficient at producing expression and inducing immunity in muscle, is much less effective at inducing immunity after intradermal injection (Raz, Carson, *et al.*, 1994). A lesson from these experiments is that expression levels in transfected tissue culture cells are poor predictors of how a plasmid will function *in vivo*. A second lesson is that although CMV promoter constructs have functioned well in a variety of tissues examined to date, one should try a number of vectors when initiating experiments in new tissues.

Small changes in the noncoding portions of the plasmid can affect protein expression (Hartikka *et al.*, 1996). Removal of a SV40 origin or replication upstream from the CMV promoter or substitution of neomycin phosphotransferase for β-lactamase as the selectable marker both cause a significant increase in luciferase expression after injection into muscle (Hartikka *et al.*, 1996).

The noncoding portions of the plasmid can also change the level of immunity induced by the vector. The same replacement of amp[r] to neo[r] that increased expression of luciferase decreased the immune response to the antigen β-galactosidase (Sato *et al.*, 1996). This finding will be discussed more thoroughly later but it illustrates the fact that even seemingly minor differences in vector sequence can lead to large differences in levels of expression or immune responses. This makes it difficult to compare results from different labs on mechanism. There have been relatively few studies in which the same plasmid has been used to vaccinate by the different delivery methods.

II. GENERATION OF IMMUNITY

Nucleic acid vaccination utilizes several different delivery methods to introduce antigen genes *in vivo*. These techniques are summarized in Table I. These methods differ in the means of plasmid DNA delivery, the cells that express antigen, and the level and duration of antigen expression. All of these methods can generate immune responses of approximately equivalent antibody titers, all generate strong cellular immune responses, and all produce protective immunity in several challenge systems. However, as we shall see, substantially different mechanisms are probably involved in generating the immunity and these will lead to a different nature of the response generated by each method. This first section examines the immune mechanisms involved in each method.

A. INTRAMUSCULAR INJECTION OF NAKED DNA

(See Wolff, 1997; Danko and Wolff, 1994.) The 1990 paper by Wolff *et al.* (1990) demonstrated that injection of plasmid expression vectors into muscle resulted in uptake and expression of reporter genes. This procedure has been called *naked DNA delivery* since it requires no compounds to facilitate DNA entry into the cell. The plasmid is simply dissolved in normal saline and injected into the muscle. Both the quadriceps (Wolff *et al.*, 1990; Rhodes *et al.*, 1993; Ulmer *et al.*, 1993) and the tibialis anterior muscle groups (Davis, Michel, *et al.*, 1993; Davis, Whalen, *et al.*, 1993) have been used for reporter gene expression and to induce immune responses.

Transgene expression can be detected within an hour of DNA injection. Ex-

Table I

Techniques of Nucleic Acid Vaccination

Delivery method	Route	Antigen expression	Cells transfected
Naked DNA	im	permanent	myotubules
Naked DNA	id	prolonged	keratinocytes, Langerhans cells
"Facilitated" DNA	im	transient	myocytes
Ballistic	id	transient	Langerhans cells, keratinocytes
Mucosal	in	?	?

Abbreviations: im, intramuscular; id, intradermal; in, intranasal.

pression peaks around 2 weeks after injection and then declines to 20–50% of the peak level where it remains for more than a year (Wolff *et al.*, 1990; Manthorpe *et al.*, 1993; Hartikka *et al.*, 1996; Doh *et al.*, 1997). Gene expression occurs in the terminally differentiated, nondividing, multinucleated myofibers. Expression is first seen in the vicinity of the tendon–muscle junction (Doh *et al.*, 1997) and later extends toward the center of the muscle myofibers (Wolff *et al.*, 1990; Doh *et al.*, 1997). The levels of transgene expression depend on the details of the expression vector (Manthorpe *et al.*, 1993; Hartikka *et al.*, 1996) but are in the range of 10 to 50 ng of luciferase per quadriceps muscle for nonsecreted reporter genes. Expression of a relatively stable secreted protein like clotting factor IX can give steady-state serum levels of 1 to 10 ng/ml in the sera and these levels are stable for several months (G. Rhodes, unpublished). There are two important properties of this method of transgene expression: (1) the plasmid DNA persists in the muscle myotubules as an unintegrated, nonreplicating plasmid episome (Wolff *et al.*, 1990) and (2) expression of the transgene continues for more than a year after injection and is probably permanent (Wolff *et al.*, 1990; Bartlett *et al.*, 1996).

A single intramuscular injection of 25 to 100 μg of a DNA expression vector gives an extremely long-lasting (probably permanent) humoral immune response in mice (Yankauckas *et al.*, 1993). IgM antibodies to the encoded antigen appear 5 to 10 days after injection (Rhodes *et al.*, 1994). IgG antibodies appear between 2 and 3 weeks after immunization and increase for 1 or 2 months (Rhodes *et al.*, 1994, 1993). Titers then stabilize or very slowly decrease over time (Rhodes *et al.*, 1994; Yankauckas *et al.*, 1993). We followed one animal for 2½ years and antibody titers were maintained at more than ½ of the peak titer over that time.

Cytotoxic T cells (CTL) that recognize the antigen are detectable 5 to 10 days after vaccination. CTL responses are long lived and have been observed 17 months after a single injection of plasmid (Yankauckas *et al.*, 1993). This long-lasting immune response could be due to the development of immunological memory or may be due to constant repriming or restimulation of the immunity due to constant exposure to the antigen.

1. Dose Response

Optimal humoral immune responses occur with doses of 25 to 100 μg of plasmid DNA, which is the same amount that gives maximal level of reporter gene expression (Manthorpe *et al.*, 1993). The dose is usually divided into two bilateral injections to reduce experimental variation. As the dose is lowered, the appearance of IgG antibodies is delayed but will ultimately reach approximately the same final titer (Rhodes *et al.*, 1994; G. Rhodes, unpublished). There appears to be a threshold effect because below a certain level of DNA, usually around 1 μg, no humoral response develops at all. A similar phenomena is seen for intradermal NAV but occurs at lower DNA concentrations (Carson, Raz, *et al.*, 1994). The dose response for CTL activity is roughly the same as that for humoral immune induction but extends to lower doses. Thus, it is possible at low DNA doses to find animals with CTL but no IgG antibodies, although the opposite has not been observed (G. Rhodes, unpublished).

Studies using reporter genes show optimal expression at 50 to 100 μg of DNA injected per muscle (Manthorpe *et al.*, 1993; Doh *et al.*, 1997). Increasing the amount of DNA injected over the optimal levels causes a decrease in the expression from the plasmid (Doh *et al.*, 1997). This factor may have played a role in the weak and transient immunity reported in some of the first primate immunization experiments as extremely large amounts of DNA were used in these injections (Liu *et al.*, 1996).

2. Multiple DNA Injections and Boosting

Repeated injection of plasmid are not required and do not appear to be needed for immune responses to most viral capsid and envelope antigens in mice. Injections of animals one, two, or three times at 2-week intervals gave the same titers within 30%, when measured 12 weeks after the first injections (Rhodes *et al.*, 1994). This concept is counterintuitive because it is opposite to our experience with protein immunization. However, protein immunization delivers antigen in pulses. Antigens levels are high after injection and then drop to very low levels until a boost. DNA immunization in muscle provides a constant, low level of antigen throughout the entire immune response. Increasing the antigen level twofold by a second injection after 2 weeks does not have much effect on the magnitude of the immune responses. However, multiple injections may be required for immune responses to some antigens, especially nonsecreted ones (see following).

An exception to the rule that multiple DNA injections do not increase immunity is provided by the malaria CSP antigen. Studies with this antigen indicate that spacing of the injections is important. Repeated injections at intervals of 28 days have little effect on the antibody titers but intervals of 45 days produce substantially higher antibody levels (Leitner *et al.*, 1997). This behavior would be expected if the antigen were toxic and slowly killed the transfected muscle cells. Immune responses to CSP are also anomalous because it is the only antigen described so far in which intramuscular NAV generates a Th2 rather than Th1 T cell response (Leitner *et al.*, 1997; see following).

Although multiple DNA injection does not have much effect on antibody levels, boosting with a small amount of antigen protein can substantially raise IgG antibody levels (Raz *et al.*, 1996). Intramuscular injection of 100 ng of antigen dissolved in saline (no adjuvant) boosted antibody titers over 50-fold in 4 days (G. Rhodes, unpublished).

3. Is Secretion Necessary for an Immune Response?

Most early NAV studies were done with viral capsid or envelope proteins. These proteins can exit the cell either by having a signal sequence or by forming particles that leave the cell. We tested five nonsecreted proteins for their ability to elicit humoral immunity after a single DNA injection. One protein, β-galactosidase, gave IgG antibodies 2 to 3 weeks after a single immunization that had titers similar to those induced by the viral proteins. Little or no immune response was obtained to the other four antigens (titers < 1/20) even several months after immunization (G. Rhodes, unpublished). Our interpretation of these data is that secretion is required for humoral immune induction and that β-galactosidase is an anomalous antigen that is somehow released from the cells and thus exposed to B cells. Immune responses to apparently nonsecreted proteins such as carcinoembryonic antigen (CEA) have been reported after multiple intramuscular injections (Conry, LoBuglio, Wright, *et al.*, 1995; conry, LoBuglio, Loeckel, *et al.*, 1995).

More recently, the question of secretion has been studied more systematically by comparing immune response to cytoplasmic, secreted, and membrane-bound forms of the same antigen. IgG responses to the different forms of a bacterial-malaria fusion antigen were comparable after multiple injections (Haddad *et al.*, 1997). Boyle *et al.* (1997) found that the humoral immune response was delayed and the titers were lower with the nonsecreted immunogen after two injections. CTL responses were also lower with the nonsecreted form after intramuscular injection. In contrast, after intradermal injection, the CTL response to the cytoplasmic form was equal or perhaps higher than that induced by the secreted protein. Antibody responses to nonsecreted forms of hepatitis C capsid antigen were observed after two injections using the "facilitated" method (Inchauspe *et al.*, 1997). Comparable responses were observed to secreted and membrane-bound antigen forms (Xiang *et al.*, 1995).

A clue as to how repeated injections could induce humoral immunity to a nonsecreted protein is provided by the finding that severe inflammation occurs when an antigen gene is injected into an animal with a preexisting immune response to the antigen (Yokoyama *et al.*, 1997). No inflammatory response is seen after a single injection into muscle (Wolff *et al.*, 1990). Thus, a likely sequence of events that leads to a humoral immune response to a nonsecreted antigen is that the initial injection induces a cellular immunity that then causes inflammation, immune activation, and antigen release after subsequent injections. Induction of immunity to nonsecreted antigens may not be such a problem with some other forms of NAV where tissue destruction and regeneration is a normal occurrence.

4. Th$_1$ Response Induced

Helper T cells can be divided into two groups based on the cytokine patterns that are made in response to antigen stimulation (Mosmann and Coffmann, 1989). Th$_1$ responses are associated with the production of IFN γ while the Th$_2$ cells secrete IL-4, IL-5, and IL-10. The cytokine patterns influence the IgG subclass usage with Th$_1$ responses favoring IgG2a while Th$_2$ responses are associated with IgG1. The immune responses induced by intramuscular NAV to a number of antigens are strongly skewed toward the Th_1 pattern. This Th_1 bias has been noted with several different antigens (Manichan *et al.*, 1995; Feltquate *et al.*, 1997; Raz *et al.*, 1996; Boyer *et al.*, 1998) and is established by most NAV methods with the exception of ballistic immunization (see following). A mechanism that accounts for the shifting of responses toward Th_1 will be discussed in a later section (see IIIA).

5. Vaccination with RNA

Injection of RNA encoding an antigen can also lead to humoral and cellular immunity (Zhou *et al.*, 1994). In contrast to vaccination with plasmid DNA, optimal immune responses with RNA vaccines require a second or boost injection 2 weeks after the initial vaccination. This again emphasizes that immune response can develop after a single injection if antigen expression is continuous but will require boosting if antigen is transiently expressed. Interestingly, Th_1 T cell immunity is induced by RNA as well as DNA after intramuscular or intradermal injection (G. Rhodes and P. Liljestrom, unpublished).

B. INTRADERMAL INJECTION OF NAKED DNA

Injection of plasmid DNA in normal saline into the dermis also induces an immune response (Raz, Carson, *et al.*, 1994; Sato *et al.*, 1996; Roman *et al.*, 1997). Transfected cells are seen in the injected area throughout the dermis and epidermis. Expression continues in the dermis for more than 3 months, whereas expression in the epidermis is seen to migrate outward with time as the cells are sloughed. The expressing cells are keratinocytes, fibroblasts, and cells with the morphological appearance of dendritic cells (Raz, Carson, *et al.*, 1994). The ability to transfect antigen-presenting cells like dendritic cells differentiate this method from muscle injection.

Immune responses induced by this method are also long lived with humoral and cellular immunity persisting for more than 70 weeks postinjection (Raz, Carson, *et al.*, 1994). The immune responses induced by NAV have a Th_1 pattern while injecting antigen protein gave Th_2 immunity (Raz *et al.*, 1996; Feltquate *et al.*, 1997). Antibody responses induced by NAV can be boosted with small amounts of protein antigen and the boosted immunity remains Th_1 (Raz *et al.*, 1996). Protection from viral challenge was also demonstrated (Raz, Carson, *et al.*, 1994).

One difference between muscle and dermal injection is that the later requires less DNA. One group has claimed that intramuscular injection gives better cellular immunity than intradermal (Shiver *et al.*, 1997, 1995) but most find that the intradermal route gives comparable or higher levels (Hinkula *et al.*, 1997; Yokoyama *et al.*, 1997; Feltquate *et al.*, 1997). In our hands, the immune responses generated by the two methods are very similar in their strength and longevity. This is somewhat surprising in that intradermal injections directly transfect antigen-presenting cells while muscle injections apparently do not. We will return to this question.

C. "FACILITATED" INTRAMUSCULAR DNA INJECTION

This method treats the muscle with a myotoxic agent such as bupivacaine (Wang *et al.*, 1993a) or cardiotoxin (Davis, Michel, *et al.*, 1993). Treatment is optimally done a day or two before the DNA injection but can also be mixed and injected with the DNA at some cost in expression levels (Davis *et al.*, 1995). These agents cause destruction of the myofiber structure in the injected muscle and promote growth of myocytes. It is the myocytes that are transfected by this method and they ultimately fuse to generate new myofibers. Expression levels of reporter genes are generally higher (4- to 40-fold or more) (Danko *et al.*, 1994) than those obtained by naked DNA injection. Expression, however, is transient and declines 2 weeks after injection (Danko *et al.*, 1994).

Macrophage and dendritic cells can be found in normal muscle but their numbers greatly increase after injury (Orimo *et al.*, 1991; Grounds, 1991; Pimorady-Esfahani, 1997). Polymorphic leukocytes peak 12 h after injury, while macrophage numbers increase for 2 days and T cell infiltration is minor (Orimo *et al.*, 1991; Pimorady-Esfahani, 1997). There are some data suggesting that different macrophage subpopulations sequentially appear at different times after the injury (McLennan, 1996). Dendritic cells increase steadily after injury and remain at elevated levels for more than 6 days (Pimorady-Esfahani, 1997). Thus, "facilitated" nucleic acid vaccination differs in at least three ways from naked injections: The antigen expressing cells are dividing myocytes instead of nondividing myofibers, antigen expression is higher but transient instead of lower and long lived, and "facilitated" vaccinations attract numerous antigen presenting cells to the site of antigen production.

The immune responses induced by "facilitated" injections are boostable by a second DNA injection since antigen levels decrease substantially from peak levels before IgG antibodies are detected. Antibody responses after a single injection are comparable after either "facilitated" or naked DNA injection. Antibody responses were boosted 200-fold by a booster-facilitated DNA injection 16 weeks after the initial immunization (Davis *et al.*, 1996). IgG levels are also boostable by injection of protein antigen (Davis *et al.*, 1996). Immune responses induced by this method give the Th_1 pattern (Leclerc *et al.*, 1997; Boyer *et al.*, 1998).

D. BALLISTIC DELIVERY

(See Swain *et al.*, 1997; Haynes *et al.*, 1996; Johnston and Tang, 1994.) In the ballistic (also called gene gun) technique, the DNA is precipitated onto small gold beads. The beads are then accelerated by one of several methods and can penetrate into the surface layers of tissues. The depth of penetration can be controlled somewhat by the amount of acceleration given to the particles. Highest levels of gene expression and antibody titers are found using conditions that maximized delivery to the epidermis (Eisenbraun *et al.*, 1993). The gold beads can be seen inside the cytoplasm as well as in the interstitial regions after bombardment. The skin is the usual tissue treated by this method although it has also been used to transfect other tissue.

Gene gun immunization requires less DNA than the other methods. Early papers reported that 15 to 80 ng of DNA was required for optimal responses (Haynes *et al.*, 1994; Pertmer *et al.*, 1995) but later papers use substantially more plasmid (Pertmer *et al.*, 1996; Iwasaki *et al.*, 1997). Antigen or reporter gene expression is transient, lasting 2 to 3 days and then declining to lower levels (Barry and Johnston, 1997; Yang *et al.*, 1990; Torres *et al.*, 1997). As expected with transient expression, antibody titers are boosted by repeated gene gun vaccinations (Le Borgne *et al.*, 1998; Pertmer *et al.*, 1995). Titers are also increased by protein boosts (Barnett *et al.*, 1997; Fuller *et al.*, 1996). This method is also very efficient at inducing cellular immunity. CTL precursor frequencies of 1/2000 have been reported after ballistic vaccination (Chen *et al.*, 1998).

The most marked difference with gene gun immunization from the other methods is that it produces Th_2 helper T cells. Early studies claimed a progression from Th_1 to Th_2 with continued immunization (Haynes *et al.*, 1994; Barry and Johnston, 1997; Fuller and Haynes, 1994; Prayaga *et al.*, 1997). The progression can be blocked by coadministration of the genes for IL-12 or IL-7 or by increasing the spacing between vaccinations (Prayaga *et al.*, 1997). Other labs see a Th_2 response from the first injection (Feltquate *et al.*, 1997). It has been proposed that the differences between gene gun and other methods can be partially accounted by the amount of DNA delivered (Barry and Johnston, 1997) but this does not seem to be the case as very similar doses given by naked DNA injection (either intramuscular or intradermal) gave Th_1 responses while the gene gun administration gave completely Th_2 (Feltquate *et al.*, 1997). The means by which ballistic immunization give a different helper T cell pattern remains unknown. It is not caused either by the amount of DNA used for immunization or by the specific antigen as the same plasmid construct gives a Th1 like response when naked DNA delivery is used and Th2 when ballistic delivery is used (Feltquate *et al.*, 1997). One possibility may be that the particulate form of the DNA bound to gold particles might influence the process by allowing transfection of different kinds of antigen–presenting cells and thereby influence the process.

Two studies have implicated cells that can rapidly migrate out of the bombardment site in the developing immune response. Klinman *et al.* (1998) examined the effect of ablating the bombarded area of the skin at various times after vaccination. They found that the magnitude of the primary immune response, measured after a single injections, slowly increased for a week after bombardment. Low anti-body levels were obtained if the skin was removed within a day of vaccination. In contrast, the memory immune response, measured after a boost injection, was established by 12 h after vaccination. The continued presence of the bombarded area of the skin for periods longer than 12 h had little effect on the memory response. T cell immunity followed these same kinetics. These results were generally confirmed in a second paper (Torres *et al.*, 1997). If the bombarded skin was transplanted to naive recipients, a memory response was observed if the skin was transplanted within 12 h of bombardment (Klinman *et al.*, 1998). These results implicate mobile skin cells, most likely dendritic cells, that migrate out of the bombarded area within 12 h and that are responsible for the development of the primary and secondary immune responses.

A novel application of this method of delivery has been expression library immunization in which random sequences from a pathogen are cloned into expression vectors and used for immunization (Johnston and Barry, 1997). This method has been shown to provide protective immunity to challenge from mycoplasma (Lai *et al.*, 1995).

E. MUCOSAL IMMUNIZATION

It is generally accepted that the mucosal immune system functions independently of the systemic immune system. Induction of a mucosal immune response requires the presence of antigen in mucosal inductor tissue that is present in several sites in the body including the nose, tonsil, and intestines. The induction of immunity in the two different compartments is quite specific. Addition of recombinant virus to nasal tissue gave mucosal but no systemic immunity while application of the same vector to the lungs produced the opposite result (Malone *et al.*, 1997). There have been several reports of application of DNA to the induction of mucosal immunity. One of the first papers published on NAV showed protection to influenza was induced by intranasal administration of the gene for hemagglutinin although no direct measurement of mucosal immunity was made (Fynan *et al.*, 1993). More recent work has demonstrated the induction of mucosal immune response following intranasal administration of DNA (Asakura *et al.*, 1997; Kuklin *et al.*, 1997) or DNA/cationic lipid complex (Klavinskis *et al.*, 1997; Okada *et al.*, 1997). The addition of cationic lipids to the DNA substantially increased the immune response (Okada *et al.*, 1997). Intravaginally administered DNA also has been reported to induce mucosal antibodies (Wang *et al.*, 1997). Thus, DNA administration to two different mucosal sites is able to generate an immune response. These studies offer a promising beginning to the development of a mucosal NAV.

F. Other Methods

Other methods have been used to deliver plasmid DNA to cell *in vivo* and to produce an immune response. These methods are either at an earlier stage of development or have only been described in a few papers. Two groups were able to deliver DNA solution to muscle using a needleless jet delivery system (Volhlsing *et al.*, 1994; Lodmell *et al.*, 1998) that propels the solution through the skin and into the underlying tissue. Both were able to detect expression of reporter genes and one was able to demonstrate an immune response after treatment.

Several groups have investigated using bacteria to deliver plasmid intracellularlly. This may be especially useful for the oral delivery of DNA to mucosal inductor tissue in the intestine (reviewed in Pascual *et al.*, 1997) in order to stimulate mucosal immunity.

Another promising approach is the use of direct DNA delivery for vaccination with live, attenuated virus vaccines. Either intramuscular or intradermal injection of feline immunodeficiency virus provirus DNA establishes a viral infection in cats (Sparger *et al.*, 1997; Rigby *et al.*, 1997). Injection of as little as 30 μg of proviral plasmid was sufficient to establish an infection. The rate of viral growth is dose dependent with larger doses giving higher immune responses and allowing virus isolation at earlier times (Sparger *et al.*, 1997). This technique promises to be very useful for delivery of attenuated viral vaccines because mutants can be easily isolated and tested since growth and purification of intact virus particles is not required.

Cationic lipids are widely used for transfection of tissue culture cells and were originally investigated for use *in vivo* for DNA delivery. These compounds inhibit expression of DNA delivered to some tissues while enhancing it in others. They inhibit expression of reporter genes in muscle (Wolff *et al.*, 1990) and completely inhibit the development of humoral immunity when used intradermally (Raz, Rhodes, *et al.*, 1994). On the other hand, these compounds can enhance expression in other tissue such as lung (Wheeler *et al.*, 1996). The reason for this differential effect could be the toxicity of these compounds. One cationic lipid has been shown to produce massive necrosis when injected into areas such as dermis or muscle where the solutions are confined after injection (Raz, Rhodes, *et al.*, 1994). Other lipids are apparently less toxic in tissue like lung where they are rapidly dispersed. Lowering the amount of lipid from the amounts used for tissue culture transfection can also reduce the *in vivo* toxicity of these compounds (Felgner *et al.*, 1994). Two recent reports claim that coinjection of lipid/DNA complexes can enhance immunity (Gregoriadis *et al.*, 1997; Ishii *et al.*, 1997). It is not known at this time whether the compounds act by increasing antigen expression, directly as adjuvants, or by a toxic mechanism similar to the compounds used for "facilitated" DNA delivery.

One potential advantage of nucleic acid vaccination is that the structure of the antigen is easily manipulated by recombinant DNA techniques. Some recent papers illustrate how this has been used to enhance immunity to an antigen. Rodriguez *et al.* (1997) have targeted the antigen for rapid degradation by engineering its ubiquitination. Antigen breakdown is so efficient that little full-length antigen is seen after transfection. They find that CTL responses are enhanced but that humoral immunity is abolished presumably because of the low levels of intact antigen protein. Another approach is to target the antigen to sites of immune induction by synthesizing the antigen as a fusion to molecules that are ligands for surface receptors on immune cells (Boyle *et al.*, 1998). The best immune enhancements were obtained when the antigen was fused to CTLA4, which binds to B7 expressing cells. Both antibody and helper T cells levels were increased by this method (Boyle *et al.*, 1998). Enhancing the expression and immune response to an antigen by changing to optimal codon usage has also been reported (Andre *et al.*, 1998).

III. IMMUNE MECHANISMS

A. Immunostimulatory Sequences (ISS)

(See Lodmell *et al.*, 1998; Carson and Raz, 1997; Krieg, 1996; Krieg *et al.*, 1998; Klinman, Takeno, *et al.*, 1997.) Immunostimulatory sequences are small DNA sequences containing an internal, unmethylated, CpG dinucleotide that have a profound stimulatory effect on most cells of the immune system. The CpG dinucleotide is underrepresented in vertebrate DNA and is usually methylated on the cytidine residue. These same sequences are unmethylated in bacterial DNA or plasmids grown in bacteria. They were first discovered during purification of water-soluble components of *Mycobacteria bovis* that induced tumor regression in mice and guinea pigs (Tokunaga *et al.*, 1989). The active purified product contained primarily DNA but no protein. This fraction was able to cause tumor rejection, stimulate natural killer cell activity, and cause the production of interferons α, β, and γ (Yamamoto *et al.*, 1988). The nature of the active DNA sequences were found by synthesizing a number of oligonucleotides from known *Mycobacteria* genes. All active oligonucleotides contained a palindromic sequence centered on a CpG dinucleotide (Kuramoto *et al.*, 1992; Yamamoto, Yamamoto, Kataoka *et al.*, 1992; Kataoka *et al.*, 1992). The synthetic oligonucleotides themselves activate NK cells and promote interferon release (Tokunaga *et al.*, 1992). The spectrum of interferons induced was different from that induced by double-stranded ribonucleotides (Tokunaga *et al.*, 1992; Yamamoto, Yamamoto, Kataoka *et al.*, 1992). The activation can be mimicked by bacterial DNA but not DNA isolated from vertebrates (Yamamoto *et al.*, 1992), presumably because the ISS sequences on the bacterial DNA are unmethylated. The immunostimulatory activity of the oligonucleotides was size dependent with optimal activity in oligonucleotides of 20 to 30 base pairs (Yamamoto *et al.*, 1994a). The active size of the ISS oligonucleotides can be shortened to hexamers and the active concentration reduced several hundredfold by adding cationic lipids to the oligonucleotides (Yamamoto *et al.*, 1994b). Cationic lipids facilitate the entry of nucleic acids into the cell, suggesting that the cellular receptor for the active ISS sequences may be located inside cells rather than on the membrane surface.

ISS sequences were discovered independently in other assays. They are able to induce murine B cells to proliferate and secrete IgM (Krieg *et al.*, 1995). The ISS oligonucleotides also directly stimulate and activate many cells of the immune system including NK (natural killer) (Yamamoto, Yamamoto, Kataoka, *et al.*, 1992; Ballas *et al.*, 1996), B cells (Krieg *et al.*, 1995), T cells (Sato *et al.*, 1996), macrophage (Sparwasser *et al.*, 1997; Chace *et al.*, 1997; Wloch *et al.*, 1998; Sweet *et al.*, 1998) and dendritic cells (Jakob *et al.*, 1998; Sparwasser *et al.*, 1998). The activated cells secrete a number of cytokines including IL-2, IL-3, IL-4, IL-5, IL-10, IL-12, IL-18, and TNF-α in addition to the interferons (Klinman *et al.*, 1996; Roman *et al.*, 1997; Chace *et al.*, 1997; Sparwasser *et al.*, 1997). ISS also cause activation and up-regulation of costimulatory molecules on dendritic cells (Jakob *et al.*, 1998; Sparwasser *et al.*, 1998). Thus, ISS have a potent activating effect on most immune effector cells as well as antigen–presenting cells.

Active, unmethylated ISS sequences in the plasmid DNA are necessary for a robust immune response after nucleic acid vaccination. One such active sequence is contained in the β–lactamase portion of the plasmid expression vector. Replacement of this portion of the plasmid with the gene that confers kanamycin resistance lowers the induced immune response (Sato *et al.*, 1996). Adding back the active ISS sequence to the plasmid, either next to the kan[r] gene or next to the eukaryotic promoter, restores the immune response (Sato *et al.*, 1996; Roman *et al.*, 1997). The immune response can also be restored *in trans* by mixing the kanamycin plasmid with an ISS containing plasmid (Sato *et al.*, 1996).

Although we know that certain CpG sequences are active and that others are inactive, there is no general way to predict which CpG sequences in DNA will be active. Early studies focused on palindromic sequences centered on a CpG (Kuramoto *et al.*, 1992; Yamamoto, Yamamoto, Kataoka *et al.*, 1992) although we now know that nonpalindromic sequences are active (Klinman *et al.*, 1997). All 64 palindromic hexamers were tested for their ability to release IFN-γ from mouse spleenocytes (Kuramoto *et al.*, 1992). Both the hexamers themselves, hexamers that were complexed to cationic lipids, or hexamers imbedded in a longer 30 nucleotide sequence were tested. There appears to be a gradation of activity with some sequences being highly active, some moderately active, and some completely inactive. Six of the 64 sequences released large amounts of cytokine, 10 released smaller amounts, and 48 were inactive. No simple consensus sequence can be drawn although most active palindromic sequences are of the form RRCGYY (R = purine, Y = pyrimidine; Kuramoto *et al.*, 1992). We now know that nonpalindromic sequences are also active but no one has yet tested the 256 nonpalindromic hexamers centered on a CpG. This lack of basic information makes it difficult to predict which CpG sequences in a plasmid will function as ISS. This is a particular problem with plasmids containing the CMV IE promoter since this region is GC rich. We also do not know whether the active CpG sequences are strain- or species-specific. One report suggests a differential effect of the various active ISS sequences with one sequence causing secretion of both IL-12 and TNF-α and another inducing only IL-12 (Lipford et al., 1997).

Methylation of the CpG sequences in a plasmid also lowers the immunity induced by NAV. One can methylate the cytosine of CpG sequences by using the Sss I methylase. As the amount of plasmid DNA methylation is increased, the immunity induced by the plasmid is lowered (Klinman, Takeno et al., 1997). Interpretation of these data is complicated by the fact that methylation also decreases expression of the antigen itself. The data are corrected for this decrease by normalizing the immune response to the amount of expression they observe after transfecting tissue culture cells. However, expression levels observed in tissue culture can be unrelated to those measured in vivo so these conclusion must be considered tentative.

Because of the many activation and stimulation effects that ISS have on immune cells, they undoubtedly play a major role in enhancing immunity after NAV. However, their exact function has not been defined and will undoubtedly be complex. One possible role is to activate the local antigen–presenting cells to accept protein antigen from the transfected cells and present it via class I MHC.

The ISS have also been postulated to play a role in producing the dominant Th_1 immunity seen after most forms of NAV. The ISS induce the production of IL-12 and IL-18, which in turn induce IFN-γ. All of these cytokines can promote the differentiation of Th_1 T cells. The concept of a cytokine milieu in which a number of Th_1 promoting cytokines contribute together has been proposed (Lodmell et al., 1998; Roman et al., 1997).

There must be other pathways that also stimulated the development of immunity and skewing to Th_1 responses. Vaccination with self-replicating RNA gives a strong Th_1 immune response (G. Rhodes and P. Liljestrom, unpublished). Animals vaccinated with stably transfected myocytes develop immune responses similar to those seen after DNA vaccination (Ulmer et al., 1996). In neither case are ISS oligonucleotides present. The same amount of DNA can give Th_2 responses when vaccinated using the gene gun and Th_1 responses when needle injected (Feltquate et al., 1997) so the delivery method is also important in determining the expansion of T cell subsets. Thus, the exact function of the ISS in NAV and their interaction with other immunostimulatory pathways remain to be defined.

B. ANTIGEN PRESENTATION

A major puzzle for intramuscular nucleic acid vaccination concerns how antigen is presented to induce immunity. Priming of a response requires the T cell to recognize two signals on the surface of the antigen presenting cell: (1) a specific recognition of the antigen peptide bound to MHC molecules by the T cell receptor and (2) the presence of costimulatory (or activation) proteins on the surface of the antigen-presenting cells. T cells that encounter the antigen in the absence of this second signal are rendered unresponsive and are incapable of activation and effector function (Quill and Schwartz, 1987; Schwartz, 1992). Most nucleated eukaryotic cells, including muscle cells, express class I MHC molecules but expression of the costimulatory proteins is thought to be restricted to professional antigen-presenting cells (APC) that are derived from bone marrow cells. Thus, the expectation from this line of argument is that intramuscular nucleic acid immunization should not be effective in inducing immunity and could perhaps even produce tolerance to the antigen. Intradermal delivery does not have this dilemma since professional APC appear to be directly transfected (see previous discussion).

Several papers have investigated the presentation of antigen by muscle cells. Three possible mechanisms have been proposed to explain how the immune system develops a class I restricted T cell response after intramuscular nucleic acid vaccination (Ulmer *et al.*, 1996):

1. The muscle cells themselves could directly present antigen to T cells by their class I MHC molecules.
2. Professional antigen-presenting cells could be transfected during the injection process.
3. The muscle cells could transfer the protein antigen to APC, which then interact with the T cells.

A number of complementary studies have shed light on these questions and make it appear very likely that the muscle cells do transfer the antigen to APC (Corr *et al.*, 1996; Doe *et al.*, 1996; Ulmer *et al.*, 1996; Iwasaki *et al.*, 1997; Casares *et al.*, 1997). All of these studies involve cross-priming of the immune response; that is, they create a mismatch between the MHC molecules on the muscle cells and the MHC of the cells of the immune system. When this occurs, muscle cells can no longer interact with the host T cells. The results of all of these studies implicate bone marrow-derived antigen-presenting cells in the immune response. Some of these papers are discussed next.

The paper by Corr *et al.* (1996) started with an F1 hybrid of $H-2^b \times H-2^d$ mice. Immune cells were ablated by radiation and the mice were reconstituted with T cell-depleted bone marrow cells from either $H-2^b$ or $H-2^d$ mice. Thus, the muscle cells of these animals contained both $H-2^b$ and $H-2^d$ haplotype as do the thymic epithelial cells while the APC and the T cell have a single haplotype. The immune system will be reconstituted by both $H-2^d$ and $H-2^b$ restricted T cells since the thymus contains both sets of class I MHC molecules. The antigen-presenting cells will be exclusively of the donor haplotype. After intramuscular vaccination, the ability to generate an $H-2^b$ or $H-2^d$ restricted immune response was measured by measuring CTL activity using synthetic peptides representing H-2^b or H-2^d restricted epitopes. The result of this experiment was that the CTL were restricted by the MHC molecules found on the bone marrow cells used to reconstitute the immune system.

A second set of cross-priming experiments was done by Doe *et al.* (1996) using mice carrying the scid mutation (sever combined immunodeficiency). Scid mice do not have functional T or B cells although they do have functional APC. In these experiments, scid mice of either H-2^d or H-2^b background were infused with cells from normal H-2^{bXd} animals and tested for their ability to mount H-2^b or H-2^d restricted CTL responses. When the mice were reconstituted with donor spleen cells, the CTL response was restricted by the haplotype of the recipient. In contrast, if animals were reconstituted with a mixture of spleen cell and bone marrow cells, CTL responses restricted by both of the donor haplotypes were observed. This is summarized in Table 2. The interpretation of these experiments is as follows: H-2^d animals reconstituted with H-2^{dXb} spleen cells will have parental muscle and APC (H-2^d) and donor B and T cells (H-2^d and H-2^b). Only the H-2^d T cells can respond if either muscle or APC are the presenting cells. When H-2^d mice are reconstituted with spleen and bone marrow cells, the muscle cells will now be H-2^d but the T cells and APC will both be H-2^d and H-2^b. Antigen presentation can now take place to T cells restricted to either haplotype. Thus, these experiments also implicate bone marrow derived APC and not muscle cells in the CTL responses generated by NAV.

These same authors also vaccinated the mice with expression vector before reconstituting the immune system. They found that animals produced similar levels of CTL even when reconstituted 3 weeks after the immunization. This experiment has two important implications. First, it makes direct transfection of

Table II

Summary of Cross-priming Experiments

Recipient	Donor	Donor cells	Muscle	APC	T cells	CTL restriction	Ref.
b X d	b	bm	b and d	b	b	b	Corr *et al.*, (1996)
b X d	d	bm	b and d	d	d	d	Corr *et al.*, (1996)
b	b X d	sc	b	b	b and d	b	Doe *et al.*, (1996)
b	b X d	sc + bm	b	b and d	b and d	b and d	Doe *et al.*, (1996)
k X d	k	transfected myoblasts	k	k and d	k and d	k and d	Ulmer *et al.*, (1996)

The haplotype of the donor and recipient animals are shown in the first two columns. The cells used for transplantation are shown in the third column, where bm means bone marrow cells and sc are spleen cells. The MHC molecules present in the muscle, antigen-presenting cells, and the T cells are shown along with the observed MHC restriction of the CTL.

APCs unlikely as a mechanism of antigen presentation. Muscle-injected DNA is very unstable with a half-life of a few minutes (Manthorpe *et al.*, 1993). Thus, there will be no DNA to transfect APC when these are introduced into the animals during reconstitution. The second implication is that the injected muscle acts as an antigen-producing factory, producing antigen and transferring it to APC to initiate immunity.

The final cross-priming experiment studied the effect of transplantation of transfected myoblasts bearing the H-2^k haplotype into H-2^{dXk} mice (Ulmer *et al.*, 1996). If muscle cells themselves present antigen, one would expect the response to be H-2^k restricted. If APC gather the antigen from muscle cells and present it to T cells, one expects both H-2^k and H-2^d restricted CTL. The later possibility is the one observed in the experiment. A summary of all of these experiments is shown in Table 2. As can be seen, the restriction pattern of CTL follows the haplotype of the antigen-presenting cells and not the muscle cells.

Thus, several lines of evidence lead to the conclusion that antigen presentation after intramuscular NAV is by bone marrow-derived APC. In addition, these experiments rule out the presentation of antigen directly by muscle cells and make it unlikely that direct transfection of APC is necessary for a class I MHC restricted immune response. It was also found that APC are not required to be present at the time of vaccination. In the absence of APC, muscle cells can synthesize antigen for 3 weeks without inducing MHC class I restricted immunity. During this time, humoral and helper T cell responses develop normally (Doe *et al.*, 1996). Three weeks after immunization, addition of APC causes the induction of CTL.

1. Transfection of APC

As discussed previously, skin dendritic cells are apparently transfected by intradermal NAV. This was directly confirmed by isolating cells from the injected dermis that were able to stimulate antigen-specific T cells. These cells were class II MHC positive and contained the injected plasmid (Casares *et al.*, 1997). Gene gun immunization also transfects dendritic cells. Antigen-expressing APC can be detected in the draining lymph nodes after vaccination by this method (Condon *et al.*, 1996). One paper has reported the detection of transfected dendritic cells after intramuscular immunization (Casares *et al.*, 1997). These results contradict the results of some of the cross-priming experiments where transfected APC would have produced immune responses that were not observed (Doe *et al.*, 1996). There were some differences in techniques used by the two groups that may explain the varying results. In one case, injection was into a smaller muscle with two immunization in two days (Casares *et al.*, 1997). It is possible that the tissue trauma from the first injection causes an influx of APC that are then transfected in the second injection.

C. Models of Immune Induction after Nucleic Acid Vaccination

1. Intramuscular Injection

The muscle cells appear to be rather passive participants in the development of immunity after intramuscular injection. Their function is to synthesize antigen protein and pass it to the APC. The APC may be activated by the ISS in the vector to take up antigen protein from the muscle cell and then migrate to the draining lymph node where they stimulate CD8$^+$ T cells. Antigen presentation to CD4$^+$ T cells may be made by the same APC or by others that pick up secreted antigen. B cells will also capture secreted antigen. This model predicts that antigen synthesis and immune induction occur at distinct and distant sites in the body.

2. Intradermal Injection

APC (Langerhans cells), keratinocytes, and fibroblasts are transfected by this method of NAV. The transfected APC can migrate to the lymph nodes and express antigen directly at the site of immune induction. Presentation can also occur by direct transfer of antigen protein to APC as in the intramuscular method.

3. "Facilitated" Intramuscular Injection

Both macrophage and dendritic cells are present in the regenerating muscle at the time of the DNA injection so in principle it is possible to transfect APC as well as the muscle myocytes. At this time there is no direct evidence for or against the transfection of APC by this method.

4. Ballistic Methods

Direct transfection of dendritic cells and their migration to the lymph nodes has been demonstrated after vaccination by ballistic methods. The transfected cells at the site of bombardment are also important in determining the level of antibody responses, most likely by providing a source of antigen for B cells (Klinman et al., 1998).

Thus, the different NAV techniques differ in their ability to transfect APC. The ballistic and intradermal methods do transfect APC while the intramuscular route appears to be unique among NAV techniques in that no direct transfection of APC occur (however, see Casares et al., 1997). Such a difference should have testable consequences, one of which is discussed next.

5. A Word of Caution

The evidence for the proposed models is fragmentary at best and they must be considered works in progress. A paper by Torres *et al.* (1997) will seriously challenge this picture if confirmed. The authors removed the injected muscle at various times after DNA vaccination (Torres *et al.,* 1997) and found that immunity develops normally even if the muscle is removed as soon as 10 minutes after injection. This implies that the muscle cells themselves do not need to synthesize antigen and may not even be involved in the immune response after intramuscular immunization. The main criticism of this paper is that the authors did not look for expression in the muscles surrounding the injected one. DNA is surprisingly facile at entering cells *in vivo*. Injection into the quadriceps muscle group usually transfects all four individual muscles even though all four are not directly injected. It would also seem that an antibody response will require a reservoir of transfected cells to provide antigen for B cells (Klinman *et al.,* 1998). However, the muscle myotubules are the dominant and almost exclusive cell type to be transfected after muscle injection.

IV. MODULATION OF IMMUNITY

(See Cohen *et al.,* 1998; Pasquini *et al.,* 1997.) There have been a number of studies that have attempted to enhance or modulate immunity by coinjection of genes for an antigen and a cytokine. Immune responses are modified in many of these experiments. However, it is uncertain whether the modulation is caused by the expression of the cytokine gene or some indirect effects are involved. As we have seen, the presence of ISS in the expression vectors can cause the release of a plethora of different cytokines and the specific action of the injected cytokine gene must be distinguished from this background. I will not attempt to summarize the studies in this area and will refer the reader to the reviews cited previously.

One method for testing the proposed models of immune responses is to look at the effect of immunomodulators on the induced immunity. Because of their rapid turnover, substances such as cytokines work locally in the area in which they are synthesized. The local action of these compounds may provide evidence for some of the notions discussed previously. For example, coexpressing a cytokine that acts on T cells such as IL-2 with an antigen in muscle would not be expected to have much effect on immune responses because IL-2 expression is distant from the developing T cells in the lymph nodes. On the other hand, coexpressing IL-2 and antigen in dermal tissue may have an effect because antigen-presenting cells are transfected and they migrate to the lymph nodes and express the cytokine locally with the antigen. Conversely, expressing a cytokine that activates APC such as gmCSF may boost immunity if introduced by either route since the APC interact with the antigen-producing muscle cells.

Coinjection of gmCSF stimulates immune responses when delivered by the intramuscular (Iwasaki *et al.,* 1997; Xiang and Ertl, 1995) or "facilitated" (Geissler *et al.,* 1997) methods. IL-2 coinjection stimulates immunity when "facilitated" injections are used (Geissler *et al.,* 1997; Kim *et al.,* 1998) but no increase is seen in two papers that used the intramuscular delivery (Abai *et al.,* 1994; Barouch *et al.,* 1998). However, a third paper does see a stimulation with IL-2 (Chow *et al.,* 1997).

This type of experiment must be done extremely carefully to avoid spurious conclusions. Since plasmid DNA can modulate immunity by itself, the total amount of DNA must be kept constant. Since the DNA sequence can also effect immunity, one must keep the sequence of the injected plasmids as close as possible. Thus, one should compare modulation by plasmids with cytokines to plasmids in which the cytokine is in the reverse orientation or a termination codon is inserted. Plasmid expression vectors that express two genes (antigen and cytokine) on the same plasmid are to be preferred to a mixture of two plasmid DNA molecules because this assures that cells that express antigen are also expressing the cytokine. Finally, one must take into account indirect effects. For example, some cytokines have been shown to inhibit expression from viral promoters including the CMV IE promoter (Harms and Splitter, 1995). In addition, many cytokines are inflammatory and can modulate immunity by attracting new cells to the site of antigen expression. While such indirect effects may be useful for vaccines, they complicate interpretation of mechanism. No paper published to date has taken all of these considerations into account and so, at this time, there are uncertainties in interpretation that will only be resolved by further experimentation.

V. SUMMARY

There are two main points to be made in this review. The first is that NAV is not one method; rather it consists of a number of techniques that differ in the cells that express the antigen, differ in the duration of antigen expression, and most importantly, differ in the way in which the antigen-producing cells interact with the immune system. There is no best method of nucleic acid vaccination. The investigator can select the procedure that provides the type of immunity that is desired.

The second object of this review is to start to define some of the immune mechanisms involved in NAV. There are large gaps in our knowledge about these mechanisms. Much of what is proposed may have to be modified. However, a model is useful for organizing data and guiding experiments and it is proposed in this spirit.

One of the aspects that make NAV research exciting is that it lies at the junction of vaccine development and basic immunological research and experiments in this field can make a contribution to both areas. In the few years since its development, nucleic acid vaccination has provided solutions to two long-standing problems in vaccine development. It provides a simple and safe way of inducing cellular immunity without using infectious virus and it has furnished a way to make a vaccine against influenza that protects across 30 years of strain variation (Rhodes *et al.*, 1993; Ulmer *et al.*, 1993). Recently, the first demonstration of protective immunity induced by NAV in primates has been reported (Lodmell *et al.*, 1998). On the immunological side, it has provided new insight into the interaction of antigen-producing and antigen-presenting cells and it has helped to demonstrate the crucial role that DNA plays in immunity. This method should continue to advance our knowledge in both fields.

ACKNOWLEDGMENTS

The author thanks David Verhoeven for editorial assistance and acknowledges the U.S. Army Medical Research and Material Command contract DAMD17-94-J-4436 and National Institute of Health grants AI 42608–02 and R01 AI44481A for salary support.

REFERENCES

Abai, A. M., Kuwahara-Rundell, A., Margalith, M., and Rhodes, G. H. (1994). Co-injection of the genes for interleukin-2 and antigen does not increase humoral immunity. In H. S. Ginsberg, F. Brown, R. M. Chanock, and R. A. Lerner (Eds), (*Vaccines 94* pp. 77–81). Cold Spring Harbor, NY: Cold Spring Harbor Press.

Andre, S., Seed, B., Eberle, J., Schraut, W., Bultmann, A., and Haas, J. (1998). Increased immune response elicited by DNA vaccination with a synthetic gp120 sequence with optimized codon usage. *J. Virol.* **72,** 1497–1503.

Asakura, Y., Hinkula, J., Leandersson, A. C., Fukushima, J., Okuda, K., and Wahren, B. (1997). Induction of HIV-1 specific mucosal immune responses by DNA vaccination. *Scand. J. Immunol.* **46,** 326–330.

Bagarazzi, M. L., Boyer, J. D., Ayyavoo, V, and Weiner, D. B. (1998). Nucleic acid-based vaccines as an approach to immunization against human immunodeficiency virus type-1. *Curr. Top. Microbiol. Immunol.* **226,** 107–143.

Ballas, Z. K., Rasmussen, W. L., Krieg, A. M. (1996). Induction of NK activity in murine and human cells by CpG motifs in oligodeoxynucleotides and bacterial DNA. *J. Immunol.* **157,** 1840–1845.

Barnett, S. W., Rajasekar, S., Legg, H., Doe, B., Fuller, D. H., Haynes, J. R., Walker, C. M., and Steimer, K. S. (1997). Vaccination with HIV-1 gp120 DNA induces immune responses that are boosted by a recombinant gp120 protein subunit. *Vaccine* **15,** 869–873.

Barouch, D. H., Santra, S., Steenbeke, T. D., Zheng, X. X., Perry, H. C., Davies, M. E., Freed, D. C., Craiu, A., Strom, T. B., Schriver, J. W., and Letvin, N. L. (1998). Augmentation and suppression of immune responses to an HIV-1 DNA vaccine by plasmid cytokine/Ig Administration. *J. Immunol.* **161,** 1875–1882.

Barry, M. A., and Johnston, S. A. (1997). Biological features of genetic immunization. *Vaccine* **15,** 788–791.

Bartlett, R. J., Secore, S. L., Singer, J. T., Bodo, M., Sharma, K., and Ricordi, C. (1996). Long-term expression of a fluorescent reporter gene via direct injection of plasmid vector into mouse skeletal muscle: Comparison of human creatine kinase and CMV promoter expression levels *in vivo*. *Cell Transplant.* **5,** 411–419.

Boyer, J., Ungen, K., Wang, B., Chattergoon, M., Tsai, A., Merva, M., and Weiner, D. B. (1998). Induction of a TH1 type cellular immune response to the human immunodeficiency type 1 virus by *in vivo* DNA inoculation. *Dev. Biol. Standard.* **92,** 169–174.

Boyle, J. S., Brady, J. L., and Lew, A. M. (1998). Enhanced responses to a DNA vaccine encoding a fusion antigen that is directed to sites of immune induction. *Nature* **392,** 408–411.

Boyle, J. S., Koniaras, C., and Lew, A. M. (1997). Influence of cellular location of expressed antigen on the efficacy of DNA vaccination: Cytotoxic T lymphocyte and antibody responses are suboptimal when antigen is cytoplasmic after intramuscular DNA immunization. *Intl. Immunol.* **9,** 1897–1906.

Carson, D. A., and Raz, E. (1997). Oligonucleotide adjuvants for T helper 1 (Th1)-specific vaccination. *J. Exp. Med.* **186,** 1621–1622.

Casares, S., Inaba, K., Brumeanu, T. D., Steinman, R. M., and Bona, C. A. (1997). Antigen presentation by dendritic cells after immunization with DNA encoding a major histocompatibility complex class II-restricted viral epitope. *J. Exp. Med.* **186,** 1481–1486.

Chace, J. H., Hooker, N. A., Mildenstein, K. L., Krieg, A. M., and Cowdery, J. S. (1997). Bacterial DNA-induced NK cell IFN-gamma production is dependent on macrophage secretion of IL-12. *Clin. Immunol. Immunopathol.* **84,** 185–193.

Chapman, B. S., Thayer, R. M., Vincent, K. A., and Haigwood, N. L. (1991). Effect of intron A from human cytomegalovirus (Towne) immediate-early gene on heterologous expression in mammalian cells. *Nucleic Acids Res.* **19,** 3979–3986.

Chen, Y., Webster, R. G., and Woodland, D. L. (1998). Induction of CD8+ T cell responses to dominant and subdominant epitopes and protective immunity to Sendai virus infection by DNA vaccination. *J. Immunol.* **160,** 2425–2432.

Chinsangaram, J., Beard, C., Mason, P. W., Zellner, M. K., Ward, G., and Grubman, M. J. (1998). Antibody response in mice inoculated with DNA expressing foot-and-mouth disease virus capsid proteins. *J. Virol.* **72,** 4454–4457.

Chow, Y. H., Huang, W. L., Chi, W. K., Chu, Y. D., and Tao, M. H. (1997). Improvement of hepatitis B virus DNA vaccines by plasmids coexpressing hepatitis B surface antigen and interleukin-2. *J. Virol.* **71,** 169–178.

Cohen, A. D., Boyer, J. D., and Weiner, D. B. (1998). Modulating the immune response to genetic immunization. *FASEB J.* **12,** 1611–1626.

Condon, C., Watkins, S. C., Celluzzi, C. M., Thompson, K., and Falo, L. D. Jr. (1996). DNA-based immunization by *in vivo* transfection of dendritic cells. *Nature Med.* **2,** 1122–1128.

Conry, R. M., LoBuglio, A. F., Loechel, F., Moore, S. E., Sumerel, L. A., Barlow, D. L., and Curiel, D. T. (1995). A carcinoembryonic antigen polynucleotide vaccine has *in vivo* antitumor activity. *Gene Ther.* **2,** 59–65.

Conry, R. M., LoBuglio, A. F., Wright, M., Sumerel, L., Pike, M. J., Johanning, F., Benjamin, R., Lu, D., and Curiel, D. T. (1995). Characterization of a messenger RNA polynucleotide vaccine vector. *Cancer Res.* **55,** 1397–1400.

Corr, M., Lee, D. J., Carson, D. A., and Tighe, H. (1996). Gene vaccination with naked plasmid DNA: Mechanism of CTL priming. *J. Exp. Med.* **184,** 1555–1560.

Danko, I., and Wolff, J. A. (1994). Direct gene transfer into muscle. *Vaccine* **12,** 1499–1502.

Danko, I., Fritz, J. D., Jiao, S., Hogan, K., Latendresse, J. S., and Wolff, J. A. (1994). Pharmacological enhancement of *in vivo* foreign gene expression in muscle. *Gene Ther.* **1,** 114–121.

Davis, H. L., Michel, M. L., and Whalen, R. G. (1993). DNA-based immunization induces continuous secretion of hepatitis B surface antigen and high levels of circulating antibody. *Human Mol. Genet.* **2,** 1847–1851.

Davis, H. L., Whalen, R. G., and Demeneix, B. A. (1993). Direct gene transfer into skeletal muscle *in vivo*: Factors affecting efficiency of transfer and stability of expression. *Human Gene Ther.* **4,** 151–159.

Davis, H. L., Michel, M. L., and Whalen, R. G. (1995). Use of plasmid DNA for direct gene transfer and immunization. *Ann. NY Acad. Sci.* **772,** 21–29.

Davis, H. L., Mancini, M., Michel, M. L., and Whalen, R. G. (1996). DNA-mediated immunization to hepatitis B surface antigen: Longevity of primary response and effect of boost. *Vaccine* **14,** 910–915.

Doe, B., Selby, M., Barnett, S., Baenziger, J., and Walker, C. M. (1996). Induction of cytotoxic T lymphocytes by intramuscular immunization with plasmid DNA is facilitated by bone marrow-derived cells. *PNAS* **93,** 8578–8583.

Doh, S. G., Vahlsing, H. L., Hartikka, J., Liang, X., and Manthorpe, M. (1997). Spatial-temporal patterns of gene expression in mouse skeletal muscle after injection of lacZ plasmid DNA. *Gene Ther.* **4,** 648–663.

Donnelly, J. J., Ulmer, J. B., Shiver, J. W., and Liu, M. A. (1997). DNA vaccines. *Annu. Rev. Immunol.* **15,** 617–648.

Eisenbraun, M. D., Fuller, D. H., and Haynes, J. R. (1993). Examination of parameters affecting the elicitation of humoral immune responses by particle bombardment-mediated genetic immunization. *DNA and Cell Biol.* **12,** 791–797.

Felgner, P. L. (1997). Nonviral strategies for gene therapy. *Scient. Am.* **276,** 102–106.

Felgner, J. H., Kumar, R., Sridhar, C. N., Wheeler, C. J., Tsai, Y. J., Border, R., Ramsey, P., Martin, M., and Felgner, P. L. (1994). Enhanced gene delivery and mechanism studies with a novel series of cationic lipid formulations. *J. Biol. Chem.* **269,** 2550–2561.

Feltquate, D. M., Heaney, S., Webster, R. G., and Robinson, H. L. (1997). Different T helper cell types and antibody isotypes generated by saline and gene gun DNA immunization. *J. Immunol.* **158,** 2278–2284.

Fuller, D. H., and Haynes, J. R. (1994). A qualitative progression in HIV type 1 glycoprotein 120-specific cytotoxic cellular and humoral immune responses in mice receiving a DNA-based glycoprotein 120 vaccine. *AIDS Res. Human Retrovir.* **10,** 1433–1441.

Fuller, D. H., Murphey-Corb, M., Clements, J., Barnett, S., and Haynes, J. R. (1996). Induction of immunodeficiency virus-specific immune responses in rhesus monkeys following gene gun-mediated DNA vaccination. *J. Med. Primatol.* **25,** 236–241.

Furth, P. A., Shamay, A., and Hennighausen, L. (1995). Gene transfer into mammalian cells by jet injection. *Hybridoma* **14,** 149–152.

Fynan, E. F., Webster, R. G., Fuller, D. H., Haynes, J. R., Santoro, J. C., and Robinson, H. L. (1993). DNA vaccines: Protective immunizations by parenteral, mucosal, and gene-gun inoculations. *PNAS* **90,** 11478–11482.

Geissler, M., Gesien, A., Tokushige, K., and Wands, J. R. (1997). Enhancement of cellular and humoral immune responses to hepatitis C virus core protein using DNA-based vaccines augmented with cytokine-expressing plasmids. *J. Immunol.* **158,** 1231–1237.

Gregoriadis, G., Saffie, R., and de Souza, J. B. (1997). Liposome-mediated DNA vaccination. *FEBS Lett.* **402** 107–110.

Grounds, M. D. (1991). Towards understanding skeletal muscle regeneration. *Pathol. Res. Pract.* **187,** 1–22.

Haddad, D., Liljeqvist, S., Stahl, S., Andersson, I., Perlmann, P., Berzins, K., and Ahlborg, N. (1997). Comparative study of DNA-based immunization vectors: Effect of secretion signals on the antibody responses in mice. *FEMS Immunol. Med. Microbiol.* **18,** 193–202.

Harms, J. S., and Splitter, G. A. (1995). Interferon-gamma inhibits transgene expression driven by SV40 or CMV promoters but augments expression driven by the mammalian MHC I promoter. *Human Gene Ther.* **6,** 1291–1297.

Hartikka, J., Sawdey, M. (1996). Cornefert-Jensen, F., Margalith, M., Barnhart, K., Nolasco, M., Vahlsing, H. L., Meek, J., Marquet, M., Hobart, P., *et al.* (1996). An improved plasmid DNA expression vector for direct injection into skeletal muscle. *Human Gene Ther.* **7,** 1205–1217.

Haynes, J. R., Fuller, D. H., Eisenbraun, M. D., Ford, M. J., and Pertmer, T. M. (1994). Accell particle-mediated DNA immunization elicits humoral, cytotoxic, and protective immune responses [Suppl 2]. *AIDS Res. Human Retrovir.* **10,** S43–S45.

Haynes, J. R., McCabe, D. E., Swain, W. F., Widera, G., and Fuller, J. T. (1996). Particle-mediated nucleic acid immunization. *J. Biotechnol.* **44**, 37–42.

Hinkula, J., Svanholm, C., Schwartz, S., Lundholm, P., Brytting, M., Engstrom, G., Benthin, R., Glaser, H., Sutter, G., Kohleisen, B., *et al.* (1997). Recognition of prominent viral epitopes induced by immunization with human immunodeficiency virus type 1 regulatory genes. *J. Virol.* **71**, 5528–5539.

Inchauspe, G., Vitvitski, L., Major, M. E., Jung, G., Spengler, U., Maisonnas, M., and Trepo, C. (1997). Plasmid DNA expressing a secreted or a nonsecreted form of hepatitis C virus nucleocapsid: Comparative studies of antibody and T-helper responses following genetic immunization. *DNA Cell Biol.* **16**, 185–195.

Ishii, N., Fukushima, J., Kaneko, T., Okada, E., Tani, K., Tanaka, S. I., Hamajima, K., Xin, K. Q., Kawamoto, S., Koff, W., *et al.* (1997). Cationic liposomes are a strong adjuvant for a DNA vaccine of human immunodeficiency virus type 1. *AIDS Res. Human Retrovir.* **13**, 1421–1428.

Iwasaki, A., Stiernholm, B. J., Chan, A. K., Berinstein, N. L., and Barber, B. H. (1997). Enhanced CTL responses mediated by plasmid DNA immunogens encoding costimulatory molecules and cytokines. *J. Immunol.* **158**, 4591–4601.

Iwasaki, A., Torres, C. A., Ohashi, P. S., Robinson, H. L., and Barber, B. H. (1997). The dominant role of bone marrow-derived cells in CTL induction following plasmid DNA immunization at different sites. *J. Immunol.* **159**, 11–14.

Jakob, T., Walker, P. S., Krieg, A. M., Udey, M. C., and Vogel, J. C. (1998). Activation of cutaneous dendritic cells by CpG-containing oligodeoxynucleotides: A role for dendritic cells in the augmentation of Th1 responses by immunostimulatory DNA. *J. Immunol.* **161**, 3042–3049.

Johnston, S. A., and Barry, M. A. (1997). Genetic to genomic vaccination. *Vaccine* **15**, 808–809.

Johnston, S. A., and Tang, D. C. (1994). Gene gun transfection of animal cells and genetic immunization. *Meth. Cell Biol.* **43**, 353–365.

Kataoka, T., Yamamoto, S., Yamamoto, T., Kuramoto, E., Kimura, Y., Yano, O., and Tokunaga, T. (1992). Antitumor activity of synthetic oligonucleotides with sequences from cDNA encoding proteins of *Mycobacterium bovis* BCG. *Jpn. Cancer Res.* **83**, 244–247.

Kim, J. J., Trivedi, N. N., Nottingham, L. K., Morrison, L., Tsai, A., Hu, Y., Mahalingam, S., Dang, K., Ahn, L., Doyle, N. K., *et al.* (1998). Modulation of amplitude and direction of *in vivo* immune responses by co-administration of cytokine gene expression cassettes with DNA immunogens. *Eur. J. Immunol.* **28**, 1089–1103.

Klavinskis, L. S., Gao, L., Barnfield, C., Lehner, T., and Parker, S. (1997). Mucosal immunization with DNA–liposome complexes. *Vaccine* **15**, 818–820.

Klinman, D. M., Sechler, J. M., Conover, J., Gu, M., and Rosenberg, A. S. (1998). Contribution of cells at the site of DNA vaccination to the generation of antigen-specific immunity and memory. *J. Immunol.* **160**, 2388–2392.

Klinman, D. M., Takeno, M., Ichino, M., Gu, M., Yamshchikov, G., Mor, G., and Conover, J. (1997). DNA vaccines: Safety and efficacy issues. *Springer Sem. Immunopathol.* **19**, 245–256.

Klinman, D. M., Yamshchikov, G., and Ishigatsubo, Y. (1997). Contribution of CpG motifs to the immunogenicity of DNA vaccines. *J. Immunol.* **158**, 3635–3639.

Klinman, D. M., Yi, A. K., Beaucage, S. L., Conover, J., and Krieg, A. M. (1996). CpG motifs present in bacteria DNA rapidly induce lymphocytes to secrete interleukin 6, interleukin 12, and interferon gamma. *PNAS* **93**, 2879–2883.

Krieg, A. M. (1996). Lymphocyte activation by CpG dinucleotide motifs in prokaryotic DNA. *Trends Microbiol.* **4**, 73–76.

Krieg, A. M., Yi, A. K., Matson, S., Waldschmidt, T. J., Bishop, G. A., Teasdale, R., Koretzky, G. A., and Klinman, D. M. (1995). CpG motifs in bacterial DNA trigger direct B-cell activation. *Nature* **374**, 546–549.

Krieg, A. M., Yi, A. K., Schorr, J., and Davis, H. L. (1998). The role of CpG dinucleotides in DNA vaccines. *Trends Microbiol.* **6**, 23–27.

Kuklin, N., Daheshia, M., Karem, K., Manickan, E., and Rouse, B. T. (1997). Induction of mucosal immunity against herpes simplex virus by plasmid DNA immunization. *J. Virol.* **71,** 3138–3145.

Kuramoto, E., Yano, O., Kimura, Y., Baba, M., Makino, T., Yamamoto, S., Yamamoto, T., Kataoka, T., and Tokunaga, T. (1992). Oligonucleotide sequences required for natural killer cell activation. *Jpn. J. Cancer Res.* **83,** 1128–1131.

Lai, W. C., Bennett, M., Johnston, S. A., Barry, M. A., and Pakes, S. P. (1995). Protection against *Mycoplasma pulmonis* infection by genetic vaccination. *DNA Cell Biol.* **14,** 643–651.

Le Borgne, S., Mancini, M., Le Grand, R., Schleef, M., Dormont, D., Tiollais, P., Riviere, Y., and Michel, M. L. (1998). *In vivo* induction of specific cytotoxic T lymphocytes in mice and rhesus macaques immunized with DNA vector encoding an HIV epitope fused with hepatitis B surface antigen. *Virology* **240,** 304–315.

Leclerc, C., Deriaud, E., Rojas, M., and Whalen, R. G. (1997). The preferential induction of a Th1 immune response by DNA-based immunization is mediated by the immunostimulatory effect of plasmid DNA. *Cell. Immunol.* **179,** 97–106.

Leitner, W. W., Seguin, M. C., Ballou, W. R., Seitz, J. P., Schultz, A. M., Sheehy, M. J., and Lyon, J. A. (1997). Immune responses induced by intramuscular or gene gun injection of protective deoxyribonucleic acid vaccines that express the circumsporozoite protein from *Plasmodium berghei* malaria parasites. *J. Immunol.* **159,** 6112–6119.

Lipford, G. B., Sparwasser, T., Bauer, M., Zimmermann, S., Koch, E. S., Heeg, K., and Wagner, H. (1997). Immunostimulatory DNA: Sequence-dependent production of potentially harmful or useful cytokines. *Eur. J. Immunol.* **27,** 3420–3426.

Liu, M. A., Ulmer, J. B., Friedman, A., Martinez, D., Hawe, L. A., DeWitt, C. M., Leander, K. R., Shi, X. P., Montgomery, D. L., Donnelly, J. J., Parker, S., Felgner, P., and Felgner, J. (1993). Immunization with DNA encoding a conserved internal viral protein results in protection from morbidity and mortality due to challenge with influenza A in mice. Immunity. In H. S. Ginsberg, F. Brown, R. M. Chanock, and R. A. Lerner (Eds). *Vaccines 93* (pp. 343–346). Cold Spring Harbor, NY: Cold Spring Harbor Press.

Lodmell, D. L., Ray, N. B., Parnell, M. J., Ewalt, L. C., Hanlon, C. A., Shaddock, J. H., Sanderlin, D. S., and Rupprecht, C. E. (1998). DNA immunization protects nonhuman primates against rabies virus. *Nature Med.* **4,** 949–952.

Lu, S., Arthos, J., Montefiori, D. C., Yasutomi, Y., Manson, K., Mustafa, F., Johnson, E., Santoro, J. C., Wissink, J., Mullins, J. I., *et al.* (1996). Simian immunodeficiency virus DNA vaccine trial in macaques. *J. Virol.* **70,** 3978–3991.

Malone, J. G., Bergland, P. J., Liljestrom, P., Rhodes, G. H., and Malone, R. W. (1997). Mucosal immune responses associated with polynucleotide vaccination. *Behring Inst. Mitteilungen* **98,** 63–72.

Manickan, E., Rouse, R. J., Yu, Z., Wire, W. S., and Rouse, B. T. (1995). Genetic immunization against herpes simplex virus. Protection is mediated by CD4+ T lymphocytes. *J. Immunol.* **155,** 259–265.

Manthorpe, M., Cornefert-Jensen, F., Hartikka, J., Felgner, J., Rundell, A., Margalith, M., and Dwarki, V. (1993). Gene therapy by intramuscular injection of plasmid DNA: Studies on firefly luciferase gene expression in mice. *Human Gene Ther.* **4,** 419–431.

Martins, L. P., Lau, L. L., Asano, M. S., and Ahmed, R. (1995). DNA vaccination against persistent viral infection. *J. Virol.* **69,** 2574–2582.

McLennan, I. S. (1996). Degenerating and regenerating skeletal muscles contain several subpopulations of macrophages with distinct spatial and temporal distributions. *J. Anat.* **188,** 17–28.

Montgomery, D. L., Leander, K. R., Shiver, J. W., Perry, H. C., Friedman, A., Martinez, D., Ulmer, J. B., Donnelly, J. J., and Liu, M. A. (1994). Nonreplicating DNA vectors designed to generate heterologous and homologous protection against influenza A. In H. S. Ginsberg, F. Brown, R. M. Chanock, R. A. Lerner (Eds), *Vaccines 94* (pp. 61–64). Cold Spring Harbor, NY: Cold Spring Harbor Press.

Mosmann, T. R., and Coffman, R. L. (1989). TH1 and TH2 cells: Different patterns of lymphokine secretion lead to different functional properties. *Annu. Rev. Immunol.* **7,** 145–173.

Okada, E., Sasaki, S., Ishii, N., Aoki, I., Yasuda, T., Nishioka, K., Fukushima, J., Miyazaki, J., Wahren, B., and Okuda, K. (1997). Intranasal immunization of a DNA vaccine with IL-12-γ and granulo-cyte-macrophage colony-stimulating factor (GM-CSF)-expressing plasmids in liposomes induces strong mucosal and cell-mediated immune responses against HIV-1 antigens. *J. Immunol.* **159,** 3638–3647.

Orimo, S., Hiyamuta, E., Arahata, K., and Sugita, H. (1991). Analysis of inflammatory cells and complement C3 in bupivacaine-induced myonecrosis. *Muscle Nerve* **14,** 515–520.

Pascual, D. W., Powell, R. J., Lewis, G. K., and Hone, D. M. (1997). Oral bacterial vaccine vectors for the delivery of subunit and nucleic acid vaccines to the organized lymphoid tissue of the intestine. *Behring Inst. Mitteilungen* **98,** 143–152.

Pasquini, S., Xiang, Z., Wang, Y., He, Z., Deng, H., Blaszczyk-Thurin, M., and Ertl, H. C. (1997). Cytokines and costimulatory molecules as genetic adjuvants. *Immunol. Cell Biol.* **75,** 397–401.

Pertmer, T. M., Eisenbraun, M. D., McCabe, D., Prayaga, S. K., Fuller, D. H., and Haynes, J. R. (1995). Gene gun-based nucleic acid immunization: Elicitation of humoral and cytotoxic T lymphocyte responses following epidermal delivery of nanogram quantities of DNA. *Vaccine* **13,** 1427–1430.

Pertmer, T. M., Roberts, T. R., and Haynes, J. R. (1996). Influenza virus nucleoprotein-specific immunoglobulin G subclass and cytokine responses elicited by DNA vaccination are dependent on the route of vector DNA delivery. *J. Virol.* **70,** 6119–6125.

Pimorady-Esfahani, A., Grounds, M. D., and McMenamin, P. G. (1997). Macrophages and dendritic cells in normal and regenerating murine skeletal muscle. *Muscle Nerve* **20,** 158–166.

Prayaga, S. K., Ford, M. J., and Haynes, J. R. (1997). Manipulation of HIV-1 gp120-specific immune responses elicited via gene gun-based DNA immunization. *Vaccine* **15,** 1349–1352.

Quill, H., and Schwartz, R. H. (1987). Stimulation of normal inducer T cell clones with antigen presented by purified Ia molecules in planar lipid membranes: specific induction of a long-lived state of proliferative nonresponsiveness. *J. Immunol.* **138,** 3704–3712.

Raz, E., Carson, D. A., Parker, S. E., Parr, T. B., Abai, A. M., Aichinger, G., Gromkowski, S. H., Singh, M., Lew, D., Yankauckas, M. A; *et al.* (1994). Intradermal gene immunization: The possible role of DNA uptake in the induction of cellular immunity to viruses. *PNAS* **91,** 9519–9523.

Raz, E., Rhodes, G. H., Bird, S. M., Abai, A. M., Tsai, Y. J., Morrow, J., Carson, D. A., and Felgner, P. L. (1994). Cationic lipids inhibit intradermal genetic vaccination. In H. S. Ginsberg, F. Brown, R. M. Chanock, and R. A. Lerner (Eds), *Vaccines 94* (pp. 71–75). Cold Spring Harbor, NY: Cold Spring Harbor Press.

Raz, E., Tighe, H., Sato, Y., Corr, M., Dudler, J. A., Roman, M., Swain, S. L., Spiegelberg, H. L., and Carson, D. A. (1996). Preferential induction of a Th1 immune response and inhibition of specific IgE antibody formation by plasmid DNA immunization. *PNAS* **93,** 5141–5145.

Rhodes, G. H., Abai, A. M., Margalith, M., Kuwahara-Rundell, A., Morrow, J., Parker, S. E., and Dwarki, V. J. (1994). Characterization of humoral immunity after DNA injection. In F. Brown (Ed.), *Recombinant vectors in vaccine development* (pp. 219–226). Basel: Karger.

Rhodes, G. H., Dwarki, V. J., Abai, A. M., Felgner, J., Felgner, P. L., Gromkoswki, S. H., and Parker, S. E. (1993). Injection of expression vectors containing viral genes induces cellular, humoral and protective immunity. In H. S. Ginsberg, F. Brown, R. M. Chanock, and R. A. Lerner (Eds), *Vaccines 93* (pp. 137–141). Cold Spring Harbor, NY: Cold Spring Harbor Press.

Rigby, M. A., Hosie, M. J., Willett, B. J., Mackay, N., McDonald, M., Cannon, C., Dunsford, T., Jarrett, O., and Neil, J. C. (1997). Comparative efficiency of feline immunodeficiency virus infection by DNA inoculation. *AIDS Res. Human Retrovir.* **13,** 405 412.

Robinson, H. L., Fynan, E. F., and Webster, R. G. (1993). Use of direct DNA inoculations to elicit protective immune responses. In H. S. Ginsberg, F. Brown, R. M. Chanock, and R. A. Lerner (Eds), *Vaccines 93* (pp. 311–315). Cold Spring Harbor, NY: Cold Spring Harbor Press.

Robinson, H. L., and Torres, C. A. (1997). DNA vaccines. *Semin. Immunol.* **9,** 271–283.

Rodriguez, F., Zhang, J., and Whitton, J. L. (1997). DNA immunization: Ubiquitination of a viral protein enhances cytotoxic T-lymphocyte induction and antiviral protection but abrogates antibody induction. *J. Virol.* **71,** 8497–8503.

Roman, M., Martin-Orozco, E., Goodman, J. S., Nguyen, M. D., Sato, Y., Ronaghy, A., Kornbluth, R. S., Richman, D. D., Carson, D. A., and Raz, E. (1997). Immunostimulatory DNA sequences function as T helper-1-promoting adjuvants [see comments]. *Nature Med.* **3,** 849–854.

Sato, Y., Roman, M., Tighe, H., Lee, D., Corr, M., Nguyen, M. D., Silverman, G. J., Lotz, M., Carson, D. A., and Raz, E. (1996). Immunostimulatory DNA sequences necessary for effective intradermal gene immunization. *Science* **273,** 352–354.

Schwartz, R. H. (1992). Costimulation of T lymphocytes: the role of CD28, CTLA-4, and B7/BB1 in interleukin-2 production and immunotherapy. *Cell* **71,** 1065–1068.

Shiver, J. W., Davies, M. E., Yasutomi, Y., Perry, H. C., Freed, D. C., Letvin, N. L., and Liu, M. A. (1997). Anti-HIV env immunities elicited by nucleic acid vaccines. *Vaccine* **15,** 884–887.

Shiver, J. W., Perry, H. C., Davies, M. E., Freed, D. C., and Liu, M. A. (1995). Cytotoxic T lymphocyte and helper T cell responses following HIV polynucleotide vaccination. *Ann. NY Acad. Sci.* **772,** 198–208.

Sparger, E. E., Louie, H., Ziomeck, A. M., and Luciw, P. A. (1997). Infection of cats by injection with DNA of a feline immunodeficiency virus molecular clone. *Virology* **238,** 157–160.

Sparwasser, T., Koch, E. S., Vabulas, R. M., Heeg, K., Lipford, G. B., Ellwart, J. W., and Wagner, H. (1998). Bacterial DNA and immunostimulatory CpG oligonucleotides trigger maturation and activation of murine dendritic cells. *Eur. J. Immunol.* **28,** 2045–2054.

Sparwasser, T., Miethke, T., Lipford, G., Erdmann, A., Hacker, H., Heeg, K., and Wagner, H. (1997). Macrophages sense pathogens via DNA motifs: Induction of tumor necrosis factor-alpha-mediated shock. *Eur. J. Immunol.* **27,** 1671–1679.

Swain, W. F., Macklin, M. D., Neumann, G., McCabe, D. E., Drape, R., Fuller, J. T., Widera, G., McGregor, M., Callan, R. J., and Hinshaw, V. (1997). Manipulation of immune responses via particle-mediated polynucleotide vaccines. *Behring Inst. Mitteilungen* **98,** 73–78.

Sweet, M. J., Stacey, K. J., Kakuda, D. K., Markovich, D., and Hume, D. A. (1998). IFN-gamma primes macrophage responses to bacterial DNA. *J. Interferon Cytokine Res.* **18,** 263–271.

Tang, D.C., DeVit, M., and Johnston, S. A. (1992). Genetic immunization is a simple method for eliciting an immune response. *Nature* **356,** 152–154.

Tighe, H., Corr, M., Roman, M., and Raz, E. (1998). Gene vaccination: Plasmid DNA is more than just a blueprint. *Immunol. Today* **19,** 89–97.

Tokunaga, T., Yamamoto, H., Shimada, S., Abe, H., Fukuda, T., Fujisawa, Y., Furutani, Y., Yano, O., Kataoka, T., Sudo, T., *et al.* (1984). Antitumor activity of deoxyribonucleic acid fraction from *Mycobacterium bovis* BCG. I. Isolation, physicochemical characterization, and antitumor activity. *J. Natl. Cancer Inst.* **72,** 955–962.

Tokunaga, T., Yano, O., Kuramoto, E., Kimura. Y., Yamamoto, T., Kataoka, T., and Yamamoto, S. (1992). Synthetic oligonucleotides with particular base sequences from the cDNA encoding proteins of *Mycobacterium bovis* BCG induce interferons and activate natural killer cells. *Microbiol. Immunol.* **36,** 55–66.

Torres, C. A., Iwasaki, A., Barber, B. H., and Robinson, H. L. (1997). Differential dependence on target site tissue for gene gun and intramuscular DNA immunizations. *J. Immunol.* **158,** 4529–4532.

Ulmer, J. B., Deck, R. R., Dewitt, C. M., Donnhly, J. I., and Liu, M. A. (1996). Generation of MHC class I-restricted cytotoxic T lymphocytes by expression of a viral protein in muscle cells: Antigen presentation by non-muscle cells. *Immunol.* **89,** 59–67.

Ulmer, J. B., Donnelly, J. J., Parker, S. E., Rhodes, G. H., Felgner, P. L., Dwarki, V. J., Gromkowski, S. H., Deck, R. R., DeWitt, C. M., Friedman, A., Hawe, L. A., Leander, K. R., Martinez, D., Perry, H. C., Shiver, J. W., Montgommery, D. L., and Liu, M. A. (1993). Heterologous protection against influenza by injection of DNA encoding a viral protein. *Science* **159**, 1745–1749.

Vahlsing, H. L., Yankauckas, M. A., Sawdey, M., Gromkowski, S. H., Manthorpe, M. (1994). Immunization with plasmid DNA using a pneumatic gun. *J. Immunol. Meth.* **175**, 11–22.

Vogel, F. R., and Sarver, N. (1995). Nucleic acid vaccines. *Clin. Microbiol. Rev.* **8**, 406–410.

Wagener, S., Norley, S., zur Megede, J., Kurth, R., and Cichutek, K. (1996). Induction of antibodies against SIV antigens after intramuscular nucleic acid inoculation using complex expression constructs. *J. Biotechnol.* **44**, 59–65.

Wang, B., Dang, K., Agadjanyan, M. G., Srikantan, V., Li, F., Ugen, K. E., Boyer, J., Merva, M., Williams, W. V., and Weiner, D. B. (1997). Mucosal immunization with a DNA vaccine induces immune responses against HIV-1 at a mucosal site. *Vaccine* **15**, 821–825.

Wang, B., Ugen, K. E., Srikantan, V., Agadjanyan, M. G., Dang, K., Refaeli, Y., Sato, A. I., Boyer, J., Williams, W. V., and Weiner, D. B. (1993). Gene inoculation generates immune responses against human immunodeficiency virus type 1. *PNAS* **90**, 4156–4160.

Wang, B., Ugen, K. E., Srikantan, V., Agadjanyan, M. G., Dang, K., Sato, A. I., Refaeli, Y., Boyer, J., Williams, W. V., and Weiner, D. B. (1993). Genetic immunization: A novel method for vaccine development against HIV. In H. S. Ginsberg, F. Brown, R. M. Chanock, and R. A. Lerner (Eds), *Vaccines 93* (pp. 143–150). Cold Spring Harbor, NY: Cold Spring Harbor Press.

Wheeler, C. J., Felgner, P. L., Tsai, Y. J., Marshall, J., Sukhu, L., Doh, S. G., Hartikka, J., Nietupski, J., Manthorpe, M., Nichols, M., *et al.* (1996). A novel cationic lipid greatly enhances plasmid DNA delivery and expression in mouse lung. *PNAS* **93**, 11454–11459.

Wloch, M. K., Pasquini, S., Ertl, H. C., and Pisetsky, D. S. (1998). The influence of DNA sequence on the immunostimulatory properties of plasmid DNA vectors. *Human Gene Ther.* **9**, 1439–1447.

Wolff, J. A. (1997). Naked DNA transport and expression in mammalian cells. *Neuromusc. Disorders* **7**, 314–318.

Wolff, J. A., Malone, R. W., Williams, P., Chong, W., Acsadi, G., Jani, A., and Felgner, P. L. (1990). Direct gene transfer into mouse muscle *in vivo*. *Science* **247**, 1465–1468.

Xiang, Z., and Ertl, H. C. (1995). Manipulation of the immune response to a plasmid-encoded viral antigen by coinoculation with plasmids expressing cytokines. *Immunity* **2**, 129–135.

Xiang, Z. Q., He, Z., Wang, Y., and Ertl, H. C. (1997). The effect of interferon-gamma on genetic immunization. *Vaccine* **15**, 896–898.

Xiang, Z. Q., Spitalnik, S. L., Cheng, J., Erikson, J., Wojczyk, B., and Ertl, H. C. (1995). Immune responses to nucleic acid vaccines to rabies virus. *Virology* **209**, 569–579.

Yamamoto, S., Kuramoto, E., Shimada, S., and Tokunaga, T. (1988). *In vitro* augmentation of natural killer cell activity and production of interferon-alpha/beta and -gamma with deoxyribonucleic acid fraction from *Mycobacterium bovis* BCG. *Jpn. J. Cancer Res.* **79**, 866–873.

Yamamoto, T., Yamamoto, S., Kataoka, T., and Tokunaga, T. (1994). Ability of oligonucleotides with certain palindromes to induce interferon production and augment natural killer cell activity is associated with their base length. *Antisense Res. Dev.* **4**, 119–122.

Yamamoto, T., Yamamoto, S., Kataoka, T., and Tokunaga, T. (1994). Lipofection of synthetic oligodeoxyribonucleotide having a palindromic sequence of AACGTT to murine splenocytes enhances interferon production and natural killer activity. *Microbiol. Immunol.* **38**, 831–836.

Yamamoto, S., Yamamoto, T., Kataoka, T., Kuramoto, E., Yano, O., and Tokunaga, T. (1992). Unique palindromic sequences in synthetic oligonucleotides are required to induce IFN [correction of INF] and augment IFN-mediated [correction of INF] natural killer activity. *J. Immunol.* **148**, 4072–4076.

Yamamoto, S., Yamamoto, T., Shimada, S., Kuramoto, E., Yano, O., Kataoka, T., and Tokunaga, T.

(1992). DNA from bacteria, but not from vertebrates, induces interferons, activates natural killer cells and inhibits tumor growth. *Microbiol. Immunol.* **36,** 983–997.

Yang, N. S., Burkholder, J., Roberts, B., Martinell, B., and McCabe, D. (1990). *In vivo* and *in vitro* gene transfer to mammalian somatic cells by particle bombardment. *PNAS* **87,** 9568–9572.

Yankauckas, M., Morrow, J. E., Parker, S. E., Rhodes, G. H., Dwarki, V. J., and Gromkowski, S. H. (1993). Long term anti-NP cellular and humoral immunity is induced by intramuscular injection of plasmid DNA containing NP gene. *DNA Cell Biol.* **12,** 771–776.

Yokoyama, M., Hassett, D. E., Zhang, J., and Whitton, J. L. (1997). DNA immunization can stimulate florid local inflammation, and the antiviral immunity induced varies depending on injection site. *Vaccine* **15,** 553–560.

Zhou, X., Berglund, P., Rhodes, G., Parker, S. E., Jondal, M., and Liljestrom, P. (1994). Self-replicating Semliki Forest virus RNA as recombinant vaccine. *Vaccine* **12,** 1510–1514.

CHAPTER 18

A Novel Gene Regulatory System

Steven S. Chua,* Mark M. Burcin,* Yaolin Wang,† and Sophia Y. Tsai*

*Department of Cell Biology, Baylor College of Medicine, Houston, Texas
†Schering-Plough Research Institute, Kenilworth, New Jersey

Many of the methodologies used to introduce genes into cells, transgenic mice, and patients are unsatisfactory due to the constitutive production of proteins and a failure in the regulation of the transferred genes. Toward the desirable goal of achieving highly regulable gene expression, we have successfully engineered a novel binary inducible system that consists of a transactivator and a target gene. The transactivator encodes a chimeric regulator that is responsive to progesterone antagonists (i.e., RU486) but not to any progesterone agonists, other hormones, or endogenous ligands for activation. The target gene can be any gene placed under the control of gal4 DNA binding sites and a minimum promoter. When the regulator is activated, it induces target gene expression by dimerizing and binding to the gal4 recognition sequences upstream of the target. The functionality of this system has been demonstrated conclusively in tissue culture and in transgenic mice. We were able to obtain dramatic levels of up to 33,000-fold activation of a human growth hormone reporter in transgenic mice. Furthermore, for applications that require higher levels of gene expression, we have also generated regulators that can induce significantly greater levels of target gene expression without compromising basal activity. In addition, we have also modified the regulator to repress both basal

Nonviral Vectors for Gene Therapy

promoter and enhancer-driven target gene expression. Since our gene regulatory system meets many of the stringent requirements stipulated for a desirable inducible system, it should prove highly useful for future applications in biology and gene therapy.

I. INDUCIBLE SYSTEMS

The process of transcription is a highly coordinated and multistep event involving the interplay of numerous factors. The first step of the process involves the recognition of specific DNA sequences by specific transcription factors. These factors then interact directly or indirectly via specific cofactors with the basal transcriptional machinery, which is composed of numerous general factors (TAFs) and RNA polymerase. Finally, through the stabilization afforded by these specific factors, transcription initiation ensues, thereby allowing a gene to be expressed.

The exact timing at which a gene is turned on or off is determined by a number of tightly regulated events. The coordinated expression of genes is crucial in development because the precocious expression of genes can be extremely lethal. In their quest toward understanding the regulation of genes in organisms, researchers have been able to duplicate some of these regulatory processes artificially in the form of inducible systems.

A. IMPORTANCE OF REGULATED EXPRESSION

Transcription is a major site of control. It regulates not only which type of cells produce a particular gene product but also the timing, duration, and level of gene expression. Based on this important tenet, many researchers have designed inducible systems to regulate the expression of a gene at the transcriptional level.

The ability to control the temporal expression of a gene is important in developmental and cancer studies because the stage in development when the causative gene is first expressed often influences the phenotypic manifestations observed (Jhappan *et al.*, 1993; Pierce *et al.*, 1993; Kordon *et al.*, 1995). Based on when the causative gene is first expressed, different functional interpretations of its action can be perceived. Inducible systems also allow investigators to bypass lethality issues in transgenic mice studies resulting from the constitutive expression of a certain transgene(s) (Quaife *et al.*, 1987; Muller *et al.*, 1988; Tepper *et al.*, 1990). This is especially useful in allowing the propagation and analyses of transgenic lines that would have been precluded otherwise. Perhaps the most superlative application of a regulatory system would be in gene therapy, to ensure that no unwarranted side effects are manifested by the uncontrolled expression of a therapeutic target gene. Thus, the justification of efficient and reliable inducible systems is highly warranted (Chua et al., 1998).

B. Types of Inducible Systems Available

A number of inducible systems have been developed over the past decade to regulate the expression of exogenously introduced genes. These include the heat-shock (Wurm et al., 1986), metallothionein (Mayo et al., 1982), steroid receptor-responsive (Hunes et al., 1981; Lee et al., 1991; Braselmann et al., 1993), lac repressor-operator (Hu and Davidson, 1987; Brown et al., 1987; Deuschle et al., 1989; Labow et al., 1990; Baim et al., 1991), tetracycline (Gossen and Bujard, 1992; Gossen et al., 1994, 1995; Shockett et al., 1995; Weinmann et al., 1994; Roder et al., 1994), and ecdysone systems (Christopherson et al., 1992; Yao et al., 1992; No et al., 1996). These systems are briefly summarized in Table 1.

In general, most of these systems are affected by one or more of the following problems (Chua et al., 1988): (1) high basal activity, (2) interference with endogenous signaling pathways, (3) undesirable pleiotropic effects, (4) toxicity of the

Table I

Summary of Inducible Systems

Inducible systems	Inducer	Applications	Induction potential
Heat shock	Heat	Tissue culture (Wurm et al., 1986)	Low
Metallothionein	Heavy metals	Tissue culture (Mayo et al., 1982)	Low
Steriod receptor-responsive	Estrogen and glucocorti-coids	Tissue culture (Hunes et al., 1981; Lee et al., 1991; Braselmann et al., 1993; Hu and Davidson, 1987)	High
Lac repressor-operator	IPTG	Tissue culture (Hu and Davidson, 1987; Brown et al., 1987; Deuschle et al., 1989; Labow et al., 1990; Baim et al., 1991)	High
Tetracycline	Tetracycline and derivatives	Tissue culture (Gossen and Bujard, 1992; Gossen et al., 1994, 1995; Shockett et al., 1995) Transgenic mice (Shockett et al., 1995) Plants (Weinmann et al., 1994; Roder et al., 1994)	Very high
Ecdysone	Muristerone	Tissue culture (Christopherson et al., 1992)	Very high

inducer, and (5) limited induction potential. However, the judicious use of any of the systems under circumstances whereby one or more of the preceding problems can be tolerated is still possible.

II. A NOVEL INDUCIBLE SYSTEM

Our binary regulatory system (Wang *et al.*, 1944), developed to overcome many of the problems inherent to preexisting inducible systems (Chua *et al.*, 1998), comprises two components: a chimeric transactivator (regulator) and a target. The regulator is first activated by an exogenously administered inducer, RU486 (Baulieu, 1989), which then primes it to induce the expression of a target gene.

A. APPROACHES

In the next section, we describe the properties of our novel gene regulatory system (Wang *et al.*, 1994) and document its potential for applications in biological systems and gene therapy.

1. System Requirements

The success of any binary inducible system is governed by the following criteria (Chua *et al.*, 1998; Wang *et al.*, 1994):

1. The regulator should not impart any phenotype or be deleterious to cells. In the absence of an inducer, it should not activate the expression of the target gene. Upon activation, it should only induce target gene expression and not interfere with endogenous pathways.

2. The inducer should be a diffusable molecule that can be easily added exogenously and at low concentrations. It should not impart any phenotype or be deleterious to cells. Upon administration, it should activate the regulator and not interfere with any of the endogenous signaling pathways. Upon removal, activation of the target gene should cease.

3. The target gene should only be induced by the activated regulator. In the absence of induction, the target gene should not impart any phenotype to the cells.

4. The induction process should be reversible and repeatable. The system as a whole should allow for a high and wide range in the induction potential of target gene expression to cater for different biological processes.

2. Design

The development of our regulator stemmed from studies aimed at dissecting the ligand-binding domain of the progesterone receptor (PR). Vegeto *et al.* (1992)

observed that the C-terminal truncation of the wild-type progesterone receptor at amino acid position 891 rendered the resulting mutant, PR-891, unresponsive to R5020 or other progesterone agonists. This also resulted in its failure to activate a progesterone receptor response element (PRE)-linked reporter by R5020. Interestingly, this mutant can now activate PRE-target genes in the presence of RU486 (Vegeto *et al.*, 1992).

Based on this finding and taking advantage of the interchangeable modular domains of transcription factors (Green and Chambon, 1986), we developed a modified transcription factor (Fig. 1) to activate genes in the presence of RU486. This regulator contains three domains: (1) a DNA-binding domain derived from the gal4 transcription factor (Giniger *et al.*, 1985), (2) a potent activation domain obtained from the herpes simplex virus protein VP16 (Triezenberg *et al.*, 1988), and (3) a ligand-binding domain derived from the PR-891 mutant (Vegeto *et al.*, 1992). By using gal4 DNA binding sites, we can exclude the possibility that our regulator would activate endogenous transcription pathways. More importantly, our regulator should now only activate genes that harbor gal4 DNA binding sites in the presence of exogenously administered ligand (Wang *et al.*, 1994).

The target gene can then be any gene placed under the control of one or more gal4 DNA binding sites and a TATA or thymidine kinase (TK) minimal promoter (Fig. 1). The use of an exogenous inducer (i.e., RU486) and unrelated

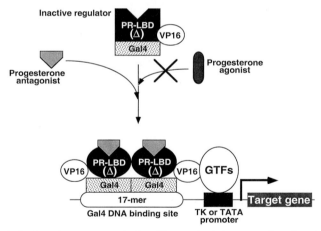

Figure 1 Mechanism of target gene activation. The constitutively expressed regulator is silent in the absence of RU486 and target gene expression will not ensue. Upon administration of the progesterone antagonist RU486, the activated regulator undergoes a conformational change, dimerizes, binds to gal4 DNA binding sites (17-mer) upstream of a target gene, and induces the transcription of the target gene. The regulator does not bind to the progesterone agonist R5020 and fails to activate target gene expression in the presence of this ligand. (GTFs are general transcription factors. PR-LBD (Δ) denotes the progesterone ligand-binding domain responsive to progesterone antagonists but not to progesterone agonists for activation.)

DNA binding sites should render our inducible system useful for most biological applications. Furthermore, using RU486 as an inducer is highly advantageous because of the following: (1) rapid delivery to cells, (2) good penetration into cells, and (3) rapid kinetics of gene activation.

3. Mechanism of Action

Since our regulator is a modified steroid receptor, its mechanism of activation mimics that of known steroid receptors (O'Malley *et al.*, 1991; Tsai and O'Malley, 1994). The regulator can be targeted for expression in any cell or tissue by utilizing the appropriate tissue-specific promoter. Although it is expressed constitutively, it is inactive in the absence of the RU486 and will not activate target gene expression (Wang *et al.*, 1994). When RU486 is added, transformation of the ligand-binding domain occurs and the regulator becomes active. The activated regulator can then dimerize, bind to the gal4 DNA binding sites located upstream of the target gene, and induce target gene expression (Fig. 1). By regulating the dose and duration of RU486 given, the level and span of target gene expression can be closely controlled, thereby expanding the repertoire of applications in which our system can be applied.

B. Verification

We have successfully applied our inducible system to tissue culture (Wang *et al.*, 1994) and transgenic mice (Wang, DeMayo, *et al.*, 1997) systems to regulate the expression of target genes. In the following sections, we present data to substantiate the functionality of our system.

1. Regulation in Cultured Cells

As a means of determining the usefulness and specificity of our system, we first tested its ability to activate a chloramphenicol acetyltransferase (CAT) reporter in transient transfections (Wang *et al.*, 1994). We placed the regulator under the control of the RSV promoter and the CAT reporter under the control of four gal4 DNA (17X4) binding sites and a minimum thymidine kinase (TK) promoter. As shown in Figure 2, we observed a robust level of induced reporter activity only in the presence of RU486. In the absence of RU486 or in the presence of the progesterone agonist R5020, we detected a low basal level of CAT reporter activity similar to that of the CAT reporter alone. A dose-dependent increase of CAT reporter activity was also observed and a concentration as low as 10^{-9}M of RU486 was sufficient to induce target gene expression (Wang *et al.*, 1994). At these levels, RU486 does not perturb PR- or glucocorticoid receptor (GR)-mediated pathways. We were also able to regulate a stably integrated tyrosine hydroxylase target gene (Wang *et al.*, 1994). Other antagonists of PR action such as the ZK and Org compounds

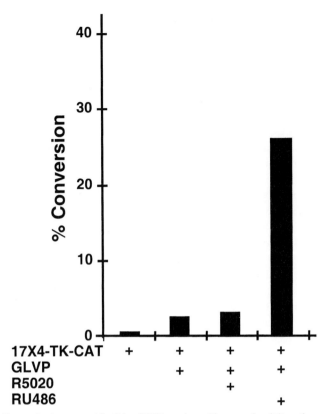

Figure 2 Transactivation potential of the GLVP regulator. To assess the ability of our regulator to induce the expression of a 17X4-TK-CAT reporter, transient transfections were performed. As shown in the graph, a robust induction of CAT activity was observed only when RU486 was administered. In the absence of RU486 or in the presence of R5020, insignificant levels of CAT activity were induced similar to that of the control reporter. "% conversion" denotes the percentage of substrate converted to acetylated forms.

(Neef *et al.*, 1984; Klein–Hitpass *et al.*, 1991) can also activate this regulator with similar potencies (Wang *et al.*, 1994). These results conclusively demonstrate that our system has a high induction potential and specificity toward the activation of specific target genes. Therefore, our system can be used successfully for transient and stable transfection assays whereby the tight regulation of a target gene is highly desirable.

2. Regulation in Transgenic Mice

To test the utility of our system for *in vivo* applications, we first assayed the ability of our regulator to induce the expression of a human growth hormone (hGH) reporter in transgenic mice (Wang, DeMayo, *et al.*, 1997). We utilize a bi-transgenic mice system similar to those of others (Khillan *et al.*, 1988; Byrne and Ruddle, 1989; Ornitz *et al.*, 1991). We targeted the expression of the regulator specifically to the liver by using the transthyretin (TTR) promoter (Yan *et al.*, 1990) and developed a target line harboring the 17X4-TK-hGH reporter. Chromosomal insulator fragments derived from the hypersensitive site IV (HS4) of the chicken β-globin gene were used to shield the regulator from positional effects of integration, possibly by establishing discrete boundaries of transcriptionally active and inactive domains (Chung *et al.*, 1993). By mating these two heterozygous lines of mice, we generated bitransgenic mice, monogenic targets, and regulator and wild-type mice. It is expected that only bitransgenic mice given RU486 should display hGH immunoreactivity in the blood when assayed by radioimmunoassays (RIA).

In the absence of RU486, undetectable amounts of hGH were observed in bitransgenic mice. However, dramatic increases in growth hormone levels (5800- to 33000-fold) were seen when bitransgenic mice were given RU486 (250 vs. 500 μg/KG body weight (Wang, DeMayo, *et al.*, 1997). We also observed that a dose of 100 μg/Kg body weight of RU486 was able to induce significant levels of hGH in bitransgenic mice (Wang, DeMayo, *et al.*, 1997). In contrast, undetectable levels of growth hormone were seen in the monogenic transactivator or target lines in the presence or absence of RU486. Oral administration of a dose of 250 μg/Kg body weight of RU486 to bitransgenic mice also yielded a robust induction of hGH immunoreactivity after 12 h (Wang, DeMayo, *et al.*, 1997). A dose-dependent increase in hGH levels was observed with higher concentrations of RU486 (Wang, DeMayo, *et al.*, 1997). Taken together, these results show that our system is silent until activated and that target gene expression can be induced over a broad range.

To understand the kinetics of growth hormone induction, we performed a time course study to monitor the induced levels of hGH in bitransgenic mice given a single dose of 100 μg/Kg body weight of RU486 (Wang, DeMayo, *et al.*, 1997). We observed that the hGH levels peak at about 8 to 12 h after RU486 administration and by 24 h, the levels of hGH start to decline. By 100 h, hGH is barely detectable. The addition of higher concentrations of RU486 also yielded a similar kinetic profile (Wang, DeMayo, *et al.*, 1997). After the RU486 has cleared sufficiently, we were still able to induce significant levels of hGH in these bitransgenic mice by repeated RU486 treatments (Wang, DeMayo, *et al.*, 1997). As expected, the control bitransgenic mice that were initially given the vehicle can also be induced to produce hGH by subsequent RU486 treatments. These results conclusively demonstrate the reversibility and reproducibility of the induction process (Wang, DeMayo, *et al.*, 1997).

To enable us to expand the utility of our system to all stages of mouse development, we first tested the effects of RU486 on pregnant females, a stage at which mice are very sensitive to RU486 levels (Wang, DeMayo, *et al.*, 1997). We injected pregnant mice with varying doses of RU486 on alternate days starting from Day 1 of mating. Our results indicated that a dose of 100 μg/Kg body weight of RU486 did not cause the pregnant mice to abort (Wang, DeMayo, *et al.,* 1997). This finding also implies that this concentration of RU486 does not impair endogenous PR and GR function (Wang *et al.,* 1994; Wang, Demayo *et al.,* 1997).

C. Improvements and Modifications

To further extend the utility of our system to meet the growing needs of many biological applications, we sought to increase the induction potential of our system without compromising basal activity.

1. Increases in Transactivation Potential

Three modifications were engineered to achieve this goal: changes in (1) activation domain, (2) length of the PR ligand binding domain, and (3) location of the VP16 activation domain (Wang, Xu *et al.,* 1997).

1. The first domain that we attempted to modify was the activation module of our first-generation regulator (Fig. 3A). It is widely believed that different activation domains interact with different coactivators of the general transcriptional

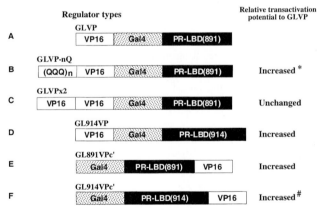

Figure 3 Diagrammatic representation of modified transactivators and their respective induction potentials compared to GLVP. **(A)** GLVP, unmodified regulator. **(B)** GLVP-nQ, regulator with glutamines (Q) added to the N-terminus. **(C)** GLVPx2, regulator with an additional VP16 activation domain at the N-terminus. **(D)** GL914VP, regulator with PR-LBD domain extended from amino acid 891 to 914. **(E)** GL891VPc', regulator with VP16 activation domain relocated to the C-terminus. **(F)** GL914VPc', regulator with extended PR-LBD and VP16 activation domain at C-terminus. (* denotes that increases in transactivation potential are seen when Q is between 10 and 34 residues long, # denotes the most potent regulator, and nQ denotes varying lengths of glutamines.)

machinery (Wang, Xu, *et al.*, 1997). The juxtaposition of these different domains in a transactivator could greatly influence its transactivation potential (Wang, Xu *et al.*, 1997). Recently, Gerber *et al.* (1994) enhanced the transactivation potential of a chimeric gal4–VP16 transcription factor by adding a stretch of polyglutamine (poly-Q) or poly-proline (poly-P) residues at the N-terminus. In a similar fashion, we added different stretches of glutamines to the N-terminal domain of our regulator (Fig. 3B). We observed that the sequential addition of between 10 and 34 glutamines increased the transactivation potential of our modified regulator gradually while further extensions from 66 to 132 glutamines actually reduced target gene expression (Wang, Xu *et al.*, 1997). These results suggested that a combination of different activation domains of an appropriate size could be mixed to improve the strength of the regulator (Wang, Xu *et al.*, 1997). To test the possibility that additional activation domains of the same type could enhance the transactivation potential of our regulator, we added a second copy of the VP16 activation domain to the N-terminus of our first-generation regulator (Fig. 3C). We observed that the addition of a second copy of the VP16 activation domain did not change the activation potential of our regulator (Wang, Xu, *et al.*, 1997.

2. Since the original PR-891 mutant was impaired in both activation function (AF2) and dimerization ability (Vegeto *et al.*, 1992; Wang, Xu, *et al.*, 1997), we wanted to address whether the extension of the progesterone ligand-binding domain can enhance the transactivation potential of our regulator. Toward this goal, we analyzed a series of c-terminal regulator deletion mutants. We found that the extension of the ligand-binding domain by 23 amino acids from position 891 to 914 (GL914VP) increased the activation potential by at least threefold (Wang, Xu, *et al.*, 1997) without affecting the basal activity (Fig. 3D). Importantly, this regulator cannot be activated by progesterone agonists (Wang, Xu, *et al.*, 1997). These findings indicate that the extension of the ligand-binding domain in our regulator most likely enhanced the weak dimerizing function, thus increasing its transactivation potential (Wang, Xu, *et al.*, 1997).

3. To investigate the possibility that the relocation of the VP16 activation domain within the context of our regulator can alter the activation of our target, we placed the VP16 activation domain at the C-terminus of our transactivator (Figs. 3E and 3F). Interestingly, we obtained an increase in transactivation potential when the VP16 activation domain was placed at the C-terminus (GL891VPc' vs. GLVP and GL914VPc' vs. GL914VP) irrespective of the length of the PR-LBD used (Wang, Xu, *et al.*, 1997). More importantly, the increases in transactivation potential were not due to higher levels of the different transactivator proteins as judged by Western analyses (Wang, Xu, *et al.*, 1997). We also found that the GL914VPc' regulator (Fig. 3F) is the most potent regulator compared to other regulators because we were able to detect higher reporter activity at lower concentrations of RU486 used (Wang, Xu, *et al.*, 1997). However, the basal activity is also

slightly enhanced (Wang, Xu, *et al.*, 1997). Nonetheless, the use of our improved transactivator should be highly applicable to systems that require higher levels of target gene expression and are more tolerable to the nominal increases in basal activity.

2. Inducible Repression

To extend the utility of our system to biological applications that require the repression of a target gene, we converted our regulator into an inducible repressor GL914KRAB (Wang, Xu, *et al.*, 1997). We replaced the VP16 activation domain of our regulator with a potent repressor module, the Krüppel associated box (KRAB) domain derived from a kidney-specific transcription factor, kid-1 (Witzgall *et al.*, 1993). The KRAB domain comprises two amphipathic helices of about 75 amino acids found at the *N*-termini of a number of the Krüppel-class transcription factors believed to be involved in protein–protein interactions (Bellefroid *et al.*, 1991). Ligation of this KRAB domain to the DNA binding domains of heterologous proteins have converted these chimeric fusions into potent repressors of reporter gene expression (Margolin *et al.*, 1994).

In transient transfections performed in HeLa cells using a basal 17X4-TK-luciferase reporter, we obtained up to an 80% reduction in luciferase activity in the presence of RU486 (Fig. 4). We were also able to repress enhancer-driven target gene expression fivefold using a 17X5-SV40-CAT reporter in the presence of RU486 (Wang, Xu, *et al.*, 1997). Taken together, these results further underscore the versatility of our novel system in the repression of target gene expression.

Other potential repressor domains that can be ligated to our regulator include the histone deacetylase (HD) module (Taunton *et al.*, 1996). Based on its ability to deacetylate histones, thereby leading to a transcriptionally inactive configuration, the HD domain when fused to our regulator should prove very applicable in the inducible repression of a stably integrated target(s).

D. Future Applications

Our current inducible system is now suited for many applications in biology. With better modifications, we can improve our gene regulatory system further and tailor its use for more specific applications including (1) transgenic disease models, (2) conditional knockout systems, (3) inducible rescue of knockouts, and (4) human gene therapy.

To further enhance the efficacy of our system for these long-term studies, we

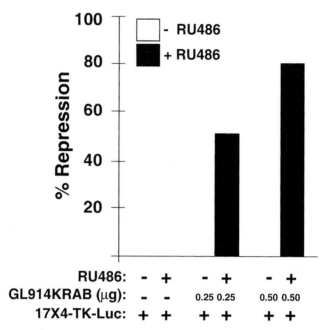

Figure 4 Inducible repression of basal target gene expression. Transient transfections were performed in HeLa cells to assess the ability of our modified transactivator, GL914KRAB, to repress the expression of a basal 17X4–TK-luciferase reporter. As shown in the graph, there is a dose-dependent repression of luciferase activity only in the presence of RU486 denoted by % repression.

can improve the delivery of RU486. This is particularly true in transgenic and knockout mice models whereby the chronic injection of RU486 can cause undue stress and trauma to the mice. To circumvent this problem, we can incorporate RU486 into slow-release pellets (Innovative Research of America, Florida, USA) that are then implanted subcutaneously into mice to allow a controlled and timed delivery of RU486 to any organ.

1. Transgenic Disease Models

In cancer studies, the expression of an oncogene at different stages in development could yield significantly diverse phenotypes (Jhappan *et al.,* 1993; Pierce *et al.,* 1993; Kordon *et al.,* 1995). Thus, our inducible system would be useful for investigators in dissecting the role a particular oncogene plays at each developmental stage to better appreciate the full spectrum of phenotypes elicited that would have been missed if constitutively expressed systems had been used. To establish an inducible oncogene system, two lines of mice must be generated: a tissue-specific regulator and a target oncogene line. Bitransgenic mice resulting from a cross between these two lines can then be subjected to analyses in the absence and presence of RU486 pellets to allow one to determine the effects of the oncogene under controlled conditions.

A variation along this theme would be to incorporate multiple levels of control by employing multiple gene regulatory systems like the tetracycline (Gossen and Bujard, 1992; Gossen *et al.*, 1994, 1995; Shockett *et al.*, 1995; Weinmann *et al.*, 1994; Roder *et al.*, 1994), ecdysone (Christopherson *et al.*, 1992; Yao *et al.*, 1992; No *et al.*, 1996), and ours (Wang *et al.*, 1994) to control the expression of multiple target oncogenes. Alternatively, dual-target (Baron *et al.*, 1995) and bicistronic target constructs (Dirks *et al.*, 1993) can be designed to allow our inducible system to regulate more than one target at the same time.

2. Conditional Knockout Systems

The applicability of regulable, tissue-specific knockout systems are also highly desirable for examining the consequences of spatiotemporal ablation of a specific gene. We can apply a variation of the Cre-Lox system (Pichel *et al.*, 1993; Orban *et al.*, 1995; Gu *et al.*, 1994) to meet this goal. Two approaches are possible.

The first approach adopts the principle of Kellendonk *et al.* (1996). Two lines of mice are needed: (1) a tissue-specific inducible Cre-PRLBDΔ fusion and (2) a floxed line containing a gene to be deleted flanked by two LoxP sites. The activity of the Cre protein is placed under the control of our modified PR-LBDΔ domain. These lines are then bred to obtain mice that harbor the tissue-specific Cre-PRLBDΔ fusion and two copies of the floxed gene. Addition of RU486 to these mice at a specific time would enable the activated Cre protein to delete the floxed gene in a tissue-specific fashion. The second approach, a modification of the first, also requires two lines of mice: (1) a bigenic line containing both the tissue-specific regulator and 17X4-TK-Cre target and (2) a floxed line. These mice are then bred to create a bigenic floxed null mice containing the regulator, 17X4-TK-Cre target, and two copies of the floxed gene. Likewise, RU486 addition would effect a spatiotemporal deletion of the floxed genes. Though the second approach appears to be more complex, it may prove to be more useful should the levels and activities of the Cre-PRLBDΔ fusion be insufficient.

Alternatively, our inducible repressor system can be used to down-regulate endogenous genes to study the effects of reduced gene levels. By exchanging the gal4 DNA binding domain of our inducible repressor for that of the DNA binding domain of a protein involved in the regulation of an endogenous target gene, we can inducibly repress that gene (Burcin *et al.*, 1998). We could also introduce gal4 DNA binding sites into the promoter of a given gene to down-regulate or ablate the expression of that gene in an inducible fashion. Both of these approaches should allow us to elucidate the physiological role of a gene of interest more precisely. Furthermore, by repressing the expression of a target gene by RU486 treatment, our system should allow investigators to bypass embryonic lethal stages so that they can assess later events in development after RU486 removal, a process not possible with traditional knockout systems.

3. Transgenic Rescue Approaches

The most convincing approach toward understanding gene function would be not only to delete that gene but also to subsequently rescue the ablated gene to fully restore its function. Our system can be used to facilitate these studies. Once again, two mice lines are needed: (1) a bigenic line harboring both the regulator and target rescue gene and (2) a heterozygote line containing only one copy of the gene. By crossing these mice, a bigenic homozygote null containing the regulator, target rescue gene, and two copies of the deleted gene can be generated. Restoration of gene function in the bigenic homozygotes can then be assessed by the administration of RU486 in a highly coordinated and controlled manner.

4. Human Gene Therapy

The application of gene therapy to cure debilitating and congenital diseases is a daunting task. The use of inducible systems should allow controlled expression of a therapeutic gene and help solve many of the unpleasant side effects associated with constitutively driven systems. To minimize the immunogenicity of the viral VP16 activation domain in our regulator, we are actively seeking to replace that domain with other potent but less immunogenic activation domains from the human SRC-1 (Onate *et al.*, 1994), CBP (Eckner *et al.*, 1994), and p65 (Schmitz and Baeuerle, 1991) proteins. However, we may still have to contend with the Ga14 DNA binding domain even though analogous domains have already been found in vertebrates.

Improved adenoviral (Bramson *et al.*, 1995) and retroviral (Gunzburg and Salmons, 1995) vectors can be used in gene therapy to enhance the delivery of the therapeutic gene. We can modify these vectors to carry both the regulator and the therapeutic gene separated by a HS4 insulator fragment (Chung *et al.*, 1993). After packaging, these recombinant vectors can then be delivered to the target tissue of choice. We are currently exploring the potential of the adenoviral system in mice.

III. CONCLUSIONS

We have successfully developed a novel gene regulatory system that can induce and repress the expression of target genes. Our system contains the following desirable characteristics: (1) low basal activity, (2) high induction potency, (3) reversibility, and (4) low toxicity. Thus, we envision that our system should be useful for a multitude of biological applications particularly in the regulation of genes involved in tumorigenesis, apoptosis, and development. An added quality that our system possesses is that we use an inducer (RU486) that is lipid soluble and can easily penetrate the blood–brain barrier. This now allows us to regulate the expression of genes in the central nervous system. Last but not least, by the appropriate exchange of DNA binding and activation domains, a variety of our inducible systems can be generated and used in future gene therapy endeavors.

ACKNOWLEDGMENTS

We are grateful to Dr. Debra Bramblett for critically reading this document.

REFERENCES

Baim, S. N., Labow, M. A., Levine, A. J., and Shenk, T. A. (1991). Chimeric mammalian transactivator based on the lac repressor that is regulated by temperature and isopropylbeta-D-thiogalactopyranoside. *Proc. Natl. Acad. Sci. USA* **88,** 5072–5076.

Baron, U., Freundlieb, S., Gossen, M., and Bujard, H. (1995). Co-regulation of two gene activities by tetracycline via a bidirectional promoter. *Nucleic Acids Res.* **23,** 3605–3606.

Baulieu, E. E. (1989). Contragestion and other clinical applications of RU 486, an antiprogesterone at the receptor. *Science* **245,** 1351–1357.

Bellefroid, E. J., Poncelet, D. A., Lecocq, P. J., Revelant, O., and Martial, J. A. (1991). The evolutionary conserved Krüppel-associated box domain defines a subfamily of eukaryotic multi-fingered proteins. *Proc. Natl. Acad. Sci. USA* **88,** 3608–3612.

Bramson, J. L., Graham, F. L., and Gauldie, J. (1995). The use of adenoviral vectors for gene therapy and gene transfer *in vivo. Curr. Opin., Biotech.* **6,** 590–595.

Braselmann, S., Graninger, P., and Busslinger, M. (1993). A selective transcriptional induction system for mammalian cells based on gal4-estrogen receptor fusion proteins. *Proc. Natl. Acad. Sci. USA* **90,** 1657–1661.

Brown, M., Figge, J., Hansen, U., Wright, C., Jeang, K. T., Khoury, G., Livingston, D. M., and Roberts, T. M. (1987). The lac repressor can regulate expression from a hybrid SV40 early promoter containing a lac operator in animal cells. *Cell* **49,** 603–612.

Burcin, M. M., O'Malley, B. W., and Tsai, S. Y. (1998). A regulatory system for target gene expression. *Frontiers in Biosci.* **3,** c1–c7.

Byrne, G. W., and Ruddle, F. H. (1989). Multiplex gene regulation: A two-tiered approach to transgene regulation in transgenic mice. *Proc. Natl. Acad. Sci. USA* **86,** 5437–5477.

Cristopherson, K. S., Mark, M. R., Bajaj, V., and Godowski, P. J. (1992). Ecdysteroid-dependent regulation of genes in mammalian cells by a *Drosphila* ecdysone receptor and chimeric transactivators. *Proc. Natl. Acad. Sci. USA* **89,** 6314–6318.

Chua, S. S., Wang, Y., DeMayo, F. J., O'Malley, B. W., and Tsai, S. Y. (1998). A novel RU486 inducible system for the activation and repression of genes. *Adv. Drug Del. Rev.* **30,** 23–31.

Chung, J. H., Whiteley, M., and Felsenfeld, G. (1993). A 5' element of the chicken β-globin domain serves as an insulator in human erythroid cells and protects against position effect in *Drosophila. Cell* **74,** 505–514.

Deuschle, U., Pepperkor, R., Wang, F. B., Giordano, T. J., McAllister, W. T., Ansorge, W., and Bujard, H. (1989). Regulated expression of foreign genes in mammalian cells under the control of coliphage T3 RNA polymerase and lac repressor. *Proc. Natl. Acad. Sci. USA* **86,** 5400–5404.

Dirks, W., Wirth, M., and Hauser, H. (1993). Dicistronic transcription units for gene expression in mammalian cells. *Gene* **128,** 247–249.

Eckner, R., Ewen, M. E., Newsome, D., Gerdes, M., DeCaprio, J. A., Lawrence, J. B., and Livingston, D. M. (1994). Molecular cloning and functional analysis of the adenovirus E1A-associated 300-kD protein (p300) reveals a protein with properties of a transcriptional adaptor. *Genes Dev.* **8,** 869–884.

Gerber, H. P., Seipel, K., Georgiev, O., Hofferer, M., Hug, M., Rusconi, S., and Schaffner, W. (1994). Transcriptional activation modulated by homopolymeric glutamine and proline stretches. *Science* **263,** 808–811.

Giniger, E., Varnum, S. M., and Ptashne, M. (1985). Specific binding of gal4, a positive regulatory protein of yeast. *Cell* **40,** 767–774.

Gossen, M., and Bujard, H. (1992). Tight control of gene expression in mammalian cells by tetracycline-responsive promoters. *Proc. Natl. Acad. Sci. USA* **89,** 5547–5551.

Gossen, M., Bonin, A. L., Freundlieb, S., and Bujard, H. (1994). Inducible gene expression systems for higher eukaryotic cells. *Curr. Opin., Biol.* **5,** 516–520.

Gossen, M., Freundlieb, S., Bender, G., Muller, G., Hillen, W., and Bujard, H. (1995). Transcriptional activation by tetracyclines in mammalian cells. *Science* **268,** 1766–1769.

Green, S., and Chambon, P. (1986). A superfamily of potentially oncogenic hormone receptors. *Nature* **324,** 615–617.

Gu, H., Marth, J. D., Orban, P. C., Mossmann, H., and Rajewsky, K. (1994). Deletion of a DNA polymerase beta gene segment in T cells using cell type-specific gene targeting. *Science* **265,** 103–106.

Gunzburg, W. H., and Salmons, B. (1995). Virus vector design in gene therapy. *Mol. Med. Today* **1,** 410–417.

Hu, M. C.-T., and Davidson, N. (1987). The inducible lac operator-repressor system is functional in mammalian cells. *Cell* **48,** 555–566.

Hynes, N. E., Kennedy, N., Rahmsdorf, U., and Groner, B. (1981). Hormone-responsive expression of an endogenous proviral gene of mammary tumour virus after molecular cloning and gene transfer into cultured cells. *Proc. Natl. Acad. Sci. USA* **78,** 2038–2042.

Jhappan, C., Geiser, A. G., Kordon, E. C., Bagheri, D., Hennighausen, L., Roberts, A. B., Smith, G. H., and Merlino, G. (1993). Targeting expression of a transforming growth factor β1 transgene to the pregnant mammary gland inhibits alveolar development and lactation. *EMBO J.* **12,** 1835–1845.

Kellendonk, C., Tronche, F., Monaghan, A. P., Angrand, P. O., Stewart, F., and Shultz, G. (1996). Regulation of Cre recombinase activity by the synthetic steroid RU 486. *Nucleic Acids Res.* **24,** 1404–1411

Khillan, J. S., Deen, K. C., Yu, S. H., Sweet, R. W., Rosenberg, M., and Westphal, H. (1988). Gene transactivation by the TAT gene of human immunodeficiency virus in transgenic mice. *Nuc. Acids Res.* **16,** 1423–1430.

Klein-Hitpass, L., Cato, A. C., Henderson D., and Ryffel, G. U. (1991). Two types of antiprogestins identified by their differential action in transcriptionally active extracts from T47D cells. *Nuc. Acids Res.* **19,** 1227–1234.

Klein-Hitpass, L., Cato, A. C., Henderson D., and Ryffel, G. U. (1991). Two types of antiprogestins identified by their differential action in transcriptionally active extracts from T47D cells. *Nuc. Acids Res.* **19,** 1227–1234.

Kordon, E., McKnight, R. A., Jhappan, C., Hennighausen, L., Merlino, G., and Smith, G. H. (1995). Ectopic TGFβ1 expression in the secretory mammary epithelium induces early senescence of the epithelial stem cell population. *Dev. Biol.* **167,** 47–61.

Labow, M. A., Baim, S. B., Shenk, T. A., and Levine, A. J. (1990). Conversion of the lac repressor into an allosterically regulated transcriptional activator for mammalian cells. *Mol. Cell. Biol.* **10,** 3343–3356.

Lee, F., Mulligan, R., Berg, P., and Ringold, G. (1991). Glucocorticoids regulate expression of dihydrofolate reductase cDNA in mouse mammary tumour virus chimaeric plasmids. *Nature* (London) **294,** 228–232.

Margolin, J. F., Friedman, J. R., Meyer, W. K.-H., Vissing, H., Thiesen, H. J., and Rauscher, F. J. III. (1994). Krüpel-associated boxes are potent transcriptional repression domains. *Proc. Natl. Acad. Sci. USA* **91,** 4509–4513.

Mayo, K. E., Warren, R., and Palmiter, R. D., (1982). The mouse metallothionein-I gene is transcriptionally regulated by cadmium following transfection into human or mouse cells. *Cell* **29,** 99–108.

Muller, W. J., Sinn, E., Patterngale, P., Wallace, R., and Leder, P. (1988). Single-step induction of mammary adenocarcinoma in transgenic mice bearing the activated c-neu oncogene. *Cell* **54,** 105–115.

Neef, G., Beier, S., Elger, W., Henderson, D., and Wiechert, R. (1984). New steroids with antiprogestational and antiglucocorticoid activities. *Steroids* **33,** 349–372.

No, D., Yao, T. P., and Evans, R. M. (1996). Ecdysone-inducible gene expression in mammalian cells and transgenic mice. *Proc. Natl. Acad. Sci. USA* **93,** 3346–3351.

O'Malley, B. W. O., Tsai, S. Y., Bagchi, M., Weigel, N. L., Schrader, W. T., and Tsai, M.-J. (1991). Molecular mechanism of action of a steroid hormone receptor. *Recent Prog. Horm. Res.* **47,** 1–24.

Onate, S. A., Tsai, S. Y., Tsai, M. J., and O'Malley, B. W. (1994). Sequence and characterization of a coactivator for the steroid hormone receptor superfamily. *Science* **270,** 1354–1357.

Orban, P. C., Chui, D., and Marth, J. D. (1995). Tissue- and site-specific DNA recombination in transgenic mice. *Proc. Natl. Acad. Sci. USA* **89,** 6861–6865.

Ornitz, D. M., Moreadith, R. W., and Leder, P. (1991). Binary system for regulating transgene expression in mice: Targeting int-2 gene expression with yeast gal4/UAS control elements. *Proc. Natl. Acad. Sci. USA* **88,** 698–702.

Pichel, J. G., Lakso, M., and Westphal, H. (1993). Timing of SV40 oncogene activation by site-specific recombination determines subsequent tumor progression during murine lens develoment. *Oncogene* **8,** 3333–3342.

Pierce, D. F., Jr., Johnson, M. D., Matsui, Y., Robinson, S. D., Gold, L. I., Purchio, A. F., Daniel, C. W., Hogan, B. L. M., and Moses, H. L. (1993). Inhibition of mammary duct development but not alveolar outgrowth during pregnancy in transgenic mice expressing active TGF-β1. *Genes Dev.* **7,** 2308–2317.

Quaife, C. J., Pinkert, C. A., Ornitz, D. M., Palmiter, R. D., and Brinster, R. L. (1987). Pancreatic neoplasia induced by ras expression in acinar cells of transgenic mice. *Cell* **48,** 1023–1034.

Roder, F. T., Schmulling, T., and Gatz, C. (1994). Efficiency of the tetracycline-dependent gene expression system—complete suppression and efficient induction of the rolB phenotype in transgenic plants. *Mol. Gen. Genet.* **243,** 32–38.

Schmitz, M. L., and Baeuerle, P. A. (1991). The p65 subunit responsible for the strong transcription activating potential of NF-kappa B. *EMBO J.* **10,** 3805–3817.

Shockett, P., Difilippantonio, M., Hellman, N., and Schatz, D. G. (1995). A modified tetracycline-regulated system provides autoregulatory, inducible gene expression in cultured cells and transgenic mice. *Proc. Natl. Acad. Sci. USA* **92,** 6522–6526.

Taunton, J., Hassig, C. A., and Schreiber, S. L. (1996). A mammalian histone deacetylase related to the yeast transcriptional regulator Rpd3p. *Science* **272,** 408–411.

Tepper, R. I., Levinson, D. A., Stanger, B. Z., Campos-Torres, J., Abbas, A. K., and Leder, P. (1990). IL-4 induces allergic-like inflammatory disease and alters T cell development in transgenic mice. *Cell* **62,** 457–467.

Triezenberg, S. J., Kingsbury, R. C., and McKnight S. L. (1988). Functional dissection of Vp-16, the transactivator of herpes simplex virus immediate early gene expression. *Genes Dev.* **2,** 718–729.

Tsai, M.-J., and O'Malley, B. W. (1994). Molecular mechanisms of action of steroid/thyroid receptor superfamily members. *Ann. Rev. Biochem.* **63,** 451–486.

Vegeto, E., Allan, G. F., Schrader, W. T., Tsai M.-J., McDonnell, D. P., and O'Malley, B. W. (1992). The mechanism of RU486 antagonism is dependent on the conformation of the carboxyl-terminal tail of the human progesterone receptor. *Cell* **69,** 703–713.

Wang, Y., DeMayo, F. J., Tsai, S. Y., and O'Malley, B. W. (1997). Ligand-inducible and liver-specific target gene expression in transgenic mice. *Nature Biotech.* **15,** 239–243.

Wang, Y., O'Malley, B. W., Jr., Tsai, S. Y., and O'Malley, B. W. (1994). A regulatory system for use in gene transfer. *Proc. Natl. Acad. Sci. USA* **91,** 8180–8184.

Wang, Y., Xu, J., Pierson, T., O'Malley, B. W., and Tsai, S. Y. (1997). Positive and negative regulation of gene expression in eukaryotic cells with an inducible transcriptional regulator. *Gene Ther.* **4,** 432–441.

Weinmann, P., Gossen, M., Hillen, W., Bujard, H., and Gatz, C. (1994). A chimeric transactivator allows tetracycline-responsive gene expression in whole plants. *Plant J.* **5,** 559–569.

Witzball, R., O'Leary, E., Gessner, R., Ouellette, A. J., and Bonventre, J. V. (1993). Kid-1, a putative renal transcription factor: Regulation during ontogeny and in response to ischemia and toxic injury. *Mol. Cell Biol.* **13,** 1933–1942.

Wurm, F. M., Gwinn, K. A., and Kingston, R. E. (1986). Inducible overexpression of the mouse c-myc protein in mammalian cells. *Proc. Natl. Acad. Sci. USA* **83,** 5414–5418.

Yan, C., Costa, R. H., Darnell, J. E., Jr., Chen, J., and Van Dyke, T. A. (1990). Distinct positive and negative elements control the limited hepatocytre and choroid plexus expression of transthyretin in transgenic mice. *EMBO J.* **9,** 869–878.

Yao, T. P., Segraves, W. A., Oro, A. E., McKeown, M., and Evans, R. M. (1992). *Drosophila ultraspiracle* modulates ecdysone receptor function via heterodimer formation. *Cell* **71,** 63–72.

INDEX

TC-chol, 317
  in lipoplexes, transgene expression and, 324, 325(figure), 326
  in reconstituted chlyomicron remnants, 305
T cells, *see also* Th1 immune response; Th2 immune response
  nucleic acid vaccination and, 394–396, 397
T-DNA, nuclear transport of, 162–163
Texas Red, 121
Th1 immune response
  immunostimulatory sequences and, 393
  nucleic acid vaccination and, 385–386, 386
Thrombocytopenia
  DOTIM/cholesterol lipoplexes and, 123
  intravenous lipoplex delivery and, 250–251
Th2 immune response, gene gun nucleic acid vaccination and, 388
Ti plasmids, 153, 162–163
Tobacco etch virus, nuclear transport by, 161–162
Toroids, 78
Toxicity
  gelatin, 276
  lipopolyplexes, 302
  PEI polyplexes, 202
  plasmid DNA, 57, 344
  in *in vivo* lipoplex delivery, 253–254, 256
Transcription, *see also* HER-2/neu oncogene, transcriptional repression
  inhibited in lipoplexes, 15
Transcriptional promoters, 185
Transfectam, 196
Transfection
  with cationic lipoplexes
    in vitro, 236–242
    in vivo, 242–260
  in cystic fibrosis gene therapy, 43, 45–46
    clinical studies in, 59–61
    efficiencies, 55, 56
    lipoplex formulation variables and, 50–54
    lipoplex structure-activity relationships and, 46–50
  early *in vitro* studies, 26–27
  effects of cell division on, 144–145
  efficiency of PEI polyplexes in, 200
  endocytosis and, 111–112
  with naked DNA, 241
  sequential injections of cationic liposomes and naked DNA, 77, 256–258
  in vivo experiments in, 28–33

Transfection efficiency
  of biopolymer-DNA nanospheres, 278, 279–280
  improving in cystic fibrosis gene therapy, 349–350
  of lipoplexes
    colloidal properties and, 77–85
    effects of neutral colipids on, 110–111
    in lung endothelium, 315, 318, 321–322
    molecular structure-activity relationships and, 73–77
    self-assembled structures and, 111–113
    size and, 245
    stability and, 84
  of lipopolyplexes, 299, 300–301
  with nonviral vectors, 92, 94–95
  of receptor-targeted polyplexes, 218
  with sequential injections of cationic liposomes and naked DNA, 77, 256–258
Transferrin
  anionic lipopolyplexes and, 304
  biopolymer-DNA nanospheres and, 273, 277–278
  PEI-transferrin polyplexes, 197
  in polycation conjugates, 211, 212(table)
Transformation, 26–27; *see also* Transfection
Transgene expression
  of biopolymer-DNA nanospheres, 279
  lipoplex-mediated, 128–131
    in cystic fibrosis gene therapy, 44, 57–58
    in lung endothelium, 315, 317–318, 323, 324–326, 329–333
    problems in, 140
    in vitro, 241, 259–260
    in vivo, 244–245, 251–252, 254–260
Transgene rescue, binary gene regulatory system and, 422
Translocation, in cystic fibrosis gene therapy, 43, 56
Tumor cell explants, canine, ballistic gene delivery, 173–177
Tumors
  lipoplex uptake, 127–128
  oral, *in vivo* ballistic transfection and, 185
Tween 80, 76

U

Ultracentrifugation, analytical, 81
Urinary bladder, lipoplex uptake, 127
Uterus, lipoplex uptake, 127